Geophysical Monograph Series

Including
IUGG Volumes
Maurice Ewing Volumes
Mineral Physics Volumes

Geophysical Monograph 164

Archean Geodynamics and Environments

Keith Benn
Jean-Claude Mareschal
Kent C. Condie
Editors

American Geophysical Union
Washington, DC

Library of Congress Cataloging-in-Publication Data

Archean geodynamics and environments / Keith Benn, Jean-Claude Mareschal, Kent C. Condie, editors.
    p.   cm. — (Geophysical monograph ; 164)
  ISBN-13: 978-0-87590-429-0
  ISBN-10: 0-87590-429-7
  1. Geodynamics. 2. Plate tectonics. 3. Earth—Mantle. 4. Earth—Crust. 5. Geology, Structural. 6.
Geology, Stratigraphic—Archaean. I. Benn, Keith, 1955- II. Mareschal, Jean-Claude, 1945- III.
Condie, Kent C. IV. Series.

  QE501.A697 2006
  551—dc22

                    2005037131

  ISBN-13: 978-0-87590-429-0
  ISBN-10: 0-87590-429-7
  ISSN 0065-8448

Front cover: White flocculent mats in and around extremely gassy, hot (>100°C) white smokers at Champagne Vent, northeast Eifuku volcano, photographed during Submarine Ring of Fire 2004: Exploration of the Mariana Arc expedition (photo courtesy of the National Oceanic and Atmospheric Administration).

Back cover: Outcrop of tonalite orthogneiss of Late Archean age, near the town of Gogama, in the Abitibi Subprovince of the southeastern Superior Province, Ontario (photo by Keith Benn; compass shows scale).

CONTENTS

Orogenies

Biogeochemical Environments and Life

PREFACE

The Archean Eon represents 1.3 Gyr of Earth's distant past, from about 3.8 Ga to 2.5 Ga—nearly one third of our planet's history. It was during the Archean that a regime of global geodynamics was established, resulting in the formation and recycling of the first lithosphere, as well as the formation, growth, deformation, differentiation, emergence, and erosion of continents. By the end of the Archean, Earth had reached a geodynamic regime dominated by plate tectonic processes. The consequent environments, at and near Earth's surface, included the different niches within which early life forms evolved. It is to the Archean evolution of Earth that we now look to better understand many of the processes that shaped the planet, as we know it.

The exposed rock record of known Archean age represents only a small fraction of the surface of the planet. Much of it is remote and difficult to access, though large regions have been mapped and studied using geological and geophysical methods. During the past two decades, the collection of geophysical, geochemical, and petrological data has confirmed the presence of deep lithospheric roots beneath the Archean cratons and yielded information on their structure and composition. These new data provide important constraints on our interpretations of the Archean Earth, and the origin and evolution of the lithosphere. The limited rock record, major differences between present and Archean thermal regimes, and, equally importantly, the need to study Archean lithosphere tectonics in the context of global geodynamics also require the application of advanced modeling methods.

In this book, we have assembled 17 papers that investigate significant aspects of Archean geodynamics, using data from the preserved Archean geological record and cutting-edge geodynamic and geophysical modeling.

The book includes studies of the geological record based on geological mapping, deep seismic-reflection, and petrological and geochemical research. Authors present new and compiled results from five Archean cratons: the Pilbara and Yilgarn cratons in Australia, the Barberton terrain in South Africa, and the Superior and Slave Provinces in Canada. Compilations of geological, geophysical, petrological, and geochemical data provide benchmark papers that bring the experienced specialist and the initiate up to date on what we

know and what questions will shape our efforts in the coming years. Other papers use modeling methods to investigate possible Archean geodynamic regimes and their effects on the global heat budget, the formation and evolution of lithosphere, and the assembly and preservation of continents.

Many of the papers include an important review component that will allow the nonspecialist reader to fully appreciate the context of the research, as well as the quality of the results and interpretations, as presented. Two papers are state-of-the-art reviews of recent and developing research into the biochemical and physical environments where early life evolved, and of the search for and documentation of early life forms.

We have organized the book into five thematic sections. Following the Introduction, the opening section includes two papers that discuss the heat budget of the Archean Earth, and consequences for mantle convection and global tectonics. It also includes a synthesis of the geological, geophysical, and petrological record of the Archean that provides an important part of the framework within which modelers pursue their efforts. The subsequent sections include papers on the origin, composition, and structure of the Archean lithosphere; tectonic regimes; the origins and nature of the early continents and their rheological evolution; orogenic styles as revealed through geological and geophysical investigations; the petrological signatures of Archean subduction processes; and the nature and evolution of biogeochemical environments and life.

The reader will find that Earth's heat budget represents a strong theme running through the book. It is a primary consideration in modeling and in interpreting the rock record petrologically and geophysically. The notion that a hotter Archean Earth may have required a geodynamic regime that differs from present-day geodynamics is certainly not new. However, several papers in this book reveal that we still have much to explore and to understand in this regard and that some previous interpretations, which have been widely accepted in recent years, do need to be revisited.

The book grew out of a session on the theme Archean Geodynamic Processes that was held at the Joint Assembly of the American Geophysical Union, in Montreal, on May 19, 2004. Holding a session on the Archean Eon in Montreal was fitting, as that city rests on the margin of the Canadian Shield, within a few hours drive of the largest Archean craton on Earth, the Superior Province. Participants at the meeting wrote some of the papers, others were written by nonparticipants who we invited to increase the scope of the book.

Archean Geodynamics and Environments
Geophysical Monograph Series 164
Copyright 2006 by the American Geophysical Union
10.1029/164GM01

The compilation of papers, which covers a broad range of topics and approaches to the Archean, allowed us to fulfill our principal aim: to present a synthesis of our present understanding of Archean processes that were fundamental to the shaping of the early Earth and its environments and of the main issues driving our research.

The editors wish to thank, first and foremost, the authors who contributed their work to this monograph. We are very grateful to them all. We also wish to whole-heartedly thank those who performed the task of reviewing the papers; their efforts contributed to making the final product of the highest quality. We also thank Allan Graubard, who first approached us about preparing the monograph; the oversight editor, Andy Nyblade, whose input made the scope of the book more comprehensive; Dawn Seigler, who worked with us during the review and editing processes; and all the AGU books staff for helping us through the publication process, beginning to end. Their efficiency and professionalism are much appreciated.

Keith Benn
Jean-Claude Mareschal
Kent C. Condie

Introduction: Archean Geodynamics and Environments

Keith Benn

Keith Benn

Ottawa-Carleton Geoscience Centre and Department of Earth Sciences, University of Ottawa, Ottawa, Ontario, Canada

Jean-Claude Mareschal

GEOTOP-UQAM-McGill Centre for Research in Geochemistry and Geodynamics, University of Québec at Montréal, Montréal, Québec, Canada

Kent C. Condie

Department of Earth and Environmental Science, New Mexico Institute of Mining & Technology, Socorro, New Mexico

Important questions remain regarding the geodynamic evolution of Earth during the Archean Eon. They include the convective regime of the mantle, the heat budget and the distribution of heat within the mantle and the crust, the nature of lithosphere tectonics, and the formation, stabilization, and preservation of continents. Archean tectonics must be considered in the context of deeper mantle processes, notably, convection. Other important issues include the petrogenesis of Archean crust in a geodynamic regime different from today's. These issues are of significance for the nature of early life as well as for the physico-chemical environments where early life evolved. In this monograph, authors address important geodynamic, tectonic, petrological, and biochemical questions, using advanced numerical and analytical modeling and extensive studies of the rock record based on new and compiled geological and geophysical data.

To study Archean geodynamics is to inquire into the ancestry of the present-day plate tectonic Earth, as well as the nature of early life and its environments. It is also to probe the birth and the early history of continental nuclei that enclose many of the large mineral deposits upon which we depend for our supplies of precious and base metals.

Our knowledge of Archean geodynamic and tectonic processes has advanced greatly in recent years, in large part due to an increase in the quantity and the quality of geological, geophysical, geochemical, and geochronological data, and also to advances in experimental and modeling methods.

Archean Geodynamics and Environments
Geophysical Monograph Series 164
Copyright 2006 by the American Geophysical Union
10.1029/164GM02

Multidisciplinary studies in several Archean provinces have provided information on the deep structure and composition of the Archean lithosphere. Still, a number of important questions regarding the Archean Earth, its physico-chemical environments, and the origins and evolution of early life remain open. For example, a robust debate continues as to the tectonic regime that governed the formation and deformation of the Archean lithosphere [*Dewit*, 1998; *Hamilton*, 1998].

Much of the debate has turned on whether plate tectonic principles can be applied to the interpretation and modeling of Archean geodynamics. There appears to be little doubt that during the Archean Eon, Earth's mantle experienced a convective regime, perhaps one more vigorous than at present. Therefore, the Archean tectonics debate can be framed as whether the Archean Earth was characterized by stagnant lid mantle convection, which would imply an immobile upper thermal boundary layer, or by active lid convection, which would imply mobile tectonic plates driven by some combination of coupling with mantle convection, and subduction. The debate can then be carried one step further, and we may ask how the kinematics and dynamics of Archean plate tectonics may have differed from those of the present day. To this we may add that the Archean Eon, which represents 1.3 Ga of early geological history, may not have known a uniform geodynamic regime but rather an evolving one.

The remaining uncertainty about global geodynamics and lithospheric tectonics in Archean time means that studying the Archean geological record presents a special challenge; we do not necessarily have a paradigm, such as modern-day plate tectonics, within which to consider our observations and data. The student of the Archean must not only understand the different components of the rock record, but also erect a tectonic framework within which to interpret them. As suggested by *Bailey* in his contribution to this book, in the absence of a generally accepted geodynamic framework, we can perhaps best frame the overriding questions regarding the Archean in this way: *What geodynamic processes can we reasonably expect to have provided the principal controls on the formation and the preservation of Archean lithosphere?*

The rather limited Archean rock record (about 5% of the surface area of the continents) and the lack of a fixed (or, at least, of a slowly evolving) tectonic framework require that we also employ quantitative modeling methods to investigate which geodynamic processes can be considered reasonable. Also, the origin and preservation, or the destruction, of Archean lithosphere cannot be completely understood unless we understand the processes that occurred in the deep mantle, specifically convection, and the related questions of mantle temperatures, viscosities, and melting. These questions must also be addressed through modeling, as well as by study of the rock record.

There is consensus that planet Earth is cooling, although proposed cooling rates vary by a factor of two. Radiogenic heat sources have been redistributed and are also running down. Earth's internal heat production was 2 to 3 times greater in the Archean, compared to that in the present day. Heat production and heat flow were likely the most important intensive parameters that controlled the Archean evolution of the mantle, the production and composition of crust, and the stabilization of continental lithosphere. The thermal evolution of Earth through the Archean represents a theme that is present, to greater and lesser degrees, in each section of this book.

GLOBAL GEODYNAMICS AND PLATE TECTONICS

Understanding the generation and workings of plate tectonics as a result of mantle convection on present-day Earth

is an exciting challenge [*Bercovici et al.*, 2000]. To understand the tectonic regime in the Archean poses some special problems, not the least of which are the prediction of the Archean thermal regime and the nature of mantle convection, which would have strongly influenced the physical properties and mechanical behavior of the lithosphere. The first section of this book includes two papers that explore the global heat budget and thermal regime of the Archean Earth.

Korenaga explores the relationships between the hotter Archean Earth, mantle convection, and tectonics through a new parameterization of mantle convection. He includes chemical changes due to high degrees of partial melting and the resulting effects on lithospheric buoyancies, thicknesses, and strengths. His proposed model also provides a solution to the *Archean thermal catastrophe*, a massive melting of the mantle that is predicted by some models for the thermal evolution of Archean Earth. An important tectonic inference is that Archean plate tectonics, while likely to have been operative, may have been more sluggish than on present-day Earth. The results also have important consequences for mantle plume activity in the Archean.

The chapter by *Lenardic* aims to test the viabilities of endmember continental growth models, "instantaneous" growth of continental crust in the Archean, and a linear growth model, by exploring their relative compatibilities with predicted thermal conditions for the Archean. One of the principal constraints is the *Archean paradox*, first raised by *Burke and Kidd* [1978], that is, the fact that Archean crustal rocks do not preserve evidence of high geothermal gradients, despite considerably higher mantle temperatures in the Archean than at present. A new heat flow scaling that includes continents may also offer a new solution to the Archean thermal catastrophe mentioned in the previous paragraph.

Of course, our interpretations of Archean lithosphere tectonics also depend on the constraints imposed by the geological record. *Condie and Benn* provide an up-to-date synthesis on our state of knowledge of the rock record and then propose a list of 10 fundamental constraints that Archean geodynamic models must address in order to satisfy our knowledge of Archean geology and geophysics. Notably, they discuss the evidence for crustal diapirism [*Collins et al.*, 1998; *Van Kranendonk et al.*, 2004] and plate tectonics recorded in the deformation of Archean terrains, and show that those two models are not mutually exclusive.

THERMOMECHANICAL EVOLUTION OF CONTINENTS

Crustal growth models suggest that 50% or more of continental crust was formed by the end of the Archean Eon [e.g., *Fyfe*, 1978; *Taylor and McLennan*, 1985; *Veizer and Jansen*,

1979], possibly with a burst of continental crust formation at about 2.7 Ga [*Condie*, 1998, 2000]. The stabilization and preservation of Archean continental crust coincided with the formation of a thick (about 200 km) buoyant, depleted lithospheric root [*Jordan*, 1978; *Lenardic and Moresi*, 1999] as revealed by seismic tomography and studies of xenolith suites [*Grand et al.*, 1997; *O'Reilly*, 2001; *Griffin et al.*, 2003; *Sleep*, 2005]. How the thick buoyant roots of the continents were formed remains an open question of obvious importance for Archean geodynamics. Existing models include the formation of highly depleted mantle residue by extensive melting in mantle plumes, thrust stacking of primitive lithosphere, and the accretion of magmatic arc-type lithosphere. The latter two models imply active plate subduction in the Archean.

Other questions regarding the evolution and stabilization of the early continents arise from the fact that radiogenic heat production in the crust must have been significantly greater, perhaps by 2 to 3 times, during the Archean. What was the effect of the hot crust on the underlying lithospheric root, and what consequences did the high crustal heat production have for the rheology and the stablization of the continental crust?

In the second section of the book, five chapters address the above questions on the formation and preservation of early continents.

The research of *Mareschal and Jaupart* addresses an apparent paradox due to the concomitant requirements of high heat production in Archean crust (suggested by radiogenic element concentrations and metamorphic geology) and the need for a strong, relatively cold lithosphere to support 40 km of crust. They address the problem by way of lithosphere-scale thermal modeling, the results of which suggest that, after the formation of Late Archean lithospheric roots, the thermal states of the crust and of the mantle lithosphere were decoupled and were in non-steady-state for perhaps 1 Gy or more.

Two papers deal directly with the formation of thick Archean continental lithosphere, taking two distinctly different approaches to the question.

Cooper et al. discuss evidence from seismic-reflection profiles that they interpret to indicate that Archean lithosphere formed by thrust stacking of primitive, *proto-cratonic* (Archean oceanic?) lithosphere. They then investigate, by way of dynamic modeling, the efficiency of the interpreted thrust stacking mechanism. The results provide important constraints on the conditions under which the mechanism may have been viable and led to the preservation of strong Archean lithospheric keels, a question of first-order importance [*King*, 2005].

Lee takes a geochemical/petrological approach to test the three models for the formation of Archean lithosphere. The chemistry of peridotite and garnet pyroxenite xenoliths from

Archean cratons is used to determine pressures (depths) at which melting occurred and to determine the protoliths of eclogites. The results are consistent with the formation of lithospheric roots by subduction-related processes, i.e., thrust stacking of oceanic lithosphere, as suggested by *Cooper et al.*, and accretion of magmatic arcs.

The last two papers in the section focus on Archean continental geotherms and the effects on the strength of crustal rocks.

Bailey extrapolates heat flow data back in time to predict possible modes of Archean continental tectonics, specifically large-scale crustal extension allowed by decoupling of upper crust from lithosphere, due to a ductile middle to lower crust. The results have implications for the evolution of crustal strength profiles with time and also for the emergence of continents from the oceans.

Bodorkos and Sandiford model the distributions of heat-producing elements in Archean crust and the resulting geothermal gradients, in order to investigate the thermal and mechanical controls on crustal deformation in Mid-Archean and Late Archean cratons of Australia. The work addresses the origins of dome and basin structures in the Mid-Archean Eastern Pilbara Craton [*Collins et al.*, 1998; *Van Kranendonk et al.*, 2004] and the linear fold-belt structural style more typical of Late Archean cratons [*Choukroune et al.*, 1997], for which they take the example of the Yilgarn Craton.

SUBDUCTION, ACCRETION, AND OROGENY

Admitting active lid mantle convection in the Archean, that is to say, mobile lithospheric plates, we must also admit consumption of lithosphere at convergent margins, as well as deformation and metamorphism compatible with accretion and collision. The nature of Archean subduction zones leaves its petrological and geochemical signatures in the tonalite–trondhjemite–granite (TTG) suites that make up much of the preserved Archean crust [*Martin and Moyen*, 2002] and in the different volcanic facies of greenstone belts.

The third and fourth sections of the book include seven papers concerning the nature of Archean subduction and tectonic accretion and the structural and metamorphic signatures of Archean accretionary and collisional tectonics.

Moyen and Stevens compile and review a large geochemical database on the experimental melting of amphibolites, the likely source material for Archean TTG magmas. They conclude that TTG were most likely formed by melting under pressures and temperatures that are most compatible with geothermal gradients in subduction environments. The finite element modeling of *O'Neill and Wyman* is designed to investigate the origins of lamprophyre-associated, Archean diamonds from Superior Province; the most likely geodynamic environment for the origin of the lamprophyre–diamond association is one of shallowly dipping subduction, followed by collisional orogeny.

The paper by *Polat and Kerrich* reviews the characteristic volcanic rock compositions that occur in high heat flow Cenozoic arcs. They then review occurrences of similar volcanic rock compositions in several Archean greenstone belts, further strengthening the case for subduction and arc magmatism in the Archean. The chapter by *Wyman and Hollings* focuses on two 2.7 Gyr greenstone belts in the Superior Province and also concludes as to the arc nature of the mafic to felsic volcanic products.

The paper by *Diener et al.* provides some exciting new metamorphic data from the high-grade portion of the Barberton terrain, in South Africa. The data indicate a clockwise pressure–temperature time path, consistent with partial subduction of an older continental fragment prior to exhumation by tectonic extension.

The paper by *van der Velden et al.* reviews deep-seismic reflection profiles from three Late Archean cratons. The authors conclude that the reflection patterns, which resemble the reflection patterns of Proterozoic and Phanerozoic orogens, are compelling indicators of terrane accretion during lithospheric plate convergence and collision events.

Benn provides a synthesis of the geology and structures of the southeastern Superior Province and a reinterpretation of published seismic reflection profiles that cross the entire region. The data provide a record of collision between terranes of different ages, which resulted in the tectonic delamination of the lower crust of the younger terrane by tectonic wedging in the plate suture. The structural signature of collision in the overriding, younger plate reveals a lithospheric strength profile with stiffer, presumably mafic granulite underlying hot, soft granitic crust.

EARLY ENVIRONMENTS AND LIFE

The origin and evolution of Earth's lithosphere have become an issue whose scope is enlarged by the realization that life on our planet began, evolved, and may have flourished, in shallow crustal and surface environments during Archean time. The two final papers deal with the origins of life on Earth and the biochemical environments within which life evolved.

Westall and Southam provide an extensive review of the evidence for, and the potential pitfalls of, identifying Mid-Archean life. The review is followed by a discussion of the metabolic strategies, environmental characteristics, and distribution of life forms, including the possible role of hydrothermalism resulting from the presumed high flow in the Archean.

Shen et al. evaluate the nature of Archean environments and of the metabolic processes of early life by studying the

fractionation of sulfur and nitrogen isotopes by biochemical, metabolic processes. That information also sheds light on the physico-chemical makeup of the environments that were influenced by Archean geodynamic processes. After reviewing isotope fractionation by biological processes in the modern oceans and hydrothermal environments, they proceed to dissect the isotopic records from Archean sedimentary rocks, in search of information on the environments of deposition and for the signatures of early life. Their review and analysis provides information on the composition of the atmosphere and points to bacteria with sulfate-reducing metabolisms having evolved by 3.5 Gy.

The papers in this book are organized according to broad themes. Many of the papers include important review components that will aid a wide range of scientists to understand the context of the different, important questions that are addressed by the authors and to fully appreciate some of the most recent advances in the study of the Archean Eon. The Editors also hope that the content and the organization of the volume will prove to be useful in the transmission of knowledge and ideas to students at the advanced undergraduate level and in graduate school. A number of the most important outstanding questions regarding the Archean Earth are explained and studied in these 17 papers. Let us now proceed to carry our understanding of the Archean much further.

REFERENCES

Bercovici, D., *et al.* (2000), The relation between mantle dynamics and plate tectonics: a primer, in *The history and dynamics of global plate motions*, edited by M.A. Richards, *et al.*, pp. 5-46, American Geophysical Union Geophysical Monograph no. 121.

Burke, K., and Kidd, W.S.F. (1978). Were Archean continental geothermal gradients much steeper than those of today, *Nature*, 272, 240-241.

Choukroune, P., *et al.* (1997), Archean crustal growth and tectonic processes: a comparison of the Superior Province, Canada and the Dharwar Craton, India, in *Orogeny through time*, edited by J.P. Burg and M. Ford, pp. 63-98, Geological Society, London, London.

Collins, W.J., *et al.* (1998), Partial Convective Overturn of Archaean Crust in the East Pilbara Craton, Western Australia - Driving Mechanisms and Tectonic Implications, *J. Struct. Geol.*, 20, 1405-1424.

Condie, K.C. (1998), Episodic continental growth and supercontinents: a mantle avalanche connection? *Earth Planet. Sci. Lett.*, 163, 97-108.

Condie, K.C. (2000), Episodic continental growth models: afterthoughts and extensions, *Tectonophysics*, 322, 153-162.

Dewit, M.J. (1998), On Archean granites, greenstones, cratons and tectonics—does the evidence demand a verdict? *Precam. Res.*, 91, 181-226.

Fyfe, W.S. (1978), Evolution of the Earth's crust: modern plate tectonics to ancient hot spot tectonics? *Chem. Geol.*, 23, 89-96.

Grand, S.P., van der Hilst, R.D., and Widiyantoro, S.R.I. (1997), Global seismic tomography: a snapshot of convection in the earth, *GSA today*, 7, 1-7.

Griffin, W.L., *et al.* (2003), The origin and evolution of Archean lithospheric mantle, *Precam. Res.*, 127, 19-41.

Hamilton, W. B. (1998), Archean magmatism and deformation were not products of plate tectonics, *Precam. Res.*, 91, 143-179.

Jordan, T.H. (1978), Composition and development of the continental tectosphere, *Nature*, 274, 544-548.

King, S.D. (2005), Archean cratons and mantle dynamics, *Earth Planet. Sci. Lett.*, 234, 1-14.

Lenardic, A., and L.N. Moresi (1999), Some thoughts on the stability of cratonic lithosphere: Effects of buoyancy and viscosity, *J. Geophys. Res.—Solid Earth.*, 104, 12747-12758.

Martin, H., and J.-F. Moyen (2002), Secular changes in tonalite-trondhjemite-granodiorite composition as markers of the progressive cooling of Earth, *Geology*, 30, 319-322.

O'Reilly, S. Y. (2001), Journey beneath southern Africa, *Nature*, 412, 777-780.

Sleep, N.H. (2005), Evolution of the continental lithosphere, *Annu. Rev. Earth Planet. Sci.*, 33, 369-393.

Taylor, S.R., and S.M. McLennan (1985), *The Continental Crust: Its Composition and Evolution*, 312 pp., Blackwell Scientific Publishers, Oxford.

Van Kranendonk, M.J., *et al.* (2004), Critical tests of vertical vs horizontal tectonic models for the Archean East Pilbara granite-greenstone terrane, Pilbara craton, Western Australia, *Precam. Res.*, 131, 173-211.

Veizer, J., and S.L. Jansen (1979), Basement and sedimentary recycling and continental evolution, *Journal of Geology*, 87, 341-370

K. Benn, Ottawa-Carleton Geosciences Centre and Department of Earth Sciences, University of Ottawa, Ottawa, Ontario K1N6N5, Canada. (kbenn@uottawa.ca)

K.C. Condie, Department of Earth & Environmental Science, New Mexico Institute of Mining & Technology, Socorro, New Mexico 87801, USA. (kcondie@nmt.edu)

J.-C. Mareschal, GEOTOP-UMAQ-McGill, University of Montréal at Québec, P.O.B. 8888, sta. "downtown", Montréal, Québec H3C3P8, Canada. (jcn@olympus.geotop.uqam.ca)

Archean Geodynamics and the Thermal Evolution of Earth

Jun Korenaga

Department of Geology and Geophysics, Yale University, New Haven, Connecticut

Possible geodynamic regimes that may have prevailed in the Archean are investigated by back-tracking the thermal history of Earth from the present-day conditions. If the temporal evolution of plate-tectonic convection is modulated by strong depleted lithosphere created at mid-ocean ridges, more sluggish plate tectonics is predicted when the mantle was hotter, contrary to commonly believed, more rapid tectonics in the past. This notion of sluggish plate tectonics can simultaneously satisfy geochemical constraints on the abundance of heat-producing elements and petrological constraints on the degree of secular cooling, in the framework of simple whole-mantle convection. The geological record of supercontinents back to ~2.7 Ga is shown to be broadly consistent with the accelerating plate motion as predicted by the new model. Furthermore, the very fact of repeated continental aggregation indicates that thicker depleted lithosphere in the past needs to move more slowly to become negatively buoyant by thermal contraction and also needs to be strong enough to support resulting thermal boundary layer. The concept of many small plates covering Archean ocean basins is thus physically implausible. As a consequence of reduced heat flux in the past, mantle plumes are expected to have been weaker in the Archean. The chemical evolution of Earth's mantle may have been encumbered by sluggish plate tectonics and weak mantle plumes, maintaining its compositional heterogeneity at various spatial scales to the present day. Internal heat production probably played an important role in controlling plate dynamics in the early Archean, for which a different mode of mantle convection is suggested.

1. INTRODUCTION

Deciphering the nature of the geodynamic regime that reigned in the Archean Earth (>2.5 Ga) is challenging owing to the paucity of unambiguous observational constraints. Exposed Archean provinces occupy only ~5% of the continental surface area [*Nisbet*, 1987]. Their great ages also imply the difficulty of preserving their primary signature without being overprinted by subsequent metamorphism and tectonic reworking. It is thus not surprising that tectonic

settings in which Archean crust was formed are still controversial [e.g., *Arndt*, 1983, *Nisbet and Fowler*, 1983, *Bickle et al.*, 1994; *Komiya et al.*, 1999; *Grove and Parman*, 2004]. Correspondingly, the interpretation of relevant geological and geochemical data in terms of Archean geodynamics is often nonunique.

Complementary to field-oriented reconstruction, a theoretical approach based on the physics of mantle convection may provide a simple yet comprehensive framework, which could facilitate to piece together fragmental geological evidence. This theoretical approach usually employs the so-called parameterized convection model, and a number of different models have been published over the last three decades. Though those models may look more or less similar, there still exist subtle differences in model predictions,

Archean Geodynamics and Environments
Geophysical Monograph Series 164
10.1029/164GM03

with substantially different implications for Archean geology. It appears that geologists tend to be discouraged by this model uncertainty and to hesitate to take the theoretical approach very seriously. Compared to the early stage of parameterized convection studies, however, we now have a much improved understanding of mantle dynamics thanks to global seismic tomography, computational fluid mechanics, and a number of laboratory experiments on mantle rheology. As I will demonstrate throughout this paper, many previous models of parameterized convection are inconsistent with our current understanding of Earth, and successful models yield rather specific predictions for the thermal and dynamic state of the Archean Earth.

My strategy is to calculate the thermal history of Earth *backward* in time: Starting from the present-day condition, which is undoubtedly the best understood part of Earth's history, to the beginning of the Proterozoic. It is reasonably safe to assume the operation of plate tectonics for this period of time [e.g., *Windley*, 1993; *Hoffman*, 1997]. The inferred state at the Archean-Proterozoic boundary is then used to speculate on possible geodynamic regimes in the Archean Earth. The overall structure of this paper is the following. I begin with the detailed account of parameterized convection models. Though this description of methodology may be somewhat redundant in part with previous studies, a self-contained description is essential to clarify the nature of various built-in assumptions and to point out the most critical aspect of this type of modeling. The preferred thermal history of Earth is then shown to be drastically different from conventional wisdom. In order to test my theoretical predictions, I explore geological, geophysical, and geochemical implications of the new evolution model, and discuss relevant observations. Finally, potential research directions are offered for unresolved issues.

2. MODELING THE THERMAL HISTORY OF EARTH

In this section, I describe physical principles and assumptions behind both conventional and new parameterized convection models in detail. First I introduce the fundamental differential equation for global heat balance in section 2.1. In order to integrate this equation with time, we have to specify the present-day conditions, so the present-day thermal budget is summarized next in section 2.2. I then explain how to parameterize the temporal variation of internal heating and convective heat flux in section 2.3. The parameterization of internal heating is straightforward, but that of convective heat flux is not. Its conventional treatment is given first in section 2.3, and in section 2.4 I show how this gives rise to the thermal catastrophe paradox. I also review existing hypotheses to reconcile this paradox. My own hypothesis is that treating plate tectonics as simple thermal convection may be the source of

all troubles. A global energy balance approach is employed in section 2.5 to construct the new parameterization of convective heat flux appropriate for plate tectonics. There I summarize the physics of multiscale mantle convection, characterized by large-scale plate-tectonic circulation and small-scale lithospheric instabilities. Finally, a new evolution model of Earth is presented in section 2.6, with predictions for internal temperature, surface heat flux, and plate velocity in the past.

2.1. Global Heat Balance

How to model the thermal history of Earth, to first order, is conceptually simple. The fundamental equation is the following global heat balance equation [*Christensen*, 1985]:

$$C\frac{dT_i(t)}{dt} = H(t) - Q(t), \qquad (1)$$

where C is the heat capacity of the whole Earth (7×10^{27} J K^{-1} [*Stacey*, 1981]) and T_i is average internal temperature. The above equation denotes that the temporal variation of internal temperature is determined by the balance between internal heating, $H(t)$, and surface heat loss, $Q(t)$. If these heat source and sink were exactly balanced, the internal temperature would remain constant with time. When surface heat loss is greater than internal heating, dT_i/dt is negative, i.e., Earth cools down with time. Equation (1) may also be written as

$$Q(t) = H(t) - C\frac{dT_i(t)}{dt}, \qquad (2)$$

which simply expresses the well-known fact that surface heat flux (the left-hand side) is composed of two heat sources: radiogenic heat generation (the first term of the right-hand side) and secular cooling (the second term). The secular cooling includes primordial heat as well as gravitational energy release by core formation at the very early history of Earth. Heating by tidal dissipation within the solid Earth is known to be insignificant [e.g., *Verhoogen*, 1980], and it is neglected here. Equation (1) is the simplest formulation of global heat balance. One may elaborate it by considering mantle and core temperatures separately [e.g., *Stevenson et al.*, 1983; *Stacey and Loper*, 1984; *Davies*, 1993], for which the parameterization of core-mantle interaction is required. The detailed modeling of the core-mantle boundary region is not attempted for now given our limited understanding of lower-mantle rheology. By employing equation (1), core cooling is assumed to follow mantle cooling, which should be valid to first order, and the internal temperature T_i is a good proxy for average mantle potential temperature, T_p (which is a hypothetical temperature of mantle adiabatically brought up to surface without melting). Since the heat capacity C is for the whole Earth, core heat flux is automatically included in the secular cooling term (the heat capacity of the core is ~1/5 of that of the entire Earth [*Stacey*, 1981]).

In order to integrate equation (1), we need to specify both $H(t)$ and $Q(t)$ over the period of interest. Determining the internal heating term $H(t)$ is relatively straightforward because the half lives of radiogenic elements and their relative abundances are well known. A real difficulty lies in how to model surface heat flux $Q(t)$, and this is where our understanding of mantle dynamics plays an essential role. In short, a heat-flow scaling law for mantle convection allows us to express Q as a function of internal temperature, which is in turn a function of time. Thus, $Q(t)$ is parameterized as $Q(T_i(t))$. The details of this parameterization are given later in this section (section 2.3 and section 2.5). The above ordinary differential equation can be integrated either forward or backward in time. A large fraction of previous studies chose to integrate forward in time, starting at 4.5 Ga [e.g., *McKenzie and Weiss*, 1975; *Schubert et al.*, 1980; *McGovern and Schubert*, 1989; *Davies*, 1993]. The initial condition is of course unknown, so a number of trial integrations are usually conducted to find an appropriate initial condition that leads to the present-day condition. As far as a single heat flow parameterization is assumed, the direction of integration does not matter. The problem is, however, that we do not know whether the entire history of Earth can be modeled by a single heat flow parameterization. The use of a *single* scaling law is a very strong assumption. Is the heat-flow scaling law appropriate for contemporary plate tectonics still valid for the early Earth dynamics? This should be left as an open question, and it is better to model the thermal history without introducing this assumption. In this regard, integrating backward in time starting from the present-day condition is more satisfactory. One can start with the present-day mantle temperature and surface heat loss, and continue to integrate as long as the operation of plate tectonics is safely assumed, for example, to the beginning of the Proterozoic. This approach can provide only a partial thermal history, but it should be viewed as a starting point to quantitatively consider Archean and Hadean geodynamics, which may be considerably different from plate tectonics. Hereinafter, I take the origin of the time axis ($t = 0$) at the present-day, and positive values denote time before present.

2.2. Present-Day Thermal Budget

Earth is currently releasing heat into the space at the rate of ~44 TW [*Pollack et al.*, 1993]. About 20% of the global heat flux (~8 TW) is estimated to originate in radiogenic isotopes in continental crust [*Schubert et al.*, 2001]; this is conductive heat flux, irrelevant to mantle convection. The present-day convective heat flux, $Q(0)$, is thus ~36 TW. On the other hand, cosmochemical and geochemical studies suggest that the radiogenic heat production of the bulk silicate Earth (i.e., mantle after core segregation but before the extraction of continental crust) is ~20 TW [*McDonough and Sun*, 1995]. Since ~8 TW must be sequestered in continental crust, however, only ~12 TW is available for convection. (Of course, continental mass and thus the fraction of heat producing elements in continents may have been smaller in the past. As discussed later, however, modeling with constant continental mass can be justified as far as the post-Archean is concerned.)

At this point, it is convenient to introduce the Urey ratio [*Christensen*, 1985], defined as $\gamma(t) = H(t)/Q(t)$, to measure the relative importance of internal heating with respect to total convective heat flux. Some literature adopts its reciprocal version as the Urey ratio, so readers must use caution when comparing this study with previous studies. My definition is the same as that of *Christensen* [1985]; $\gamma = 0$ corresponds to the case of no internal heating whereas $\gamma = 1$ denotes an exact balance between internal heating and surface heat loss. The Urey ratio is a time-dependent quantity as H and Q can vary independently to each other. From the above global heat budget, we can see that the present-day (cosmochemical) Urey ratio $\gamma(0)$ is ~0.3. I note that there is some confusion in literature when calculating the Urey ratio. *Schubert et al.* [2001], for example, arrive at the Urey ratio of ~0.6 by dividing the bulk-Earth heat production of ~20 TW by convective heat flux of ~36 TW. This is clearly a mistake; the numerator H in this case includes heat production in continental crust, which is, however, not considered as a part of convective heat flux Q.

One problem with the above estimation of the Urey ratio (~0.3) is that it is inconsistent with petrological observations. When the mantle rises beneath mid-ocean ridges, it starts to melt typically at the depth of 60-80 km [*McKenzie and Bickle*, 1988; *Langmuir et al.*, 1992], and the product of this mantle melting is known as mid-ocean ridge basalts (MORB). Terrestrial magmatism is dominated by mid-ocean ridge magmatism [*Crisp*, 1984], so the bulk of petrologically-observable mantle is restricted to the MORB source mantle. It has long been known that the MORB source mantle is depleted in heat-producing elements, i.e., K, U, and Th [*Jochum et al.*, 1983]. If the MORB source mantle constitutes the entire mantle, for example, its heat production would be ~6 TW at most (based on 8 ppb U, 16 ppb Th, and 100 ppm K), corresponding to the Urey ratio of less than 0.17. (Note: If we also include source mantle creating hotspots and large igneous provinces, this value may increase by ~10%.) Although the preferred value of heat production according to *Jochum et al.* [1983] is much lower than this (~2.4 TW), ~6 TW appears to be reasonable because a recent estimate of

MORB primary melt composition implies ~100 ppm K in the source mantle [*Korenaga and Kelemen*, 2000]. This is also supported by a more comprehensive compilation of global MORB database [*Su*, 2002]. At any rate, the cosmochemical Urey ratio should be regarded as the *upper* bound, and I will consider the range of the Urey ratio from 0.15 (petrological) to 0.3 (cosmochemical) in my modeling.

2.3. Parameterization of H(t) and Q(t)

Internal heat production in Earth is provided by the following four radiogenic isotopes: ^{238}U, ^{235}U, ^{232}Th, and ^{40}K. Heat production in the past was of course greater than the present because of radiogenic decay, which also changes the relative abundance of those four isotopes owing to differences in their half lives. The heat source term may thus be modeled as:

$$H(t) = H(0) \sum_{n=1}^{4} h_n \exp(\lambda_n t),$$ (3)

where

$$H(0) = \gamma(0)Q(0),$$ (4)

and

$$h_n = \frac{c_n p_n}{\sum_n c_n p_n}.$$ (5)

Present-day relative concentrations of the radiogenic isotopes are denoted by c_n, and their heat generation rates by p_n. Values used for these parameters are summarized in Table 1.

The parameterization of surface heat flux $Q(t)$ is more involved. Some earlier studies assumed that $Q(t)$ should behave similarly as $H(t)$ [e.g., *McKenzie and Weiss*, 1975; *Bickle*, 1986], but such a simple relation does not hold for a system in non-steady state (e.g., the cooling Earth) [*Daly*, 1980]. A conventional approach is explained first here, to be compared later with a more recent parameterization proposed for plate-tectonic convection. The conventional parameterization is based on the following scaling law for convection:

$$Nu \propto Ra^\beta$$ (6)

where Nu is the Nusselt number, which is heat flux normalized by conductive heat flux, and Ra is the Rayleigh number, which is a measure of convective potential for a given fluid system. The exponent β in express on (6) determines the sensitivity of surface heat flux with respect to a change in the vigor of convection, and a number of experimental and numerical studies shows that this exponent is ~0.3 for *thermal* convection [e.g., *Turcotte and Oxburgh*, 1967; *Gurnis*, 1989; *Davaille and Jaupart*, 1993; *Korenaga*, 2003]. In terms of internal temperature T_i, these two nondimensional parameters may be expressed as

$$Nu = \frac{Q}{kA(T_i / D)},$$ (7)

where A, D, and k denote surface area, the system depth, and thermal conductivity, respectively, and

$$Ra = \frac{\alpha \rho g T_i D^3}{\kappa \eta(T_i)},$$ (8)

where α, ρ, g, and κ denote thermal expansivity, density, gravitational acceleration, thermal diffusivity, respectively. Viscosity, $\eta(T_i)$, indicates its dependence on temperature.

From equations (6)-(8), one can see that

$$Q \propto \frac{T_i^{1+\beta}}{(\eta(T_i))^\beta}$$ (9)

The rheology of Earth's mantle is known to be strongly temperature-dependent [e.g., *Weertman*, 1970], and hotter mantle has lower viscosity. Thus, the above scaling indicates that hotter mantle convects faster, releasing correspondingly higher heat flux. More precisely, the temperature dependency takes the following Arrhenius form:

Table 1. Radiogenic Heat Production.

Isotope	c_n^a	p_n [W/kg][b]	h_n	$\tau_{1/2}$[Gyr][b]	λ_n[1/Gyr][b]
^{238}U	0.9927	9.37×10^{-5}	0.372	4.47	0.155
^{235}U	0.0072	5.69×10^{-4}	0.0164	0.704	0.985
^{232}Th	4.0	2.69×10^{-5}	0.430	14.0	0.0495
^{40}K	1.6256	2.79×10^{-5}	0.181	1.25	0.555

[a]Relative concentration normalized by the abundance of total U, with U:Th:K = 1:4:(1.27×10^4), $^{238}U/U = 0.9927$, $^{235}U/U = 0.0072$, and $^{40}K/K = 1.28 \times 10^{-4}$. All Th is ^{232}Th.
[b]From *Turcotte and Schubert* [1982].

$$\eta(T_i) \propto \exp\left(\frac{E}{R(T_i + T_{\text{off}})}\right), \quad (10)$$

where E is activation energy, R is universal gas constant, and T_{off} is 273 K to convert T_i (which is defined with respect to Earth's surface temperature) to absolute temperature. The functional dependence of surface heat flux on internal temperature is controlled by the activation energy, and for each chosen value of E, one can derive (by least-squares fit) the following heat-flow scaling law:

$$Q \approx a' T_i^{\beta'}. \quad (11)$$

Examples with $E = 0, 300$, and 600 kJ/mol are shown in Figure 1. The scaling constant a' is determined to yield Q of 36 TW at T_i of 1350°C. The case of $E = 0$ corresponds to constant viscosity irrespective of internal temperature. Diffusion creep is typically characterized by $E \sim 300$ kJ/mol, whereas dislocation creep by $E \sim 500\text{-}600$ kJ/mol [*Karato and Wu*, 1993]. Within the Newtonian approximation, effective activation energy for nonlinear (dislocation) creep is reduced by a factor of ~ 2 [*Christensen*, 1984]. Thus, regardless of the type of microscopic deformation mechanism involved, $E \sim 300$ kJ/mol is probably appropriate to describe the temperature dependency of mantle viscosity.

2.4. Thermal Catastrophe and Common Resolutions

The result of backward integration of equation (1) from the present-day mantle temperature of 1350°C [*Langmuir et al.*, 1992] is shown in Figure 2, with the Urey ratio of 0.15-0.3 and with the activation energy of 300 kJ/mol. Internal temperature quickly rises and diverges toward infinity before reaching 2 Ga. This is known as thermal catastrophe, a first-order paradox in reconstructing Earth's cooling history, because mantle temperature is believed to have been lower than ~1800°C even in the Archean [e.g., *Abbott et al.*, 1994]. Note that this catastrophic thermal history is inconsistent with the petrological constraints on the Archean thermal state, well beyond the uncertainty associated with the genesis of komatiites (i.e., ~1800°C for dry melting and ~1500°C for wet melting [*Grove and Parman*, 2004]). The reason for this rapid increase in temperature is a positive feedback between secular cooling and heat flux. The Urey ratio of 0.3, for example, means that 70% of convective heat flux must come from secular cooling, which results in a sharp increase in temperature back in time. Higher temperature, in turn, corresponds to higher convective heat flux (equation (11)),

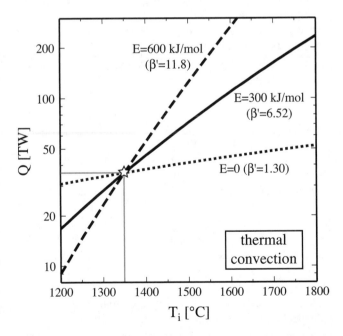

Figure 1. "Conventional" heat flux parameterization based on $Nu\text{-}Ra$ scaling and single mantle rheology (equation (11)). Three cases are shown: (1) $E = 0$ (dotted, $a' = 3.07 \times 10^{-3}$, $\beta' = 1.30$), (2) $E = 300$ kJ/mol (solid, $a' = 1.35 \times 10^{-19}$, $\beta' = 6.52$), and (3) $E = 600$ kJ/mol (dashed, $a' = 5.95 \times 10^{-36}$, $\beta' = 11.8$). Star denotes the present-day mantle condition, to which the scaling law is calibrated.

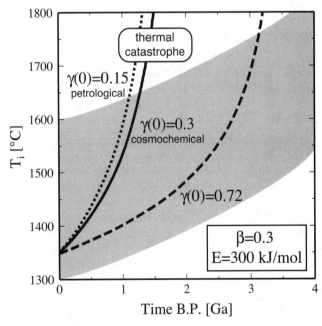

Figure 2. Thermal evolution modeling with conventional heat flux scaling of equation (11). The activation energy is set to 300 kJ/mol. Three present-day Urey ratios are used: 0.15 (petrological, dotted), 0.3 (cosmochemical, solid), and 0.72 (superchondritic, dashed). Gray shading denotes the range of potential temperatures recorded in MORB-like suites including greenstone belts and ophiolite suites Figure 5 [*Abbott et al.*, 1994, their Figure 5].

but because radiogenic heat production increases much more slowly, the Urey ratio further decreases, i.e., even greater secular cooling is required in the past. This thermal catastrophe paradox is one of long-standing issues in global geodynamics and geochemistry: A simple theory of convective cooling, when combined with geochemical constraints on the heat production budget, cannot reproduce a reasonable cooling history of Earth.

One common resolution to thermal catastrophe is to use the Urey ratio greater than 0.7. This high heat production reduces the degree of secular cooling and pushes back the timing of thermal catastrophe into the early Earth history (Figure 2). Almost all of parameterized whole-mantle convection studies in the past took this approach [e.g., *Schubert et al.*, 1980; *Turcotte*, 1980; *Stevenson et al.*, 1983; *Richter*, 1985; *McGovern and Schubert*, 1989; *Davies*, 1993]. Some considered that geochemistry did not tightly constrain the abundance of radiogenic isotopes in the mantle, and treated the Urey ratio as a free parameter to be determined by parameterized convection models. This type of resolution is not particularly meaningful, however, as long as whole-mantle convection is assumed. As already noted, the heat production of petrologically observable mantle is even lower than the cosmochemical estimate, corresponding to the Urey ratio of ~0.15. If mid-ocean ridges randomly sample the mantle that is mixed by whole-mantle convection, it is difficult to explain this low observed Urey ratio while claiming that the true Urey ratio is actually as high as 0.7. This discrepancy in the Urey ratio is simply too high to be reconciled, even by whole-mantle convection with distributed blobs of enriched composition [*Helffrich and Wood*, 2001] or with transition-zone water filter [*Bercovici and Karato*, 2003], because source mantle for ocean island basalts (OIB) does not have such high concentrations of radiogenic elements (i.e., the melting of 'primitive' mantle is often used to model the trace-element geochemistry of OIB). Thus, although increasing the Urey ratio does provide a reasonable thermal history, this is not an attractive solution if our goal is to understand both the physics and chemistry of Earth. Accordingly, predictions for heat flux and plate velocity based on this very high Urey ratio must be viewed with suspicion.

The other common resolution is to invoke layered-mantle convection. Traditionally, the lower mantle below the 660-km seismic discontinuity had often been considered to be isolated from upper-mantle convection. Layered-mantle convection can avoid thermal catastrophe because the "effective" Urey ratio for upper-mantle convection is high owing to heat input from the lower convection system [e.g., *Richter*, 1985]. In other words, heat is released less efficiently if convection is layered. The concept of a layered mantle has also an advantage of explaining the depleted nature of the upper mantle. Geophysical evidence for whole-mantle convection, however,

has steadily been growing in the last decade; in addition to subducting slabs reaching deep mantle [*van der Hilst et al.*, 1997; *Fukao et al.*, 2001], we now have seismic evidence for mantle plumes rising from the core-mantle boundary [*Romanowicz and Gung*, 2002; *Montelli et al.*, 2004]. Confronted with these geophysical observations, *Allegre* [1997] proposed that the mode of mantle convection may have changed from layered-mantle to whole-mantle recently, at ~1 Ga. However, this scenario would not resolve thermal catastrophe because mantle temperature is already high (~1600°C) at 1 Ga (Figure 2) and the Urey ratio is too low (<0.1). Others have tried to modify the traditional model of layered-mantle convection to be more consistent with recent geophysical observations, such as deep layering with an irregular and nearly invisible interface [*Kellogg* et al., 1999]. It must be noted, however, that the layered mantle may resolve thermal catastrophe only in the upper mantle; *Spohn and Schubert* [1982] pointed out that the lower mantle still becomes extremely hot in the past. Recent variants of the layered convection model are characterized by complex thermochemical convection allowing finite mass transfer between layers [e.g., *Davaille*, 1999; *Gonnermann et al.*, 2002], which may have a potential to avoid thermal runaway in the lower mantle. The dynamical behaviors of such elaborate models are, however, currently poorly constrained by observations because the layering is not associated with major seismic discontinuities in the mantle.

The positive feedback that leads to thermal runaway is effected by the heat-flow scaling law of equation (11). Though the original *Nu-Ra* relationship (equation (6)) is firmly established for thermal convection, it is not immediately obvious how valid it would be for plate-tectonic convection in Earth's mantle. A frequently used theoretical argument for $\beta \sim 0.3$ in equation (6) is that heat flux Q becomes independent of the system height D when $\beta = 1/3$ (Nu is proportional to QD, and Ra is proportional to D^3). When a fluid is vigorously convecting, how heat is released from its surface is expected to depend only on the top boundary layer, not on the entire system, so this theoretical justification appears reasonable [e.g., *Howard*, 1966]. However, it is also true that, when the material properties of a fluid are not constant, we need more than one nondimensional parameter (i.e., Ra) to describe the convection system [*Buckingham*, 1914; *Barenblatt*, 1996]. In other words, Nu is no longer a simple function of Ra only, and there is no reason to expect that the exponent β is in the neighborhood of 1/3. One may still hope to cast the heat flow scaling law in the form of equation (6), and indeed it is possible for the case of stagnant-lid convection with temperature-dependent viscosity [e.g., *Davaille and Jaupart*, 1993; *Solomatov*, 1995]. As it will be discussed later, however, the spatial variation of mantle rheology may considerably be affected by chemical differentiation associated with plate tectonics, and adhering to the conventional *Nu-Ra* relationship becomes awkward.

We must remember that plate tectonics is more than thermal convection with simple material properties. If mantle rheology characterized by purely temperature-dependent viscosity as in equation (10), mantle convection should be in the regime of stagnant-lid convection [*Solomatov*, 1995]. The surface layer becomes too rigid to deform (surface viscosity is greater than interior viscosity by >15 orders of magnitude), and only the hot interior convects beneath a single plate covering the entire surface; this is the situation believed to be in action for other terrestrial planets like Venus and Mars [*Schubert et al.*, 2001]. Strictly speaking, therefore, equation (11) is valid only when $E \approx 0$ (i.e., isoviscous convection). Realistic activation energy corresponds to stagnant-lid convection, in which surface plate velocity is zero. There must be some mechanism on Earth that could reduce the strength of the cold boundary layer so that the entire boundary layer can sink into the interior, though there is currently no consensus on the actual mechanism. Understanding the generation of plate tectonics from first principles is still at the frontier of geodynamics [e.g., *Bercovici*, 2003].

We can, however, attempt to construct a heat-flow scaling law appropriate for plate tectonics, by examining how energy is created and consumed in convecting mantle. How plate tectonics can be initiated is not questioned in this approach. What concerns us is how plate tectonics is maintained and regulated when it is already taking place. The boundary layer theory, based on this energy balance on a global scale, provides an invaluable guideline to infer how plate tectonics may have operated when the mantle was hotter. The energetics of plate-tectonic convection is thus described in the following.

2.5. Energetics of Plate-Tectonic Convection

2.5.1. Large-scale flow. A conceptual mantle system is depicted in Figure 3. Plate tectonics is a surface manifestation of this large-scale mantle convection, whose velocity scale is denoted by U. The mantle is so viscous that inertia effects are virtually absent. Starting with the conservation of momentum for this zero-Reynolds-number system, and incorporating free-slip boundary conditions appropriate for whole-mantle convection, one can readily arrive at the following integral relationship [*Chandrasekhar*, 1981]:

$$\int_V \sigma_{ij} \frac{\partial u_i}{\partial x_j} dV = \int_V u_i f_i dV, \qquad (12)$$

where integration is over the entire model domain V, σ_{ij} is stress, u_i is velocity, and f_i is external force. This global balance is exact. Einstein summation convention is assumed for indices i and j. With the constitutive relation and the

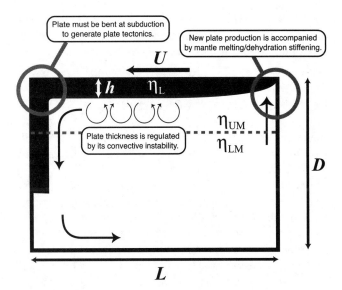

Figure 3. Conceptual flow geometry for plate-tectonic convection. Plate is moving with velocity U, and surface cooling combined with this horizontal motion results in the development of a cold boundary layer as denoted by solid shading. The thickness of this boundary layer is limited to h by its convective instability. Cold and strong plate, whose characteristic viscosity is given by η_L, is bent at subduction zone, and continues to sink into the mantle with vertical dimension of D. Lower-mantle viscosity η_{LM} is higher than upper-mantle viscosity η_{UM}. The aspect ratio of this large-scale convection cell is L/D.

incompressible fluid approximation, and also noting that gravity is the only external force, one can further simplify this to

$$2\int_V \eta \dot{\varepsilon}_{ij} \dot{\varepsilon}_{ij} dV = \int_V u_z \rho g dV, \qquad (13)$$

where $\dot{\varepsilon}_{ij}$ is strain rate. The left hand side corresponds to the rate of viscous dissipation, Φ_{vd}, whereas the right hand side expresses the rate of potential energy release, Φ_{pe}.

Negative buoyancy associated with subducting plate originates in thermal contraction, and it is proportional to $\alpha \rho T_i g$. The dimension of the subducting part of plate is proportional to Dh, where D is the height of whole mantle and h is the thickness of plate. Thus, the ratio of potential energy release may be expressed as [*Turcotte and Schubert* (1982)]:

$$\Phi_{pe} = C_{pe} \alpha \rho_0 g T_i D h U, \qquad (14)$$

where ρ_0 is reference density and C_{pe} is a scaling constant to be discussed later.

This energy input is consumed by at least two kinds of viscous dissipation mechanisms [*Solomatov*, 1995; *Conrad* and Hager, 1999]. The first one is mantle-wide dissipation, in which strain rate is proportional to U/D. The rate of this type of dissipation is thus given by

$$\Phi_{vd}^M = C_{vd}^M \eta_M \left(\frac{U}{D} \right)^2 D^2 = C_{vd}^M \eta_M U^2, \qquad (15)$$

where η_M is average mantle viscosity. Since lower-mantle viscosity is believed to be greater than upper-mantle viscosity by ~2 orders of magnitude [e.g., Hager, 1991; King, 1995], I assume that $\eta_M \approx \eta_{LM}$. The important aspect of plate-tectonic convection is the deformation of strong surface boundary layer at subduction; otherwise plate tectonics does not take place. This subduction-zone dissipation is proportional to lithospheric viscosity, η_L. Simply applying temperature-dependent viscosity would predict unrealistically high lithospheric viscosity (thus stagnant-lid convection), so some kind of "effective" lithospheric viscosity, which takes into account the effects of various weakening mechanisms such as brittle failure, must be used instead. Based on the numerical studies of mantle convection with strong plates [Zhong and Gurnis, 1995; Gurnis et al., 2000], a reasonable range for η_L appears to be 10^{22}-10^{24} Pa s. Though lithospheric viscosity has a large uncertainty, the main conclusions of this paper are not substantially affected as long as η_L is greater than η_M by more than one order of magnitude. Bending strain rate is proportional to Uh/R^2 [Turcotte and Schubert, 1982; Conrad and Hager, 1999], where R is the radius of curvature, and the volume of plate being bent scales with Rh, so the rate of subduction-zone dissipation may be expressed as

$$\Phi_{vd}^S = C_{vd}^S \eta_L \left(\frac{Uh}{R^2} \right)^2 Rh = C_{vd}^S \eta_L \left(\frac{h}{R} \right)^3 U^2 \qquad (16)$$

Another potentially important energy sink is fault-zone dissipation at subduction zones. Conrad and Hager [1999] suggested that this may add up to ~10% of total dissipation, but also that it is not constrained well because of its trade-off with mantle-wide dissipation. For the sake of simplicity, this type of dissipation is ignored here, so the global energy balance is given by

$$\Phi_{pe} = \Phi_{vd} = \Phi_{vd}^M + \Phi_{vd}^S. \qquad (17)$$

Because the rate of potential energy release is proportional to U and the rate of viscous dissipation is proportional to U^2, the above energy balance can be solved for the velocity scale U. Using the scaling relation between plate velocity and surface heat flux [Turcotte and Schubert, 1982],

$$Q \propto T_i \sqrt{U}, \qquad (18)$$

we finally arrive at the parameterization of plate-tectonic heat flux as

$$Q = a \left(\frac{C_{pe} \alpha \rho_0 g T_i^3 Dh}{C_{vd}^M \eta_M + C_{vd}^S \eta_L (h/R)^3} \right)^{\frac{1}{2}}, \qquad (19)$$

where the constant a is to be calibrated so that surface heat flux for present-day mantle temperature equals to 36 TW. Based on a series of numerical experiments and comparing them to present-day plate velocities, Conrad and Hager [1999] estimated the scaling coefficients for energy source and sink as

$$C_{pe} \sim \frac{1}{\sqrt{\pi}}, \qquad (20)$$

$$C_{vd}^M \sim 3(L/D + 2.5), \qquad (21)$$

$$C_{vd}^S \sim 2.5. \qquad (22)$$

Uncertainty associated with these coefficients is probably less than an order of magnitude. This does not affect the main conclusion of this paper, which is based on the observation that, under certain mantle conditions, Φ_{vd}^S can become greater than Φ_{vd}^M by a few orders of magnitude.

To complete the heat flux parameterization, one must determine the implicit temperature dependency of parameters involved if any. Average mantle viscosity η_M is assumed to take the Arrhenius-type temperature dependency. Lithospheric viscosity η_L is, on the other hand, most likely independent of internal temperature, because the surface boundary layer is always cold (by definition). The radius of curvature for plate bending is assumed to be constant at the current value of ~200 km [Bevis, 1986]; it might have changed through time, though no scaling relation is available at present. According to a global study of trench topography by Levitt and Sandwell [1995], however, plate bending at subduction zone is characterized by a roughly constant spatial scale (termed 'flexural parameter') regardless of the age of subducting plate. This observation may justify the use of the constant R.

Plate thickness at subduction zone, h, on the other hand, could strongly depend on mantle temperature, at least in two different ways. Surface cooling gives rise to the gradual growth of thermal boundary layer, but this plate thickening does not continue indefinitely because plate eventually becomes convectively unstable. When mantle is hotter, its viscosity is lower, enhancing this convective instability. Thus plate thickness is expected to decrease with increasing mantle temperature. When mantle is hotter, however, it also starts to melt deeper beneath mid-ocean ridges, leaving thicker dehydrated (thus stiff) lithosphere. This dehydration stiffening suppresses convective instability and helps the growth of thermal boundary layer. In this case, plate thickness is expected to increase with increasing mantle temperature. As evident from equation (19), plate thickness essentially determines the relative importance of subduction-zone dissipation over mantle-wide dissipation, so it is important to know how plate thickness may be regulated by the above two competing processes. This is a problem known as the onset of sublithospheric convection or the convective instability of oceanic lithosphere, which will be discussed next.

2.5.2. Convective stability of oceanic lithosphere. The role of small-scale convection in modulating plate thickness has long been speculated on the basis of physical reasoning. Recent high-resolution seismological studies of suboceanic upper mantle [*Katzman et al.*, 1998; *Montagner*, 2002, *Ritzwoller et al.*, 2004] suggest that such sublithospheric dynamics may indeed be taking place. The convective stability of oceanic lithosphere has been investigated by a number of studies [e.g., *Parsons and McKenzie*, 1978; *Yuen and Fleitout*, 1984; *Buck and Parmentier*, 1986; *Davaille and Jaupart*, 1994; *Korenaga and Jordan*, 2003]. Obviously, the onset of small-scale convection is controlled primarily by mantle rheology, but *how* it is controlled was not properly understood until recently; it was considered to be extremely sensitive to the activation energy of temperature-dependent viscosity [*Davaille and Jaupart*, 1994; *Choblet and Sotin*, 2000]. On the basis of carefully designed numerical experiments as well as a new scaling analysis with the differential Rayleigh number, *Korenaga and Jordan* [2003] demonstrated that the onset of convection is much less sensitive to the activation energy than previously thought. *Huang et al.* [2003] further confirmed this by numerical simulation with a wider parameter range. *Korenaga and Jordan* [2002a] also extended the scaling theory for the case of temperature- and depth-dependent viscosity, and the nondimensionalized onset time of convection, t_c^*, is given in the following form:

$$Ra_c = \frac{\pi^2}{2^5} Ra \cdot F(t_c^*), \qquad (23)$$

where Ra_c is the critical Rayleigh number ($\sim 10^3$) and $F(t^*)$ is a functional that depends on time-dependent buoyancy distribution (see *Korenaga and Jordan* [2002a] and *Korenaga and Jordan* [2003] for its full account). This scaling law forms an important theoretical basis for the parameterization of plate thickness in terms of mantle temperature, as discussed below.

One critical difference between classical thermal convection and plate-tectonic convection is that the latter involves chemical differentiation during the formation of the top boundary layer (Figure 3). Adiabatic upwelling beneath

divergent plate boundaries results in decompressional melting, the product of which is observed as oceanic crust. The MORB source mantle is slightly hydrated with ~ 100-200 ppm H_2O [*Michael*, 1988; *Dixon et al.*, 1988], and with the present-day potential temperature, it first crosses wet solidus at ~ 100 km and starts to get dehydrated, followed by more copious melting when it crosses dry solidus at ~ 60 km [*Hirth and Kohlstedt*, 1996] (Figure 4). Upwelling mantle is assumed here to continue to melt to the surface with the maximum degree of melting of 20%, to have the degree of mantle depletion consistent with the thickness of normal oceanic crust (i.e., 10% melting on average of 60 km mantle column produces 6-km-thick oceanic crust). In reality, mantle melting continues only to the base of oceanic crust, and such details will be considered later when discussing the net buoyancy of oceanic lithosphere. In terms of convective instability of lithosphere, this complication does not concern us because the instability is a relatively local phenomenon, taking place near the base of lithosphere.

This mantle melting upon the formation of new plate at mid-ocean ridges has two important consequences. First, residual mantle after melt extraction is less dense than unmelted mantle. This compositional effect can be quantified as

$$\left(\frac{d\rho}{dF}\right)_C = \frac{d\rho}{d\text{Mg}\#}\frac{d\text{Mg}\#}{dF} \qquad (24)$$

where F is degree of melting, and Mg# is defined as $100\times$ molar Mg/(Mg+Fe) of mantle composition. *Jordan* [1979] suggested $(d\rho/d\text{Mg}\#)$ to be -15 kg/m^3/Mg#, which is recently confirmed by *Lee* [2004] based on the extensive compilation of mantle xenolith data. Mantle melting models generally show that $(d\text{Mg}\#/dF) \sim 0.08$ Mg#/% [*Langmuir et al.*, 1992; *Kinzler and Grove*, 1992], thus $(d\rho/dF)_C \approx -1.2$ kg/m^3/%. Residual mantle also cools down because of the latent heat of fusion, and this thermal effect tends to offset the above compositional effect on density. It may be expressed as

$$\left(\frac{d\rho}{dF}\right)_T = \alpha\rho\frac{dT}{dF} = \frac{\alpha\rho H_f}{100c_v}, \qquad (25)$$

where H_f and c_p are the latent heat of fusion and specific heat of mantle. With the uncertainty associated with H_f [*Langmuir et al.*, 1992] $(d\rho/dF)_T \sim 0.3$–0.6 kg/m^3/%, thus compensating the compositional effect by 50% at most (Figure 4).

Second, dehydration caused by mantle melting can considerably stiffen residual mantle. The presence of impurities such as defects is essential for the subsolidus deformation of silicate mantle, and the most important impurity in this regard is the trace amount of water in nominally anhydrous minerals [*Karato*, 1986; *Hirth and Kohlstedt*, 1996]. This dehydration-induced stiffening has been confirmed by a

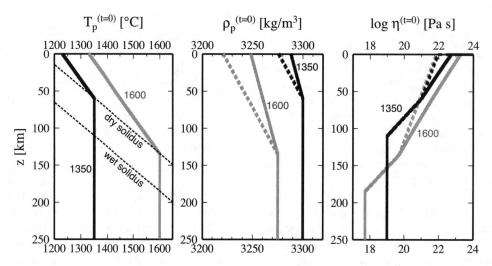

Figure 4. Zero-age oceanic mantle profiles for two potential temperatures of unmelted mantle: 1350°C (solid) and 1600°C (gray). (left) Potential temperature. Dry solidus is based on *Takahashi and Kushiro* [1983], and wet solidus is placed 50 km below dry solidus. A constant melting rate (after crossing dry solidus) of 10%/GPa is assumed, and H_f/c_p is set to 600 K. (middle) Potential density. Dashed lines denote the case of melt depletion effect only. The effect of cooling due to the latent heat of fusion is incorporated in solid lines. (right) Viscosity. Reference viscosity (for oceanic upper mantle) is 10^{19} Pa s at 1350°C [*Hager*, 1991; *Karato and Wu*, 1993; *Hirth and Kohlstedt*, 1996; *Korenaga and Jordan*, 2002b], and activation energy for temperature-dependent viscosity is set to 300 kJ/mol. Dehydration stiffening is modeled as 100-fold viscosity increase during the wet-to-dry transition, followed by 3-fold increase per every additional 10% of melting. Dashed lines show the effect of dehydration whereas solid lines incorporate cooling upon melting.

number of laboratory experiments in mineral physics [e.g., *Hirth and Kohlstedt*, 2003]. When mantle melts during its ascent, water is strongly partitioned into the melt phase, so residual mantle becomes much more viscous than unmelted mantle. Following *Hirth and Kohlstedt* [1996], I assume 100-fold viscosity increase during the wet-to-dry transition, followed by 3-fold increase per every another 10% of melting (Figure 4).

These changes in the density and viscosity of residual mantle can support the growth of thermal boundary layer. It can be quantified by calculating the onset time of small-scale convection in the presence of chemically depleted lithosphere. To evaluate the relative importance of these two effects, I first consider the influence of dehydration stiffening. When the mantle was hotter in the past, it must cross solidus at a deeper level, producing more melt (i.e., thicker crust) and thicker depleted lithosphere (Figure 4). Therefore, even though the viscosity of asthenosphere is lower, this alone cannot limit the growth of thermal boundary layer, which is stabilized by stiff depleted lithosphere. Figure 5 shows predictions for maximum plate thickness based on the onset-time scaling law of *Korenaga and Jordan* [2002a]. When potential temperature is 1600 °C, for example, plate can grow as thick as ~170 km; if dehydration stiffening is not taken into account, plate thickness is limited to only ~40 km. This result is basically the same as that of *Korenaga* [2003],

in which a slightly different formulation of stiffening is employed.

Adding the effect of compositional buoyancy is not straightforward, because we do not yet have a scaling law (verified by finite-amplitude convection experiments) for the onset time when both compositional and viscous stratifications are present. Probably the closest one available is that of *Zaranek and Parmentier* [2004], who studied the influence of linear compositional stratification on the onset of convection with temperature-dependent (but not composition-dependent) viscosity. They showed that chemical buoyancy hardly affects the onset time when the stratification is restricted to a rigid boundary layer. This is almost exactly the case for depleted lithosphere (Figure 4), so compositional buoyancy is not expected to modify predicted plate thickness if dehydration stiffening is already present. In order to verify this, I approximate the effect of compositional buoyancy using "effective temperature," T_{eff}, which is defined through

$$\rho(t,z) = \rho_p(z)(1 - \alpha(T(t,z) - T_o))$$
$$= \rho_o(1 - \alpha(T_{eff}(t,z) - T_o)), \qquad (26)$$

where $\rho_p(z)$ is potential density at $t = 0$ (Figure 4), and ρ_o is reference density of unmelted mantle at $T = T_o$. Predicted plate thickness with this approximated compositional buoyancy is also shown in Figure 5. For potential temperature less

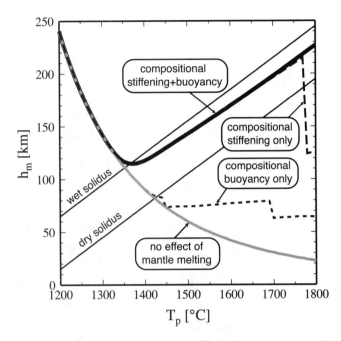

Figure 5. Maximum plate thickness as a function of mantle potential temperature. As in Figure 4, reference viscosity is 10^{19} Pa s at 1350°C, and activation energy for temperature-dependent viscosity is set to 300 kJ/mol. Gray curve corresponds to this purely temperature-dependent viscosity. Dotted line shows the effect of added compositional buoyancy, dashed line shows the effect of added compositional stiffening, and solid line shows the effects of adding both. For comparison, dry and wet solidi are also shown. Plate thickness for $T_p > 1350$°C roughly follows the wet-to-dry transition.

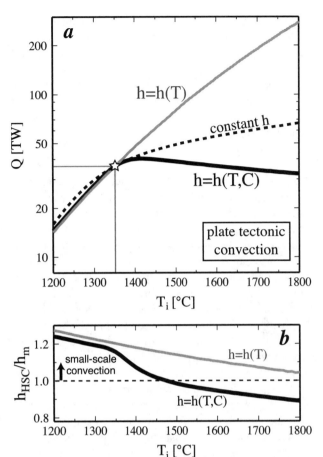

Figure 6. (a) Heat flux parameterization for plate-tectonic convection (equation (19)). Dotted curve is for the case of constant plate thickness of 100 km. The case of purely temperature-dependent viscosity is denoted by gray lines, and that of compositional stiffening and buoyancy is by solid lines. (b) The role of small-scale convection in regulating maximum plate thickness (h_m) is shown by comparing with hypothetical thickness by half-space cooling only (h_{HSC}). See text for discussion.

than ~1750°C, there is no noticeable effect of compositional buoyancy if lithosphere is already stiffened by dehydration, thus confirming the above speculation. When the mantle is as hot as 1800°C, thermal weakening becomes so significant to almost cancel dehydration stiffening (Figure 5, dashed). Compositional buoyancy does help to stabilize in this extreme case. Also note that the effect of compositional buoyancy alone is insufficient to maintain 100-km-thick plate when mantle becomes hotter than present (Figure 5, dotted).

The most likely variation of plate thickness in response to a change in mantle temperature is shown as solid curve in Figure 5, which takes into account both dehydration stiffening and compositional buoyancy. Interestingly, its minimum roughly corresponds to the present-day condition. When mantle is cooler, plate thickens owing to temperature-dependency of mantle viscosity, and when it is hotter, plate thickens owing to deeper melting with dehydration stiffening. Thus, as already pointed out by *Korenaga and Jordan* [2002b], the effect of mantle melting is hard to distinguish from that of temperature-dependent viscosity for present-day

(normal) oceanic lithosphere. It becomes important only when mantle was hotter in the past. Note that, even with the added chemical factors, plate thickness can still be modeled as a function of a single parameter, T_p, which simplifies its use in parameterized convection modeling. This is because mantle melting is also a function of mantle temperature, just as mantle viscosity.

2.5.3. Heat-flow scaling law. Using the plate thickness as shown in Figure 5 for equation (19), we can finalize the heat-flow scaling law for plate-tectonic convection (Figure 6a). Average mantle viscosity is assumed to be 10^{21} Pa s under present-day conditions, and lithospheric viscosity is set to 10^{23} Pa s. The aspect ratio of large-scale convection cell, L/D, is set to 2, which is representative of present-day plate geometry. When plate thickness is controlled only by temperature-dependent viscosity ($h = h(T)$), the corresponding scaling law is similar to the conventional one with $E = 300$ kJ/mol (Figure 1, solid). In this case, plate is thinner for hotter mantle, and correspondingly, subduction-zone dissipation is also smaller. Thus, when only thermal effects are considered, plate-tectonic convection behaves like classical thermal convection *even with this extra dissipation* [*Korenaga*, 2003]. The effects of mantle melting substantially modifies this scaling, and because of thicker depleted lithosphere, plate tectonics becomes less efficient in heat transport when mantle is hotter (Figure 6a, solid). Traditionally, hotter mantle in the past is almost always *assumed* to convect faster, so this scaling law may appear counter-intuitive. However, this conventional wisdom is based on the notion that hotter mantle is less viscous. Though this temperature dependency is correct in most cases, it may not hold for the most critical part of plate-tectonic convection, i.e., oceanic lithosphere. It is a well-established fact in petrology and mineral physics that plate tectonics involves chemical differentiation, and that this differentiation stiffens the residual mantle (oceanic lithosphere). In other words, effective mantle viscosity in terms of large-scale circulation may be *higher* for *hotter* mantle, reducing its strength of convection and thus heat transport. In some sense, the classical *Nu-Ra* relation (equation (6)) with a positive exponent still holds; only the relation between *Ra* and mantle temperature is modified by the effects of mantle melting.

One can also predict plate velocity U from heat flux through equation (18) and calculate hypothetical plate thickness by half-space cooling (i.e., without the onset of convection) at subduction. This plate thickness, h_{HSC}, is compared with the maximum plate thickness allowed by convective stability in Figure 6b. Plate velocity is scaled so that present-day value is equal to 4 cm/yr, and the age of subducting plate is calculated with the assumption of $L/D = 2$ (i.e., ~6000 km away from a ridge axis). For the case of purely temperature-dependent viscosity, h_{HSC} is always greater than the maximum thickness, meaning that plate thicknesses is always regulated by small-scale convection. With depleted lithosphere, the role of convective instability becomes insignificant when mantle is hotter than ~1450°C. Depleted lithosphere becomes so thick that thermal conduction does not fully propagate down to its base before subduction. This justifies the simple scaling of plate velocity and surface heat flux expressed by equation (18). Note that difference in these two thicknesses is only < 10%, which also justifies the use of maximum plate thickness in equation (6). That is, plate thickness, plate velocity, and aspect ratio are all self-consistent in the new heat-flow scaling law.

2.6. Evolution of Plate Tectonics

A predicted thermal history based on the energetics of plate-tectonic convection are presented in Figure 7a. Reduced heat flux for convection with hotter mantle breaks the positive feedback loop that leads to thermal catastrophe, and whole-mantle convection can reproduce a moderate cooling history with the Urey ratio of 0.15-0.3. Compared to thermochemical layered convection models, this resolution may be more appealing because it is based on shallow mantle processes and does not attempt to hide discrepancies between geodynamics and geochemistry in deep mantle phenomena. Decompressional melting beneath mid-ocean ridges, the role of water in upper-mantle rheology, and plate bending at subduction are all well-known concepts. It is the self-consistent combination of these components that gives rise to this (only apparently) counter-intuitive behavior of plate tectonics. A corresponding history of plate velocity is shown in Figure 7b. Thicker depleted lithosphere in the past retards plate motion; plate tectonics at the Archean-Proterozoic boundary may have been more sluggish than present by a factor of ~2.

The model results are shown back to 4 Ga, only to facilitate discussion on Archean geodynamic regimes, and not to imply that Archean tectonics is similar to contemporary plate tectonics. It will be shown later (section 4.2) that the energetics of plate-tectonic convection as derived above is likely to become invalid in the early Archean. Thus, sluggish plate tectonics in the post-Archean does not readily indicate that the present-day thermal state of Earth is sensitive to early Earth conditions; different modes of mantle convection in the Hadean and Archean could erase the signature of initial conditions. One robust feature of this new evolution model is, however, that the role of internal heating becomes progressively more important back in time (Figure 7c,d). This is a natural corollary of nearly constant convective heat flux over time, because the history of internal heat production is determined solely by radiogenic decay. The degree of secular cooling, therefore, is more subdued in the past, and so is core heat flux.

In what follows, I will explore various geological and geochemical implications of this new evolution model. In section 3, the geological record of past plate motion is compared with the predicted history of plate velocity. In section 4, the nature of oceanic lithosphere in past plate tectonics is closely examined, first for its subductability (section 4.1) and then for its bearing on the mode of mantle convection (section 4.2). The implication of sluggish plate tectonics for ancient plume dynamics will also be discussed (section 4.3). Finally, a potential link between sluggish plate tectonics and global geochemistry is discussed in section 5.

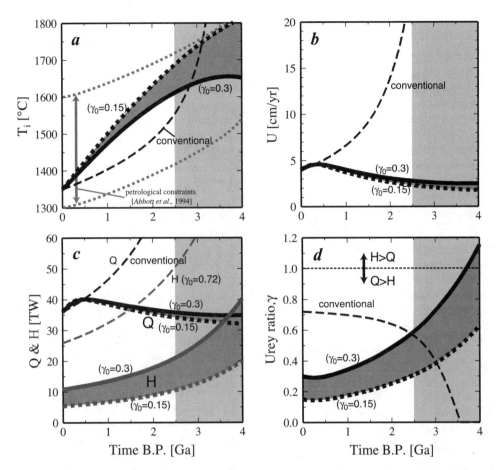

Figure 7. Thermal evolution modeling incorporating the energetics of plate-tectonic convection, with the Urey ratio of 0.15-0.3. For comparison, results from the conventional model with the Urey ratio of 0.72 (Figure 2) are also shown as dashed curve. (a) Mantle internal temperature. (b) Plate velocity. (c) Surface heat flux Q and internal heating H. (d) Urey ratio. Gray shading for >2.5 Ga indicates large uncertainty involved in predicted thermal history, i.e., heat flux scaling to be used may be considerably different in this time range. In (a), the range of potential temperatures recorded in MORB-like suites [*Abbott et al.*, 1994] is also shown as dotted lines.

3. GEOLOGICAL RECORD OF PAST PLATE MOTION

Whereas it can resolve the thermal catastrophe paradox, the new evolution model of plate tectonics is drastically different from what is commonly believed, in terms of its prediction for plate velocity (Figure 7b). Thus, the most critical test of this new model may be done by comparing its prediction with the geological record of plate motion. More sluggish plate tectonics when the mantle was hotter is, indeed, the essence of plate-tectonic convection regulated by depleted oceanic lithosphere.

The ocean floor provides the history of plate motion since the breakup of the supercontinent Pangea at ~200 Myr. Though the rate of seafloor spreading is commonly believed to be higher in the Cretaceous than today [*Sprague and Pollack*, 1980; *Larson*, 1991; *Lithgow-Bertelloni and Richards*, 1998], the rapid seafloor spreading is more likely an artifact resulting from overlooking plate reorganizations and using an inaccurate geological time scale [*Heller et al.*, 1996]. A more robust estimate based on the area-age distribution of the ocean floor yields a nearly constant rate of plate creation during the last 180 Myr [*Parsons*, 1982; *Rowley*, 2002]. This record is, however, too short to discriminate between the new and conventional models. On the other hand, the estimate of continental drift rates can go back to ~3.5 Ga [*Ullrich and der Voo*, 1981; *Kröner and Layer*, 1992], but its accuracy is often severely limited by the quality of paleopole determinations, inaccurate age data, and sparse sampling in time [*der Voo and Meert*, 1991]. In addition, the possibility of true polar wander [e.g., *Evans*, 2003] tends to blur its relevance to plate motion. Though this may simply be owing to the paucity of data, it is often claimed that there is no geological evidence supporting rapid plate tectonics in the Archean [*Kröner and Layer*, 1992; *de Wit*, 1998].

I suggest that the history of supercontinents may provide a straightforward constraint on global plate motion. Why a supercontinent forms and breaks up is still a matter of debate [*Storey*, 1995], but the Wilson cycle is essentially the opening and closure of large ocean basins. The frequency of supercontinental formation, therefore, could be a good proxy for a global as well as time average of plate velocity, with local irregularities automatically smoothed out. In fact, *Hoffman* [1997] already pointed out that the recurrence interval for supercontinent assemblies appeared to have become shorter with time (Figure 8): Kenorland (2.7-2.6 Ga), Nuna (1.8-1.7 Ga), Rodinia (1.1-1.0 Ga), Gondwanaland (0.6-0.5 Ga), and Pangea (0.3-0.2 Ga). The first three are still hypothetical because their configurations are yet to be settled, but the distribution of greenstone ages indicates that

there were prominent peaks of global tectonism, most likely the assembly of a supercontinent, at ~2.7 Ga, ~1.9 Ga, and ~1.1 Ga [*Condie*, 1995]. In light of the new evolution model, the implication of these accelerating Wilson cycles is no longer puzzling. By assuming that the recurrence interval is inversely proportional to average plate velocity, those decreasing intervals correspond to gradually speeding-up plate tectonics, broadly consistent with the new model (Figure 8, case A). If Gondwanaland is viewed not as a stand-alone supercontinent, but rather merely a building block of the more complete Pangea landmass, then the Wilson cycles are seen to have occurred at nearly constant intervals of ~800 Myr (Figure 8, case B). In either case, the supercontinental record appears to contradict the standard model of substantially faster Precambrian plate tectonics, yet it is in broad agreement with the new thermal evolution model presented here.

The above constraint on past plate motion by the Wilson cycles is admittedly simplistic and has unquantifiable uncertainties. One may also claim that the motion of continents may be irrelevant to the dynamics of oceanic plates. On the present-day Earth, in fact, continental plates move at much slower rates than oceanic plates [*Forsyth and Uyeda*, 1975]. Slow continental motion today, however, may simply reflect that the history of seafloor spreading recorded in ocean floor is restricted to the opening mode of the Wilson cycle [*Gordon et al.*, 1979]. To create a supercontinent, an ocean basin between continental plates must vanish, and in this closing mode, the motion of continents is tightly coupled with subducting plates. A rapid motion of India before colliding to Eurasia is a modern (albeit small-scale) example of this. Plate reconstruction over the Phanerozoic [*Jurdy et al.*, 1995] suggests that all continents exhibit fluctuating plate motion with a time scale of ~100 Myr, and that they were often moving at the velocity of >4–6 cm/yr. Thus, using the Wilson cycle to infer the motion of oceanic plates is not so far off the mark. What drives plate tectonics is ultimately the release of gravitational potential energy (equation (13)), i.e., subduction. The motion of continents is always coupled with that of subducting plates, either directly (subducting plate attached to continent) or indirectly (viscous drag via wedge mantle) [e.g., *Hager and O'Connell*, 1981].

One subtle feature of the predicted history of plate motion is the presence of the velocity maximum (Figure 8); plate tectonics was accelerating until ~0.3 Ga, and then started to slow down. This velocity maximum corresponds to the plate thickness minimum, which happens to take place under nearly present-day conditions (Figure 5). This currently decelerating plate tectonics might explain why a new supercontinent has not been formed if the case-A interpretation of the Wilson cycles is correct. The exact timing of this velocity maximum is, however, controlled by a delicate balance

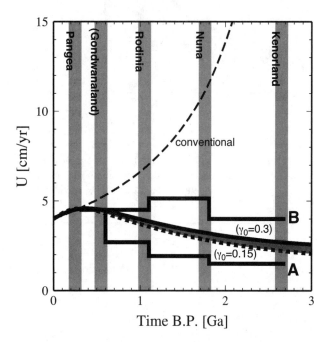

Figure 8. Comparison of predicted plate velocity with the geological estimate based on the history of supercontinents. Legend for model prediction is the same as in Figure 7b. Gray bars denote the periods of supercontinent assemblies based on *Hoffman* [1997]. Case A is a geological estimate based on five "supercontinents": Pangea, Gondwanaland, Rodinia, Nuna, and Kenorland. Plate velocity is calculated as the reciprocal of the recurrence interval, and is scaled with 4.5 cm yr^{-1} for the Pangea-Gondwanaland interval to line up with the model prediction at 0.3 Ga for comparison. Case B is based on four supercontinents, Pangea, Rodinia, Nuna, and Kenorland. The same velocity scaling is applied with the Pangea-Rodinia interval.

between temperature-dependent viscosity and melting-induced viscosity stratification, so any coincidence with geology is probably fortuitous. A more important point is that the existence of the plate thickness minimum prevents ever-accelerating plate tectonics while Earth is cooling down. It stabilizes mantle convection over a wide range of internal temperature, resulting in a nearly steady heat flux over a major fraction of Earth's history.

4. IMPLICATIONS FOR ARCHEAN MAGMATISM AND TECTONICS

4.1. Subductability of Oceanic Lithosphere

Hotter mantle in the past must have created thicker oceanic crust as well as thicker depleted lithospheric mantle, both of which are less dense than unmelted mantle. As far as the energy balance of plate-tectonic convection is concerned, this compositional buoyancy does not matter because basaltic crust is transformed into denser eclogitic crust at depths less than 100 km, so the chemical density defect of subducting plate as a whole is insignificant. However, it is still an important issue for the initiation of subduction [e.g., *Bickle*, 1986; *Davies*, 1992], which is essential to close ocean basins and amalgamate continents. As a necessary (but probably not sufficient) condition for the initiation of subduction, the net buoyancy of oceanic lithosphere must be negative, i.e., plate must become denser than underlying asthenosphere to sink through it. Denoting the thickness of oceanic crust by d_c and its density by ρ_c, and similarly using d_l and ρ_l for depleted lithospheric mantle, compositional buoyancy may be defined as

$$B_C \equiv (\rho_m - \rho_c)d_c + (\rho_m - \rho_l)d_l, \quad (27)$$

where ρ_m is the density of unmelted mantle. This buoyancy is fixed when plate is created at mid-ocean ridges. On the other hand, thermal (negative) buoyancy continues to increase as thermal boundary layer thickens, and it is defined as

$$B_T \equiv \rho_m \alpha \bar{T} h, \quad (28)$$

where $\bar{T}$ is the average temperature within the thermal boundary layer, whose thickness is denoted by h. By modeling the growth of boundary layer as $h = 2\sqrt{\kappa t}$ (i.e., half-space cooling), then, the average temperature is given by

$$\bar{T} \approx \frac{\int_0^1 T_p \text{erfc}(x)dx}{\int_0^1 dx} \approx 0.52 T_p, \quad (29)$$

where erfc(x) is the complementary error function. For plate to be negatively buoyant with respect to underlying mantle, sufficient cooling must be attained so that $B_T/B_C > 1$. The

critical thickness of thermal boundary layer to exceed this threshold is denoted by h_{crit} and the corresponding plate age by t_{crit}.

To calculate h_{crit} as a function of mantle temperature, we need to model d_c, d_l, ρ_c, and ρ_l as a function of mantle temperature. The mantle melting model of *Korenaga et al.* [2002] can be exploited for this purpose. The initial pressure (in GPa) of dry melting is related to potential temperature as

$$P_0 = (T_p - 1150)/100 \quad (30)$$

and the final pressure of melting is given by

$$P_f = P_0 + \left(\frac{\partial F}{\partial P}\right)_S^{-1}\left\{1 - \left[1 + 2P_0\left(\frac{\partial F}{\partial P}\right)_S\right]^{\frac{1}{2}}\right\}, \quad (31)$$

where $(\partial F/\partial P)_S$ is a change in melt fraction with a change in pressure above the solidus during adiabatic decompression. It is set here as 15 GPa/% so that $T_p = 1350°C$ produces 7-km-thick crust. The mean degree of melting is then given by

$$\bar{F} = 0.5(P_0 - P_f)\left(\frac{\partial F}{\partial P}\right)_S, \quad (32)$$

and the mean pressure of melting is defined as

$$\bar{P} = 0.5(P_0 + P_f). \quad (33)$$

The thickness of depleted mantle is equivalent to the zone of melting, so we have

$$d_l = 30(P_0 - P_f), \quad (34)$$

and crustal thickness is then given by

$$d_c = d_l \bar{F}/100. \quad (35)$$

Note that the final pressure of melting is not zero; this is because mantle can rise only up to the base of crust.

The density of depleted mantle can be calculated from $(d\rho/dF)_C$ (equation (24)), using the mean degree of melting $\bar{F}$. To calculate crustal density, we first compute crustal P-wave velocity from $\bar{P}$ and $\bar{F}$ using the following empirical relation [*Korenaga et al.*, 2002]:

$$V_p = a_0 + \\ W_L(P,F)(b_0 + b_1P + b_2F + b_3P^2 + b_4PF + b_5F^2) + \\ W_H(P,F)(c_0 + c_1P + c_2F + c_3P^2 + c_4PF + c_5F^2) \quad (36)$$

where P is the pressure of melting in GPa, F is melt fraction, and $a_0 = 7.52$, $b_0 = -1.73$, $b_1 = -0.55$, $b_2 = 7.71$, $b_3 = -0.11$,

$b_4 = 8.87$, $b_5 = -146.11$, $c_0 = -0.35$, $c_1 = 0.034$, $c_2 = 0.51$, $c_3 = 0.0016$, $c_4 = -0.040$, and $c_5 = 0.046$. (Note: The coefficient b_1 was incorrectly published in *Korenaga et al.* [2002].) $W_L(P,F)$ and $W_H(P,F)$ are hypertangent window functions defined as

$$W_L(P,F) = \frac{1}{4}\{1 - \tan h[p(P - P_t)]\}\{1 - \tan h[q(F - F_t)]\}$$

and

$$W_H(P,F) = \frac{1}{4}\{1 + \tan h[p(P - P_t)]\}\{1 + \tan h[q(F - F_t)]\}$$

where $p = 0.6$, $q = 8.4$, $P_t = 1.0$, and $F_t = 0.05$. Equation (36) is calibrated over a wide range of pressure and melt fraction, based on high-quality experimental data [*Kinzler and Grove*,

1993; *Baker and Stolper*, 1994; *Kinzler*, 1997; *Walter*, 1998], and despite its rather complicated form, its behavior is stable for mantle temperatures of our interest here. *Korenaga et al.* [2002] also showed that the relation between P-wave velocity and crustal density is best approximated by Birch's law for rocks with mean atomic weight of ~21 [*Birch*, 1961], so we adopt

$$\rho_c = 0.77 + 0.3V_p, \tag{37}$$

which completes the parameterization of compositional buoyancy as a function of mantle potential temperature (Figure 9a,b).

The critical thickness of oceanic lithosphere is shown in Figure 9c, and the corresponding critical age is shown in Figure 9d. At the present-day condition, plate must be older than only 20 Ma to become negatively buoyant, but when the

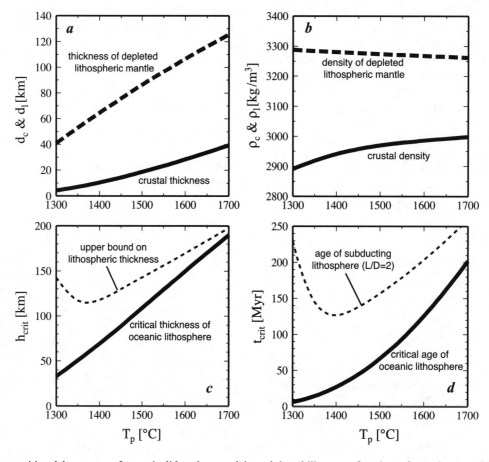

Figure 9. Compositional buoyancy of oceanic lithosphere and its subductability as a function of mantle potential temperature. (a) Thickness of oceanic crust (solid) and depleted mantle (dashed). (b) Density of oceanic crust (solid) and depleted mantle (dashed). Crustal density increases with potential temperature because higher degree of melting increases the olivine content of resulting crust. (c) Critical thickness for net negative buoyancy (solid). Upper limit for the thickness of oceanic lithosphere when dehydration stiffening is present (Figure 5) is also shown as dashed. (d) Critical plate age for subduction (solid). Dashed curve denotes the age of subducting plate assumed in the heat-flow parameterization for plate-tectonic convection.

mantle was hotter, say, 1600°C, it must be older than ~120 Ma. However, because of more sluggish plate tectonics in the past, a typical age of subducting plate is expected to be greater than present (Figure 9d, dashed). Thus attaining negative buoyancy is not a major problem. This is quite contrary to previous suggestion on the emergence of plate tectonics based solely on compositional buoyancy [e.g., *Davies*, 1992], in which the conventional scaling is assumed for heat flow so that plates are moving faster in the past.

It is interesting to note that, if plate thickness is controlled only by temperature-dependent viscosity and compositional buoyancy (Figure 5, dotted curve), the critical thickness for subduction can never be achieved for $T_p > {\sim}1400$ °C; small-scale convection limits the maximum plate thickness to ~70 km. Thus, without dehydration stiffening, the initiation of subduction could take place within a fairly limited temperature range, and accordingly, for a very limited time period (< 0.5 Ga, Figure 7a). The geological observations of continental aggregation back to ~2.7 Ga (Figure 8) thus strongly support the notion of strong depleted lithosphere.

4.2. Different Modes of Mantle Convection?

The aspect ratio of large-scale mantle circulation, L/D, is assumed to be 2 throughout this paper. Changing this ratio by a factor of < 2 (which is an acceptable variation for whole-mantle convection) does not affect the main conclusions of this study. On the other hand, one of prevailing concepts in Archean geology is convection cells with much smaller aspect ratios (i.e., L ~ 200–300 km). *Hargraves* [1986], for example, proposed that Archean plate tectonics is characterized by a large number of slowly-moving small plates. His argument is, however, based entirely on the premise of *higher* heat flux in the Archean. Though internal heat production is indeed higher in the past, surface heat flux, which is the sum of heat production and secular cooling (equation (2)), does not necessarily correlate with internal heat production, as demonstrated in Figure 7c. On the same (and most likely incorrect) premise, *Abbott and Menke* [1990] applied the size-number statistics of the current plate configuration to Archean cratons to estimate the length of global plate boundaries, and their result (> 60 plates at 2.4 Ga) appears to be consistent with the model of *Hargraves* [1986]. Their approach is, however, fundamentally biased because predicting a large number of plates in the past is simply based on the fact that Archean cratons are much smaller than present-day continents. Surface heat flux does not have to be higher in the past, and a need for greater ridge lengths would evaporate in light of the new heat-flow scaling law for plate-tectonic convection. *Hargraves* [1986] calculated the maximum age of subduction for hypothesized small plates as ~20 Ma. Those plates cannot be negatively buoyant (Figure 9d), so plate

tectonics would halt. Thus, the Archean tectonics with numerous small plates is dynamically implausible.

How the aspect ratio should scale with other parameters of plate-tectonic convection is not understood well; we cannot explain even the plate size distribution on the present-day Earth. It is, however, difficult to expect smaller aspect ratios for hotter mantle because the subduction of thicker depleted lithosphere requires a longer duration of cooling. If plate velocity is fixed, therefore, a larger aspect ratio may be more physically reasonable for hotter mantle. Plate velocity, however, probably decreases with increasing temperature, so aspect ratio may stay relatively constant. In fact, *Sleep* [1992] suggested that plate dimension in the Archean are probably similar to modern examples based on the duration of a transpressive strike-slip episode in the Superior province [*Card*, 1990], though this type of inference may be too indirect.

Another popular thinking regarding Archean geodynamics is a change in the mode of convection from whole-mantle to layered-mantle [e.g., *Breuer and Spohn*, 1995]; that is, the endothermic phase change at the 660 km boundary may have been more important when mantle was hotter, so convection could be layered then whereas it is now in the regime of whole-mantle convection. This notion is, however, based on numerical studies on convection with unrealistically simple mantle rheology [e.g., *Tackley et al.*, 1993; *Solheim and Peltier*, 1994; *Yuen et al.*, 1994]. Models with variable viscosity are usually limited to either depth-dependent viscosity or weakly temperature-dependent viscosity. Even constant viscosity is employed in some cases. In those studies, convection with hotter mantle (i.e., with lower viscosity) is characterized by small-scale flows typical to high-Rayleigh-number convection. The effect of an endothermic phase boundary increases with smaller-scale flows [*Tackley*, 1995], and convection can be layered if the Rayleigh number becomes sufficiently high. More realistic numerical models with strong plates show a much weaker influence of endothermic phase change on mantle dynamics [e.g., *Zhong and Gurnis*, 1994]. The presence of strong plates can support large-scale circulation, even with increasing mantle temperature, and the transition to layered-mantle convection is unlikely to take place. The influence of phase changes on the energetics of whole-mantle convection is also expected to be small, partly because the effect of endothermic phase change at 660 km is largely compensated by the opposite effect of exothermic phase change at 410 km. Given that density jumps associated with these phase changes are less than 10% and that both phase changes have Clapeyron slopes of a few MPa/K, a correction to be made for the rate of potential energy release can be shown to be on the order of only a few percent.

Discussion so far may imply that there could not be any major change in the mode of convection in the past, but there is one thing that suggests a different kind of mantle convection in the Archean. Even though stagnant-lid convection is not taking place on Earth today, we can still calculate the hypothetical thickness of stagnant-lid in equilibrium with internal heat production H as

$$\bar{h}_{SL} \sim \frac{kAT_i}{H},\qquad(38)$$

where A is the total surface area of Earth. (Note: This is a crude estimate because the internal temperature T_i is given by Figure 7a, instead of calculated self-consistently with assumed temperature-dependent viscosity. The most important variation of $\bar{h}_{SL}$ is, however, caused by the secular change of internal heating H, so this detail does not matter.) On the other hand, mean plate thickness averaged from the ridge axis to subduction is given by

$$\bar{h}_{PT} \sim \frac{2}{3}h_m,\qquad(39)$$

where h_m is the maximum plate thickness as in Figure 5. These two mean thicknesses are compared in Figure 10, which shows that plate thickness for plate-tectonic convection could become greater than that for stagnant-lid convection at >3 Ga. This implies that, instead of delamination by small-scale convection, thermal erosion by intense internal heating may control plate thickness. The currently available scaling

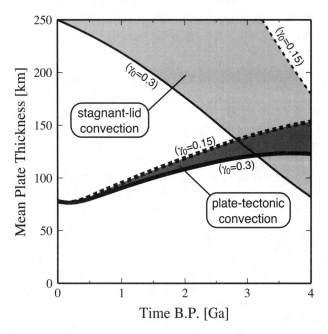

Figure 10. Mean plate thickness for plate-tectonic convection and stagnant-lid convection. Stagnant-lid convection is assumed to be in equilibrium with internal heat production. Thermal conductivity k is set to 4 W m^{-1} K^{-1} in equation (38).

law for the convective instability of oceanic lithosphere (equation (23)) does not take into account the effect of internal heating. Strong internal heating may halt the growth of thermal boundary layer, regardless of the rheological structure of oceanic mantle.

This is similar to what puzzled Richter regarding the survival of thick cratonic lithosphere [*Richter*, 1985, 1988]. The problem with cratonic lithosphere may be solved by its intrinsic rheological strength or by its tendency to be surrounded by cold downing, but these *local* solutions cannot be applied here. The impact of thermal erosion on oceanic lithosphere is on a global scale. Subducting plate may have been thinner than predicted in Figure 5, so it may have been moving faster. Alternatively, thermal erosion could prevent oceanic lithosphere to become negatively buoyant, and stagnant-lid convection instead of plate tectonics may characterize the early Archean Earth. In either case, the geodynamic regime of the early Archean is expected to be different from contemporary plate tectonics. It remains to be seen how enhanced heat production can affect global plate dynamics, thereby the thermal history of the early Earth.

4.3. Core Heat Flux and Origins of Archean Komatiites

The petrogenesis of komatiites has been paid special attention in Archean geology because their field occurrence is mostly restricted to Archean provinces, even though their abundance is relatively minor, compared with more common tholeiitic rocks in Archean greenstone belts. Major competing hypotheses for the origin of Archean komatiites include (1) the melting of very hot mantle (the potential temperature of >1800°C) originating in mantle plumes [*Bickle et al.*, 1977; *Herzberg*, 1992; *Walter*, 1998; *Arndt*, 2003], (2) the wet melting of moderately hot mantle (~1500°C) in arc magmatism [*Allegre*, 1982; *Parman et al.*, 1997; *Grove and Parman*, 2004], and (3) the dry polybaric melting with the mean pressure of melting of ~3 GPa in mid-ocean ridge magmatism [*Kelemen et al.*, 1998] (which translates to the potential temperature of ~1650°C with the melting model of §4.1). Mantle plumes are needed in the first hypothesis because more common tholeiitic samples constrain average mantle temperature as ~1500–1600°C (Figure 2). The rarity of komatiites requires hotter mantle to be present only locally, which conforms to the behavior of mantle plumes [e.g., *Campbell and Griffiths*, 1992].

How could the new thermal history contribute to this debate? As a consequence of nearly constant surface heat flux, the Urey ratio was higher in the past (Figure 7d), that is, the contribution of secular cooling to surface heat flux was smaller [*Sleep et al.*, 1988]. Core cooling comprises ~1/5 of whole-Earth secular cooling [*Stacey*, 1981], so the new model implies lower core heat flux (i.e., lower plume flux) in the Archean (Figure 11). The present-day core heat flux is estimated to be ~5-10 TW, based on a likely temperature gradient across the D″ layer and its thermal conductivity [*Buffett*, 2002; *Labrosse*, 2002]. This is larger than plume-related heat flux, the global total of which is only ~2 TW [*Sleep*, 1990]. This discrepancy is not surprising because heat can be extracted from the core not only by removing hot boundary layer materials as plumes but also by warming up cold subducted slabs. Thus, the plume flux is expected to be only a fraction of the core heat flux, and the new core cooling history implies much subdued plume activities in the Archean. Some geologists seem to take the presence of mantle plumes for granted throughout Earth's history [e.g., *Condie*, 2001; *Ernst and Buchan*, 2003], but this uniformitarian view should be treated with caution. Plume activities have an intimate relation with secular cooling, so one cannot freely speculate on their strength independently of the global cooling history of Earth.

The existence of mantle plumes in the Archean is not guaranteed as discussed above. The new evolution model thus may prefer the arc or mid-ocean ridge hypotheses for the origin of komatiites. The required mantle potential temperature in these hypotheses is ~1500-1650 °C, which is consistent with the predicted thermal history in the Archean (Figure 7a) given that mantle near subduction zones are relatively cooler than mantle beneath mid-ocean ridges. Moreover, slower plate motion may facilitate the heating of subducting slab, which

may have led to the enhanced activity of arc magmatism in the Archean.

Finally, it is noted that the conventional model with a high present-day Urey ratio is not consistent with the estimated core heat flux based on the characteristics of the D″ layer (Figure 11, dashed); the high Urey ratio leaves little (only ~2 TW) for core heat flux. The diminishing core heat flux in the past as predicted by the new model has intriguing implications for the age of the inner core and the history of geodynamo, which will be explored elsewhere.

5. MANTLE MIXING AND CHEMICAL GEODYNAMICS

More sluggish plate tectonics in the past implies less efficient mantle mixing than previously thought. It is widely believed that whole-mantle convection can mix the mantle very efficiently [e.g., *Kellogg and Turcotte*, 1990; *Ferrachat and Ricard*, 1998; *van Keken et al.*, 2002] (though some argue against this notion [e.g., *Davies*, 2002]). To preserve chemical heterogeneities observed in hotspot magmatism, therefore, some kind of special mechanism such as layered convection is usually invoked [e.g., *Hofmann*, 1997]. Numerical models of mantle mixing are often scaled with the present-day vigor of convection, and the degree of mixing acquired after running models for ~4 Gyr is usually quoted as a "conservative" estimate [e.g., *van Keken et al.*, 2002], because convection is commonly believed to be more vigorous in the past.

Though the dynamics of chemical mixing is a complicated problem, mixing should be more or less proportional to the time-integrated history of plate velocity. The thermal history of the core (Figure 11) suggests reduced core heat flux and thus weaker mantle plumes in the past, supporting that mixing is predominantly controlled by plate dynamics. Mantle is cooled from above, and not very much heated from below. Furthermore, the role of small-scale convection is also expected to be smaller or even negligible in the past (Figure 6b). Based on the predicted velocity (Figure 7b), therefore, one can compare the new and conventional models in terms of mixing efficiency (Figure 12). As a measure of mixing, I use the processing rate of the mantle at mid-ocean ridges. With the current rate of plate construction of 3.4 km^2/yr [*Parsons*, 1982] and the initial depth of mantle melting of ~60 km, the present-day processing rate is ~200 km^3/yr or 6.7×10^{14} kg/yr. As going back in time, the depth of mantle melting increases owing to higher potential temperature, but plate velocity decreases as well. These two effects tend to cancel out, so the processing rate is nearly constant through time. The global ridge length is assumed here to be constant through time, which is probably reasonable as discussed in section 4.2.

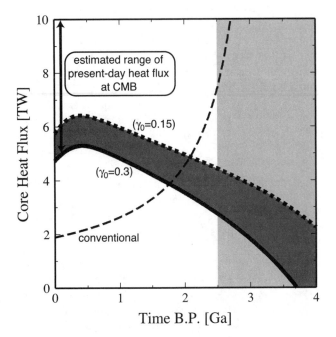

Figure 11. The first-order thermal history of the core. Legend is the same as in Figure 7. Core heat flux is calculated as ~19% of the whole-Earth secular cooling [*Stacey*, 1981].

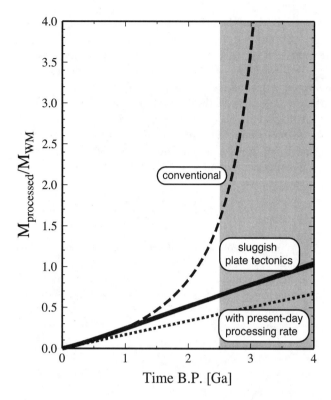

Figure 12. Efficiency of mantle mixing in terms of processed mantle mass normalized by the mass of the whole mantle. Because of the trade-off between plate velocity and the initial depth of mantle melting, the difference in the assumed Urey ratio (i.e., 0.15 or 0.3) results in negligible difference in the processing rate (solid curves). Dashed curve denotes the prediction based on the convection model with the Urey ratio of 0.72. A hypothetical case with the constant processing rate of the present-day value is also shown as dotted. Sluggish plate tectonics is characterized by low processing rates similar to this extreme case.

The new evolution model suggests that it would take ~4 Gyr for the entire mantle to be processed at mid-ocean ridges (Figure 12, solid). Because the mode of mantle convection was probably different in the early Archean (section 4.2), this estimate does not necessarily imply the preservation of primordial mantle. For chemical heterogeneity created after ~3 Ga, however, it appears plausible that such old heterogeneity can remain intact to the present time. This is in a stark contrast with what the conventional model predicts; the entire mantle would have been processed four times at mid-ocean ridges since 3 Ga owing to rapid convection in the past (Figure 12, dashed). Note that the low processing rate means inefficient degassing of noble gas such as argon. The ^{40}Ar budget of Earth's atmosphere is often quoted as supporting evidence for layered mantle [e.g., *Allégre et al.*, 1996; *Hofmann*, 1997]. However, this geochemical argument heavily relies on the conventional notion of vigorous convection

in the past; all of previously produced argon in the convection mantle is assumed to degas in case of whole-mantle convection. On the other hand, exactly the same argon budget has also been used to claim that plate motion must have been nearly constant (with the assumption of whole-mantle convection) [*Sleep*, 1979; *Tajika and Matui*, 1993]. The energetics of plate-tectonic convection described in this paper suggests that such speculation based on a mass-balance argument now has a reasonable physical basis.

This slow mixing, combined with the fact that chemical heterogeneities are constantly injected by subduction, may explain why mantle chemistry can remain heterogeneous at various spatial scales with whole-mantle convection. After all, the chemical state of the mantle may be well described by the distributed blob model [*Helffrich and Wood*, 2001], provided that mantle dynamics is governed by the new convection energetics. How such blobs can maintain their integrity has been a concern [*Manga*, 1996; *Becker et al.*, 1999], but mantle mixing itself could be very inefficient in the first place; there is no need to invoke higher viscosity for those enriched blobs to survive. Flood basalt and hotspot magmatism may simply be tapping blob-like heterogeneities from time to time [e.g., *Korenaga*, 2005b], not being fed by a hidden deep reservoir. Intrinsic density anomaly associated with those enriched components [*Christensen and Hofmann*, 1994; *Korenaga*, 2004] may explain preferential sampling of depleted MORB source mantle in terrestrial magmatism.

6. DISCUSSION AND CONCLUSION

Several issues, with various degrees of complexity, have been left out from consideration, including (1) the effect of continental growth on internal heating, (2) the role of core-mantle boundary processes in the thermal evolution, (3) the discrepancy between the petrological and cosmochemical Urey ratios, (4) the validity of the new energetics of multi-scale convection, and (5) the style of pre-plate-tectonic convection in the early Archean. They are discussed below in order.

Because the present-day continental crust is responsible for ~40% of the bulk Earth heat production, its growth in the past could considerably impact the heat budget of mantle convection. The growth rate of continental crust in the post-Archean is, however, believed to be low; a recent geochemical study suggests that ~75% of the present-day continental mass already existed at ~3 Ga [*Campbell*, 2003]. Thus, in terms of back-tracking the thermal history to the beginning of the Proterozoic, incorporating the history of continental growth would result in minor differences. If the early Archean is characterized by much smaller continental mass, the effect of internal heating on lithospheric dynamics becomes even more important than suggested by Figure 10. A likely change in the mode of mantle convection is expected to be strongly coupled with the growth of continents. A more refined history of continental growth is much desired from geochemistry, and different modes of mantle convection should be explored for their implications for the petrogenesis of continental crust.

I have adopted the simplest approach to incorporate the effect of core cooling in thermal evolution (equation (1)). Though this is better than completely ignoring it as sometimes done in previous studies [e.g., *Jackson and Pollack*, 1984; *Williams and Pan*, 1992], we could still benefit from explicitly modeling the core-mantle interaction to check how good this simple approach could be. The parameterization of core-mantle interaction, however, may not be so simple as assumed in previous coupled core-mantle evolution models [*Stevenson et al.*, 1983; *Stacey and Loper*, 1984; *Davies*, 1993; *Yukutake*, 2000; *Nimmo et al.*, 2004]. Traditionally, the Rayleigh-Taylor instability of a hot bottom boundary layer is thought to produce the upwelling of a less viscous plume through a more viscous overlying fluid. Viscosity contrast between a plume and the ambient mantle is typically assumed to be on the order of $10^2 - 10^3$, and this contrast results in the formation of a large spherical head followed by a narrow conduit [e.g., *Richards et al.*, 1989; *Griffiths and Campbell*, 1990]. It is thus quite surprising that a recent finite-frequency tomography has resolved quite a few deep mantle plumes with very large radii, typically ranging from 200 to 400 km [*Montelli et al.*, 2004]. Recently, *Korenaga* [2005a] suggested that, to reconcile with the geodynamical estimate of plume buoyancy flux based on hotspot swell topography [*Sleep*, 1990], the viscosity of those thick plumes must be as high as $10^{21} - 10^{23}$ Pa s (i.e., comparable or greater than lower-mantle viscosity), and that the temperature dependency of lower-mantle rheology is probably dominated by the grain size-dependent part of diffusion creep, i.e., hotter mantle has higher viscosity [*Solomatov*, 1996, 2001]. The impact of this new kind of plume dynamics should be considered in future studies on the coevolution of the core-mantle system.

As discussed in section 2.2, the reasonable range of the Urey ratio goes from ~0.15 (petrological) to ~0.3 (cosmochemical), and all calculations have been done with this range. Though this uncertainty in the Urey ratio does not lead to substantial difference in predicted mantle temperature and plate velocity, compared with uncertainties in corresponding geological constraints (Figures 2, 7, and 8), it becomes important when discussing the timing of a change in the mode of mantle convection (Figure 10). The factor-of-two discrepancy between the cosmochemical and petrological values is also a concern because it may imply the presence of a hidden reservoir enriched in trace elements. We must remember, however, that the heat production of 20 TW for the bulk silicate Earth is not an axiom. *McDonough and Sun*

[1995], for example, report the nominal uncertainty of ~20% for the concentration of heat-producing elements in their pyrolite mantle model. If the heat production is as low as 16 TW, then, the cosmochemical Urey ratio would be only ~0.2, after carving off continental heat production. A better quantification of uncertainties involved in the composition of the bulk silicate Earth may be a key to resolving this "missing heat source" paradox.

To overcome the compositional buoyancy of thick depleted lithosphere for the initiation of plate tectonics, depleted lithosphere must also be strong enough to support the growth of thermal boundary layer (section 4.1). This is exactly what is predicted by dehydration stiffening (Figure 5). Moreover, to become negatively buoyant, the minimum age of subducting plate (at the initiation of subduction at least) must be greater for thicker plate (i.e., hotter potential temperature, Figure 9d), which implies that plate must move slower for thicker depleted lithosphere. This is also what is predicted by the new heat-flow scaling law for plate-tectonic convection (Figure 6). Thus, based on this line of reasoning and on the geological observation of repeated continental aggregation back to ~2.7 Ga (Figure 8), it is plausible that plate tectonics may have been more sluggish in the past. Nevertheless, the heat-flow scaling law is constructed by combining the scaling law for large-scale convection with that for small-scale convection, and it remains to be seen how valid this composite scaling law would be as a whole, using the numerical modeling of mantle convection. Such modeling may also be able to investigate how the radius of curvature for slab bending is controlled by other model parameters, and more importantly, how the aspect ratio of convection is determined in evolving plate tectonics characterized by the Wilson cycle. This is a challenging problem, because modeling plate-tectonic convection itself is still considered to be difficult, let alone its sensitivity to a change in mantle temperature. We may have to wait for our understanding of the generation of plate tectonics to become mature enough. Successful numerical modeling is essential to go beyond global-average characteristics based on parameterized convection models and investigate regional-scale dynamics in a quantitative manner. Without this, connection between field geology and theoretical geophysics would remain loose.

It is suggested that the style of mantle convection may have been different from contemporary plate tectonics in the early Archean. *How* different it could be is an open question, and to answer this, we need to first understand how enhanced internal heating could affect plate dynamics. Important observational constraints on this early Archean dynamics may come from isotope geochemistry. As already discussed, mantle mixing since the late Archean is probably inefficient. However, some radiogenic isotope systematics (e.g., Sm-Nd) of Archean rocks suggest much faster convective motion during the Archean [*Blichert-Toft and Albarède*, 1994]. The Pb model age (~2 Ga) of present-day MORBs and OIBs also seems to place the upper bound on the oldest chemical heterogeneities survived through convective mixing [*Christensen and Hofmann*, 1994]. Though the interpretation of those geochemical signatures is often nonunique, we may better quantify their implications by coupling thermal evolution modeling with mantle mixing (e.g., Figure 12).

In conclusion, plate-tectonic convection characterized by strong depleted lithosphere is shown to be able to simultaneously satisfy major geophysical, petrological, geochemical, and geological constraints on the thermal evolution of Earth, at least back to the late Archean. Only whole-mantle convection is involved, without calling for compounded layered-mantle convection. Even with a relatively low concentration of heat-producing elements as imposed by geochemistry, a reasonable thermal history can be reconstructed because thermal catastrophe is avoided by reduced heat flux in the past. More sluggish plate tectonics is consistent with the history of supercontinents. Furthermore, the assembly of supercontinents requires the initiation of subduction between continents, and the history of continental aggregation back to ~2.7 Ga strongly supports the notion of sluggish plate tectonics with thick plate because depleted lithosphere must be sufficiently cooled to overcome its compositional buoyancy. Reduced surface heat flux in the past also suggests that plumes were weaker or even absent in the Archean, and thick depleted lithosphere may have been sufficiently stabilized to suppress small-scale convection. Mantle mixing is thus controlled largely by the history of plate motion, and sluggish plate tectonics may explain the presence of long-lived chemical heterogeneities as well as inefficient noble gas degassing, in the context of whole-mantle convection; layered mantle is no longer required. Early Archean dynamics is suggested to be in a different mode of mantle convection, with internal heating playing an important role in controlling plate dynamics.

Acknowledgments. This paper has benefited from discussion with Dave Evans, Peter Olson, Dave Stevenson, and Peter Kelemen. The author also thanks Geoff Davies and Norm Sleep for official reviews, which helped to improve the clarity of the manuscript.

REFERENCES

Abbott, D., and W. Menke, Length of the global plate boundary at 2.4 ga, *Geology*, 18, 58-61, 1990.

Abbott, D., L. Burgess, J. Longhi, and W.H.F. Smith, An empirical thermal history of the Earth's upper mantle, *J. Geophys. Res.*, 99, 13, 835-13, 850, 1994.

Allegre, C.J., Genesis of Archaean komatiites in a wet ultramafic subducted plate, in *Komatiites*, edited by N.T. Arndt and E.G. Nisbet, pp. 495-500, George Allen & Unwin, London, 1982.

Allegre, C.J., Limitation on the mass exchange between the upper and lower mantle: the evolving convection regime of the Earth, *Earth Planet. Sci. Lett.*, 150, 1-6, 1997.

Allègre, C.J., A. Hofmann, and K. O'Nions, The argon constraints on mantle structure, *Geophys. Res. Lett.*, 23, 3555-3557, 1996.

Arndt, N., Komatiites, kimberlites, and boninites, *J. Geophys. Res.*, 108(*B6*), 2293, doi:10.1029/2002JB002,157, 2003.

Arndt, N.T., Role of a thin, komatiite-rich oceanic crust in the Archean plate-tectonic process, *Geology*, 11, 372-375, 1983.

Baker, M.B., and E.M. Stolper, Determining the composition of high-pressure mantle melts using diamond aggregates, *Geochim. Cosmochim. Acta*, 58, 2811-2827, 1994.

Barenblatt, G.I., *Scaling, Self-similarity, and Intermediate Asymptotics: Dimensional Analysis and Intermediate Asymptotics*, Cambridge, New York, 1996.

Becker, T.W., J. Kellogg, and R.J. O'Connell, Thermal constraints on the survival of primitive blobs in the lower mantle, *Earth Planet. Sci. Lett.*, 171, 351-365, 1999.

Bercovici, D., The generation of plate tectonics from mantle convection, *Earth Planet. Sci. Lett.*, 205, 107-121, 2003.

Bercovici, D., and S. Karato, Whole-mantle convection and the transition-zone water filter, *Nature*, 425, 39-44, 2003.

Bevis, M., The curvature of Wadati-Benioff zones and the torsional rigidity of subducting plates, *Nature*, 323, 52-53, 1986.

Bickle, M.J., Implications of melting for stabilisation of the lithosphere and heat loss in the Archean, *Earth Planet. Sci. Lett.*, 80, 314-324, 1986.

Bickle, M.J., C.E. Ford, and E.G. Nisbet, The petrogenesis of peridotitic komatiites: Evidence from high-pressure melting experiments, *Earth Planet. Sci. Lett.*, 37, 97-106, 1977.

Bickle, M.J., E.G. Nisbet, and A. Martin, Archean greenstone belts are not oceanic crust, *J. Geol.*, 102, 121-138, 1994.

Birch, F., The velocity of compressional waves in rocks to 10 kilobars, part 2, *J. Geophys. Res.*, 66, 2199-2224, 1961.

Blichert-Toft, J., and F. Albarède, Short-lived chemical heterogeneities in the Archean mantle with implications for mantle convection, *Science*, 263, 1593-1596, 1994.

Breuer, D., and T. Spohn, Possible flush instability in mantle convection at the Archaean-Proterozoic transition, *Nature*, 378, 678-610, 1995.

Buck, W.R., and E.M. Parmentier, Convection beneath young oceanic lithosphere: Implications for thermal structure and gravity, *J. Geophys. Res.*, 91, 1961-1974, 1986.

Buckingham, E., On physically similar systems; illustrations of the use of dimensional equations, *Phys. Rev.*, 4, 345-376, 1914.

Buffett, B.A., Estimates of heat flow in the deep mantle based on the power requirements for the geodynamo, *Geophys. Res. Lett.*, 29(12), doi:10.1029/2001GL014, 649, 2002.

Campbell, I.H., Constraints on continental growth models from Nb/U ratios in the 3.5 ga Barberton and other Archaean basalt-komatiite suites, *Am. J. Sci.*, 303, 319-351, 2003.

Campbell, I.H., and R.W. Griffiths, The changing nature of mantle hotspots through time: Implications for the chemical evolution of the mantle, *J. Geol.*, 92, 497-523, 1992.

Card, K.D., A review of Superior Province of the Canadian shield, a product of Archean accretion, *Precambrian Res.*, 48, 99-156, 1990.

Chandrasekhar, S., *Hydrodynamic and hydromagnetic stability*, Dover, New York, 1981.

Choblet, G., and C. Sotin, 3D thermal convection with variable viscosity: can transient cooling be described by a quasi-static scaling law?, *Phys. Earth Planet. Int.*, 119, 321-336, 2000.

Christensen, U., Convection with pressure- and temperature-dependent non-Newtonian rheology, *Geophys. J. R. Astron. Soc.*, 77, 343-384, 1984.

Christensen, U.R., Thermal evolution models for the Earth, *J. Geophys. Res.*, 90, 2995-3007, 1985.

Christensen, U.R., and A.W. Hofmann, Segregation of subducted oceanic crust in the convecting mantle, *J. Geophys. Res.*, 99, 19, 867-19, 884, 1994.

Condie, K.C., Episodic ages of greenstones: A key to mantle dynamics?, *Geophys. Res. Lett.*, 22, 2215-2218, 1995.

Condie, K.C., *Mantle Plumes and Their Record in Earth History*, Cambridge, New York, 2001.

Conrad, C.P., and B.H. Hager, Effects of plate bending and fault strength at subduction zones on plate dynamics, *J. Geophys. Res.*, 104, 17, 551-17, 571, 1999.

Crisp, J.A., Rates of magma emplacement and volcanic output, *J. Volcanol. Geotherm. Res.*, 20, 177-211, 1984.

Daly, S., Convection with decaying heat sources: constant viscosity, *Geophys. J. R. Astron. Soc.*, 61, 519-547, 1980.

Davaille, A., Simultaneous generation of hotspots and superswells by convection in a heterogeneous planetary mantle, *Nature*, 402, 756-760, 1999.

Davaille, A., and C. Jaupart, Transient high-Rayleigh-number thermal convection with large viscosity variations, *J. Fluid Mech.*, 253, 141-166, 1993.

Davaille, A., and C. Jaupart, Onset of thermal convection in fluids with temperature-dependent viscosity: Application to the oceanic mantle, *J. Geophys. Res.*, 99, 19, 853-19, 866, 1994.

Davies, G.F., On the emergence of plate tectonics, *Geology*, 20, 963-966, 1992.

Davies, G.F., Cooling the core and mantle by plume and plate flows, *Geophys. J. Int.*, 115, 132-146, 1993.

Davies, G.F., Stirring geochemistry in mantle convection models with stiff plates and slabs, *Geochim. Cosmochim. Acta*, 66, 3125-3142, 2002.

der Voo, R.V., and J.G. Meert, Late Proterozoic paleomagnetism and tectonic models: a critical appraisal, *Precambrian Res.*, 53, 149-163, 1991.

de Wit, M.J., On Archean granites, greenstones, cratons and tectonics: does the evidence demand a verdict?, *Precambrian Res.*, 91, 181-226, 1998.

Dixon, J.E., E. Stolper, and J.R. Delaney, Infrared spectroscopic measurements of CO_2 and H_2O in Juan de Fuca Ridge basaltic glasses, *Earth Planet. Sci. Lett.*, 90, 87-104, 1988.

Ernst, R.E., and K.L. Buchan, Recognizing mantle plumes in the geological record, *Annu. Rev. Earth Planet. Sci.*, 31, 469-523, 2003.

Evans, D.A.D., True polar wander and supercontinents, *Tectonophysics*, 362, 303-320, 2003.

Ferrachat, S., and Y. Ricard, Regular vs. chaotic mantle mixing, *Earth Planet. Sci. Lett.*, 155, 75-86, 1998.

Forsyth, D., and S. Uyeda, On the relative importance of the driving forces of plate motion, *Geophys. J. R. Astron. Soc.*, 43, 163-200, 1975.

Fukao, Y., S. Widiyantoro, and M. Obayashi, Stagnant slabs in the upper and lower mantle transition region, *Rev. Geophys.*, 39, 291-323, 2001.

Gonnermann, H.M., M. Manga, and A. Jellinek, Dynamics and longevity of an initially stratified mantle, *Geophys. Res. Lett.*, 29(10), 10.1029/2002GL01, 485, 2002.

Gordon, R.G., M.O. McWilliams, and A. Cox, Pre-Tertiary velocities of the continents: A lower bound from paleomagnetic data, *J. Geophys. Res.*, 84, 5480-5486, 1979.

Griffiths, R.W., and I.H. Campbell, Stirring and structure in mantle plumes, *Earth Planet. Sci. Lett.*, 99, 66-78, 1990.

Grove, T.L., and S.W. Parman, Thermal evolution of the Earth as recorded by komatiites, *Earth Planet. Sci. Lett.*, 219, 173-187, 2004.

Gurnis, M., A reassessment of the heat transport by variable viscosity convection with plates and lids, *Geophys. Res. Lett.*, 16, 179-182, 1989.

Gurnis, M., S. Zhong, and J. Toth, On the competing roles of fault reactivation and brittle failure in generating plate tectonics from mantle convection, in *The History and Dynamics of Global Plate Motions*, pp. 73-94, American Geophysical Union, Washington, DC, 2000.

Hager, B.H., Mantle viscosity: A comparison of models from postglacial rebound and from the geoid, plate driving forces, and advected heat flux, in *Glacial Isostasy, Sea-Level and Mantle Rheology*, edited by R. Sabatini, K. Lambeck, and E. Boschi, pp. 493-513, Kluwer Academic Publishers, Dordrecht, 1991.

Hager, B.H., and R.J. O'Connell, A simple global model of plate dynamics and mantle convection, *J. Geophys. Res.*, 86, 4843-4867, 1981.

Hargraves, R.B., Faster spreading or greater ridge length in the Archean?, *Geology*, 14, 750-752, 1986.

Helffrich, G.R., and B.J. Wood, The Earth's mantle, *Nature*, 412, 501-507, 2001.

Heller, P.L., D.L. Anderson, and C.L. Angevine, Is the middle Cretaceous pulse of rapid sea-floor spreading real or necessary?, *Geology*, 24, 491-494, 1996.

Herzberg, C., Depth and degree of melting of komatiites, *J. Geophys. Res.*, 97, 4521-4540, 1992.

Hirth, G., and D. Kohlstedt, Rheology of the upper mantle and the mantle wedge: A view from the experimentalists, in *Inside the Subduction Factory*, edited by J. Eiler, pp. 83-105, American Geophysical Union, Washington, D.C., 2003.

Hirth, G., and D.L. Kohlstedt, Water in the oceanic mantle: Implications for rheology, melt extraction, and the evolution of the lithosphere, *Earth Planet. Sci. Lett.*, 144, 93-108, 1996.

Hoffman, P.F., Tectonic genealogy of North America, in *Earth Structure: An Introduction to Structural Geology and Tectonics*, edited by B.A. van der Pluijm and S. Marshak, pp. 459-464, McGraw-Hill, New York, 1997.

Hofmann, A.W., Mantle geochemistry: the message from oceanic volcanism, *Nature*, 385, 219-229, 1997.

Howard, L.N., Convection at high Rayleigh number, in *Proceedings of the Eleventh International Congress of Applied Mechanics*, edited by H. Gortler, pp. 1109-1115, Springer-Verlag, New York, 1966.

Huang, J., S. Zhong, and J. van Hunen, Controls on sublithospheric small-scale convection, *J. Geophys. Res.*, 108(B7), 2405, doi:10.1029/2003JB002, 456, 2003.

Jackson, M.J., and H.N. Pollack, On the sensitivity of parameterized convection to the rate of decay of internal heat sources, *J. Geophys. Res.*, 89, 10, 103-10, 108, 1984.

Jochum, K.P., A.W. Hofmann, E. Ito, H.M. Seufert, and W.M. White, K, U and Th in mid-ocean ridge basalt glasses and heat production, K/U and K/Rb in the mantle, *Nature*, 306, 431-436, 1983.

Jordan, T.H., Mineralogies, densities and seismic velocities of garnet lherolites and their geophysical implications, in *The Mantle Sample: Inclusions in Kimberlites and Other Volcanics*, edited by F. Boyd and H. Meyer, pp. 1-14, American Geophysical Union, Washington, D.C., 1979.

Jurdy, D.M., M. Stefanick, and C.R. Scotese, Paleozoic plate dynamics, *J. Geophys. Res.*, 100, 17, 965-17, 975, 1995.

Karato, S., Does partial melting reduce the creep strength of the upper mantle?, *Nature*, 319, 309-310, 1986.

Karato, S., and P. Wu, Rheology of the upper mantle: A synthesis, *Science*, 260, 771-778, 1993.

Katzman, R., L. Zhao, and T.H. Jordan, High-resolution, two-dimensional vertical tomography of the central Pacific mantle using ScS reverberations and frequency-dependent travel times, *J. Geophys. Res.*, 103, 17, 933-17, 971, 1998.

Kelemen, P.B., S.R. Hart, and S. Bernstein, Silica enrichment in the continental upper mantle via melt/rock reaction, *Earth Planet. Sci. Lett.*, 164, 387-406, 1998.

Kellogg, L.H., and D.L. Turcotte, Mixing and the distribution of heterogeneities in a chaotically convecting mantle, *J. Geophys. Res.*, 95, 421-432, 1990.

Kellogg, L.H., B.H. Hager, and R.D. van der Hilst, Compositional stratification in the deep mantle, *Science*, 283, 1881-1884, 1999.

King, S.D., Models of mantle viscosity, in *Mineral Physics and Crystallography: A Handbook of Physical Constants*, edited by T.J. Ahrens, pp. 227-236, American Geophysical Union, Washington, D.C., 1995.

Kinzler, R.J., Melting of mantle peridotite at pressures approaching the spinel to garnet transition: Application to mid-ocean ridge basalt petrogenesis, *J. Geophys. Res.*, 102, 852-874, 1997.

Kinzler, R.J., and T.L. Grove, Primary magmas of mid-ocean ridge basalts, 2, applications, *J. Geophys. Res.*, 97, 6907-6926, 1992.

Kinzler, R.J., and T.L. Grove, Corrections and further discussion of the primary magmas of mid-ocean ridge basalts, 1 and 2, *J. Geophys. Res.*, 98, 22, 339-22, 347, 1993.

Komiya, T., S. Maruyama, T. Masuda, S. Nohda, M. Hayashi, and K. Okamotmo, Plate tectonics at 3.8-3.7 Ga: Field evidence from the Isua accretionary complex, southern West Greenland, *J. Geol.*, 107, 515-554, 1999.

Korenaga, J., Energetics of mantle convection and the fate of fossil heat, *Geophys. Res. Lett.*, 30(8), 1437, doi:10.1029/2003GL016, 982, 2003.

Korenaga, J., Mantle mixing and continental breakup magmatism, *Earth Planet. Sci. Lett.*, 218, 463-473, 2004.

Korenaga, J., Firm mantle plumes and the nature of the core-mantle boundary region, *Earth Planet. Sci. Lett.*, 232, 29-37, 2005a.

Korenaga, J., Why did not the Ontong Java Plateau form subaerially?, *Earth Planet. Sci. Lett.*, 234, 385-399, 2005b.

Korenaga, J., and T.H. Jordan, Onset of convection with temperature- and depth-dependent viscosity, *Geophys. Res. Lett.*, 29(19), 2002a.

Korenaga, J., and T.H. Jordan, On 'steady-state' heat flow and the rheology of oceanic mantle, *Geophys. Res. Lett.*, 29(22), 2056, doi:10.1029/2002GL016, 085, 2002b.

Korenaga, J., and T.H. Jordan, Physics of multiscale convection in Earth's mantle: Onset of sublithospheric convection, *J. Geophys. Res.*, 108(B7), 2333, doi:10.1029/2002JB001, 760, 2003.

Korenaga, J., and P.B. Kelemen, Major element heterogeneity in the mantle source of the North Atlantic igneous province, *Earth Planet. Sci. Lett.*, 184, 251-268, 2000.

Korenaga, J., P.B. Kelemen, and W.S. Holbrook, Methods for resolving the origin of large igneous provinces from crustal seismology, *J. Geophys. Res.*, 107, 2178, 2002.

Kröner, A., and P.W. Layer, Crust formation and plate motion in the Early Archean, *Science*, 256, 1405-1411, 1992.

Labrosse, S., Hotspots, mantle plumes and core heat loss, *Earth Planet. Sci. Lett.*, 199, 147-156, 2002.

Langmuir, C.H., E.M. Klein, and T. Plank, Petrological systematics of mid-ocean ridge basalts: Constraints on melt generation beneath ocean ridges, in *Mantle Flow and Melt Generation at Mid-Ocean Ridges*, edited by J. Phipps Morgan, D.K. Blackman, and J.M. Sinton, vol. 71 of *Geophys. Monogr. Ser.*, pp. 183-280, AGU, Washington, D. C., 1992.

Larson, R.L., Geological consequences of superplumes, *Geology*, 19, 963-966, 1991.

Lee, C.-T.A., Compositional variation of density and seismic velocities in natural peridotites at STP conditions: implications for seismic imaging of copmositional heterogeneities in the upper mantle, *J. Geophys. Res.*, 108, 2441, doi:10.1029/2003JB002, 413, 2004.

Levitt, D.A., and D.T. Sandwell, Lithospheric bending at subduction zones based on depth soundings and satellite gravity, *J. Geophys. Res.*, 100, 379-400, 1995.

Lithgow-Bertelloni, C., and M.A. Richards, The dynamics of Cenozoic and Mesozoic plate motions, *Rev. Geophys.*, 36, 27-78, 1998.

Manga, M., Mixing of heterogeneities in the mantle: Effect of viscosity differences, *Geophys. Res. Lett.*, 23, 403-406, 1996.

McDonough, W.F., and S.-S. Sun, The composition of the Earth, *Chem. Geol.*, 120, 223-253, 1995.

McGovern, P.J., and G. Schubert, Thermal evolution of the Earth: effects of volatile exchange between atmosphere and interior, *Earth Planet. Sci. Lett.*, 96, 27-37, 1989.

McKenzie, D., and M.J. Bickle, The volume and composition of melt generated by extension of the lithosphere, *J. Petrol.*, 29, 625-679, 1988.

McKenzie, D.P., and N. Weiss, Speculation on the thermal and tectonic history of the Earth, *Geophys. J. R. Astron. Soc.*, 42, 131-174, 1975.

Michael, P.J., The concentration, behavior and storage of H_2O in the suboceanic upper mantle: Implications for mantle metasomatism, *Geochim. Cosmochim. Acta*, 52, 555-566, 1988.

Montagner, J., Upper mantle low anisotropy channels below the Pacific plate, *Earth Planet. Sci. Lett.*, 202, 263-374, 2002.

Montelli, R., G. Nolet, F.A. Dahlen, G. Masters, E.R. Engdahl, and S.-H. Hung, Finite-frequency tomography reveals a variety of plumes in the mantle, *Science*, 303, 338-343, 2004.

Nimmo, F., G.D. Price, J. Brodholt, and D. Gubbins, The influence of potassium on core and geodynamo evolution, *Geophys. J. Int.*, 156, 363-376, 2004.

Nisbet, E.G., *The Young Earth: An introduction to Archaean geology*, Allen&Unwin, Boston, 1987.

Nisbet, E.G., and C.M.R. Fowler, Model for Archean plate tectonics, *Geology*, 11, 376-379, 1983.

Parman, S.W., J.C. Dann, T.L. Grove, and M.J. de Wit, Emplacement conditions of komatiite magmas from the 3.49 Ga Komati Formation, Barberton Greenstone Belt, South Africa, *Earth Planet. Sci. Lett.*, 150, 303-323, 1997.

Parsons, B., Causes and consequences of the relation between area and age of the ocean floor, *J. Geophys. Res.*, 87, 289-302, 1982.

Parsons, B., and D. McKenzie, Mantle convection and the thermal structure of the plates, *J. Geophys. Res.*, 83, 4485-4496, 1978.

Pollack, H.N., S.J. Hurter, and J.R. Johnston, Heat loss from the earth's interior: analysis of the global data set, *Rev. Geophys.*, 31, 267-280, 1993.

Richards, M.A., R.A. Duncan, and V.E. Courtillot, Flood basalts and hot-spot tracks: Plume heads and tails, *Science*, 246, 103-107, 1989.

Richter, F.M., Models for the Archean thermal regime, *Earth Planet. Sci. Lett.*, 73, 350-360, 1985.

Richter, F.M., A major change in the thermal state of the Earth at the Archean-Proterozoic boundary: Consequences for the nature and presevation of continental lithosphere, *J. Petrol.*, *Special Lithosphere Issue*, 39-52, 1988.

Ritzwoller, M.H., N.M. Shapiro, and S. Zhong, Cooling history of the Pacific lithosphere, *Earth Planet. Sci. Lett.*, 226, 69-84, 2004.

Romanowicz, B., and Y.C. Gung, Superplumes from the core-mantle boundary to the lithosphere: implications for heat flux, *Science*, 296, 513-516, 2002.

Rowley, D.B., Rate of plate creation and destruction: 180 Ma to present, *Geol. Soc. Am. Bull.*, 114, 927-933, 2002.

Schubert, G., D. Stevenson, and P. Cassen, Whole planet cooling and the radiogenic heat source contents of the Earth and Moon, *J. Geophys. Res.*, 85, 2531-2538, 1980.

Schubert, G., D.L. Thurcotte, and P. Olson, *Mantle Convection in the Earth and Planets*, Cambridge, New York, 2001.

Sleep, N.H., Thermal history and degassing of the Earth: Some simple calculations, *J. Geol.*, 87, 671-686, 1979.

Sleep, N.H., Hotspots and mantle plumes: Some phenomenology, *J. Geophys. Res.*, 95, 6715-6736, 1990.

Sleep, N.H., Archean plate tectonics: what can be learned from continental geology?, *Can. J. Earth Sci.*, 29, 2066-2071, 1992.

Sleep, N.H., M.A. Richards, and B.H. Hager, Onset of mantle plumes in the presence of preexisting convection, *J. Geophys. Res.*, 93, 7672-7689, 1988.

Solheim, L.P., and W.R. Peltier, Avalanche effects in phase transition modulated thermal convection: A model of Earth's mantle, *J. Geophys. Res.*, 99, 6997-7018, 1994.

Solomatov, V.S., Scaling of temperature- and stress-dependent viscosity convection, *Phys. Fluids*, 7, 266-274, 1995.

Solomatov, V.S., Can hotter mantle have a larger viscosity?, *Geophys. Res. Lett.*, 23, 937-940, 1996.

Solomatov, V.S., Grain size-dependent viscosity convection and the thermal evolution of the Earth, *Earth Planet. Sci. Lett.*, 191, 203-212, 2001.

Spohn, T., and G.Schubert, Modes of mantle convection and the removal of heat from the Earth's interior, *J. Geophys. Res.*, 87, 4682-4696, 1982.

Sprague, D., and H.N. Pollack, Heat flow in the Mesozoic and Cenozoic, *Nature*, 285, 393-395, 1980.

Stacey, F.D., Cooling of the earth - a constraint on paleotectonic hypotheses, in *Evolution of the Earth*, edited by R.J. O'Connell and W.S. Fyfe, pp. 272-276, American Geophysical Union, Washington, DC, 1981.

Stacey, F.D., and D.E. Loper, Thermal histories of the core and the mantle, *Phys. Earth Planet. Int.*, 36, 99-115, 1984.

Stevenson, D.J., T.Spohn, and G.Schubert, Magnetism and thermal evolution of the terrestrial planets, *Icarus*, 54, 466-489, 1983.

Storey, B.C., The role of mantle plumes in continental breakup: case histories from Gondwanaland, *Nature*, 377, 301-308, 1995.

Su, Y.-J., Mid-ocean ridge basalt trace element systematics: constraints from database management, icpms analyses, global data compilation, and petrologic modeling, Ph.D. thesis, Columbia University, New York, 2002.

Tackley, P.J., On the penetration of an endothermic phase transition by upwellings and downwellings, *J. Geophys. Res.*, 100, 15, 477-15, 488, 1995.

Tackley, P.J., D.J. Stevenson, G.A. Glatzmaier, and G.Schubert, Effects of an endothermic phase transition at 670 km depth in a spherical model of convection in the Earth's mantle, *Nature*, 361, 699-704, 1993.

Tajika, E., and T.Matui, Evolution of seafloor spreading rate based on ^{40}Ar degassing history, *Geophys. Res. Lett.*, 20, 851-854, 1993.

Takahashi, E., and I.Kushiro, Melting of a dry peridotite at high pressures and basalt magma genesis, *Am. Mineral.*, 68, 859-879, 1983.

Turcotte, D.L., On the thermal evolution of the Earth, *Earth Planet. Sci. Lett.*, 48, 53-58, 1980.

Turcotte, D.L., and E.R. Oxburgh, Finite amplitude convective cells and continental drift, *J. Fluid Mech.*, 28, 29-42, 1967.

Turcotte, D.L., and G. Schubert, *Geodynamics: Applications of continuum physics to geological problems*, John Wiley & Sons, New York, 1982.

Ullrich, L., and R.V. der Voo, Minimum continental velocities with respect to the pole since the Archean, *Tectonophysics*, 74, 17-27, 1981.

Van der Hilst, R.D., S.Widiyantoro, and E.R. Engdahl, Evidence for deep mantle circulation from global tomography, *Nature*, 386, 578-584, 1997.

van Keken, P.E., E.H. Hauri, and C.J. Ballentine, Mantle mixing: The generation, preservation, and destruction of chemical heterogeneity, *Annu. Rev. Earth Planet. Sci.*, 30, 493-525, 2002.

Verhoogen, J., *Energetics of the Earth*, National Academy of Sciences, Washington, D.C., 1980.

Walter, M.J., Melting of garnet peridotite and the origin of komatiite and depleted lithosphere, *J. Petrol.*, 39, 29-60, 1998.

Weertman, J., The creep strength of the Earth's mantle, *Rev. Geophys. Space Phys.*, 8, 146-168, 1970.

Williams, D.R., and V.Pan, Internally heated mantle convection and the thermal and degassing history, *J. Geophys. Res.*, 97, 8937-8950, 1992.

Windley, B.F., Uniformitarianism today: plate tectonics is the key to the past, *J. Geol. Soc. London*, 150, 7-19, 1993.

Yuen, D.A., and L. Fleitout, Stability of the oceanic lithosphere with variable viscosity: an initial-value approach, *Phys. Earth Planet. Int.*, 34, 173-185, 1984.

Yuen, D.A., D.M. Reuteler, S.Balachandar, V. Steinbach, A.V. Malevsky, and J.J. Smedsmo, Various influences on 3-dimensional mantle convection with phase transitions, *Phys. Earth Planet. Int.*, 86, 185-203, 1994.

Yukutake, T., The inner core and the surface heat flow as clues to estimating the initial temperature of the Earth's core, *Phys. Earth Planet. Int.*, 121, 103-137, 2000.

Zaranek, S.E., and E.M. Parmentier, Convective cooling of an initially stably stratified fluid with temperature-dependent viscosity: Implications for the role of solid-state convection in planetary evolution, *J. Geophys. Res.*, 109, B03, 409, doi:10.1029/2003JB002, 462, 2004.

Zhong, S., and M.Gurnis, Role of plates and temperature-dependent viscosity in phase change dynamics, *J. Geophys. Res.*, 99, 15, 903-15, 917, 1994.

Zhong, S., and M.Gurnis, Mantle convection with plates and mobile, faulted plate margins, *Science*, 267, 838-843, 1995.

J. Korenaga, Department of Geology and Geophysics, P.O. Box 208109, Yale University, New Haven, Connecticut 06520-8109, USA. (jun.korenaga@yale.edu)

Continental Growth and the Archean Paradox

A. Lenardic

Department of Earth Science, Rice University, Houston, Texas

Proposed continental growth curves, based on geochemical and/or petrologic data, have fallen into two broad camps, instantaneous versus progressive growth. Archean thermal constraints can potentially help to elucidate which end-member is more viable as the scenarios show greatest divergence in the Archean. Toward this end, we review a heat flow scaling theory for mantle convection with continents and use the theory to explore the thermal implications of different end-member models of continental growth. Both instantaneous and progressive growth models predict that the ratio of oceanic to subcontinental mantle heat flux should have increased in the Earth's past, when the planet had more heat to lose. This allows both scenarios to match the observation-based constraint that heat flow through continents was not much greater in the Archean than it is at present. A progressive growth scenario is mildly more effective at matching this constraint under a wider range of parameter conditions and is, thus, somewhat more robust than the instantaneous growth end-member. An added separation between the end-members comes from the fact that a progressive continental growth scenario predicts a buffering of mantle temperature over time. This buffering has the potential to limit wide-scale mantle melting in the Archean. The lack of evidence of any such melting favors a progressive growth scenario although the constraint is not strong as several other physical mechanisms, not associated with continental growth, can also limit Archean melting. Collectively, the thermal constraints discussed cannot rule out either end-member. A progressive continental growth model is more robust in terms of parameter uncertainties which provides mild support for a continental growth curve nearer to this end-member.

INTRODUCTION

Figure 1 shows several continental growth curves that have been proposed over the years based on geochemical and/or petrologic data [*Fyfe*, 1978; *Armstrong*, 1981; *McCulloch and Bennett*, 1994; *Taylor and McLennen*, 1995; *Kramers and Tolstikhin*, 1997; *Sylvester et al.*, 1997; *Collerson and Kamber*, 1999]. The figure shows that the disagreement as to how continents have grown can be broadly

broken down into two camps: instantaneous versus progressive growth. The differences each end-member scenario would imply for the Earth's thermal state become most pronounced in the Archean, where the end-member curves reach maximum divergence. The purpose of this contribution is to explore the implications of the end-member growth curves for the thermal state of the Earth in the Archean. The goal is to determine if one or the other of the end-members can better fit Archean thermal constraints. If so, then these thermal constraints, which were not a priori considered in determining any of the curves of Figure 1, can provide added, independent support for a specific class of growth models. The thermal issues to be addressed are discussed in subsections below. As we will be dealing with the effect of

Archean Geodynamics and Environments
Geophysical Monograph Series 164
Copyright 2006 by the American Geophysical Union
10.1029/164GM04

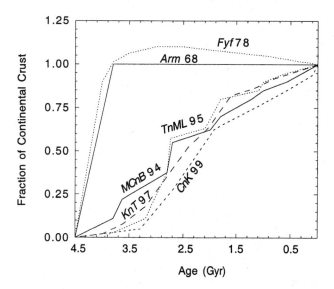

Figure 1. Continental growth curves. Notation: Fyf78—*Fyfe*, 1978; Arm68—*Armstrong*, 1981; TnML95—*Taylor and McLennen*, 1995; MCnB94—*McCulloch and Bennett*, 1994; KnT97—*Kramers and Tolstikhin*, 1997; CnK99—*Collerson and Kamber*, 1999. More recent work than that of Armstrong and Fyfe has also been shown to be consistent with models of instantaneous growth [e.g., *Sylvester et al.*, 1997].

continents, and continental growth, on the thermal state of the Earth, its mantle in particular, a third introductory sub-section on continent-mantle coupling is also included.

The Archean Paradox

The Archean paradox stems from the fact that while certain lines of evidence suggest that Archean continental geotherms were similar to those at present [*Burke and Kidd*, 1978; *England and Bickle*, 1984; *Boyd et al.*, 1985], radiogenic mantle heat production must have been higher, suggesting an average mantle heat flux significantly greater than at present [*McKenzie and Weiss*, 1975; *Davies*, 1980; *Richter*, 1984]. Over the years several potential solutions have been offered for the apparent contradiction that arises [*Burke and Kidd*, 1978; *Bickle*, 1978; 1986; *Davies*, 1979; *Jarvis and Campbell*, 1983; *Richter*, 1985; *Pollack*, 1986; *Ballard and Pollack*, 1988; *Lenardic*, 1998]. One of the more popular holds that creation and subduction of oceanic plates carried a greater proportion of the Earth's heat loss load in the Archean relative to the present. This, it is argued, would allow continental geotherms to remain relatively unchanged over time. Such a solution hinges on the ratio of continental to oceanic mantle heat flux. Specifically, the ratio must decline as one goes back in time, i.e., as the vigor of mantle convection increases. The dynamic plausibility of this was tested using

numerical simulations that made variable assumptions [*Davies*, 1979; *Lenardic*, 1998] but, to the best of my knowledge, the issue of how different continental growth scenarios would affect this ratio has not been explored to date. Doing so would offer an added means of discriminating between the two end-member growth scenarios if it turned out that one was more robust in its ability to address the Archean paradox.

The Archean Thermal Catastrophe

There is no evidence that the majority of the Earth's mantle was molten in the Archean. Indeed, the preservation of sections of Archean continents argues against this. A molten mantle would be a "thermal catastrophe" and the lack of evidence for it means that the Archean thermal catastrophe is a non-event. It may seem odd to discuss a non-event in the context of thermal constraints on continental growth. However, the non-event has proven to be a valuable constraint in thermal history modeling as models of the Earth's cooling that predict wide-scale mantle melting in the Archean can be considered far more problematic than those that do not. That certain classes of thermal history models predict unsupported thermal catastrophes in the Archean was pointed out by *Christensen* [1985]. A brief review of thermal history modeling will help to highlight the significance of his observation.

Classic thermal history models are based on a volume-averaged energy conservation equation for the Earth's mantle [e.g., *Davies*, 1980; *Schubert et al.*, 1980]. The equation balances the heat flow from the Earth's surface with the heat generated by the decay of radioactive elements within the solid Earth and the Earth's primordial heat, i.e., the heat associated with formation and early differentiation. Key quantities predicted by thermal history models, that can be compared to data constraints, are the variation of Earth's average internal temperature over time, the variation of surface heat flow, the variation of average seafloor spreading rate, and the ratio of radiogenic heat production to heat loss, a quantity referred to as the Urey ratio. The key assumption needed to close thermal history models is a relationship between the Earth's surface heat flow and the vigor of convection within the mantle. A measure of convective vigor is given by the mantle Rayleigh number, Ra, which is a nondimensional ratio of the forces driving convection relative to those resisting convection. A measure of the efficiency of convective heat transport is the Nusselt number, Nu, which is defined as the surface heat flux divided by the heat flux that would result from pure conduction. All classic thermal history models have parameterized convective heat transport using a scaling relationship of the form

$$\mathrm{Nu} = a_0 \mathrm{Ra}^\beta \qquad (1)$$

where a_0 is a geometric scaling constant and β is a scaling exponent. The higher the value of the scaling exponent, the more efficient convection is in cooling the mantle. For convecting fluids in which inertial forces are negligibly small, like the Earth's mantle, the absolute upper bound on β is 1/3 [*Chan*, 1971; *Constantin and Doering*, 1999]. The high viscosity of the mantle is the reason inertial forces are so low and this is often expressed in the literature by stating that the mantle is effectively an infinite Prandtl number fluid, where the Prandtl number is the ratio of the kinematic viscosity to the thermal diffusivity.

As *Christensen* [1985] noted, thermal history studies traditionally employed a scaling exponent, β, between 1/4 to 1/3, a range that is well confirmed by laboratory, numerical, and theoretical work for the case of constant viscosity, Rayleigh-Bérnard convection [e.g., *Busse*, 1985; 1989; and references therein]. Fluid dynamic confirmation for simple convecting systems led early thermal history modelers to adopt the 1/3 value and the thermal history models that resulted were compatible with added data constraints provided that the present-day Urey ratio was 0.7 or greater [e.g., *Schubert et al.*, 1980]. This is where *Christensen* [1985] noted a problem.

Based on estimates of mantle heat production and loss, *Christensen* [1985] derived a value for the present-day Urey ratio of 0.4. There is uncertainty but what is key is that the value is relatively low. Subsequent estimates have lead to even lower Urey ratios [e.g., *Kellogg et al.*, 1999; *Korenaga*, 2003]. A low Urey ratio means that a significant component of the Earth's present-day heat loss is coming from primordial heat and this presents a problem for thermal history models that invoke relatively efficient convective cooling. The problem is that if convective cooling is efficient, and if the Earth still contains a large proportion of primordial heat, then one is led to the prediction that the Earth should have contained enough heat in the Archean to completely melt the mantle. This is the essence of *Christensen's* [1985] argument that the value of β must be < 0.1, significantly lower than previously assumed. Stated another way, the cooling efficiency of the Earth's mantle over geologic time must be considerably less than the maximum value for a high Prandtl number convecting fluid. *Christensen* [1985] himself suggested that the temperature dependence of mantle viscosity could lead to less efficient mantle cooling because the high-viscosity upper boundary layer of the mantle would move sluggishly at best and thus not cool the interior efficiently [*Christensen*, 1984]. This suggestion went against generally accepted thoughts at the time [e.g., *Tozer*, 1972] and was subsequently criticized because, as it was specifically stated, it sacrificed active oceanic plate creation and subduction, a solution that was not deemed reasonable [*Gurnis*, 1989]. Subsequently, several alternative solutions have been proposed [*Becker et al.*, 1999; *Kellogg et al.*, 1999; *Conrad and Hager*, 2001;

Helfrich and Wood, 2001; *Butler and Peltier*, 2002; *Korenaga*, 2003]. None of those cited specifically considered how continents might affect mantle cooling and whether the effects could or could not lead to thermal history models that do not predict unreasonable thermal conditions in the Archean. As with the Archean paradox, this will be a focus of this contribution.

Continent-Mantle Thermal Coupling

Oceanic lithosphere is relatively short-lived, attesting to the fact that it is the active upper thermal boundary layer of mantle convection [e.g., *Turcotte and Oxburgh*, 1967]. Accordingly, its thermal structure has been modeled within a thermal convection framework [e.g., *Parsons and Sclater*, 1977; *Parsons and McKenzie*, 1978; *Stein and Sten*, 1992; *Doin and Fleitout*, 1996]. Unlike oceanic lithosphere, the longevity of continental crust and deeper continental lithosphere [e.g., *Boyd et al.*, 1985] has led to continental thermal structure being treated within a thermal conduction framework. The bulk of continental thermal models to date are based on solving a conduction equation within a column meant to represent the vertically layered structure of continental lithosphere [e.g., *Pollack and Chapman*, 1977; *Morgan*, 1984; *Cermak*, 1993; *Balling*, 1995; *Zhou*, 1996; *Jaupart et al.*, 1998; *Rudnick et al.*, 1998]. Mantle convection is introduced indirectly as a free lower thermal boundary condition applied to the conduction solution.

As conduction is less efficient in transferring heat than convection, the differences between oceanic and continental lithosphere, noted above, lead to the idea that continents act as thermal insulation about the convecting mantle. Indeed, the bulk of studies that have explored the thermal coupling between continents and the convecting mantle have been founded on this idea [e.g., *Elder*, 1967; *Whitehead*, 1972; *Busse*, 1978; *Gurnis*, 1988; *Zhong and Gurnis*, 1993; *Lowman and Jarvis*, 1993; 1995; *Guillou and Jaupart*, 1995; *Lenardic and Kaula*, 1995; 1996]. The major focus of these studies has been to explore how the presence of continents can alter patterns of convective flow within the mantle and, more specifically, how the interaction between insulating continents and the convecting mantle can drive continental aggregation, dispersal, and drift. The issue of how the presence of continental lithosphere, of potentially time variable area, affects the global thermal balance of our planet over geologic time has received less attention. Intuitively, the fact that continents can act as insulation above the mantle suggests that they should lower the Earth's global rate of internal heat loss. This assumption has generally been adopted for global thermal history studies that have sought to incorporate continents [e.g., *Schubert and Reymer*, 1985; *Spohn and Breuer*, 1993; *Grigne and Labrosse*, 2001]. Global mantle heat flow is treated as a simple area-weighted average of the heat flow through oceanic lithosphere and the heat flow into the base of continents. As subcontinental mantle heat flux is very low [*Ashwal et al.*, 1987; *Pinet et al.*, 1991; *Guillou-Frottier et al.*, 1995; *Ketchman*, 1996] relative to oceanic heat flux [*Pollack et al.*, 1993], this implies that increased continental area should decrease global mantle heat flow and retard mantle cooling. It is worth stating the key assumption noted above in another way: The assumption being made is that an increase or decrease in the amount of continental insulation does not affect the convective heat flux from oceanic lithosphere. That is, the presence of insulating continents is assumed to not significantly affect the rate of oceanic plate creation and subduction.

The assumption above, although intuitively reasonable, has only recently been tested using numerical simulations of mantle convection below conducting continental lithosphere [*Lenardic and Moresi*, 2003; *Lenardic et al.*, 2005]. For an isoviscous mantle, the assumption is compatible with the results of numerical simulations. However, for a mantle with temperature-dependent viscosity the assumption falls short. The reason for this is because a key effect of continental insulation is to increase the internal temperature and, hence, to reduce the viscosity of the bulk mantle, which leads, in turn, to higher convective velocities and enhanced heat flow through ocean basins. The enhanced oceanic heat flow can outweigh the lowered subcontinental mantle heat flow, due to continental insulation, provided the surface area covered by oceans is sufficiently large. Thus, depending on the ratio of ocean to continent surface area, continental insulation can have no effect on and can even increase global mantle heat flow. This result is consistent with a theoretical analysis of partially insulated thermal convection and with the results of both numerical simulations and laboratory experiments that explore the heat transfer properties of partially insulated convecting layers [*Lenardic and Moresi*, 2003; *Lenardic et al.*, 2005]. The consistency between the fore mentioned theory and both simulations and experiments suggests that the theory can have utility for exploring the interaction between mantle heat loss and increasing volumes of insulating continental lithosphere, i.e., continental growth. This will be the main methodology of this contribution. The heat flow scaling theory will be used to address how end-member growth curves affect mantle heat loss and predictions will be compared to the thermal constraints discussed in the previous subsections. The theory will be reviewed first as will the work used to test it.

THEORY

We begin by considering the heat transfer properties of a mantle lacking continents. In a plate tectonic mode of convection, oceanic lithosphere subducts and cools the interior mantle, i.e., the oceanic lithosphere actively participates in convective overturn. It has long been hypothesized that convection in a mantle with temperature-dependent viscosity, and an active lithosphere, should behave as an equivalent isoviscous system with a viscosity equal to that of the average internal viscosity of the temperature-dependent system [*Tozer*, 1972, *Gurnis*, 1989]. If this is true, then the theoretically expected form of the mantle heat flux scaling should be equivalent to Eq (1) with the constant property Rayleigh number, Ra, replaced by a Rayleigh number defined in terms of average internal viscosity.

The validity of the argument above has been confirmed via numerical simulations of mantle convection with a viscoplastic rheology [*Moresi and Solomatov*, 1998]. A viscoplastic mantle rheology allows for an active lid mode of convection for oceanic lithosphere, i.e., lithospheric recycling, and for an internal mantle viscosity that depends strongly on temperature [*Moresi and Solomatov*, 1998]. The rheology law remains on a temperature-dependent viscous branch for stresses below a specified yield stress, τ_{yield}. Along this branch, the viscosity function is given by

$$\mu = A\exp[-\theta T], \qquad (2a)$$

where A and θ are material parameters, T is temperature, and μ is mantle viscosity. For stresses above a yield stress, the flow law switches to a plastic branch with a nonlinear, effective viscosity given by

$$\mu_{plastic} = T_{yield}/I, \qquad (2b)$$

where I is the second strain-rate invariant. This rheologic formulation allows zones of localized lithospheric failure, analogs to weak plate boundaries, to form in a self-consistent way. The failure zones allow for lithospheric subduction and mantle stirring akin to plate tectonics.

Moresi and Solomatov [1998] explored numerical convection simulations employing the rheologic formulation above. Bottom heating was assumed and no continents were present. The best fit scaling for mantle heat flux in the active lid regime was found to be

$$q_o = 0.385 Ra_{io}^{0.293}, \qquad (3)$$

where q_o and Ra_{io} are, respectively, the nondimensional surface heat flux and the Rayleigh number defined in terms of average internal (subscript '*i*') viscosity (subscript 'o' indicates that this scaling applies to the oceanic lithosphere only end-member case). This is reasonably close to the theoretically expected scaling if the temperature-dependent, active lid system behaves as an equivalent isoviscous system [*Tozer*, 1972, *Howard*, 1966]. For high degrees of convective vigor, the average internal temperature, T_{io}, was shown to approach the mean of the surface and base temperatures, as is theoretically expected [*Tozer*, 1972, *Howard*, 1966]. For lower degrees of convective vigor, a relationship for T_{io} as a function of known system parameters was presented [*Moresi and Solomatov*, 1998]. This allows Ra_{io} to be determined and closes the heat flux scaling for the oceanic lithosphere only end-member case.

Notice that the scaling exponent in Eq (3) is comparable to the value used by classic thermal history studies and, thus, is in the range *Christensen* [1985] found unacceptable as it led to thermal histories that predicted global melting in the Archean. The issue we will address is whether the effects of continental lithosphere can alleviate this problem and, as such, we must incorporate the effects of continents into a mantle heat flow scaling. To do so, we employ a thermal network analysis [*Seely*, 1964; *Incropera and DeWitt*, 1996]. The solid Earth is modeled as a thermal network composed of three primary heat transfer components (Figure 2). One component is associated with conductive heat transfer within chemically buoyant, non-subductable continental lithosphere of average thickness d. The second component is associated with heat transfer from the convecting mantle into the base of long-lived continental lithosphere. Mantle heat is transferred across the thermal sublayer, of thickness δ_c, that forms below chemically buoyant continental lithosphere. The thickness of the sublayer, and thus the heat transfer efficiency of this second component, is determined by the vigor of mantle convection. Components one and two, linked in series, form the continental heat transfer path. This path is linked in parallel to the final thermal component associated with mantle heat transfer through oceanic lithosphere. For an active lid, i.e., plate tectonic-like, mode of convection, heat will be transferred to the base of the ocean floor across the active upper thermal boundary layer of the convecting mantle. This boundary layer, i.e., the oceanic thermal lithosphere, has a thickness of δ_o, which is determined by the vigor of mantle convection.

End-member cases of a planet covered by, or devoid of, continents can be associated with thermal resistances of R_c and R_o, respectively. Intermediate cases will have a total network resistance, R_t. Total system heat flux, q_t, can thus be expressed as $q_t = \Delta T_i / R_t$, where ΔT_i is the average temperature drop associated with heat transfer from the interior mantle to the surface. Similarly, for the end-members, $q_c = \Delta T_c / R_c$ and $q_o = \Delta T_o / R_o$. We assume thorough thermal mixing such that the average internal temperature of the full thermal network, T_i, can be taken to be a volume-weighted average of the internal temperatures for each end-member, T_{ic} and T_{io}. Considering surface and volume ratios to be proportional leads to

$$T_i = A_c T_{ic} + A_o T_{io} \qquad (4)$$

where A_c and A_o are defined as the ratio of the planet's surface covered by continents and oceans, respectively (i.e., $A_c + A_o = 1$).

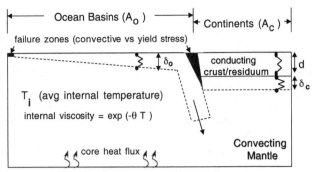

Model of Conduction-Convection Interaction

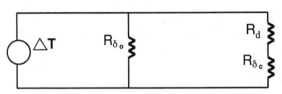

Figure 2. Cartoon of the solid Earth heat transfer system (top) together with a thermal network model (bottom). The simplest relevant network is composed of three components. The first represents conductive heat transfer within continental lithosphere of average thickness d, surface area A_c, and thermal conductivity K_c. The thermal resistance of this component depends directly on d and inversely on K_c and A_c. A second component represents heat transfer from the convecting mantle into the base of continents across a convective sub-layer. The average thickness of this boundary layer, δ_c, and thus the effective resistance of this component depends inversely on convective mantle vigor and, by association, directly on internal mantle viscosity. In series, components one and two form the continental path. This path is linked in parallel to an oceanic component. For oceanic regions, internal heat is transferred across an active thermal boundary layer of surface area A_o. The average thickness of this boundary layer, δ_o, and thus the effective resistance of this path depend inversely on convective mantle vigor and, by association, directly on internal mantle viscosity. The average internal temperature of the mantle, T_i, minus the temperature at the system surface, T_s, is the temperature drop, ΔT_i, driving heat transfer across the network.

Oceanic and continental paths are considered to be linked in parallel and the effective resistance of each path is assumed to decrease as its relative surface area increases (Figure 2). As well as surface area effects on local resistance, the effects of internal temperature must also be considered. The temperature of the full network will be greater than that of the oceanic end-member and less than that of the continental end-member. Thus, the internal viscosity of the system will differ from that of either end-member. This, in turn, can alter the local resistance of the oceanic and continental path within the full system as compared to end-member cases. This will not have a large effect on the continental path provided that the thermal resistance of stable continental lithosphere is much greater than that of the mantle sublayer below. This is because the continental path is itself a composite of two resistance in series (Figure 2). Changes in internal mantle viscosity will effect the thickness of the mantle sublayer but, for relatively thick continental lithosphere, this will have a negligible effect on the composite continental resistance. We will assume this to be the case as, for high Rayleigh numbers, the ratio d/δ_c will be greater than one. The oceanic path, on the other hand, is not a composite and changes in internal temperature can have large effects on its effective thermal resistance. Higher temperature reduces internal mantle viscosity, which facilitates more rapid recycling of oceanic lithosphere. Thus, the resistance of the oceanic path in the network will be less than its value for the oceanic lithosphere only end-member case. To account for this, we consider the internal viscosity for the oceanic end-member, $\mu_{io} = A\exp[-\theta T_{io}]$, and for the network, $\mu_i = A\exp[-\theta(A_c T_{ic} + A_o T_{io})]$. In an active lid regime, surface heat flux for the oceanic end-member was found to scale inversely with the internal viscosity to the 0.293 power [*Moresi and Solomatov*, 1998]. This scaling, together with the ratio of μ_i to μ_{io}, allows a weighting factor to be defined by $(\exp[\theta A_c(T_{io} - T_{ic})])^{0.293}$. This nondimensional factor, multiplied by R_o, accounts for the reduction in the thermal resistance of the oceanic path due to the higher internal mantle temperature associated with continental insulation.

Combining the ideas of the previous two paragraphs leads to an expression for the inverse resistance of the system given by

$$R_t^{-1} = R_c^{-1}A_c + R_o^{-1}A_o(\exp[EA_c(T_{ic} - T_{io})])^{0.293}. \quad (5)$$

The average system heat flux is then given by

$$q_t = A_c q_c \left[A_c + A_o \frac{T_{io}}{T_{ic}} \right] + A_o q_o \left[A_c \frac{T_{ic}}{T_{io}} + A_o \right]$$
$$\times (\exp[\theta A_c(T_{ic} - T_{io})])^{0.293} \quad (6)$$

where the surface temperature is set to zero. Notice that an increase in total heat flow, with added continental insulation, can occur if the increase in oceanic heat flux is great enough and if there is sufficient oceanic surface area. Two effects allow for increased oceanic heat flux with increased continental area. The first is an increased temperature drop across the oceanic lithosphere. The second is reduced sublithospheric viscosity allowing for faster plate speeds and thinner oceanic lithosphere. The previously discussed results of *Moresi and Solomatov* [1998] provide the scalings for the oceanic lithosphere only end-member, i.e., the scalings for q_o and T_{io}. To close the full scaling we must derive expressions

for the equivalent quantities in the end-member limit of a mantle completely covered by continents.

We consider a mantle completely covered by stable continental lithosphere of average thickness d. Stability of continental lithosphere is most likely achieved through a combination of chemical buoyancy and intrinsic strength [*Jordan*, 1981; *Pollack*, 1986; *Lenardic et al.*, 2003]. Thus, d parameterizes continental lithosphere that is chemically buoyant and strong to the degree that it is not recycled into the convecting mantle nor is it deformed by convection. The rheology for the mantle is given by Eq (2a) and (2b). As the continental lithosphere does not participate in mantle overturn, its presence can reduce the effective Rayleigh number driving mantle convection, Ra_{eff}, by reducing the temperature difference across, and the overall depth of, the convecting layer. Conversely, the insulating effect of the continental lithosphere can increase Ra_{eff} by decreasing internal mantle viscosity, μ_{ic}. We thus define the Rayleigh number driving convection as $Ra_{eff} = \rho_0 g\alpha(T_b - T_c)(D - d)^3/\mu_{ic}\kappa$ where ρ_0 is reference mantle density, g is gravitational acceleration, α is the thermal expansion coefficient, T_b is the temperature at the base of the mantle, T_c is the average temperature at the base of the continental lithosphere, D is mantle depth, and κ is thermal diffusivity. As Ra_{eff} is not known a priori, it will be useful to consider a Rayleigh number, Ra_s, based on the system temperature drop, ΔT, and the mantle viscosity defined using the surface temperature, μ_s. The two Rayleigh numbers are related by

$$Ra_{eff} = \frac{(T_b - T_c)}{\Delta T}(1 - d/D)^3 \frac{\mu_s}{\mu_{ic}} Ra_s. \quad (7)$$

Heat source enrichment within continental lithosphere can be approximated by allowing the continental lid to have a relatively low thermal conductivity [*Busse*, 1978]. Heat flux into the continental base, q_c, can then be equated to surface heat flux, leading to $q_c = K_c T_c/d$, where K_c is the thermal conductivity of continental lithosphere. A linear thermal gradient is assumed to hold across the active mantle boundary layer of thickness δ_c that forms below continents. Mantle heat flux, q_m, can be written as $q_m = [K_m(T_{ic} - T_c)]/\delta_c$, where K_m is the thermal conductivity of the mantle and T_{ic} is the average internal temperature of the mantle. Equating q_m and q_c leads to

$$\delta_c = \frac{(T_{ic} - T_c)K_m d}{T_c K_c}. \quad (8)$$

We introduce a local boundary layer Rayleigh number, Ra_{δ_c}, and assume it to remain near a constant value [*Howard*, 1966]. The critical Rayleigh number for convective onset is $\approx 10^3$ for a range of boundary conditions [*Sparrow et al.*, 1964]. To allow for variation [*Sotin and Labrosse*, 1999], we consider $\mathrm{Ra}_{\delta_c} = a_1 10^3$ where a_1 is a scaling constant. We thus have a second, independent, expression for δ_c given by

$$\delta_c^3 = \frac{a_1 10^3 D^3 \Delta T}{(T_{ic} - T_c)\mathrm{Ra}_{\mathrm{eff}}}. \qquad (9)$$

The problem can be closed by expressing T_{ic} in terms of T_c and T_b. The result depends on the boundary conditions at the top and bottom of the convecting region. The mantle below stable continents is in contact with a rigid boundary. In contrast, the core-mantle boundary presents a free-slip mechanical boundary condition to mantle convection. Thus, an asymmetry is imposed on the mantle even in the limit of d going to zero and this will affect internal mantle temperature. For a convecting mantle layer with a rigid surface and a free slip base it can be shown that the internal temperature of the mantle will be 0.64 times its base temperature plus 0.36 times its surface temperature [*Lenardic et al.*, 2004; 2005]. Continental lithosphere does not participate in convection so the surface temperature of the convecting layer is the lid base temperature, T_c, and the internal temperature of the convecting layer is then given by

$$T_{ic} = 0.64T_b + 0.36T_c. \qquad (10)$$

Equations (7)-(10), together with rheologic assumptions of Eq (2a) and (2b), lead to an expression for T_c. The expression is simplified by nondimensionalizing with D as the length scale, K_m as the conductivity scale, and ΔT as the temperature scale (dimensionless surface and base temperatures are set to zero and one). The final expression is

$$\frac{(1 - T_c)^5}{T_c^3 \exp[-0.36\theta T_c]} = \frac{a_1 5960.46 K_c^3 \exp[-0.64\theta]}{(d - d^2)^3 \mathrm{Ra}_s}. \qquad (11)$$

This allows us to solve for surface heat flux using the expression $q_c = K_c T_c / d$. We thus have scalings for q_c and T_{ic}, which can be used with Eq (6) and the previous discussed scalings for q_o and T_{io} to solve for the total system heat flow as a function of known parameters. As well as predicting global heat flow, which will be key for application to the Archean thermal catastrophe, Eq (6) also contains within it expressions for the local heat flux within continental and oceanic lithosphere, which will be key for application to the Archean paradox.

The scaling theory reviewed above has been tested using in excess of 400 numerical simulations and a suite of laboratory experiments [*Lenardic and Moresi*, 2003; *Lenardic et al.*, 2004; 2005]. For vigorous convection, i.e., high Rayleigh numbers, the consistency between theory predictions and both simulation and experimental results was good. Figures 3 and 4 show representative results.

The simulations used for testing (Figure 3) are similar to those of *Moresi and Solomatov* [1998] except that a conducting surface layer is included to represent continental lithosphere [*Lenardic and Moresi*, 2003]. The yield stress was set to a value low enough to allow for an active lid mode of convection, i.e., a plate tectonic-like regime [*Moresi and Solomatov*, 1998]. For any simulation suite, the thermal field from a simulation with no continent present was used as the initial condition for a case that imposed a continent with a nondimensional lateral extent of 0.05. The calculation was run to a statistically steady-state and then used as the initial condition for the next case, which increased the extent by 0.05. Simulation suites were performed that imposed continents above either a zone of mantle upflow or a zone of downflow with total heat flux varying by less than 5% as a result. For a compact description of results, it is useful to define an effective thermal thickness of stable continental lithosphere, d_T, as $(dK_m)/(DK_c)$. Fixed parameters for any suite of simulations are d_T; the yield stress; the basal Rayleigh number, Ra_b; and the degree of temperature-dependence of mantle viscosity, θ (cf. Eq 2a).

Figure 4 compares theory predictions to simulation results. The prediction that total mantle heat flow should progressively flatten and then increase with continental extent, as mantle viscosity becomes more temperature-dependent, is borne out by simulation results. This result was also confirmed using laboratory experiments, and added theory predictions were also tested against simulations with results comparable to those shown in Figure 4 [*Lenardic and Moresi*,

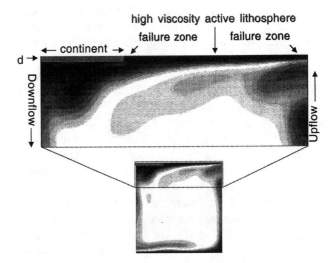

Figure 3. Image plots from numerical simulations. The blown-up top image shows the upper 1/3 of the full modeling domain, which is shown in the lower image.

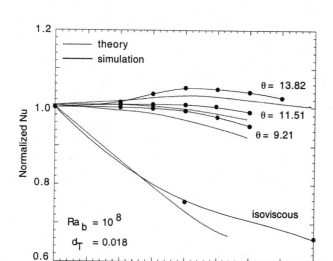

Figure 4. Normalized system heat flux versus percentage of continental lithosphere for four numerical simulation suites with different degrees of temperature-dependent viscosity compared to theory predictions.

2003; *Lenardic et al.*, 2004; 2005]. In addition, the robustness of key predictions, for different mantle heating modes (bottom versus internal) was also tested with good results [*Lenardic et al.*, 2005]. The interested reader is referred to the cited papers for added discussions of theory testing.

APPLICATION

At present, total mantle heat flux is ≈ 60 mW/m². Estimates of the heat flux in the earliest Archean, i.e., ≈ 4 bya, are not as certain and range from 2 to 4 times higher than the present-day value [*Davies*, 1980; *Schubert et al.*, 1980; *Christensen*, 1985]. The present-day subcontinental mantle heat flux is ≈ 15–20 mW/m² [*Ashwal et al.*, 1987; *Pinet et al.*, 1991; *Guillou-Frottier et al.*, 1995; *Ketchman*, 1996]). Based on the rarity of minimum-melting granites in the Superior Province of North America, *Burke and Kidd*, [1978] concluded that average continental geothermal gradients in the Archean were no more that 1.3 times greater than those of the present day. Accounting for the fact that crustal heat generation was greater in the past, this implies that subcontinental mantle heat flux must have remained constant over the bulk of geologic history. This conclusion was borne out by subsequent investigations that used metamorphic-based constraints [*England and Bickle*, 1985]. The latter authors restated the paradox this result leads to, given that the Earth had more heat to lose in the Archean.

To satisfy the constraint offered by the Archaen paradox, a continental growth model must allow subcontinental mantle

heat flux to remain constant as the mantle Rayleigh number increases. The heat flow scaling theory of the previous section can be used to guage whether different end-member growth scenarios are more or less effective at matching this constraint. In principal, the scalings could be used to explore a range of thermal history scenarios. However, at this stage we will not undertake a large suite of thermal history calculations. Rather, we will explore scenarios that maximize the potential thermal differences between end-member growth models. This is done with a degree of hindsight as we will see that even for such cases, the difference between the growth models in terms of matching Archean paradox constraints is not large.

Figure 5 shows theory predictions of global and subcontinental mantle heat flux as a function of the basal mantle Rayleigh number, Ra_b. The range of Rayleigh numbers shown is at the high end of likely range over the geologic age of the Earth. This large range will act to maximize any potential differences between growth models. For either growth end-member, the surface area of continental lithosphere, at the lowest Rayleigh number plotted, is taken to be 40% of the entire Earth's surface. For any assumed continental thickness, the constant continental volume models fix the surface area to be constant as the Rayleigh number is increased. The progressive growth models assumes that crustal volume grows in a linear manner as the Rayleigh number moves from the high to low values. The effects of different average thermal thicknesses, d_T, of continental lithosphere are also displayed, as are the effects of varying degrees of temperature-dependence of mantle viscosity, which is expressed via the nondimensional activation energy, θ (cf. Eq 2a).

The simple approach of Figure 5 is not a substitute for full thermal history calculations but it is useful in highlighting the maximum degree of variations that end-member growth models could lead to in terms of subcontinental and total mantle heat flux. What is immediately clear is that both growth end-members lead to an increase in the ratio of oceanic to continental heat flux as the vigor of convection increases. As convective vigor increases, the thickness of the thermal sublayer below continents decreases. This, in turn, increases the relative insulating effect of continental lithosphere, which leads to higher mantle temperatures (Eqs 10 and 11). This effect is not tied to continental area, which is why both growth models allow for it. Higher mantle temperatures increase the temperature drop across oceanic lithosphere, which increases oceanic heat flux regardless of mantle viscosity. This will lead to an increase in the ratio of oceanic to continental heat flux, provided that oceanic lithosphere is thinner than chemically stabilized continental lithosphere, which is assumed to remain at a constant thickness over geologic time for any individual model suite of Figure 5. The efficiency of the oceanic lithosphere in transfering heat allows

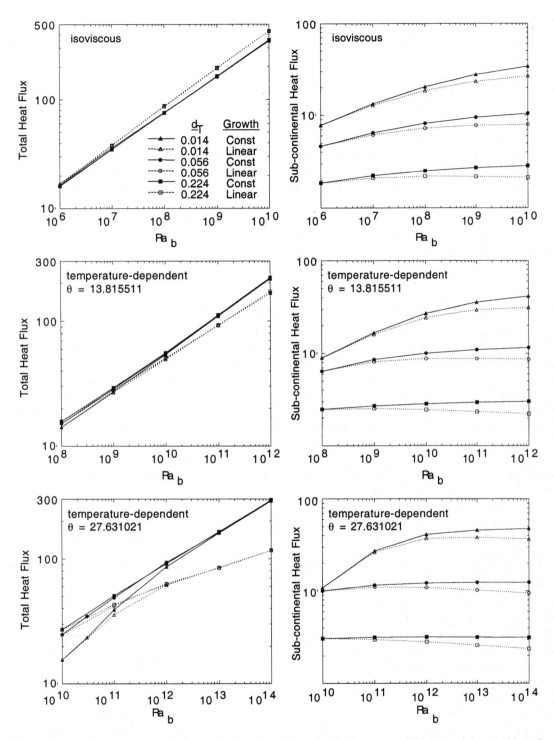

Figure 5. Predicted non-dimensional global (left graphs) and non-dimensional subcontinental (right graphs) mantle heat flux versus basal mantle Rayleigh number for different end-member continental growth scenarios and a range of thermal and rheological parameter values. Note that both growth models allow for relatively mildly varying subcontinental mantle heat flux with increasing Rayleigh numbers. The separation between the models in terms of global mantle heat flux seen in the lower left-hand graph is due to the fact that the models predict different average internal mantle temperatures for equivalent basal Rayleigh numbers, the result of differing amounts of continental insulation. Different internal temperatures lead to different bulk mantle viscosities. For the assumptions of our theory, internal mantle viscosity has a strong effect on mantle heat flux, which leads to the divergent trends in the lower left graph of the figure.

subcontinenal mantle heat flux to vary mildly while the global mantle heat flux increases more substantially with increasing Rayleigh number (Figure 5). Allowing for temperature-dependent mantle viscosity leads to a further increase in the proportion of heat loss carried by oceanic lithosphere because a lower viscosity, due to a higher mantle temperature, allows for more rapid overturn of oceanic lithosphere (cf. Eq 6). A decrease in continental area with increasing Rayleigh number leads to more constant temperatures at the base of continents, which, in turn, leads to milder variations of subcontinental mantle heat flux (cf. Eq 6). It is this latter area effect that allows the progressive growth model to be mildly more effective in buffering subcontinental mantle heat flux.

Figure 5 shows that assumptions regarding average continental thickness have relatively small effects on global mantle heat flux but can have large effects on subcontinental heat flux. More elaborate models that allow for changing depths and extents of continents over geologic time can be explored but Figure 5 already shows that if continental thickness has increased significantly as the Earth has cooled, then any growth scenario will have difficulties in addressing the Archean paradox. There is no direct evidence that the thickness of continental crust has changed appreciably over geologic time [*England and Bickle*, 1985] and there is good evidence that the thickness of continental lithosphere formed in the Archean has not changed over geologic time [*Jordon*, 1975; 1978; *Richardson et al.*, 1984; *Boyd et al.*, 1985; *Boyd*, 1989].

Although the progressive model is slightly more effective in regulating continental thermal conditions over time (Figure 5, right column), this constraint cannot rule out an instantaneous growth model. This is re-inforced by the fact that our simple approach maximizes any potential differences in terms of subcontinental mantle heat flux. That variations between the models remain mild shows that constraints offered by the Archean paradox cannot discriminate between growth models. A potentially larger separation between the two growth scenarios comes from their predictions regarding total mantle heat flux. As the temperature dependence of mantle viscosity is increased to one representative of the Earth, the predicted global mantle heat flux over time becomes a stronger function of the mode of continental growth (Figure 5, left column). The effect can become more extreme if the temperature dependence of mantle viscosity is increased beyond the values shown. For the Earth, the value of θ is estimated to be between 20 and 45, for a range of potential creep mechanisms [*Moresi and Solomatov*, 1998; *Solomatov*, 1995]. The higher values, however, are for non-linear flow laws. As our theory does not account for non-newtonian viscosity, the bottom graphs of Figure 5 must be considered as representing the maximum effect that can be predicted by the theory as it stands in its present form.

As *Christensen* [1985] showed, in order to prevent wide-scale melting in the Archean, total mantle heat flux must depend weakly on the mantle Rayleigh number. The progressive model does predict a weaker dependence, although it should be noted that the degree of this effect is not large enough on its own to prevent some degree of mantle melting in the Archean (this can be determined by using the theory to predict mantle temperature over time; cf. Eqs 4 and 10).

The results of Figure 5 for the progressive growth scenario may seem initially counterintuitive, based on the idea that continental lithosphere should insulate the mantle. That is, if there was less continental lithosphere in the past, when the mantle Rayleigh number was higher, and if insulating lithosphere leads to less efficient mantle cooling, then one might expect that a progressive continental growth model would, if anything, compound the problem associated with the Archean thermal catastrophe as less insulation in the past should have led to more efficiently convective cooling. The theory reviewed in this paper reveals why this is not the case for a mantle with a strongly temperature-dependent viscosity and an oceanic lithosphere that actively partakes in convective mantle overturn. For such conditions, local insulation of the mantle does not decrease global heat loss. It leads instead to lower mantle viscosity, which allows for the more rapid overturn of oceanic lithosphere and, provided the area of ocean basins is always at or greater than its present-day value, this can increase global mantle heat flux.

The progressive growth heat flux trends can be better understood by assuming present-day values of key parameters and considering what would happen if we took portions of the continental lithosphere and put them back into the mantle. This would increase the mantle Rayleigh number as more radiogenic elements would be put back into it. On its own this would increase global heat flow. This is rather intuitive and we will refer to it as the Ra-effect. However, the area of continents would also decrease and, as the heat flow scaling shows, global heat flow would scale with continental area in a manner that is not initially intuitive. That is, smaller continental area, on its own, leads to lower mantle temperature which leads, in turn, to higher mantle viscosity, less efficient overturn of oceanic lithosphere, and decreased global heat flow. We will refer to this as the area-effect. For a strongly temperature-dependent viscosity, the area-effect can compete with the Ra-effect which leads to the flattening of the Nu-Ra relationship for the progressive growth model in the lower left-hand graph of Figure 5.

Figure 5 shows that, for the assumptions we have made, a progressive continental growth model predicts a relationship between total mantle heat flux and convective vigor that trends in the right direction to limit mantle melting in the Archean. This is not meant to imply that progressive continental growth is the only factor that can prevent, or has prevented, wide-scale melting in the Archean. As was noted in the introductory sections, a range of hypothesis have been proposed that can also accomplish this. This last point makes it difficult to use the Archean thermal catastrophe as a hard discriminate between growth models.

CONCLUSION

The thermal implications of end-member continental growth models have been explored using a heat flow scaling for mantle convection with continents. The scaling applies for a temperature-dependent viscosity mantle that maintains a plate tectonic-like mode of convection allowing for the continued creation and subduction of oceanic lithosphere over geologic time. Both instantaneous and progressive continental growth models can account for mildly varying thermal conditions within continents over geologic time that allow them to match constraints imposed by the Archean paradox. The progressive growth model also has the potential to limit mantle-wide melting in the Archean. This separation between the two scenarios is not a strong one as several other physical mechanisms, not tied to continental growth, can also limit Archean melting. Collectively, the constraints that continental thermal structure was not greatly different in the Archean than it is at present and that no mantle-wide melting event occurred in the Archean cannot rule out instantaneous continental growth but they do mildly favor continental growth as closer to a progressive end-member.

Acknowledgments. This work has been supported by NSF Grant EAR-0448871 and NASA-MDAP Grant NAG5-12166. Thanks to an anonymous referee for a critical and very helpful review that stopped me from pushing the theory discussed beyond the limits of what it might actually be able to say about the Earth. A related thanks goes to Kelly Russell, whose insightful question after a seminar also made me rethink how strong the constraints discussed in this paper really were for discriminating between growth models. A final thanks goes to the editors for equally helpful reviews and for patience.

REFERENCES

Anderson, D.L., Hotspots, polar wander, Mesozoic convection and the geoid, *Nature*, 297, 391-393, 1982.

Armstrong, R.L., Radiogenic isotopes: the case for crustal recycling on a near-steady-state no-continental growth Earth, *Philos. Trans. Roy. Soc. Lond. A*, 301, 443-472, 1981.

Ashwal, L.D., P. Morgan, S.A. Kelley, and J.A. Percival, Heat production in an Archean crustal profile and implications for heat flow and mobilization of heat-producing elements, *Earth Planet. Sci. Lett*, 439-450, 1987.

Ballard, S., and H.N. Pollack, Modern and ancient geotherms beneath southern Africa, *Earth Planet. Sci. Lett*, 88, 132-142, 1988.

Becker, T.W., J.W. Kellogg, and R.J. O'Connell, Thermal constraints on the survival of primitive blobs in the lower mantle, *Earth Planet. Sci. Lett*, 171, 351-365, 1999.

Bickle, M.J., Heat loss form the Earth: a constraint on Archean tectonics from the relation between geothermal gradients and the rate of plate production, *Earth Planet. Sci. Lett*, 40, 301-315, 1978.

Bickle, M.J., Implications of melting for stabilization of the lithosphere and heat loss in the Archean, *Earth Planet. Sci. Lett*, 80, 314-324, 1986.

Boyd, F.R., Compositional distinction between oceanic and cratonic lithosphere, *Earth Planet. Sci. Lett*, 96, 15-26, 1989.

Boyd, F.R., J.J. Gurney, and S.H. Richardson, Evidence for a 150-200 km thick Archean lithosphere from diamond inclusion thermobarometry, *Nature*, 315, 387-389, 1985.

Burke, K., and W.S.F. Kidd, Were Archean continental geothermal gradients much steeper than those of today?, *Nature*, 272, 240-241, 1978.

Busse, F.H., A model of time-periodic mantle flow, *J. R. Astron. Soc.*, 52, 1-12, 1978.

Busse, F.H., Transitions to turbulence in Rayleigh-Bénard convection, in *Hydrodynamic Instabilities and the Transition to Turbulence*, edited by H.L. Swinney and J.P. Gollub, pp. 97-137, Springer-Verlag, Berlin, 1985.

Busse, F.H., Fundamentals of thermal convection, in *Mantle Convection: Plate Tectonics and Global Dynamics*, edited by W.R. Peltier, 23-95, Gordan and Breach Science Publishers, New York, 1989.

Butler, S.L., and W.R. Peltier, Thermal evolution of the Earth: Models with time-dependent layering of mantle convection which satisfy the Urey ratio constraint, *J. Geophys. Res.*, 107, 10.1029/2000JB000018, 2002.

Chan, S.-K., Infinite Prandtl number turbulent convection, *Stud. Appl. Maths.*, 50, 13-49, 1971.

Christensen, U.R., Heat transport by variable viscosity convection and implications for the Earth's thermal evolution, *Phys. Earth Planet. Inter.*, 35, 264-282, 1984.

Christensen, U.R., Thermal evolution models for the Earth, *J. Geophys. Res.*, 90, 2995-3007, 1985.

Collerson, K.D., and B.S. Kamber, *Science*, 283, 1519-1522, 1999.

Conrad, C.P., and B.H. Hager, Mantle convection with strong subduction zones, *Geophys. J. Int.*, 144, 271-288, 2001.

Constantin, P., and C. Doering, Infinite Prandtl number convection, *J. Statist. Phys.*, 94, 159-172, 1999.

Davies, G.F., Thickness and thermal history of continental crust and root zones, *Earth Planet. Sci. Lett*, 44, 231-238, 1979.

Davies, G.F., Thermal histories of convective earth models and constraints and constraints on radiogenic heat production in the Earth, *J. Geophys. Res.*, 85, 2517-2530, 1980.

Elder, J., Convective self-propulsion of continents, *Nature*, 214, 657-660, 1967.

England, P., and M. Bickle, Continental thermal and tectonic regimes during the Archean, *J. Geol.*, 92, 353-367, 1984.

Fyfe, W.S., Evolution of the Earth's crust: modern plate tectonics to ancient hot spot tectonics?, *Chem. Geol.*, 23, 89-96, 1978.

Grigne, C., and S. Labrosse, Effects of continents on Earth cooling: thermal blanketing and depletion in radioactive elements, *Geophys. Res. Lett.*, 28, 2707-2710, 2001.

Guillou-Frottier, L., J.-C. Mareschal, C. Jaupart, C. Gariepy, R. Lapointe, and G. Bienfait, Heat flow variations in the Grenville Province, Canada, *Earth Planet. Sci. Lett*, 136, 447-460, 1995.

Gurnis, M., Large-scale mantle convection and the aggregation and dispersal of supercontinents, *Nature*, 332, 695-699, 1988.

Gurnis, M., A reassessment of the heat transport by variable viscosity convection with plates and lids, *Geophys. Res. Lett.*, 16, 179-182, 1989.

Helfrich, G.R., and B.J. Wood, The Earth's mantle, *Nature*, 412, 501-507, 2001.

Howard, L.N., Convection at high Rayleigh number, in *Applied Mechanics, Proceedings of the 11th Congress of Applied Mechanics, Munich (Germany)*, edited by H. Gortler, pp. 1109-1115, Springer-Verlag, Berlin, 1966.

Incropera, F.P., and D.P. DeWitt, *Fundamentals of Heat and Mass Transfer*, John Wiley and Sons, New York, 1996.

Jarvis, G.T., and I.H. Campbell, Archean komatiites and geotherms: Solution to an apparent contradiction, *Geophys. Res. Lett.*, 10, 1133-1136, 1983.

Jordan, T.H., The continental tectosphere, *Rev. Geophys.*, 13, 1-12, 1975.

Jordan, T.H., Composition and development of the continental tectosphere, *Nature*, 274, 544-548, 1978.

Jordan, T.H., Continents as a chemical boundary layer, *Phil. Trans. R. Soc. Lond., A*, 301, 359-373, 1981.

Kellogg, L.H., B.H. Hager, and R.D. van der Hilst, Compositional stratification in the deep mantle, *Science*, 283, 1881-1884, 1999.

Ketchman, R.A., Distribution of heat-producing elements in the upper and middle crust of southern and west central Arizona: evidence from the core complexes, *J. Geophys. Res.*, 101, 13, 611-13,632, 1996.

Korenaga, J., Energetics of mantle convection and the fate of fossil heat, *Geophys. Res. Lett.*, 30, 1437, doi:10.1029/2003GL016982, 2003.

Kramers, J.D., and N. Tolstikhin, *Chem. Geol.*, 139, 5-15, 1997.

Lenardic, A., On the partitioning of mantle heat loss below oceans and continents over time and its relationship to the Archean paradox, *Geophys. J. Int.*, 134, 706-720, 1998.

Lenardic, A., and L.-N. Moresi, Thermal convection below a conducting lid of variable extent: heat flow scalings and two-dimensional, infinite Prandtl number numerical simulations, *Phys. Fluids*, 15, 455-466, 2003.

Lenardic, A., L.-N. Moresi, and H. M.hlhaus, Longevity and stability of cratonic lithosphere: insights from numerical simulations of coupled mantle convection and continental tectonics, *J. Geophys. Res.*, 108, 10.1029/2002JB001859, 2003.

Lenardic, A., L.-N. Moresi, A.M. Jellinek, and M. Manga, Continental insulation, mantle cooling, and the surface area of oceans and continents, *Earth Planet. Sci. Lett.*, In Press, 2005.

Lenardic, A., F. Nimmo, and L.-N. Moresi, Growth of the hemispheric dichotomy and the cessation of plate tectonics on Mars, *J. Geophys. Res.*, 109, 10.1029/2003JE002172, 2004.

Lowman, J.P., and G.T. Jarvis, Mantle convection models of continental collision and breakup incorporating finite thickness plates, *Phys. Earth Planet. Int.*, 88, 53-68, 1995.

McCulloch, M.T., and V.C. Bennett, *Geochim. Cosmochim Acta*, 58, 197-214, 1994.

McKenzie, D.P., and N.O. Weiss, Speculations on the thermal and tectonic history of the Earth, *J. R. Astron. Soc.*, 42, 131-174, 1975.

Moresi, L.-N., and V.S. Solomatov, Mantle convection with a brittle lithosphere: thoughts on the global tectonic styles of the Earth and Venus, *Geophys. J. Int.*, 133, 669-682, 1998.

Morgan, P., The thermal structure and thermal evolution of the continental lithosphere, *Physics and Chemistry of the Earth*, 15, 107-193 1984.

Nyblade, A.A., and H.N. Pollack, A comparative study of parameterized and full thermal convection models in the interpretation of heat flow from cratons to mobile belts, *Geophys. J. Int.*, 113, 747-751, 1993.

Parsons, B. and D.P. McKenzie, Mantle convection and the thermal structure of the plates, *J. Geophys. Res.*, 83, 4485-4496, 1978.

Parsons, B., and J.G. Sclater, An analysis of the variation of ocean floor bathymetry and heat flow with age, *J. Geophys. Res.*, 82, 803-827, 1977.

Pinet, C., C. Jaupart, J.-C. Mareschal, C. Gariepy, G. Bienfait, and R. Lapointe, Heat flow and structure of the lithosphere in the eastern Canadian Shield, *J. Geophys. Res.*, 96, 19,941-19,963, 1991.

Pollack, H.N., Cratonization and thermal evolution of the mantle, *Earth Planet. Sci. Lett.*, 80, 175-182, 1986.

Pollack, H.N., and D.S. Chapman, On the regional variation of heat flow, geotherms, and the thickness of the lithosphere, *Tectonophysics*, 38, 279-296, 1977.

Pollack, H.N., S.J. Hurter, and J.R. Johnson, Heat flow from the Earth's interior: analysis of the global data set, *Rev. Geophys.*, 31, 267-280, 1993.

Richardson, S.H., J.J. Gurney, A.J. Erlank, and J.W. Harris, Origin of diamonds in old enriched mantle, *Nature*, 310, 198-202, 1984.

Richter, F.M., Regionalized models for the thermal evolution of the Earth, *Earth Planet. Sci. Lett.*, 68, 471-484, 1984.

Rudnick, R.L., W.F. McDonough, and R.J. O'Connell, Thermal structure, thickness and composition of continental lithosphere, *Chemical Geology*, 145, 395-411, 1998.

Schubert, G., and A. Reymer, Continental volume and free-board through geologic time, *Nature*, 316, 336-339, 1985.

Schubert, G., D. Stevenson, and P. Cassen, Whole mantle cooling and the radiogenic heat source content of the Earth and moon, *J. Geophys. Res.*, 85, 2531-2538, 1980.

Seely, S., *Dynamic Systems Analysis*, Reinhold Publishing, New York, 1964.

Solomatov, V.S., Scaling of temperature- and stress-dependent viscosity convection, *Phys. Fluids*, 7, 266-274, 1995.

Sotin, C., and S. Labrosse, Three-dimensional thermal convection in an iso-viscous, infinite Prandtl number fluid heated from within and from below: applications to the transfer of heat through planetary mantles, *Phys. Earth Planet. Int.*, 112, 171-190, 1999.

Sparrow, E.M., R.J. Goldstein, and V.K. Jonsson, Thermal instability in a horizontal fluid layer: effect of boundary condition and nonlinear temperature profile, *J. Fluid Mech.*, 18, 513-528, 1964.

Spohn, T., and D. Breuer, Mantle differentiation through continental crust growth and recycling and the thermal evolution of the Earth, in *Evolution of the Earth and Planets, Geophys. Monogr. Ser., 74*, edited by E. Takahashi, R. Jeanloz, and D.C. Rubie, pp. 55-71, AGU, Washington D.C., 1993.

Sylvester, P.J., I.H. Campbell, and D.A. Bower, Niobium/uranium evidence for early formation of the continental crust, *Science*, 275, 521-523, 1997.

Taylor, S.R., and S.M. McLennan, *The Continental Crust: Its Composition and Evolution*, Blackwell Science Publ., Oxford, 1985.

Tozer, D.C., The present thermal state of the terrestrial planets, *Phys. Earth Planet. Inter.*, 6, 182-197, 1972.

Turcotte, D.L., and E.R. Oxburgh, Finite amplitude convection cells and continental drift, *J. Fluid. Mech.*, 28, 29-42, 1967.

Whitehead, J.A., Moving heaters as a model of continental drift, *Phys. Earth Planet. Int.*, 5, 199-212, 1972.

Zhong, S., and M. Gurnis, Dynamic feedback between a continent-like raft and thermal convection, *J. Geophys. Res.*, 98, 12,219-12,232, 1993.

A. Lenardic, Department of Earth Science MS 126, P.O. Box 1892, Rice University, Houston, Texas 77251-1892, USA. (Adrian@geophysics.rice.edu)

Archean Geodynamics: Similar to or Different From Modern Geodynamics?

Kent C. Condie

Department of Earth & Environmental Science, New Mexico Institute of Mining & Technology, Socorro, New Mexico

Keith Benn

Ottawa-Carleton Geoscience Centre and Department of Earth Sciences,
University of Ottawa, Ottawa, Ontario, Canada

There is a wealth of geologic, geochemical, structural, volcanologic, and sedimentologic data that are consistent with Archean plate tectonics, especially after 3.0 Ga. Neither the eruption of submarine basalts onto thinned continental crust nor the existence of ductile or viscous diapirism precludes the existence of plate tectonics during the Archean. Some "missing indicators" of plate tectonics are found in Archean terranes (probable oceanic crust, melange, possible passive margin sequences, boninite), whereas the absence of others (such as blueschists) can be explained by a higher Archean mantle geotherm. Bimodal magmatism is not limited to the Archean but occurs in several modern tectonic settings. The relative abundance of komatiites in the Archean reflects hotter Archean mantle and possibly widespread mantle plume activity. Any viable model for Archean geodynamics must accommodate the following 10 constraints: During the Archean, the mantle was hotter than it is today; there are two styles of crustal deformation in the Archean; komatiite is proportionally more abundant in Archean greenstones than in younger greenstones; tonalite–trondhjemite–granodiorite depleted in heavy rare earth elements is more widespread in the Archean than afterwards; thick lithosphere underlies many Archean cratons; portions of the mantle were strongly depleted in large ion lithophile elements during the Archean; many Archean greenstones comprise arc-like rock assemblages; a significant proportion of Archean greenstones contain volcanic rocks with geochemical characteristics similar to modern plume-derived basalts; paleomagnetic data indicate that apparent polar wandering occurred during the Archean; and a large volume of continental crust was produced about 2.7 Ga.

INTRODUCTION

Were Archean tectonic regimes similar to or different from present tectonic regimes? This question has been with us for many years, and despite the large amount of high-precision data that has become available from Archean rocks, the question still lingers. From the very onset of modern studies of Archean greenstone belts [*Langford and Morin*, 1976; *Blackburn*, 1980], plate tectonics has been the favored paradigm. And from that time onwards, few seemed to have looked back: Plate tectonics carried the Archean banner [*Smithes and Champion*, 2000; *Chen et al.*, 2003; *Percival et al.*, 2004].

Archean Geodynamics and Environments
Geophysical Monograph Series 164
Copyright 2006 by the American Geophysical Union
10.1029/164GM05

At the same time, not all investigators have accepted that modern plate tectonics operated in the Archean. For instance, *Goodwin* [1977] suggested that granite-greenstone belts formed by "sagduction" (sagging of volcanic–sedimentary basins and remelting of the lower crust in response to plume-induced rifting; *Goodwin and Smith* [1980]). More recently, other investigators have challenged the plate tectonic paradigm, in whole or in part, as applied to the Archean [*Davies*, 1992; *Hamilton*, 1998; *Bleeker*, 2002; *Trendall*, 2002], with some suggesting that modern-style plate tectonics did not appear until the Late Archean [*Davies*, 1995; *Smithes et al.*, 2003; *Peschler et al.*, 2004].

In fact, a proper understanding of Archean geodynamics should not necessitate a choice between two end-member models (e.g., vertical or horizontal deformation). It is more likely that there was an evolution of geodynamic settings as Earth cooled during the Archean [*Vlaar et al.*, 1994]. In this short contribution, we address some of the questions that have been raised about plate tectonics in the Archean, chiefly from a geological and geochemical point of view. Geophysical considerations, such modes of continental accretion and rheological behavior of the the lithosphere, are more fully discussed in other papers in this volume. We present a list of 10 constraints that any model for Archean geodynamics must accommodate. This is not to say these are the only 10 constraints, but in our opinion, they are 10 of the most robust constraints.

PROBLEMS WITH ARCHEAN PLATE TECTONICS

Problems presented for the existence of Archean plate tectonics generally fall into one of three categories: eruption of greenstones onto felsic crust; the existence of vertical crustal tectonics in the Archean; and the absence, or apparent absence, of key plate tectonic indicators from the Archean geological record.

Early Felsic Crust

Archean greenstones that appear to have been erupted onto or through felsic crust have now been described from many cratons [*Martin et al.*, 1993; *Swager*, 1997; *Bleeker et al.*, 1999]. In some cases actual contacts are exposed such as at the classic site in Zimbabwe [*Bickle et al.*, 1994]. In other cases, Nd isotopic data or inherited components in igneous zircons indicate contamination of mafic and komatiitic magmas with felsic crust [*Stevenson*, 1995; *Lesher and Arndt*, 1995; *Tomlinson et al.*, 2004]. Although it is not yet clear what proportion of Archean greenstones are crustally contaminated, it is probably less than 50%. In cases where komatiitic and mafic lavas were erupted onto felsic Archean crust, the felsic component must have been relatively thin so as not to isostatically rise above sea level, or sea level must

have been high to flood all or most of the continents. We can consider three possibilities to explain low-standing continental crust: (1) the production rate of felsic crust was very low prior the Late Archean; (2) Archean submarine volcanics were erupted on thin felsic crust around the margins of microcontinents, which were flooded in part or entirely with seawater; and (3) Archean felsic crust was thinned due to rifting or lateral flow of weak lower crust during or prior to submarine volcanism. Because Earth was hotter in the Archean [*Bickle*, 1978], it is unlikely that the production rate of continental crust was low—in fact, it should have been high. Perhaps the continental crust production rate was high, but the recycling rate into the mantle was even higher, such that only a small amount of this early crust survived. If so, we are faced with the question of why such high recycling rates?

Eruption of greenstone volcanics on thin felsic crust around the margin of flooded microcontinents is feasible as pointed out by *Arndt* [1999]. Perhaps high sea level in the Archean was caused by high-standing ocean ridges or by numerous oceanic plateaus. Still another possibility is that Archean continental crust was subjected to intense rifting, possibly caused by mantle plumes. In the Slave and Superior provinces, there is evidence for rifting of the felsic basement upon which greenstone volcanics were erupted [*Percival*, 2001; *Bleeker*, 2002]. If mantle plumes were more widespread in the Archean, both possibilities two and three may have been responsible for eruption of thick submarine basalt successions onto thinned continental crust. Continental crust produced in the Early Archean by subduction-related processes may have been thinned by later rifting processes, such that it did not emerge above sea level. Magmas derived from mantle plumes, ocean ridges, or subduction zones could have been erupted on to such crust. As a Phanerozoic example, for instance, during early stages of the opening of the Red Sea and the Gulf of California in the Tertiary, pillow basalts may have erupted on extended continental crust. During the opening of back arc basins, such as the Rocas Verdes ophiolites in the southern Andes during the Cretaceous [*Stern and de Wit*, 2003], submarine basalts can be erupted onto extended continental crust. The Kerguelen oceanic plateau may have been erupted, at least in part, onto thin continental crust around the margin of Antarctica.

So what is the bottom line? Eruption of submarine basalts onto submerged continental crust, whether thinned by rifting or not, does NOT preclude the existence of plate tectonics in the Archean.

Vertical Tectonics in the Archean

Although structural, geochemical, and reflection seismology studies of Archean cratons have shown that horizontal tectonic regimes, much like modern plate tectonic regimes at convergent margins, dominated in the Late Archean [*Ludden et al.*, 1993; *Polat and Kerrich*, 2001; *Percival et al.*, 2004], there is evidence to support vertical tectonics, i.e., crustal diapirism in some Archean cratons [*Chardon et al.*, 1996; *Van Kranendonk et al.*, 2004; *Bedard et al.*, 2004]. The best documented case for vertical tectonics is in the Eastern Pilbara craton during the Mid-Archean [*Collins et al.*, 1998; *Hickman*, 2004; *Van Kranendonk et al.*, 2004]. The dome and keel pattern of the Eastern Pilbara has long been recognized as its salient feature. A similar pattern has been described in the Archean Slave province [*Bleeker*, 2002]. The greenstone successions in the Eastern Pilbara (3.7–3.5 Ga) contain no low-angle thrusts, and stratigraphic units can be correlated between greenstone belts. Intrusion of syntectonic tonalite–trondhjemite–granodiorite (TTG) sheets and laccoliths initiated doming (3.49–3.43 Ga) followed by gravity-driven diapiric doming and sinking of denser greenstones into adjacent synclines [*Hickman*, 2004]. Unlike the Eastern Pilbara, the Western Pilbara comprises a collage of three separate, fault-bounded terranes, each of which has a unique stratigraphic section and structural history. No diapiric domes are present in the Western Pilbara, where all deformation appears to result from episodes of compressive deformation, consistent with a plate tectonic regime [*Hickman*, 2004].

Should we abandon plate tectonics in the Archean because some examples of regions with crustal diapirism have been found? Is it not possible that crustal diapirism can operate on a plate tectonic Earth? The main question concerning crustal diapirism is that of the rheological behavior of continental crust and its response to body forces, tectonic stresses, and different geothermal gradients. Ductile or viscous diapirism in the Archean need not be inconsistent with global plate tectonics, as discussed below.

Apparent Absence of Key Plate Tectonic Indicators

One has to be very careful in using "apparent absence" of specific plate tectonic indicators in ancient rocks as evidence for the absence of plate tectonics. Some components, especially those that are subject to being subducted, such as melange, are not universally preserved even in the Phanerozoic record, yet we don't question the applicability of plate tectonics in arc successions where melange is missing. Ophiolites are commonly cited as examples of missing plate tectonic indicators in the Archean (a possible 2.5-Ga ophiolite has recently been described in China [*Kusky et al.*, 2001], but also see *Zhai et al.* [2002]; *Kusky and Li* [2002]). There are several problems with the recognition and preservation of complete ophiolite successions in the Archean. For instance, if Archean oceanic crust was significantly thicker than modern oceanic crust [*Sleep and Windley*, 1982; *Foley et al.*, 2003], deep portions of

oceanic crust that contain sheeted dykes, layered gabbros, and ultramafic cumulate rocks may not be accreted to the continents in subduction zones but instead recycled into the mantle. The rheological profiles of thick oceanic crust could have resulted in tectonic delamination of the middle to lower crustal sections and mantle lithosphere during collisional tectonics, and the emplacement of only the upper, volcanic crust within Archean orogens ("oceanic flake tectonics" [*Hoffman and Ranalli*, 1988]). One of the great challenges of the future is that of recognizing Archean oceanic crust when and if it is preserved in greenstones. Perhaps many of the mafic plain greenstones are remnants of oceanic crust rather than of oceanic plateaus [*Condie*, 1994; *Tomlinson and Condie*, 2001].

The apparent absence of high-pressure, low-temperature metamorphic rocks in the Archean has also been used to argue against plate tectonics [*Hamilton*, 1998; *Bleeker*, 2002]. Yet the absence of these rocks in the Archean is predicted from the probable higher temperature regimes in Archean subduction zones as convincingly shown by *Komiya et al.* [2002]. In fact, it would be a difficult problem if we found blueschists in the Archean!

Some have cited the apparent absence of passive margin successions in the Archean as evidence against plate tectonics [*Hamilton*, 1998; *Bleeker*, 2002]. To identify a passive margin sequence in the geologic record requires excellent preservation of rocks recording the opening of an ocean basin, something we do not have in the Archean (or at least have not yet recognized). There are, however, Archean sedimentary successions composed of shales and quartz arenites resting uncomfortably on older basement that could be remnants of passive margin successions, such as the Moodies Supergroup in South Africa [*Eriksson*, 1978; *Lowe*, 1994], remnants in the Narryer Complex in Western Australia [*Maas and McCulloch*, 1991], the Mount Bruce Supergroup in Western Australia [*Barley et al.*, 1992], the Bababudan Group in southern India [*Srinivasan and Ojakangas*, 1986], and sites in the Canadian shield [*Donaldson and de Kemp*, 1998] and in southern Zimbabwe [*Fedo and Eriksson*, 1996]. Certainly, lithologic packages characteristic of passive margins occur in the Archean.

The absence of melange and accretionary prisms in the Archean also has been used to question the existence of plate tectonics [*Hamilton*, 1998]. But are they really absent? First of all, melange is a rock type easily subducted into the mantle, and even pre-Mesozoic melanges are not universally preserved. *McCall* [2003] discussed the strong contrasts between the accretionary complexes of the convergent Makran margin, Iran, and Archean greenstone and metasedimentary belts, and concluded that zones of Archean plate convergence must have differed significantly from what we know to exist at present. However, a tectonic melange resembling a classic Mesozoic-style accretionary melange has been described from the Schreiber greenstone belt in the Superior Province of Canada [*Polat and Kerrich*, 1999]. Like Phanerozoic counterparts, this Archean melange is characterized by varying degrees of fragmentation and mixing, ranging from intact sedimentary layers and volcanic flows to intensely sheared, transposed, and mixed tectonic melange. It differs from Phanerozoic examples in the composition of the blocks that are included within the melange, and also in metamorphic facies. The total package of rocks in the Schreiber greenstone, however, is comparable to those found in young arc successions that evolved through compressional to transpressional deformation. In addition, there are several Archean greenstones that have been described that contain large blocks ranging from dismembered to intact, which are similar to blocks in young accretionary prisms [*Thurston*, 1994; *Mueller et al.*, 1996; *Stott*, 1997; *de Wit and Ashwal*, 1997; *de Wit*, 1998; *Kusky*, 1998]. The work of *Polat and Kerrich* [1999] underscores one major challenge in applying the plate tectonic model to the Archean: how to recognize Archean subduction–accretion complexes that may differ significantly from modern analogs.

TEN CONSTRAINTS FOR ARCHEAN TECTONIC REGIMES

A step in the right direction for a better understanding of the Archean is to come up with a list of robust constraints that must be accommodated into any viable tectonic scenario for the Archean. As a beginning to such a list, below we present 10 constraints, which seem to be well established from published data.

1. During the Archean, the mantle was hotter than it is today.

Several lines of evidence indicate that Earth's mantle was hotter in the Archean than afterwards [*Bickle*, 1978; *Hoffman and Ranalli*, 1988; *Davies*, 1992]. The higher concentrations of radiogenic isotopes (principally U, Th, and K) in the mantle prior to 2.5 Ga strongly suggest that radioactive heat production in the mantle was two to three times that of today [*Pollack*, 1997], although the ratio of heat loss to heat production in the early Earth is not well constrained.

2. There are two contrasting styles of Archean crustal deformation.

Structural signature may provide insights to geodynamic settings and rheological state of the lithosphere during the Archean. Two end-member structural types are documented by mapping in Archean terranes: dome and keel style and linear, upright fold belts. The latter are most common, but not restricted to the Late Archean (≤2.7 Ga).

The Eastern Pilbara province in Western Australia is the best documented example of a pre-2.9-Ga province with dome and keel structural architecture. Extensive structural, metamorphic, and geophysical modeling have provided abundant and compelling evidence for crustal diapirism in the Eastern Pilbara during the Mid-Archean [*Collins et al.*, 1998; *Van Kranendonk et al.*, 2004; *Peschler et al.*, 2004], resulting in the classical pattern of granite-gneiss domes that is strikingly displayed on satellite images of the region (Figure 1 in *Van Kranendonk et al.* [2004]). Diapirism occurred in response to the accumulation of thick mafic and komatiitic volcanic packages that sank into the felsic middle crust, resulting in "partial convective overturn" [*Collins et al.*, 1998]. The implication is that the middle crust was softened by thermal blanketing, conductive incubation [*Sandiford et al.*, 2004], and/or heating by a mantle plume, allowing the crust to flow under the weight of a 10–18-km-thick volcanic pile.

The Dharwar craton in southern India is another Archean province where studies have confirmed the role of crustal diapirism related to body forces. This interpretation is based on extensive mapping of structures and finite strain patterns in the Dharwar craton that appear to be related to diapirism that occurred ca. 3.0 Ga [*Bouhallier et al.*, 1993] or possibly later, between 2.8 and 2.6 Ga [*Chardon et al.*, 1998]. Diapiric tectonics between 2.9 and 2.7 Ga has also been proposed for the plutonic-gneissic Minto Block of northern Quebec [*Bédard et al.*, 2003], although a plate tectonic model has also been proposed for this region [*Percival et al.*, 1994].

The evidence provided by studies in the Pilbara and Dharwar cratons strongly supports the interpretation that crustal doming and diapiric deformation of continental crust occurred during the Archean, in particular in provinces older than 2.9 Ga. In contrast, post-2.8-Ga Archean provinces are generally elongate belts with strong linear structural trends parallel to the belts. For the most part, they are upright fold belts that may be bounded by subparallel fault lineaments and elongate intrusive complexes.

As examples, in both the southern Superior and Yilgarn provinces, upper, unreflective crust is deformed by upright folds with wavelengths on the order of tens of kilometers, and by fold-parallel fault zones continuous for hundreds of kilometers. The more reflective middle crust is also folded and cross-cut by ductile shear zones. In the Abitibi granite-greenstone belt, the amplitude of folds is sufficient to expose amphibolite-grade orthogneiss in anticlinal cores [*Benn*, 2004]. Upright folding of the middle crust is also interpreted from seismic profiles [*Wilde et al.*, 1996; *Benn*, this volume] and documented in the field in the amphibolite-grade Opatica and Pontiac belts that bound the Abitibi belt

[*Benn*, 1992, 1994]. Some shear zones are crustal-scale features interpreted as sutures, and elongate belts may be accreted tectono-stratigraphic terranes [*Calvert and Ludden*, 1999; *Wilde et al.*, 1996; *Goleby et al.*, 2004; *Benn*, this volume].

It is difficult to attribute the structural style of Late Archean crust, especially the upright regional-scale fold belts and locally transpressive shear zones, to geodynamic processes other than horizontal displacements and collisions of relatively rigid lithospheric plates [*Choukroune et al.*, 1997]. Disagreement may exist as to the identification of individual terrane sutures in Late Archean cratons, but a plate tectonic model is invoked by most workers to explain the first-order structures. Although elongate upright fold belts are less typical of pre-2.8-Ga crust, this does not preclude horizontal activity by tectonics before 2.8 Ga. As mentioned previously, in the Western Pilbara province, thrust stacking of identifiable lithotectonic terranes, the presence of elongate fold belts and plutonic complexes, and documented large-scale transcurrent shear zones provide compelling evidence for tectonic plate collisions prior to 2.9 Ga [*Hickman*, 2004]. The apparent plate collisions in the Western Pilbara would have occurred during a period between two crustal diapirism events in the Eastern Pilbara.

Thus, it appears that prior to 2.9 Ga, the rheology of continental crust was such that crustal diapirism was an efficient process for the redistribution of mass and heat. This imposes a constraint of relatively weak and hot middle and perhaps lower crust compared to the Phanerozoic, where such large-scale vertical tectonism is unknown. Several questions arise. Does crustal diapirism record transient thermal anomalies or a high steady-state geotherm in Early Archean continents? Was the "extra" heat in the pre-2.9-Ga continental crust internally generated [*Sandiford et al.*, 2004], in which case it might have been an intrinsic property of the early continental crust? Or was it derived from an external source, such as mantle plumes, which may have been more common in the Archean [*Stein and Hofmann*, 1994; *Choukroune et al.*, 1997; *Tomlinson and Condie*, 2001], or catastrophic mantle overturn events in a layered mantle [*Davies*, 1995]?

Our existing database confirms that tectonic terrane accretion occurred in the western Pilbara at least as early as 2.9 Ga [*Hickman*, 2004] and in the Barberton granite-greenstone belt as early as 3.2 Ga [*Diener et al.*, this volume]. Therefore, horizontal displacements of relatively rigid lithospheric plates (i.e., plate tectonics) must have occurred at that time, and we expect that constructive, destructive, and transcurrent terrane boundaries existed as well. This leads to two important questions: (1) How did tectonic forces generated by plate convergence and collision interact with body forces in weak crust, inboard from plate boundaries where crustal diapirism occurred, and (2) what was the dynamic and thermal nature of early Archean subduction zones?

3. Komatiite is proportionally more abundant in Archean greenstones than in younger greenstones.

As more quantitative data have become available from greenstones, it is clear that relative to basalt, komatiite is more abundant in Archean greenstones than it is in either Proterozoic or Phanerozoic greenstones [*de Wit and Ashwal*, 1995]. The significance of this observation, however, remains elusive [*Arndt et al.*, 1998; *Grove and Parman*, 2004]. Most Archean komatiites occur in "mafic plain" type greenstones, which appear to represent remnants of oceanic plateaus or oceanic crust [*Condie*, 1994, 2003a]. If this is the case, the relative abundance of Archean komatiite may indicate that mantle plumes were more abundant before 2.5 Ga [*Xie and Kerrich*, 1994; *Kerrich et al.*, 1999; *Galer and Mezger*, 1998; *Campbell*, 1998; *Tomlinson and Condie*, 2001]. If so, this is important because it may mean that plume cooling of the mantle was more important in the Archean than afterwards. However, evidence from experimental petrology, geochemistry, and field studies suggest that at least some komatiites crystallized from water-rich magmas, perhaps associated with subduction [*Grove and Parman*, 2004]. Although both dry and hydrous komatiite magmas demand high temperatures in their mantle sources, mantle plume sources suggest temperatures of 300–500°C higher than at present, whereas subduction sources require temperatures only about 100°C higher than at present. Perhaps komatiite magma is produced in more than one mantle source; one of our future challenges is to come up with criteria to distinguish komatiites from different sources, if they exist.

So at present all we can conclude is that the relative abundance of Archean komatiites reflects higher mantle temperatures in the Archean, but it may not mean that any one tectonic setting, such as mantle plumes, was more widespread in the Archean.

4. TTG depleted in heavy rare earth elements (REE) is more widespread in the Archean than at later times.

Because TTGs are the most voluminous rock type in the preserved Archean crust, it is critical to understand their source and origin to better understand the origin and early evolution of continents. Even in the Tertiary, TTGs are an important juvenile component added to the continental crust in the Andes and other continental-margin arcs. Because of geochemical similarities of TTGs to modern high-silica adakites, they are commonly assumed to have similar origins [*Drummond and Defant*, 1990; *Martin*, 1999; *Martin et al.*, 2005], and in some cases, the two terms are used interchangeably. Both rock types exhibit strong enrichment in the most incompatible elements, with notable negative anomalies at Nb-Ta and Ti and strong depletion in heavy REE and Y [*Martin*, 1993]. Hence, as with high-silica adakites, TTGs are commonly assumed to represent melts of descending slabs, a conclusion with important implications for tectonic settings in the Archean [*Martin et al.*, 2005]. However, some investigators have pointed to dissimilarities between high-silica adakites and TTGs, and hence to different tectonic settings [*Smithes*, 2000; *Kamber et al.*, 2002; *Condie*, 2004a]. It is important to emphasize that geochemical data can be used to help constrain tectonic setting, but it cannot be used alone to reconstruct ancient tectonic settings.

Although high-Al TTG and high-silica adakite show strongly fractionated REE and incompatible element patterns, many TTGs have lower Sr, Mg, Ni, Cr, and Nb/Ta than many high-silica adakites [*Condie*, 2004a]. These compositional differences cannot be easily related by shallow fractional crystallization. While high-silica adakites are probably slab melts that have reacted with ultramafic rocks in the mantle wedge during ascent, many TTGs may be produced by partial melting of hydrous mafic rocks in the lower crust in arc systems, or in the Archean, perhaps in the root zones of oceanic plateaus [*White et al.*, 1999; *Smithes*, 2000; *Whalen et al.*, 2002; *Condie*, 2004a]. Depletion in heavy REE and low Nb/Ta ratios in Archean TTGs require both garnet and low-Mg amphibole in the restite, whereas moderate to high Sr values allow little if any plagioclase in the restite [*Foley et al.*, 2003]. To meet these requirements requires melting in the hornblende eclogite stability field between 40 and 80 km deep and between 700°C and 800°C. The thick mafic root and the descending slab models have in common the requirement of garnet and hornblende in the source restite.

If many Archean TTGs are not adakites, as appears to be the case, how do we produce a thick mafic crust during the Archean in which TTG magmas are generated? At present, two models are in the running: (1) beneath oceanic plateaus, and (2) underplated mafic crust from subduction zone magmas. Our challenge in the near future is to find geochemical and isotopic approaches to further evaluate these two models and see if either can withstand the test of time.

5. A large volume of continental crust was produced about 2.7 Ga.

Nd isotopes and detrital zircon age distributions strongly suggest a peak in production of continental crust at 2.7 Ga [*Condie*, 1998, 2000; *Bennett*, 2003; *Rino et al.*, 2004]. This peak has been related to a catastrophic mantle overturn event, which gave rise to a large number of mantle plumes [*Condie*, 1998]. Supporting a global mantle plume event at 2.7 Ga is a large peak in the frequency of plume proxies, such as komatiites, oceanic plateau basalts, flood basalts, giant dyke

swarms, and large layered intrusions [*Ernst and Buchan*, 2003; *Condie*, 2003b]. In addition, flood basalt eruptions occur at this time on both the Kaapvaal (Ventersdorp, 2725–2710 Ga) and Pilbara cratons (Fortescue, 2764–2756 Ga). The formation of perhaps the first supercontinent or supercratons [*Aspler and Chiarenzelli*, 1998; *Bleeker*, 2003] followed peak crustal production by 10 to 50 My as indicated by U/Pb zircon chronology of Archean TTG-greenstones [*Condie*, 2000]. A peak in gold production between 2700 and 2650 Ma also correlates with supercontinent formation. Also consistent with a worldwide mantle plume event at 2.7 Ga is a peak in La/Yb ratios of TTGs, reflecting perhaps an increased production rate of garnet-rich, mafic crust that served as TTG magma sources. In addition, a possible sudden increase in Nb/Th ratio of plume-related basalts after 2.7 Ga may reflect rapid extraction of juvenile crust from deep mantle sources via plumes during the 2.7-Ga event [*Condie*, 2003a, b].

A catastrophic mantle plume event at 2.7 Ga also should affect the ocean/atmosphere/biosphere systems and may leave a permanent imprint in the geologic record. Possible examples include the following [*Condie et al.*, 2001; *Condie*, 2003b]: (1) Evidence for low sea level 2.7 Ga may reflect direct hits of mantle plumes beneath the only two large cratons, Kaapvaal and Pilbara; (2) relatively high Chemical Index of Alterations values in shales of this age may record global warming caused by increased input of CH_4 and/or CO_2 into the atmosphere; (3) a corresponding peak in black shale deposition may reflect increased input of nutrients into the oceans and increased anoxia related to CH_4 input; (4) a small peak in deposition of banded iron formation at 2.7 Ga may be related to enhanced hydrothermal spring activity pumping more Fe into the oceans; and (5) a well-defined decrease in $\delta^{13}C$ of kerogens at 2.7 Ga could reflect enhanced activity of methanogenic bacteria due to greater input of CH_4 into the oceans.

6. Thick lithosphere underlies many Archean cratons.

Seismic velocity distribution beneath Archean cratons suggests that the lithosphere is often unusually thick, as great s 300 km or more in some cratons [*James et al.*, 2001; *Gung et al.*, 2003]. In addition, surface heat flow is typically low within these cratons, averaging about 40 mW m^{-2} [*Nyblade and Pollack*, 1993]. Mantle xenoliths from beneath Archean cratons are chiefly lherzolites that are highly refractory, having undergone significant degrees of melt extraction [*Boyd*, 1989]. Most have been partially refertilized by one or more stages of penetration with metasomatic fluids [*Griffin et al.*, 1999, 2003]. Furthermore, Re/Os model ages indicate that the deep depleted keels of Archean cratons formed at or before the overlying crust, perhaps as part of the same event or events that formed the crust [*Pearson et al.*, 1995; *Chesley et al.*, 1999; *Irvine et al.*, 2001].

The origin and survival of these deep depleted roots of some Archean cratons has attracted much attention in recent years. They appear to have survived because they are cool and refractory, and thus less dense than surrounding asthenosphere. Just how they formed, however, remains controversial. Two models have been suggested. In the first, the thick keels are formed from the collection of buoyant oceanic plates beneath continental lithosphere of normal thickness. Because Earth was hotter in the Archean, oceanic plates were warmer at the time of subduction and should undergo flat subduction and perhaps accumulate as thick keels beneath the continents. One of the problems with this model is that large volumes of oceanic crust should have been plastered beneath Archean continents, and yet deep-seated mafic xenoliths (eclogite, garnet, granulite) are not common in Archean xenolith suites.

In contrast, if the depleted lherzolites carried to the surface in kimberlite pipes are representative of the deep Archean roots, and if they formed as restites from one or more dry-melting events in mantle plumes, up to 40% komatiite melt must have been extracted [*Walter*, 1999]. For a keel 200 km thick, the resulting succession of komatiites would exceed 70 km in thickness [*Rudnick*, 2004]. Because such thicknesses are not found in Archean greenstones, if this mechanism of keel formation is viable, large quantities of komatiite melt must never have reached the surface, and/or must have rapidly been recycled back into the mantle, while the restite accumulated beneath the continents. One possibility is that the melts were more dense than the restite and thus never made it to the surface [*Boyd*, 1989]. Another is that the melts somehow traveled laterally and were erupted in ocean basins to become part of the oceanic crust, which was rapidly recycled into the mantle. And still another, favored by the Fe/Mg ratio of mantle xenoliths, is that the keels represent the restite remaining after extraction of plume-type basaltic magma rather than komatiitic magma [*Herzberg*, 2004].

7. Portions of the mantle were strongly depleted in large ion lithophile elements during the Archean.

Large positive E_{Nd} values in Early Archean igneous rocks require depletion of large ion lithophile elements in at least one mantle reservoir, probably caused by extraction of continental crust [*Bennett*, 2003]. Although E_{Nd} values ≥+3 may reflect post-crystallization disturbance in the Sm/Nd system, values up to +2 were prevalent in the Early Archean upper mantle. If E_{Nd} values ≥+3 at 3.8 Ga are real, however, this is very important in that it requires extreme fractionation in the mantle within the first 400 My of Earth history.

The question of where this depleted mantle resides and how large a volume it represents is still one of disagreement.

In terms of high field strength element (HFSE) ratios Nb/Th, Zr/Nb, Nb/Y, and Zr/Y, most basalts from mafic plain type Archean greenstones are similar to oceanic plateau basalts derived from mantle plume sources [*Condie*, 2003a, 2004b). The near absence of Archean greenstone basalts with HFSE ratios similar to NMORB is puzzling, and may mean that a "shallow" depleted mantle source was absent or volumetrically small in the Archean. In that case, the Archean depleted mantle source may have resided in the deep mantle, thus contributing to mantle plumes. Such a source has been documented today for at least some mantle plumes, such as Iceland [*Fitton et al.*, 1997; *Kempton et al.*, 2000]. If this is the case, we are immediately faced with the question of how this deep depleted reservoir formed. Perhaps it formed at shallow depths from extraction of early crust, and then it was recycled into the deep mantle by some catastrophic event in the first few hundred million years of Earth history?

High field strength element ratios in Archean basalts also indicate the existence of recycled components HIMU, EM1, and EM2 in the mantle by the Late Archean [*Condie*, 2003a]. This suggests that oceanic lithosphere as well as perhaps continental lithosphere were recycled into the deep mantle and became incorporated in some mantle plumes by the Late Archean. High field strength element ratios also indicate an important contribution of continental crust and/or subcontinental lithosphere to some mafic plain type Archean greenstone basalts. This implies that at least thin continental lithosphere was relatively widespread in the Archean. Such a conclusion is also supported by widespread isotopic evidence of contamination of Archean greenstone basalts by continental crust [*Tomlinson et al.*, 2004].

8. Many Archean greenstones comprise arc-like rock assemblages.

Archean greenstones can roughly be divided into two major types: those that comprise chiefly basalt and komatiite, the "mafic plain type", and those that contain calc-alkaline volcanics and related sediments, the "arc type". The arc-type greenstones are widespread in the Archean Superior province in Canada and also occur in the Yilgarn province in western Australia [*Condie and Harrison*, 1976; *Davis and Condie*, 1976; *Hallberg et al.*, 1976; *Giles and Hallberg*, 1982; *Cousens*, 2000; *Polat and Kerrich*, 2001; *Percival et al.*, 2004]. They appear to be the most important type of greenstone in the Late Archean terranes of Zimbabwe and Kenya [*Condie and Harrison*, 1976]. Although composed dominantly of pillow basalts, arc-type greenstones include up to 50% calc-alkaline volcanics, dominantly andesites and dacites [*Boily and Dion*, 2002]. In addition, these greenstones contain appreciable amounts of graywacke and felsic tuff. The Raquette Lake Formation in the Archean Slave

Province is a superb example of a volcanic–sedimentary succession that appears to represent a fragment of a continental margin arc system [*Mueller and Corcoran*, 2001]. Detailed mapping and sedimentologic studies show facies of high-energy clastic sedimentation contemporary with explosive volcanism, much like that found along modern continental margin arcs like Japan and the Andes. In recent years, the occurrence of boninite has been documented in several Archean greenstones, some as old as 3 Ga [*Boily and Dion*, 2002; *Smithes et al.*, 2004].

Most Archean terranes are bimodal in composition, as pointed out long ago by *Barker and Arth* [1976]. Some recent investigators have cited this bimodality as another argument against plate tectonics in the Archean [*Hamilton*, 1998]. However, bimodal magmatism is widespread in the Phanerozoic record, found in both continental rift systems and in many island arcs [*Yoder*, 1973; *Feiss*, 1982; *Nakajima and Arima*, 1998; *Smith et al.*, 2003; *Leat et al.*, 2003). Thus, bimodal magmatism is clearly not an Archean phenomenon and it occurs in at least two major plate tectonic settings.

To summarize: The striking similarity of lithologic associations in arc-type Archean greenstones, including boninite in some, argues strongly for a subduction-related origin for these greenstones.

9. A significant proportion of Archean greenstones contain volcanic rocks with geochemical characteristics similar to modern plume-derived basalts.

In terms of the worldwide database for greenstone belts, it would appear that about 35% of Late Archean (3.0–2.5 Ga) greenstone belts have plume affinities [*Condie*, 1994; *Tomlinson and Condie*, 2001]. Possibly as much as 80% of Early Archean greenstone belts may have plume affinities, although there are relatively few examples to base this on. These proportions of Archean plume-related greenstones are greater than those found in most post-Archean greenstones [*Condie*, 1994].

The observation that plume-related basalts and komatiites are more frequent in the Archean than subsequently rests in part on two major assumptions: (1) These volcanics are truly part of oceanic plateaus and not crust generated at ocean ridges, and (2) the relative abundance of greenstone belts with plume affinities is not a relict of preservational bias in the Archean.

Because lithologic assemblages formed at ocean ridges and mantle plumes may be similar (i.e., submarine basalts and intrusive equivalents together with minor chemical sediments), ocean ridge and oceanic plateau basalts may be difficult to distinguish from each other, especially in the Archean, when the production rate of oceanic crust was probably greater than today [*Abbott et al.*, 1994]. If Archean oceanic crust was the order of 20 km thick [*Sleep and Windley*, 1982; *Galer and Mezger*, 1998], the oceanic lithosphere would

have been more buoyant and could have behaved much like an oceanic plateau. So where do we draw the line between oceanic plateau and ocean ridge basalt in greenstone successions? The following two observations favor, but do not prove, plume affinities for many Archean mafic plain greenstones [*Tomlinson and Condie*, 2001]. First of all, komatiites, which reflect higher mantle temperatures and greater melting depths than basalts [*Herzberg*, 1995] occur almost exclusively in mafic plain greenstone belts. If komatiites require a mantle plume source as previously mentioned, then komatiite-bearing greenstones should reflect mantle plume sources. The problem is that some komatiites may be formed in subduction zones, also as mentioned previously. Secondly, the Th/Ta, Zr/Nb, and Nb/Th ratios in Archean mafic plain basalts differ from these ratios in NMORB yet are similar to ratios in young oceanic plateau basalts [*Condie*, 2003a, 2004b].

If plume-related basalts are selectively preserved in the Archean, their higher frequency may reflect a preservational bias rather than a greater abundance of Archean mantle plumes. What may lead to increased preservation of plume-related greenstones in the Archean? If oceanic plateaus were more buoyant in the Archean, they would be less subductable, and during plate collisions they may have accreted to the growing continental crust [*Xie and Kerrich*, 1994; *Condie*, 1997; *Kerrich et al.*, 1999; *Polat and Kerrich*, 2001], thus selectively preserving volcanic rocks with plume sources compared to volcanic rocks with arc sources. After the Archean, when mantle temperatures fell, oceanic plateaus may have been more subductable and thus the frequency of preservation of oceanic plateau type greenstones decreased.

In any case, we are then faced with the difficult question of what proportion of the Archean ocean floor was plume-generated and what proportion was generated at ocean ridges.

10. Paleomagnetic data indicate that apparent polar wandering occurred during the Archean.

Paleomagnetic data from the Archean of the Kaapvaal craton in southern Africa indicate that there has been significant polar wandering during the Late Archean [*Layer et al.*, 1989; *Kroner and Layer*, 1992]. Results suggest that between 3450 and 2700 Ma the Kaapvaal craton moved from polar to equatorial latitudes at least twice for an average latitudinal plate velocity of 17 mm/y [*Kroner and Layer*, 1992]. Pole positions estimated from Late Archean flood basalts from the Pilbara craton in Western Australia also suggest plate movements, some at a considerably faster rate than present-day rates [*Strik et al.*, 2003; *Blake et al.*, 2004]. The minimum movement rate of plates deduced from flood basalts from Nullagine synclinorum is 10 cm/y. If such fast rates are verified by future studies, this implies that of the two mechanisms suggested for escape of the greater Archean heat flux—longer ocean ridges or faster plate motions—the latter may have been very important [*Blake et al.*, 2004].

However, perhaps the most important conclusion from the paleomagnetic data (assuming that true polar wander was minor) is that plates were moving about on Earth's surface by 3.5 Ga, and hence plate tectonics *in some form* must have been operative by this time.

WHAT'S ON THE ARCHEAN HORIZON?

With all the renewed interest in the Archean, we might ask where do we stand and where are we going? Plate tectonics, which is assumed by most investigators to have operated in the Archean, is now being challenged. And indeed, such a challenge should bring about renewed efforts to understand if and how the Archean differs from later times in Earth history. Despite the challenges, however, a wealth of geologic, geochemical, structural, volcanologic, and sedimentologic data are, at very least, consistent with plate tectonics, at least in the Late Archean. Yes, there are some differences, most of which, however, can be explained by a hotter mantle along with, at least periodically, more elevated crustal geothermal gradients, in the Archean. It would seem to us that rather than abandon plate tectonics, which so well explains so much observational data, it is better to look for differences in dynamic processes and tectonic regimes to explain the "unusual" tectonic signatures we see in the Archean geological record.

Clearly we need more precise information to answer some of the problems with Archean plate tectonics. Why is evidence not found for a shallow depleted mantle similar to that from which NMORB is derived? If the major depleted reservoirs in the Archean mantle were at great depth, how did they form? If diapirism tectonics is important on some Archean cratons but not on others, what does this mean for evolving plate tectonic regimes? If the Archean lower crust was hotter and less viscous in some cratons, so that it convected [*Collins et al.*, 1998], how did other (most?) Archean cratons escape such heating? Perhaps a thick depleted mantle keel is the answer. In that case, we must further ask why some cratons developed keels and others did not. And finally what really happened 2.7 Ga in Earth's mantle? If it was a mantle overturn event, what caused it and what effect did it have on evolving Archean tectonic regimes?

Acknowledgments.   This paper has been substantially improved from a review by Jean Bedard.

REFERENCES

Abbott, D., L. Burgess, J. Longhi, and W.H.F. Smith, (1994), An empirical thermal history of the Earth's upper mantle. *J. Geophys. Res.*, 99, 13, 835-850.

Arndt, N. (1999), Why was flood volcanism on submerged continental platforms so common in the Precambrian, *Precamb. Res.*, 97, 155-164.

Arndt, N., C. Ginibre, C. Chauvel, F. Albarede, M. Cheadle, C. Herzberg, G. Jenner, and Y. Lahaye, (1998), Were komatiites wet? *Geology*, 26, 739-742.

Aspler, L.B., and J.R. Chiarenzelli, (1998), Two Neoarchean supercontinents? Evidence from the Paleoproterozoic. *Sediment. Geol.* 120, 75-104.

Barker, F., and J.G. Arth, (1976), Generation of trondhjemitic-tonalitic liquids and Archean bimodal trondhjemite-basalt suites, *Geology*, 4, 596-600.

Barley, M.E., T.S. Blake, and D.I. Groves, (1992), The Mount Bruce Megasequence set and eastern Yilgarn craton: examples of Late Archean to Early Proterozoic divergent and convergent craton margins and controls on mineralization, *Precambrian Res.*, 58, 55-70.

Bedard, J.H., P. Brouillette, L. Madore, and A. Berclaz, (2003), Archean cratonization and deformation in the northern Superior province, Canada: an evaluation of plate tectonic versus vertical tectonic models. *Precamb. Res.*, 127, 61-87.

Benn, K. (2004), Late Archean Kenogamissi complex, Abitibi Subprovince, Ontario: doming, folding and deformation-assisted melt remobilization during syntectonic batholith emplacement. Transactions of the Royal Society of Edinburgh: *Earth Sciences*, 95, 297-307.

Benn, K., W. Miles, M.R. Ghassemi, and J. Gillett, (1994), Crustal structure and kinematic framework of the northwestern Pontiac Subprovince, Quebec: an integrated structural and geophysical study. *Canadian Journal of Earth Sciences*, 31, 271-281.

Bennett, V.C. (2004), Compositional evolution of the mantle. Treatise on Geochemistry, Chapt. 2.13, Elsevier, Amsterdam, p. 493-519.

Bickle, M.J. (1978), Heat loss from the Earth: a constraint of Archean tectonics from the relation between geothermal gradients and the rate of plate production. *Earth Planet. Sci. Lettr.*, 40, 301-315.

Bickle, M.J., E.G. Nisbet, and A. Martin, (1994), Archean greenstone belts are not oceanic crust, *Jour. Geol.*, 102, 121-138.

Blackburn, C.E. (1980), Towards a mobilist tectonic model for part of the Archean of NW Ontario, *Geosci. Canada*, 7 (2), 64-72.

Blake, T.S., R. Buick, S.J.A. Brown, and M.E. Barley, (2004), Geochronology of a Late Archean flood basalt province in the Pilbara craton, Australia: constraints on basin evolution, volcanic and sedimentary accumulation, and continental drift rates, *Precamb. Res.*, 132, 143-173.

Bleeker, W. (2002), Archean tectonics: a review, with illustrations from the Slave craton, *Geol. Soc. London, Spec. Publ.*, 199, 151-181.

Bleeker, W. (2003), The late Archean record: a puzzle in ca. 35 pieces, *Lithos*, 71, 99-134.

Bleeker, W., J.W.F. Ketchum, V.A. Jackson, and M.E. Villeneuve, (1999), The central Slave basement complex, Part I: it structural topology and autochthonous cover. *Canad. Jour. Earth Sci.*, 36, 1083-1109.

Boily, M., and C. Dion, (2002), Geochemistry of boninite-type volcanic rocks in the Frotet-Evens greenstone belt, Opatica subprovince, Quebec: implications for the evolution of Archean greenstone belts. *Precamb. Res.*, 115, 349-371.

Bouhallier, H., P. Choukroune, and M. Ballèvre, (1993), Diapirism, bulk homogeneous shortening and transcurrent shearing in the

Archean Dharwar craton: the Holenarsipur area, southern India. *Precambrian Research*, 63, 43-58.

Boyd, F.R. (1989) Compositional distinction between oceanic and cratonic lithosphere. *Earth Planet. Sci. Lett.*, 96, 15-26.

Calvert, A.J., and J.N. Ludden, (1999), Archean continental assembly in the southeastern Superior Province of Canada. *Tectonics*, 18, 412-429.

Chardon, D., P.L. Choukroune, and M. Jayananda, (1996), Strain patterns, decollement and incipient sagducted greenstone terrains in the Archean Dharwar craton South India. *Jour. Struct. Geol.*, 18, 991-1004.

Chardon, D., P.L. Choukroune, and M. Jayananda, (1998), Sinking of the Dharwar Basin (South India): implications for Archean tectonics. *Precambrian Research*, 91, 15-39.

Chen, S.F., R. Riganti, S. Wyche, J.E. Greenfield, and D.R. Nelson, (2003), Lithostratigraphy and tectonic evolution of contrasting greenstone successions in the central Yilgarn craton, Western Australia. *Precamb. Res.*, 127, 249-266.

Chesley, J.T., R.L. Rudnick, and C.T. Lee, (1999), Re-Os systematics of mantle xenoliths from the East African rift: age, structure, and history of the Tanzanian craton, Geochim. *Cosmochim. Acta*, 63, 1203-1217.

Choukroune, P., J. Ludden, A.J. Calvert, H. Bouhallier, (1997), Archean crustal growth and tectonic processes: a comparison of the Superior Province, Canada and the Dharwar Craton, India. in: Orogeny through time, Burg, J.P. and Ford, M. (eds.), Geological Society Special Publication No. 121, London, p. 63-98.

Collins, W.J., M.J. Van Kranendonk, and C. Teyssier, (1998), Partial convective overturn of Archean crust in the east Pilbara craton, Western Australia: driving mechanisms and tectonic implications. *Jour. Struct. Geol.*, 20, 1405-1424.

Condie, K.C. (1994), Greenstones through time. In: K.C. Condie (editor), Archean Crustal Evolution, Elsevier, Amsterdam, pp. 85-120.

Condie, K.C. (1997), Contrasting sources for upper and lower continental crust: the greenstone connection. *J. Geol.*, 105, 729-736.

Condie, K.C. (1998), Episodic continental growth and supercontinents: a mantle avalanche connection? *Earth Planet. Sci. Lettr.*, 163, 97-108.

Condie, K.C (2000), Episodic continental growth models: afterthoughts and extensions. *Tectonophys.*, 322, 153-162.

Condie, K.C. (2003a), Incompatible element ratios in oceanic basalts and komatiites: tracking deep mantle sources and continental growth rates with time. *Geochem. Geophys. Geosyst.*, 4 (1), 10.1029/2002GC000333.

Condie, K.C. (2003b), What on earth happened 2.7 billion years ago? Amer. Geophys. Union Abstract vol., Nice, France, EAE3-A-01269.

Condie, K.C. (2004a), TTGs and adakites: are they both slab melts? Lithos (in press).

Condie, K.C. (2004b), High field strength element ratios in Archean basalts: a window to evolving sources of mantle plumes, *Lithos* (in press).

Condie, K.C., and N.M. Harrison, (1976), Geochemistry of the Archean Bulawayan Group, Midlands Greenstone belt, Rhodesia, *Precambrian Res.*, 3, 253-271.

Condie, K.C., D.J. Des Marais, and D. Abbott, (2001), Precambrian superplumes and supercontinents: a record in black shales, carbon isotopes, and paleoclimates? *Precambrian Research*, 106, 239-260.

Cousens, B.L. (2000), Geochemistry of the Archean Kam Group, Yellowknife greenstone belt, Slave province, Canada, *Jour. Geol.*, 108, 181-197.

Davies, G.F. (1992), On the emergence of plate tectonics. *Geology*, 20, 963-966.

Davies, G.F. (1995), Puncuated tectonic evolution of the earth. *Earth and Planetary Science Letters*, 136, 363-379.

Davis, Jr., P.A., and K.C. Condie, (1977), Trace element model studies of Nyanzian greenstone belts western Kenya. *Geochim. Cosmochim. Acta*, 41, 271-277.

de Wit, M.J. (1998), On Archean granites, greenstones, cratons and tectonics: does the evidence demand a verdict? *Precambrian Res.*, 91, 181-226.

de Wit, M.J., and L.D. Ashwal, (1997), Greenstone Belts. Oxford Univ. Press, Oxford, 809 pp.

Donaldson, J.A., and E.A. de Kemp, (1998), Archean quartz arenites in the Canadian shield: examples from the Superior and Churchill provinces. *Sediment. Geol.*, 120, 153-176.

Drummond, M.S., M.J. Defant, (1990), A model for trondhjemite-tonalite-dacite genesis and crustal growth via slab melting: Archean to modern comparisons. *Journ. Geophys. Res.*, 95, 21503-21521.

Eriksson, K.A. (1978), Alluvial and destructive beach facies from the Archean Moodies Group, Barberton Mountain Land, South Africa and Swaziland, Canad. Society Petrol. *Geol. Memoir* 5, 287-311.

Ernst, R.E., and K.L. Buchan, (2003) Recognizing mantle plumes in the geological record. *Ann. Rev. Earth Planet. Sci.* 2003, 31, 469-523.

Fedo, C.M., and K.A. Eriksson, (1996), Stratigraphic framework of the 3 Ga Buhwa greenstone belt: a unique stable shelf succession in the Zimbabwe Archean craton. *Precamb. Res.*, 77, 161-178.

Feiss, P.G. (1982), Geochemistry and tectonic setting of the volcanics of the Carolina slate belt, *Econ. Geol.*, 77, 273-293.

Fitton, J.G., A.D. Saunders, M.J. Norry, B.S. Hardarson, and R.N. Taylor, (1997), Thermal and chemical structure of the Iceland plume. *Earth Planet. Sci. Lettr.*, 153, 197-208.

Foley, S.F., S. Buhre, and D.E. Jacob, (2003), Evolution of the Archean crust by delamination and shallow subduction, *Nature*, 421, 249-252.

Galer, S.J.G., and K. Metzger, (1998), Metamorphism, denudation and sea level in the Archean and cooling of the Earth, *Precamb. Res.*, 92, 389-412.

Giles, C.W., and J.A. Hallberg, (1982), The genesis of the Archean Welcome Well volcanic complex, Western Australia, *Contrib. Mineral. Petrol.*, 80, 307-318.

Goleby, B.R., R.S. Blewett, R.J. Korsch, D.C. Champion, K.F. Cassidy, L.E.A. Jones, P.B. Groenewald, and P. Henson, (2004), Deep seismic reflection profiling in the Archaean northeastern Yilgarn Craton, Western Australia: implications for crustal architecture and mineral potential. *Tectonophysics*, 388, 119-133.

Goodwin, A.M. (1977), Archean basin-craton complexes and the growth of Precambrian shields, *Canadian Journal of Earth Sciences*, 14, 2737-2759.

Goodwin, A.M., and J.E.M. Smith, 1980, Chemical discontinuities in Archean volcanic terrain and the development of Archean crust, *Precambrian Research*, 10, 301-311.

Griffin, W.L., S.Y. O'Reilly, and C.G. Ryan, (1999), The composition and origin of subcontinental lithospheric mantle, Geochemical Society Spec. Publ. No. 6, 13-45.

Griffin, W.L., S.Y. O'Reilly, L.M. Natapov, and C.G. Ryan, (2003), the evolution of lithospheric mantle beneath the Kalahari craton and its margins, *Lithos*, 71, 215-241.

Grove, T.L., and S.W. Parman, (2004), thermal evolution of the Earth as recorded by komatiites, *Earth Planet. Sci. Lettr.*, 219, 173-187.

Gung, Y., M. Panning, and B. Romanowicz, (2003), Global anisotropy and the thickness of continents, *Nature*, 422, 707-710.

Hallberg, J.A., D.N. Carter, and K.N. West, (1976), Archean volcanism and sedimentation near Meekatharra, Western Australia, *Precambrian Res.*, 3, 577-595.

Hamilton, W.B. (1998), Archean magmatism and deformation were not products of plate tectonics, *Precambrian Res.*, 91, 143-179.

Herzberg, C. (1995), Generation of plume magmas through time: An experimental perspective. *Chem. Geol.*, 126, 1-16.

Herzberg, C. (2004), Geodynamic information in peridotite petrology, *Jour. Petrology*, 45, 2507-2530.

Hickman, A.H. (2004), Two contrasting granite-greenstone terranes in the Pilbara craton, Australia: evidence for vertical and horizontal tectonic regimes prior to 2900 Ma, *Precamb. Res.*, 131, 153-172.

Hoffman, P.F., and G. Ranalli, (1988), Archean oceanic flake tectonics, *Geophys. Res. Lettr.*, 15, 1077-1080.

Irivine, G.J., D.G. Pearson, and R.W. Carlson, (2001), Lithospheric mantle evolution of the Kaapvaal craton: a Re-Os isotope sudy of peridotite xenoliths from Lesotho kimberlites, *Geophys. Res. Lettr.*, 28, 2505-2508.

James, D.E., M.J. Fouch, J.C. VanDecar, and S. van der Lee, (2001), Tectospheric structure beneath southern Africa, *Geophys. Res. Lettr.*, 28(13), 2485-2488.

Kamber, B.S., A. Ewart, K.D. Collerson, M.C. Bruce, G.D. McDonald, (2002), Fluid-mobile trace element constraints on the role of slab melting and implications for Archean crustal growth models. *Contrib. Mineral. Petrol.*, 144, 38-56.

Kempton, P.D., J.G. Fitton, A.D. Saunders, G.M. Nowell, R.N. Taylor, B.S. Hardarson, and G. Pearson, (2000), The Iceland plume in space and time: a Sr-Nd-Pb-Hf study of the North Atlantic rifed margin. *Earth Planet. Sci. Lettr.*, 177, 255-271.

Kerrich, R., A. Polat, D. Wyman, and P. Hollings, (1999), Trace element systematics of Mg- to Fe-tholeiitic basalt suites of the Superior Province: implications for Archean mantle reservoirs and greenstone belt genesis. *Lithos*, 46, 163-187.

Komiya, T., M. Hayashi, S. Maruyama, and H. Yurimoto, (2002), Intermediate-P/T type Archean metamorphism of the Isua supracrustal belt: implications for secular change of geothermal gradients at subduction zones and for 'Archean plate tectonics, *Amer. Jour. Sci.*, 302, 806-826.

Kroner, A., and P.W. Layer, (1992), Crust formation and plate motion in the Early Archean. *Science*, 256, 1405-1411.

Kusky, T.M. (1998), Tectonic setting and terrane accretion of the Archean Zimbabwe craton. *Geology*, 26, 163-166.

Kusky, T.M., and J.H. Li, (2002), Is the Dongwanzi complex an Archean ophiolite? Response. *Science*, 295, no. 5557, DOI: 10.1126/science.295.5557.923a.

Kusky, T.M., J-H. Li, and R.D. Tucker, (2001), The Archean Dongwanzi ophiolite complex, North China craton: 2.505 billion year old oceanic crust and mantle. *Science*, 292, 1142-1145.

Langford, F.F., and J.A. Morin, (1976), The development of the Superior province of NW Ontario by merging island arcs. *Amer. Jour. Sci.*, 276, 1023-1034.

Layer, P.W., A. Kroner, M. McWilliams, and D. York, (1989), Elements of the Archean thermal history and apparent polar wander of the eastern Kaapvaal craton, Swaziland, from single grain dating and paleomagnetism, *Earth Planet. Sci. Lettr.*, 93, 23-34.

Leat, P.T., J.L. Smellie, I.L. Millar, and R.D. Larter, (2003), Magmatism in the South Sandwich arc, *Geol. Soc. London, Spec. Publ.* 219, 285-313.

Lesher, C.M., and N.T. Arndt, (1995), REE and Nd isotope geochemistry, petrogenesis and volcanic evolution of contaminated komatiites at Kambalda, Western Australia, *Lithos*, 34, 127-157.

Lowe, D.R. (1994), Archean greenstone-related sedimentary rocks. In: K.C. Condie (editor), Archean Crustal Evolution, Elsevier, Amsterdam, pp. 121-170.

Ludden, J., C. Hubert, A. Barnes, B. Milkereit, and E.W. Sawyer, (1993), A 3-dimensional perspective on the evolution of Archean crust - Lithoprobe seismic-reflection images in the southwestern Superior Province. *Lithos*, 30, 357-372.

Maas, R., and M.T. McCulloch, (1991), The provenance of Archean clastic metasediments in the Narryer Gneiss complex, Western Australia: trace element geochemistry, Nd isotopes, and U-Pb ages for detrital zircons, *Geochim. Cosmochim. Acta*, 55, 1915-1932.

Martin, A., E.G. Nisbet, M.J. Bickle, and J.L. Orpen, (1993), Rock units and stratigraphy of the Belingwe greenstone belt: the complexity of the tectonic setting. In: M.J. Bickle and E.G. Nisbet (eds), the Geology of the Belingwe Greenstone Belt, Zimbabwe: A Study of the Evolution of Archean Continental Crust. Geol. Soc. Zimbabwe, Spec. Publ., 2, Balkema, Rotterdam, pp. 13-37.

Martin, H. (1993), The mechanisms of petrogenesis of the Archean continental crust-comparison with modern processes. *Lithos*, 30, 373-388.

Martin, H. (1999), Adakitic magmas: modern analogues of Archean granitoids, *Lithos*, 46, 411-429.

Martin, H., R.H. Smithies, R. Rapp, J.-F. Moyen, and D. Champion, (2005), An overview of adakite, tonalite-trondhjemite-granodiorite (TTG), and sanukitoid: relationships and some implications for crustal evolution, *Lithos*, 79, 1-24.

McCall, G.J.H. (2003), A critique of the analogy between Archaean and Phanerozoic tectonics based on regional mapping of the Mesozoic-Cenozoic plate convergent zone in the Makran, Iran, *Precambrian Research*, 127, 5-17.

Mueller, W.U., and P.L. Corcoran, (2001), Volcano-sedimentary processes operating on a marginal continental arc: the Archean

Raquette Lake Formation, Slave province, Canada, Sediment, *Geology*, 141-142, 169-204.

Mueller, W.U., R. Daigneault, J.K. Mortensen, and E.H. Chown, (1996), Archean terrane docking: upper crust collision tectonics, Abitibi greenstone belt, Quebec, Canada, *Tectonophys.*, 265, 127-150.

Nakajima, K., and M. Arima, (1998), Melting experiments on hydrous low-K tholeiite: implications for the genesis of tonalitic crust in the Izu-Bonin-Mariana arc, *The Island Arc*, 7, 359-373.

Nyblade, A.A., and H.N. Pollack, (1993), a global analysis of heat flow from Precambrian terrains: implications for the thermal structure of Archean and Proterozoic lithosphere, *Jour. Geophys. Res.*, 98, 12207-218.

O'Reilly, S.Y., W.L. Griffin, Y.H. Poudjom Djomani, and P. Morgan, (2001), Are lithospheres forever? Tracking changes in subcontinental lithospheric mantle through time. *GSA Today*, 11 (4), 4-10.

Pearson, D.G., R.W. Carlson, S.B. Shirey, F.R. Boyd, and P.H. Nixon, (1995), The stabilization of Archean lithospheric mantle: a Re-Os isotope study of peridotite xenoliths from the Kaapvaal craton, *Earth Planet. Sci. Lettr.*, 134, 341-357.

Percival, J.A., R.A. Stern, T. Skulski, K.D. Card, J.K. Mortensen, and N.J. Begin, (1994), Minto Block, Superior Province - missing link in deciphering assembly of the craton at 2.7 Ga. *Geology*, 22, 839-842.

Percival, J.A., V. McNicoll, J.L. Brown, and J.B. Whalen, (2004), Convergtent margin tectonics, central Wabigoon subprovince, Superior Province, Canada, *Precamb. Res.*, 132, 213-244.

Peschler, A.P., K. Benn, and W.R. Roest, (2004), Insights on Archean continental geodynamics from gravity modeling of granite-greenstone terranes, *Jour. Geodynamics*, (in press).

Polat, A., and R. Kerrich, (1999), Formation of an Archean tectonic melange in the Schreiber-Hemlo greenstone belt, Superior province, Canada: implications for Archean subduction-accretion process, *Tectonics*, 18, 733-755.

Polat, A., and R. Kerrich, (2001), Geodynamic processes, continental growth, and mantle evolution recorded in late Archean greenstone belts of the southern Superior province, Canada, *Precamb. Res.*, 112, 5-25.

Pollack, H.N. (1997), Thermal characteristics of the Archean. In: M.J. de Wit and L.D. Ashwal (eds.), Greenstone Belts. Clarendon Press, Oxford, pp. 223-293.

Rino, S., K. Tsuyoshi, B.F. Windley, I. Katayama, A. Motoki, and T. Hirata, (2004), Major episodic increases of continental crustal growth determined from zircon ages of river sands: implication for mantle overturns in the Early Precambrian, *Phys. Earth Planet. Inter.*, 146, 369-394.

Rudnick, R.L. (2004), Cratons and contiental evolution, 2004 Intern. Basement Tectonics Assoc. Conference, Oak Ridge, TN, Abstract Volume, p. 44-45.

Sandiford, M., M.J. Van Kranendonk, and S. Bodorkos, (2004), Conductive incubation and the origin of dome-and-keel structure in Archean granite-greenstone terranes: a model based on the eastern Pilbara Craton, Western Australia. *Tectonics*, 23, doi:10.1029/2002TC001452.

Sleep, N.H., and B.F. Windley, (1982), Archean plate tectonics: constraints and inferences. *Jour. Geol.*, 90, 363-379.

Smith, I.E.M., T.J. Worthington, R.B. Stewart, R.C. Price, and J.A. Gamble, (2003), Felsic volcanism in the Kermadec arc, SW Pacific: crustal recycling in an oceanic setting, Geol. Soc. London, Spec. Publ. 219, 99-118.

Smithies, R.H. (2000), The Archean tonalite-tondhjemite-granodiorite (TTG) series is not an analogue of Cenozoic adakite. *Earth Planet. Sci. Lettr.*, 182, 115-125.

Smithies, R.H., and D.C. Champion, (2000), The Archean high-Mg diorite suite: links to tonalite-trondhjemite-granodiorite magmatism and implications for early Archean crustal growth. *Jour. Petrol.*, 41, 1653-1671.

Smithes, R.H., D.C. Champion, and K.F. Cassidy, (2003), Formation of Earth's early Archean continental crust. *Precamb. Res.*, 127, 89-101.

Smithes, R.H., D.C. Champion, and S-S. Sun, (2004), the case for Archean boninites, *Contrib. Mineral. Petrol.*, 147, 705-721.

Srinivasan, G., and R.W. Ojakangas, (1986), Sedimentology of quartz-pebble conglomerates and quartzites of the Archean Bababudan Group, Dharwar craton, South India: evidence for early crustal stability. *Jour. Geol.*, 94, 199-214.

Stein, M., and A.W. Hofmann, (1994), Mantle plumes and episodic crustal growth. *Nature*, 372, 63-69.

Stern, C.R., and M.J. de Wit, (2003), The role of spreading centre magma chambers in the formation of Phanerozoic oceanic crust: evidence from Chilean ophiolites, in A. Panayiotou (editor), Ophiolites: Proceedings Intern. Ophiolite Sympos., Cyprus 1979, Nocosia, Geol. Survey Dept., p. 497-506.

Stevenson, R. (1995), Crust and mantle evolution in the Late Archean: evidence from a Sm-Nd isotopic study of the North Spirit Lake greenstone belt, NW Ontario, Canada, *Geol. Soc. America Bull.*, 107, 1458-1467.

Stott, G.M. (1997), The Superior province, Canada. In: M. J. de Wit and L. D. Ashwal (eds.), Greenstone Belts. Clarendon Press, Oxford, pp. 480-507.

Strik, G., T.S. Blake, T.E. Zegers, S.H. White, and C.G. Langereis, (2003), Paleomagnetism of flood basalts in the Pilbara craton, Western Australia: Late Archean continental drift and the oldest known reversal of the geomagnetic field, *Jour. Geophys. Res.*, 10.1029/2003JB002475.

Swager, C.P. (1997), Tectonostratigraphy of Late Archean greenstone terranes in the southern Eastern Geoldfields, Western Australia. *Precamb. Res.*, 83, 11-42.

Thurston, P.C. (1994) Archean volcanic patterns. In: K.C. Condie (editor), Archean Crustal Evolution, Elsevier, Amsterdam, pp. 45-84.

Tomlinson, K.Y., and K.C. Condie, (2001), Archean mantle plumes: evidence form greenstone belt geochemistry. *Geol. Soc. America Memoir*, 352, 341-357.

Tomlinson, K.Y., G.M. Stott, J.A. Percival, and D. Stone, (2004), Basement terrane correlations and crustal recycling in the western Superior province: Nd isotopic character of granitoid and felsic volcanic rocks in the Wabigoon subprovince, N Ontario, Canada, *Precamb. Res.*, 132, 245-274.

Trendall, A.F. (2002), The significance of iron formation in the Precambrian stratigraphic record. In: W. Altermann and P. L. Corcoran (eds), Precambrian Sedimentary Environments:

A Modern Approach to Depositional Systems. IAS Spec. Paubl., 44, Blackwell, Oxford, pp. 33-66.

Van Kranendonk, M.J., W.J. Collins, A. Hickman, and M.J. Pawley, (2004), Critical tests of vertical vs horizontal tectonicmodels for the Archean East Pilbara granite-greenstone terrane, Pilbara craton, Western Australia, *Precamb. Res.*, 131, 173-211.

Vlaar, N.J., P.E. van Keken, and A.P. van den Berg, (1994), Cooling of the Earth in the Archean: consequences of pressure-release melting ni a hotter mantle, *Earth and Planetary Science Letters*, 121, 1-18.

Walter, M.J. (1999), Melting residues of fertile peridotite and the origin of cratonic lithosphere. In: Y. Fei, M. Bertka and B.O. Mysen (editors), Mantle Petrology: Field Observations and High-Pressure Experimentation. Geochem. Society, Spec. Publ., Houston, TX, p. 225-240.

Whalen, J.B., J.A. Percival, V.J. McNicoll, F.J. Longstaffe, (2002), A mainly crustal origin for tonalitic granitoid rocks, Superior province, Canada: implications for Late Archean tectonomagmatic processes. *Journ. Petrol.*, 43, 1551-1570.

White, R.V., J. Tarney, A.C. Kerr, A.D. Saunders, P.D. Kempton, M.S. Pringle, and G.T. Klaver, (1999), Modification of an oceanic plateau, Aruba, Dutch Caribbean: implications for the generation of continental crust. *Lithos*, 46, 43-68.

Wilde, S.A., M.F. Middleton, and B.J. Evans, (1996), Terrane accretion in the southwestern Yilgarn Craton: evidence from a deep seismic crustal profile. *Precambrian Research*, 78, 179-196.

Xie, Q., and R. Kerrich, (1994), Silicate-perovskite and majorite signature komatiites from the Archean Abitibi greenstone belt: implications for early mantle differentiation and stratification. *J. Geophys. Res.*, 99, 15,799-812.

Yoder, Jr., H.S. (1973), Contemporaneous basaltic and rhyolitic magmas, *Amer. Mineralog.*, 58, 153-171.

Zhai, M.G., G.C. Zhao, and Q. Zhang, (2002), Is the Dongwanzi complex an Archean ophiolite? *Science*, 295, no. 5557, DOI: 10.1126/science.295.5557.923a.

K. Benn, Ottawa-Carleton Geoscience Centre and Department of Earth Sciences, University of Ottawa, Ottawa, Ontario K1N 6N5, Canada. (kbenn@uottawa.ca)

K. C. Condie, Department of Earth & Environmental Science, New Mexico Institute of Mining & Technology, Socorro, New Mexico 87801, USA. (kcondie@nmt.edu)

Archean Thermal Regime and Stabilization of the Cratons

Jean-Claude Mareschal

GEOTOP-UQAM-McGill, Centre for Research in Geochemistry and Geodynamics, University of Québec at Montréal, Montréal, Québec, Canada

Claude Jaupart

Institut de Physique du Globe de Paris, Paris, France

Archean provinces are presently characterized by low heat flow, with an average of 41 mW m^{-2}, less than the global continental average (56 mW m^{-2}). The range of regionally averaged heat flow values in Archean Provinces (18–54 mW m^{-2}) is narrower than in younger terranes. At the end of the Archean, when crustal heat production was double the present, surface heat flow varied over a range ($\approx$45-90 mW m^{-2}) as wide as that presently observed in Paleozoic Provinces. For the present-day vertical distribution of radio-elements, high heat production during the Archean is insufficient to account for elevated lower crustal temperatures. High temperature-low pressure metamorphism conditions require additional heat input, for example by emplacement of large volumes of basaltic melts, crustal thickening or larger concentrations of radio-elements in the lower crust. In the Archean, with a crust thicker than 40 km or with the radio-elements uniformly distributed throughout a 40 km thick crust, the lower crust was near melting and, with an effective viscosity $\approx$10^{19}Pa s, it could not sustain the stress due to crustal thickening. Long-term crustal stability requires the enrichment of the upper crust in radio-elements through melt extraction from the lower crust. After root emplacement, thermal conditions in cratons remained far from equilibrium for 1-2 Gy. Depending on the mechanism of root formation, temperature at 150 km might only be 150 K higher than present, implying that the lithospheric mantle remained sufficiently cold and strong to preserve Archean features. In thick continental lithosphere, temperatures in the crust and deep in the continental root are effectively decoupled for a long time.

INTRODUCTION

Record of the tectonics of early Earth has been preserved in many ancient terranes now amalgamated in cratons. Crustal and mantle records seem to be in conflict. On the

Archean Geodynamics and Environments
Geophysical Monograph Series 164
Copyright 2006 by the American Geophysical Union
10.1029/164GM06

one hand, Archean crustal metamorphism appears to be biased towards high temperature-low pressure conditions, in contrast to modern analogs. In Canada's Slave Province, for example, Archean tectonics involved extensive rifting, deformation and the intrusion of large volumes of granites. On the other hand, seismic studies show preserved Archean features in the lithospheric mantle [*Bostock*, 1998], which suggests that the mantle was strong, and hence cold. Some authors [*Sandiford & Hand*, 1998] have argued that, because heat production was higher than today, high crustal temperatures were due to elevated heat production and did not reflect transient thermal events triggered by mantle processes. What is at stake, therefore, is the ability to use the Archean record to reconstruct tectonic processes at the lithospheric scale and to assess whether or not they were different from today. Also at stake is the growth of thick and buoyant continental lithosphere, which occurs at different temperatures depending on the mechanisms of melting and buoyant residue formation (i.e. mantle plumes or subduction-related processes). These questions can only be addressed in the framework of thermal models at the lithospheric scale.

Today, heat from the Earth's interior escapes mostly through the seafloor. The flow of heat from the continents represents about 20% of the total heat flow from the Earth. Most of the continental heat flow is accounted for by radiogenic heat production in the crust and very little heat from the Earth's interior escapes through the continents. During the Archean, part of Earth surface was covered by continents with a thick (> 35 km) crust, part of which was stabilized and preserved in the cratons. Presently, the cratons are characterized by low surface heat flow, and by a thick and cold lithosphere. Because radiogenic heat production during the Archean was higher than today, the cratons were hotter when they stabilized than they are today. How much hotter and what was the role of the thermal regime in their stabilization remains debated. The intrusion of granites without any subsequent crustal melting event implies that the hot crust cooled down and remained cold after the end of the Archean. The presence of inclusions of Archean age in diamonds has been interpreted by some [*Richardson et al.*, 1984; *Boyd et al.*, 1985] to imply that the diamonds are Archean but this interpretation has been disputed [*Shimizu & Sobolev*, 1995]. If the diamonds are Archean, it implies that the lithospheric mantle was sufficiently cold to be in the diamond stability field. Several researchers estimated an Archean steady state temperature profile and claimed that the temperature in the lithospheric mantle was above the stability field for diamonds; According to them, diamonds could form only during the Proterozoic after the lithosphere cooled down due to the rundown of radioactivity [*Morgan*, 2004]. Also, geodynamic models have been used to show that because of temperature fluctuations, the sublithospheric mantle is unlikely to remain in the diamond stability field for 3 Gy [*O'Neill & Moresi*, 2003].

The mechanism of formation of the cratonic roots is still widely debated. On one side, some authors favor plate tectonic like processes and call for under-thrusting and stacking of subducted slabs beneath the crust to form the root [e.g. *Abbott*, 1991; *Bostock*, 1999; *Carlson et al.*, 2005]. On the other side, some authors invoke a completely different tectonic style during the Archean and propose that the sublithospheric mantle was extracted from hot rising mantle diapirs [e.g. *Griffin et al.*, 2003]. The very different thermal conditions at the time of stabilization implied by these models must be included in discussions of their feasibility.

In this paper, we discuss thermal models for Archean cratons with emphasis on two key points. One is the large time-constant ($\approx$1-2 Gy) for diffusive equilibration in thick lithosphere and the other is the amount and vertical distribution of radio-elements in the crust. We review the present heat flow and heat production data from Archean provinces and examine their implications for the stabilization of the cratons. Detailed analysis of the heat flow and heat production data in stable continents shows that the mantle component of the heat flow is weak ($\approx$10-15 mW m^{-2}) and that radiogenic heat production in the crust accounts for most of the heat flow and explains its variations [*Pinet et al.*, 1991; *Roy and Rao*, 2000; *Jaupart et al.*, 1998; *Jaupart & Mareschal*, 1999]. From the heat flow and heat production data, determining the crustal heat production after the crust became stable is straightforward, but there is no direct way to determine what was the mantle contribution to the surface heat flow during and at the end of the Archean. Thermal models must satisfy the conditions that temperature was near melting in the lower crust before it became stable at the end of the Archean and that the lithospheric mantle remained strong and buoyant. After concentration of the heat producing elements in the upper crust, steady state temperatures in the lower crust remained below melting provided that the heat flow from the mantle was not higher than at present. For the lithosphere to be sufficiently strong and the cratons to remain stable, the mantle temperatures must be lower than in the steady state regime with basal heat flow not much higher at the end of the Archean than it is today. This might be achieved in transient regime with an initial "cold" condition (for instance, following accretion of the cratonic roots by under-thrusting of cold subducting plates).

2. HEAT FLOW AND HEAT PRODUCTION IN THE ARCHEAN CRATONS

Many studies have documented that heat flow is lower in Archean cratons than in younger provinces [*Polyak & Smirnov*, 1968; *Jessop & Lewis* 1978; *Morgan*, 1985]. *Nyblade & Pollack* [1993] have reported that the average heat flow for all Archean provinces is 41 mW m^{-2} compared to a global continental average of 56 mW m^{-2}.

Studies of continental heat flow have often relied on the linear heat flow-heat production relationship [*Birch et al.*, 1968]:

$$Q = Q_r + D \times H \qquad (1)$$

where Q is the measured surface heat flow and H the heat production at the heat flow site. The parameters Q_r and D are used to characterize the vertical distribution of radiogenic heat production and to elaborate thermal models for the crust and lithosphere. In the simplest model, D is the depth scale of a surficial layer enriched in radio-elements, and the reduced heat flow Q_r is the heat flow from below that layer. However such considerations do not apply in Archean provinces because the linear heat flow-heat production relationship is not observed [*Kutas*, 1984; *Jaupart & Mareschal*, 1999].

The average heat flow and heat production vary between Archean cratons and average geothermal characteristics might not be representative of any of the Archean provinces [*Mareschal & Jaupart*, 2004]. Therefore, using the global average to infer an "Archean thermal regime" may be misleading and we prefer to analyze separately the heat flow and heat production in the main cratons where sufficient data are available.

2.1 Heat Flow in the Archean Provinces

Heat flow and heat production data from several Archean Provinces and subprovinces are sufficient for meaningful statistics to be made. These statistics presented in Table 1 illustrate that heat flow varies between Archean subprovinces.

The Superior Province. The Superior province is the largest of the Archean provinces and it is often considered the archetype of an Archean craton. The heat flow data have been analyzed to determine the scales of their variability [*Mareschal & Jaupart*, 2004]. Differences in composition between the main belts of the Superior are reflected in the heat flow. For instance, the Abitibi greenstone belt has a lower heat flow than the entire Superior Province (37 vs 42 mW m^{-2}). The trend of lower heat flow in the greenstone belts than in the meta-sedimentary belts is confirmed in the western part of the Superior Province where the heat flow is lower in the

greenstone than in meta-sedimentary belts [*Rolandone et al.*, 2003]. Conventional methods of analysis such as the linear heat flow heat production relationship do not apply in the Superior province and crustal heat production has been estimated from extensive sampling of the main rock units. Within the individual belts of the Superior Province, there are also important variations. For instance, the heat flow increases from 28 mW m^{-2} to 45 mW m^{-2} between the eastern and the western Abitibi [*Pinet et al.*, 1991] resulting in ≈150 K temperature differences at Moho depth. After removing the crustal contribution, mantle heat flow was estimated to be 11-15 mW m^{-2} throughout the Canadian shield [*Jaupart & Mareschal*, 1999]. *Rudnick & Nyblade* [1999] have suggested a higher mantle heat flow (≈18 mW m^{-2}) in the Superior from the P-T array for the Kirkland Lake xenoliths.

Kaapvaal Province. The mean heat flow in the Kaapvaal craton appears higher than that of the Superior (51 vs 42 mW m^{-2}) and that of the other Archean Provinces. This result must be qualified because most of the heat flow values have been estimated from bottom hole temperature in the Witwatersrand Basin where the contribution of the sediments to the heat flow was calculated to be 7 mW m^{-2} [*Jones*, 1988]. After removing the effect of the sediments, the mean heat flow for the Kaapvaal is only marginally higher than in the Superior Province. The few heat flow values outside the Witwatersrand Basin are low (33 mW m^{-2}) although they were obtained on granite domes with a high heat production. On the southern edge of the Archean craton, the high heat flow anomaly (61 mW m^{-2}) observed in Lesotho is accounted for by high crustal heat production [*Jones*, 1992]. Crustal heat production for the Kaapvaal craton was estimated from sampling in the Vredefort structure

Table 1. Mean Heat Flow and Heat Production in Different Archean Provinces in the World. $\langle Q \rangle$ is the average heat flow ± the standard error, σ_Q is the standard deviation, N_Q number of sites, $\langle A \rangle$ is the average heat production, σ_A the standard deviation, N_A the number of sites where samples were analyzed.

	$\langle Q \rangle$ mW m^{-2}	σ_Q	N_Q	$\langle A \rangle$ μW m^{-3}	σ_A	N_A
Superior Province	42 ± 2.0	12	57	0.95 ± 0.15	1.	44
(excl. Abitibi)	45 ± 2.4	12	31	1.4 ± 0.26	1.2	22
Abitibi	37 ± 1.0	7	26	0.41 ± 0.07	0.33	22
Western Australia	39 ± 2.1	7	27	2.7	2.5	13
Yilgarn Block	39 ± 1.5	7	23	2.7	2.5	13
Pilbara Block	43 ± 1.5	3	4			
South Africa						
Kaapvaal granite domes	33 ± 0.8	2	7	2.3	0.8	7
Witwatersrand	45 ± 1.8	6	10			
(Witwatersrand)	51 ± 0.7	6	81[†]			
Lesotho	61 ± 5.3	16	9			
Siberian Shield	38 ± 1.2	14	143			
Baltic Shield	34 ± 1.5	12	60			
Ukraine	36 ± 2.4	8	12	0.9	0.2	7
Dharwar craton (India)	36 ± 2.0	6	8			

[†]Including heat flow estimates from bottom hole temperature.

[*Nicolaysen et al.*, 1981]. After removing the crustal contribution, the mantle heat flow was estimated to be 18 mW m^{-2}, slightly higher than in the Superior Province. An independent estimate of the mantle heat flow for the Kaapvaal craton was obtained from geothermobarometry on mantle xenoliths. It has yielded a value of 15 mW m^{-2} slightly less than the estimate above [*Rudnick & Nyblade*, 1999; *Michaut & Jaupart*, 2004].

Western Australian craton: Yilgarn and Pilbara. The western Australian craton contains two distinctive blocks: the Yilgarn and Pilbara. Heat flow is slightly lower in the Yilgarn than in the Pilbara (39 vs 43 mW m^{-2}) but the average surface heat production is high for the Yilgarn (2.5μW m^{-3}) with some extremely high local values (> 8μW m^{-3}) [*Cull*, 1991]. Because of this high heat production in the shallow crust, the Yilgarn is one of the few Archean provinces where the linear heat flow heat production relationship appears verified. The low heat flow and the small slope in the linear relationship (D = 4 km) imply that the layer enriched in radio-elements is thin.

Dharwar craton (India). In the Dharwar craton, where average heat flow is 36 mW m^{-2}h *Roy and Rao* [2000] have estimated the crustal contribution from systematic sampling and derived the mantle component to be 11 mW m^{-2}.

Eurasia: Baltic, Ukrainian, and Siberian cratons. These three Archean cratons in the Eurasian plate have lower than average heat flow. Heat flow values, after correction for Holocene climate variations remain very low particularly in the Baltic and Siberian Shield. Heat flow values are low throughout the Siberian craton, with very low values (15-25 mW m^{-2}) consistently observed in the area surrounding the Anabar massif [*Duchkov*, 1991]. In the Baltic Shield, even if some suspiciously low heat flow values [*Kukkonen et al.*, 1998] are not included, the average heat flow is low (Table 1). Geothermobarometry on xenoliths from the Baltic Shield yield a low value (11 ± 4 mW m^{-2}) for the mantle heat flow [*Kukkonen & Peltonen*, 1999]. In the Ukrainian Shield, the heat flow shows regional variations from < 30 to > 50 mW m^{-2}, reflecting the variability of crustal heat production [*Galushkin et al.*, 1991].

Regional variations. In order to determine heat flow variations on the scale of lithospheric thickness, we have applied an averaging window of 2° × 2° to the distribution of heat flow values in the regions where sufficient data are available. The results in Table 2 show that regional averages can be very low (< 20 mW m^{-2}) and that the long wavelength variability of heat flow is ≈25 mW m^{-2} within each province. This variability is accounted for by the variations in crustal heat production [*Jaupart & Mareschal*, 1999; *Mareschal & Jaupart*, 2004].

2.2. Heat Production in the Archean Crust

A crude estimate of the heat production of the bulk Archean crust can be made by removing the mantle heat flow

Table 2. Regional Variations of the Heat Flow in Different Cratons. Minimum and maximum values obtained by averaging over 200 km × 200 km windows.

	minimum mW m^{-2}	maximum mW m^{-2}
Superior	22	48
Australia	34	54
Baltic	15	39
Siberia	18	46

from the mean surface heat flow. The area weighted average heat flow for all Archean Provinces is 41 mW m^{-2}. Estimates of the mantle heat flow by various authors vary between 11 and 18 mW m^{-2} [*Jaupart & Mareschal*, 1999]. After removing the mantle heat flow, the average crustal component of the surface heat flow in Archean Provinces is between 23 and 30 mW m^{-2}. This implies that the average heat production for the bulk Archean crust is 0.65 ± 0.1 μW m^{-3} [*Jaupart & Mareschal*, 2004]. This is within the range expected from global geochemical models [*Allégre et al.*, 1988]. There are large variations in crustal heat production between cratons and within each craton.

2.3 Variability of the Crustal Heat Production

In the Superior province where the total contribution from the crust to the heat flow was estimated 26-31 mW m^{-2}, the mean crustal heat production is 0.68-0.76 μ W m^{-3}. In the Kaapvaal craton, with mantle heat flow 18 mW m^{-2}, crustal heat production is in the same range. In Siberia and the Baltic Shield, the average heat flow is low, local values in the 15 mW m^{-2} range have consistently been obtained, and regional (2° × 2°) avrages can be as low as 15-18 mW m^{-2}. In these regions, the mantle heat flow can not exceed 10–11 mW m^{-2}, and the regionally averaged crustal heat production is extremely low < 0.25 μW m^{-3}. After removing the mantle component from the surface heat flow, the averaged heat production of the crust is estimated to be between 0.55 μW m^{-3} and 0.7 μW m^{-3} for the whole Baltic and Siberian Shield.

In addition to heat production measurements at heat flow sites, systematic sampling of the heat production in the crust has been done in several Archean provinces (Table 3). In several Archean regions (Yilgarn craton, Australia; Slave Province, Canada; Lesotho belt, South Africa) high heat generation anomalies have been detected by surface sampling. In western Australia, the low average surface heat flow requires that the enriched layer is very thin. In Lesotho, the average surface heat flow is markedly higher than in the adjacent Kaapvaal craton and implies that the whole crust is enriched. Because the Lesotho belt is narrow, mantle temperatures are

Table 3. Average Surface Heat Production for Different Geological Provinces Obtained by Systematic Regional Sampling Over Large Areas. $\langle A \rangle$ is the average heat production, σ_A is the standard deviation, N is the number of samples.

	$\langle A \rangle \; \mu$W m^{-3}	σ_A	N	References
Yilgarn (Archean, Western Australia)	3.3	3.3	540	*Heier & Lambert* [1978]
Superior (Archean, Canada)				
New Québec region	1.22	†	3085	*Eade & Fahrig* [1971]
Wawa Gneiss Terrance	1.01	‡	56	*Shaw et al.* [1994]
Sachigo subprovince	1.04	‡	20	*Fountain et al.* [1987]
Slave (Archean, Canada)				
Yellowknife region	2.3	‡	85	*Thompson et al.* [1995]
Winter-Lac de Gras region	1.7	‡	85	*Thompson et al.* [1995]

†Analyses were made on mixed powders, implying that the standard deviation of the analyses underestimates the true spread of values for individual rock samples.

‡Average calculated by weighting according to the abundances of the different rock types.

only marginally higher than in surrounding subprovinces. In the Slave Province, few heat flow measurements are available [*Mareschal et al.*, 2004] but they are consistent with the high values of heat production that have been measured [*Thompson et al.*, 1995]. In this province, the bulk crustal heat production is higher than that of the average Archean crust. Surface heat flow thus reflects the variability of the average crustal heat production in the Archean crust that ranges from < 0.2 to > 1.1 μ W m^{-3}.

3. THE ARCHEAN THERMAL REGIME

3.1. Crustal Heat Production

Crustal heat production is less today than it was during the Archean because of the rundown of the radio-elements. For a fixed present ratio of the heat producing elements (Th/U = 4; K/U = 12,000), the past crustal heat production can be calculated (Figure 1). Crustal heat production was double the present value at 2.55 Ga and triple at 3.55 Ga. The mantle heat flow component during the Archean can not be directly constrained. Nevertheless, because crustal heat production

accounts for most of the present heat flow, the average heat flow during the Archean was 20–50 mW m^{-2} higher than present. The differences in mantle temperatures were not proportional to heat flow because the shallower the heat sources, the smaller the increase in temperature.

3.2. Temperature in the Lithosphere During the Archean

The differences in regional heat flow and heat production were larger during the Archean than at present, implying differences in lithospheric temperatures. Crustal heat production models are needed to calculate the temperature in the lithosphere. We shall first consider two end models of heat source distribution from the Canadian Shield: the eastern Abitibi, and the Slave Province. The present heat flow (28 mW m^{-2}) is lower than average in the eastern Abitibi. The Slave Province is unusual because of its higher than average heat flow ($\approx$50 mW m^{-2}) due to the very high heat generation in the upper crust [*Thompson et al.*, 1995]. For each region, we have calculated two geotherms with two different values for the mantle heat flow: same as today (13 mW m^{-2}) or 50% higher than today (20 mW m^{-2}), yielding a surface heat flow of 45 (52)mW m^{-2} for the eastern Abitibi and 87 (94)mW m^{-2} for the Slave Province. The vertical distribution of heat production for both regions is detailed in Table 4. Over the range of crustal temperatures, the thermal conductivity can vary by as much as 50%. We have thus used the following law for the thermal conductivity of the crust [*Durham et al.*, 1987]:

$$K = 2.26 - \frac{618.241}{T} + K_0 \left(\frac{255.576}{T} - 0.30247 \right) \quad (2)$$

where K is thermal conductivity (in W m^{-1} K^{-1}), T is the absolute temperature and K_0 is the thermal conductivity at the surface (for $T = 273$ K). When calculated with temperature dependent conductivity, Moho temperatures are $\approx$150 K higher than for constant conductivity. Present Moho

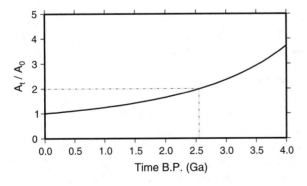

Figure 1. Variation in crustal heat production with time for Th/U = 4 and K/U = 12,000. Crustal heat production is relative to present.

Table 4. Heat Production Models for the Archean Crust. H: layer thickness, A_0: present heat production, $A_{2.6}$: Archean heat production, Q_0: present surface heat flow, $Q_{2.6}$: Archean surface heat flow (with 20 mW m^{-2} mantle heat flow).

	H km	A_0 μW m^{-3}	$A_0 \times H$ mW m^{-2}	Q_0 mW m^{-2}	$A_{2.6}$ μW m^{-3}	$A_{2.6} \times H$ mW m^{-2}	$Q_{2.6}$ mWm^{-2}
Eastern Abitibi							
	40	0.4	16		0.8	32	
				29			45 (52)
Western Abitibi							
	20	1.2	24		2.4	48	
	20	0.4	8		0.8	16	
				45			77 (84)
Slave (Lac de Gras)							
	10	1.7	17		3.4	34	
	10	1.2	12		2.4	24	
	20	0.4	8		0.8	16	
				50			87 (94)

temperatures calculated with conductivity following equation 2 are consistent with P_n velocities observed in Shield areas [*Shapiro & Ritzwoller*, 2004]. In the mantle, we include a radiative component to the thermal conductivity [*Schärmeli*, 1979]:

$$K_r = 0.37 \times 10^{-9} T^3 \qquad (3)$$

The resulting steady-state geotherms (Figure 2) show that mantle temperatures remain low for all the models without increase in mantle heat flow. The Moho temperature varies between 500 and 700°C. Regardless of the assumed mantle heat flow, steady state temperatures are always below melting temperature at the base of the crust. With mantle heat flow 20 mW m^{-2}, Moho temperatures are 100 K higher than with 13 mW m^{-2}, but the temperature difference between the two models increases in the mantle, and the lithosphere is thin (< 100 km) when mantle heat flow is high. High mantle heat flow does not result in a hot crust and cold mantle. For this reason, several authors have invoked a thermal pulse that affected the lower crust but not the mantle [e.g. *Davis et al.*, 2003].

We shall now examine whether high temperature in the lower crust can be accounted for by heat generation only. Because of the high heat production in the Slave crust during the Archean, Moho temperature was extremely sensitive to the vertical distribution of the radio-elements and to crustal thickness. For instance, with the same crustal heat production (i.e. same surface and mantle heat flow) and crustal thickness, the Moho temperature is 200 K higher with the radio-elements uniformly distributed in the crust than with the upper crust enriched in radio-elements. A comparable increase in Moho temperature is obtained by thickening the Slave crust by 25% from 40 to 50 km. Uniform vertical distribution of radio-elements in the crust or a thickened crust bring the lower crust near the melting conditions. In steady-state, mantle temperatures are raised by the same amount as the lower crust (Figure 3).

These steady-state geotherms do not consider the time needed for equilibrium to be reached. Crustal radioactivity heats the crust in a geologically short time, but a much longer time is required to affect the lower lithosphere. If the lithospheric mantle is formed by the under-thrusting of subducted slabs beneath the crust, it will initially be colder than in steady state. Because the time for thermal equilibrium in a thick lithosphere is of the same order as the half-life of some of radio-elements, the crustal radioactivity will have decreased by ≈30% before the lithosphere has been heated [*Michaut & Jaupart*, 2004]. In the Appendix, we determine how the temperature at the base of the lithosphere varies in response to crustal radioactivity (decreasing with time). The calculations show that because of the time delay and the rundown of the radio-elements, the maximum temperature increase at the base of the lithosphere will be 70% that expected for steady state with the initial crustal heat production. The mantle is thermally decoupled from the crust unless the lithospheric mantle stabilized with its temperature in quasi-steady state, which is implausible. If the root forms with its initial temperature below steady-state, the mantle temperature will never be as elevated as implied by the high crustal production.

3.3. Rheology of the Crust and Mantle

It has been noted that the Archean Moho seems to be flat [*Cook et al.*, 1999; *James et al.*, 2003]. The temperature profiles are useful to determine the strength of the Archean lithosphere and the conditions for relaxation of crustal roots. Crustal rocks usually deform by power law creep [*Ranalli*, 1995]:

$$\dot{\varepsilon} = A\sigma^n \exp(-(E + PV^*)/RT) \qquad (4)$$

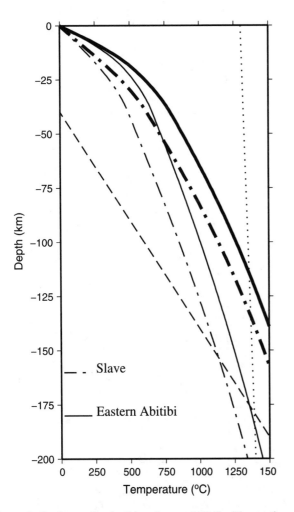

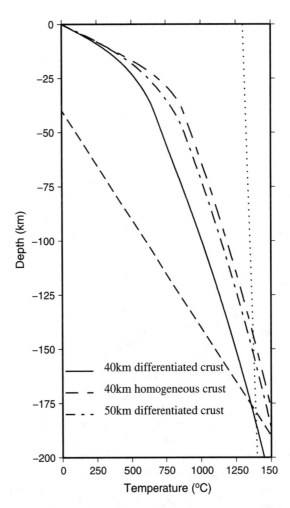

Figure 2. Geotherms for the lithosphere at 2.55 Ga. The geotherms are calculated for the crustal production of the Slave and and eastern Abitibi at 2.55 Ga, mantle heat flow 13 W m^{-1} K^{-1} (thin lines) and 20 mW m^{-2} (thick lines), thermal conductivity (given by equation 2) is 2.8 W m^{-1} K^{-1} at the surface. The dotted line is the 1300°C adiabat. The dashed line represents the graphite-diamond transition.

Figure 3. Effect of a different distribution of radioelements and of crustal thickening on lithospheric geotherms at 2.55 Ga. The geotherms are calculated for a vertical distribution of radio-elements similar to the present in the Slave Province back calculated at 2.55 Ga and with mantle heat flow 13m W m^{-2}, with the radio-elements distributed uniformly throughout the crust, or with a 50 km thick crust. For the crustal thickening model, the crust with vertical radio-elements profile is uniformly thickened from 40 to 50 km. The dotted line is the 1300°C adiabat. The dashed line represents the graphite-diamond transition.

Where $\dot{\varepsilon}$ is the strain rate, σ the deviatoric stress, A and n are constants characteristic of the material, E the activation energy, V^* the activation volume, R the gas constant and T thermodynamic temperature. Usually, the strength is defined as the stress required to maintain a fixed rate of deformation, typically 10^{-15}s^{-1}. We have used rheological parameters corresponding to dry rheologies for the crust and mantle (Table 5). We have also calculated an "effective viscosity" μ_{eff} for a fixed deviatoric stress of 10 MPa.

$$\mu_{eff} = A^{-1}\sigma^{1-n}\exp\left((E + PV^*)/RT\right) \qquad (5)$$

The effective viscosity can be used to determine the order of a time of relaxation for topography on an interface between

two layers with a density difference $\Delta\rho$ [*Chandrasekhar*, 1961]:

$$\tau \approx \frac{8\pi\mu_{eff}}{g\Delta\rho\lambda} \qquad (6)$$

where g is the acceleration of gravity and λ is the wavelength of the interface topography. Equation 6 yields only an order of magnitude because the relaxation time is modulated by a function depending on the geometry and the boundary

Table 5. Creep Parameters for Lithospheric Materials Used in Calculating the Strength of the Lithosphere [*Ranalli* 1995; *Carter and Tsenn* 1989].

	A (MPa^{-n} s^{-1})	n	H (kJ mol^{-1})
upper crust			
(dry granite)	$1. \times 10^{-7}$	3.2	144.
lower crust			
(mafic granulites)	1.4×10^4	4.2	445.
mantle			
(dry dunite)	$3. \times 10^4$	3.6	535.

conditions [*Chandrasekhar*, 1961]. For a differentiated 40 km crust, the relaxation time is larger than 1 Gy. Strength profiles for a 40 km thick crust in the eastern Abitibi and the Slave are shown on Figure 4. Table 6 gives the effective viscosity for a stress of 10 MPa, the size of the deviatoric stress caused by variations in crustal thickness under isostatic conditions [*Jeffreys*, 1959; *Artyushkov*, 1972]. If the temperature was close to steady-state, the effective viscosity of the crust and mantle would have been sufficiently high ($\approx 10^{24}$ Pa s) at the end of the Archean to inhibit lower crustal flow in response to stresses due to density heterogeneities or differences in crustal thickness. The effective viscosity of the mantle is $>10^{21}$Pa s above 100 km. Because of the nonlinearity of the creep law, the effective viscosity drops by 2-3 orders of magnitude when the stress increases by one order. Nevertheless, it appears that if the thermal regime was near steady-state with the mantle heat flow and the vertical distribution of radio-elements the same as today, a 40 km thick dry crust was strong at the end of the Archean.

Figure 5 shows that, if the crust was thicker than 40 km or if the radio-elements had not been re-mobilized in the upper crust, the lower crust was weak and the lithosphere was thin. With a uniform distribution of the heat generating elements within the crust, the effective viscosity of the lower crust at 40 km depth is 10^{19}Pa s for 10 MPa deviatoric stress (Table 7). Thickening of the crust by 10 km yields an effective viscosity 10^{19}Pa s at Moho depth. The strength profiles in the mantle are almost identical with a thick crust or a uniform crustal heat generation, the effective viscosity reaching 10^{21}Pa s at 60-70 km depth. For given surface heat flow, the vertical distribution of the radiogenic elements is one key factor controlling the strength of the lithosphere and the maximum thickness that the crust can maintain without spreading laterally.

Because of the long thermal relaxation time for the lithosphere, the steady state calculations give only a lower bound for the strength of the mantle. We show in the appendix that, with "cold" initial conditions for the root, the lithospheric mantle was colder than steady state during the Archean, and

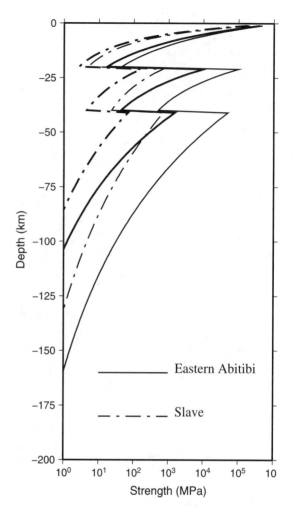

Figure 4. Strength of the lithosphere at 2.55 Ga for the temperature profiles in Figure 2. Stress required to maintain a strain rate $\dot{\varepsilon} = 10^{-15} s^{-1}$. The rheological parameters are those for dry granite, mafic granulite, and dry dunite for the upper crust, lower crust, and mantle, respectively. Values for the creep parameters are given in Table 5. Thin lines 13 mW m^{-2} mantle heat flow, thick lines 20 mW m^{-2}.

Table 6. Temperature and Rheology in the Archean Lithosphere. Temperature is calculated for varying thermal conductivity in the crust and mantle (equations 2 and 3). The effective viscosity for 10 MPa deviatoric stress is calculated for mafic granulites in the lower crust and dry dunite in the mantle. Two values for Moho heat flow are used with each province.

	Eastern Abitibi		Slave Province	
q_m (mW m^{-2})	13	20	13	20
Moho temperature (°C)	490	620	660	790
μ_{eff} (Pa s)lower crust	10^{29}	10^{25}	10^{24}	10^{21}
μ_{eff} (Pa s) mantle	10^{34}	10^{30}	10^{29}	10^{25}
150 km temperature (°C)	1120	1460	1250	1560
$\mu_{eff} = 10^{21}$Pa s	125 km	82 km	98 km	64 km

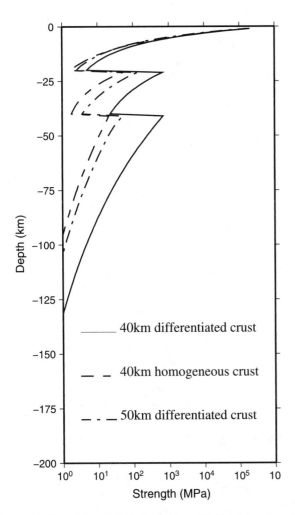

Figure 5. Effect of the vertical distribution of radio-elements and of crustal thickening on the strength of the lithosphere at 2.55 Ga for the temperature profiles in the Slave Province with mantle heat flow 13 mW m^{-2} (Figure 3). Stress required to maintain a strain rate $\dot{\varepsilon} = 10^{-15}\,\mathrm{s}^{-1}$. The rheological parameters are those for dry granite, mafic granulite, and dry dunite for the upper crust, lower crust, and mantle, respectively. Values for the creep parameters are given in Table 5.

temperatures in the mantle might have been higher during the Proterozoic than at the end of the Archean.

4. DISCUSSION

There is an apparent inconsistency in the geological record of the Archean thermal regime. There is good evidence that the lower crust was hot at the end of the Archean. There is also evidence that the lithosphere was thick and that the cratonic roots have preserved Archean features, suggesting that the mantle was cold. Heat flow and heat production data from Archean provinces are not sufficient to determine the

Table 7 Temperature and Rheology for the Slave Province During the Archean With Vertical Distribution of Radio-Elements in a 40 km Crust, in a 50 km Crust, and With the Radio-Elements Uniformly Distributed in a 40 km Crust. Temperature is calculated for thermal conductivity given by equations 2 and 3 in the crust and mantle and mantle heat flow 13 mW m^{-2}. The effective viscosity for 10 MPa deviatoric stress is calculated for mafic granulites in the lower crust and dry dunite in the mantle.

	40 km	50 km	40 km (uniform)
Moho temperature (°C)	514	880	875
μ_{eff} (Pa s) lower crust	10^{24}	10^{19}	10^{19}
μ_{eff} (Pa s) mantle	10^{29}	10^{23}	10^{23}
150 km temperature (°C)	1250	1360	1400
$\mu_{eff} = 10^{21}$Pa s	98 km	70 km	61 km

Archean thermal regime but they provide some important parameters for thermal models. Steady-state thermal models can be used to estimate upper bounds for the temperature and lower bounds for the strength of the lithosphere at the end of the Archean, but it must be emphasized that the Archean lithosphere could not have been in steady state. Because the thermal time constant for a thick lithosphere is > 1 Gy and on the same order as the half-life of crustal radioactivity, it is still not in steady state today [*Michaut & Jaupart*, 2004].

Crustal structure and composition indicate a hot lower crust at the end of the Archean. Seismic reflection profiles show that the Archean Moho is flat in the Canadian Shield [*Cook et al.*, 1999] and in the Kaapvaal craton [*James et al.*, 2003]. This suggests that the lower crust was ductile and could flow at geological time scale. Another evidence for elevated Moho temperatures is provided by the abundance of late granite intrusions, for example in the Superior Province [*Davis et al.*, 2000]. In the Kaapvaal craton, there is also evidence for a late Archean Ventersdorp thermal event when lower crustal flow and re-equilibration of the Moho could have occurred [*Schmitz & Bowring*, 2003]. The present structure of the crust was achieved after the final episode of granite genesis and emplacement in the upper crust and, obviously, is not relevant to the thermal structure of the crust before.

The present vertical distribution of radio-elements implies that the temperatures at the base of a 40 km crust were not high enough for the lower crust to flow. The steady-state models examined above show that there are two ways to have a hot crust without invoking an input of heat from the mantle: crustal thickening, or redistributing the radio-elements through the crust. In regions with high crustal heat generation, crustal thickening to about 50 km gives lower crustal temperatures sufficiently high for a crustal root to relax in a short time. Alternatively, with the heat producing elements uniformly distributed within the crust, temperature was sufficiently high to bring the lower crust very near melting conditions. Following the emplacement of granite intrusions and the subsequent re-distribution of the radio-elements in the crust, the steady state temperature decreased by ≈200 K at Moho depth. This re-equilibration could have coincided with the removal of the dense mafic residue left after the extraction of the silicic rocks from the lower crust [Kay & Kay, 1993; Jull & Kelemen, 2001].

There is good evidence that the age of the cratonic roots is Archean [Moser et al., 2001; Carlson et al., 2005]. Also, deep seismic reflection profiles have shown Archean features in the mantle beneath the cratons [Calvert et al., 1995; Bostock, 1998] and their preservation requires that the mantle down to >150 km does not flow on geological time scales. Inclusions of Archean age have been identified in diamonds, implying that thermal conditions appropriate for diamond stability were established relatively early, perhaps 100-200 My after the last major tectono-metamorphic event recorded in the crust [Richardson et al., 2001]. All these data suggest that the lithospheric mantle was cold. In steady state, this cannot be reconciled with a hot crust. If the lithospheric root is emplaced cold, the time needed for crustal radioactivity to heat the lithospheric mantle is such that crustal heat production decreased before the cratonic root in the mantle was heated up. It has been argued, on the basis of steady state thermal models, that the lithospheric mantle could not be in the diamond stability field during the Archean and that the diamonds formed during the Proterozoic [Thompson et al., 1995; Morgan, 2004]. Transient calculations show that thermal models can not rule out an Archean age for the diamonds. These calculations also show that, in transient regime, the high surface heat flow during the Archean in provinces such as the Slave (90 mW m^{-2} at 2.55 Ga) does not preclude the mantle to be cold and strong.

APPENDIX: TRANSIENT EFFECTS

In this section, we shall estimate the time needed for (time dependent) crustal radioactivity to heat up the entire lithosphere. The temperature depends on initial conditions, which we do not know, but the time needed to reach a pseudo steady state does not. So we shall consider the one dimensional heat equation in the crust and lithospheric mantle:

$$\frac{1}{\kappa}\frac{\partial T_c}{\partial t} = \frac{\partial^2 T_c}{\partial z^2} + \frac{H\exp(-\lambda t)}{K} \quad 0 < z < h \quad (A.1)$$

$$\frac{1}{\kappa}\frac{\partial T_m}{\partial t} = \frac{\partial^2 T_m}{\partial z^2} \quad h < z < L \quad (A.2)$$

where κ is the thermal diffusivity, K is thermal conductivity, H is the volumetric heat production. The boundary conditions are $T_c = 0$ at the surface and $\partial T/\partial z = 0$ at the base of the lithosphere. Heating the lithosphere from below is a distinct problem which has been addressed by many studies and it can be calculated separately.

The solution is found with the Laplace transform of the equations above:

$$\frac{s}{\kappa}T_c = \frac{d^2T_c}{dz^2} + \frac{H}{K(s+\lambda)} \quad 0 < z < h \quad (A.3)$$

$$\frac{s}{\kappa}T_m = \frac{d^2T_m}{dz^2} \quad h < z < L \quad (A.4)$$

where s is variable of the Laplace transform defined by:

$$T_i(s,z) = \int_0^\infty \exp(-st)T_i(t,z)dt \quad (A.5)$$

$$T_i(t,z) = \frac{1}{2\pi i}\int_{\gamma-i\infty}^{\gamma+i\infty} T_i(s,z)\exp(st)ds \quad (A.6)$$

with $i = c$ or m, and γ is a real number such that the contour of integration lies to the right of all the singularities of the Laplace transform in the complex s plane. The solution for the mantle temperature is obtained as:

$$T_m(s,z) = \frac{H\kappa}{sK(s+\lambda)}\frac{\cosh\sqrt{\tau s}(z-L)/L}{\cosh\sqrt{\tau s}}$$
$$\times\left(\cosh\left(\sqrt{\tau s}\frac{h}{L}\right) - 1\right) \quad (A.7)$$

where $\tau = L^2/\kappa$. The inverse Laplace transform gives:

$$T_m(z,t) = \frac{HL^2}{\lambda\tau K}\exp(-\lambda t)\frac{\cos(\sqrt{\lambda\tau}(z-L)/L)}{\cos(\sqrt{\lambda\tau})}$$
$$\times\left(1 - \cos\left(\sqrt{\lambda\tau}\frac{h}{L}\right)\right)$$
$$- \frac{4HL^2}{\pi^2 K}\sum_{n=0}^\infty \frac{(-)^n}{(2n+1)}\frac{1}{(2n+1)^2\pi/4 - \lambda\tau}$$
$$\times(1 - \cos\left((2n+1)\frac{\pi h}{2L}\right)$$
$$\times\cos\left((2n+1)\frac{\pi(z-L)}{2L}\right)$$
$$\times\exp\left(\frac{-(2n+1)^2\pi^2 t}{4\tau}\right) \quad (A.8)$$

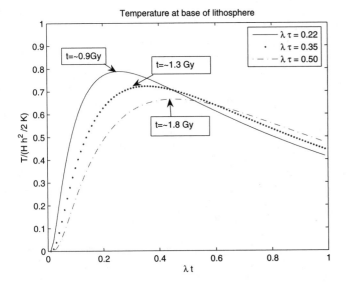

Figure A1. Temperature variation at the base of the lithosphere following heating by crustal radioactivity. Temperature is divided by the steady state temperature that would be observed when the root is emplaced. Time is multiplied by the average decay constant of crustal radio-activity $\lambda \approx 2.7 \times 10^{-10} Gy^{-1}$. The characteristic time $\tau = L^2/\kappa \approx 1.3$ Gy for a 200 km thick lithosphere.

Figure A1 shows the time variation of the crustal radioactivity component of the temperature at the base of the lithosphere after the emplacement of the root. The self-evident conclusion is that, when the half-life of crustal radioactivity is of the same order as the thermal time of the lithosphere, the temperature increase at the base of the lithosphere is less than the steady state. The temperature at the base of the lithosphere reaches a maximum after 1-2 Gy, depending on lithospheric thickness and the peak temperature is ≈70% what one would infer from steady state models.

Acknowledgments. The authors are grateful to Norm Sleep, co-editor Kent Condie, and an anonymous reviewer for their constructive comments. This work was supported by NSERC (Canada) and INSU (CNRS, France).

REFERENCES

Abbott, D., The case for accretion of the tectosphere by buoyant subduction, *Geophys. Res. Lett.*, 18, 585-588, 1991.

Allégre C.J., E. Lewin, and B. Dupré, A coherent crust-mantle model for the Uranium-Thorium-Lead isotopic system. *Chem. Geol.*, 70, 211-234, 1988.

Artyushkov, E.V., Convective instability in geotectonics, *J. Geophys. Res.*, 76, 1197-1211, 1971.

Birch F., R.F. Roy, and E.R. Decker, Heat flow and thermal history in New England and New York. In *Studies of Appalachian Geology* (ed. E. An-Zen), pp. 437-451. Wiley-Interscience, 1968.

Bostock, M.G., Mantle stratigraphy and evolution of the Slave Province, *J. Geophys. Res.*, 103, 21,183-21,200, 1998.

Bostock, M.G., Seismic imaging of lithospheric discontinuities and continental evolution, *Lithos*, 48, 1-16, 1999.

Boyd, F.R., J.J. Gurney, and S.H. Richardson, evidence for a 150-200 km thick Archean lithosphere from diamond inclusions thermobarometry, *Nature*, 315, 387-389, 1985.

Calvert, A.J., E.W. Sawyer, W.J. Davis, J.N. Ludden, Archean subduction inferred from seismic images of a mantle suture in the Superior Province, *Nature*, 375, 670-673.

Carlson, R.W., D.G. Pearson, D.E. James, Physical, chemical, and chronological characteristics of continental mantle, *Rev. Geophys.*, 43, RG1001, doi:10.1029/2004RG000156, 2005.

Carter, N.L., and M.C. Tsenn, Flow properties of continental lithosphere, *Tectonophys.*, 136, 27-63, 1987.

Chandrasekhar, S., *Hydrodynamic and hydromagnetic stability*, Oxford University Press, pp. 654., 1961.

Cook, F.A., A.J. van der Velden, K.W. Hall, and B.J. Roberts, Frozen subduction in Canada's Northwest Territories; Lithoprobe deep lithospheric reflection profiling of the western Canadian Shield. *Tectonics*, 18, 1-24, 1999.

Cull, J.P., Heat flow and regional geophysics in Australia, in *Terrestrial heat flow and the lithosphere structure*, edited by V. Cermak and L. Rybach. Springer-Verlag. 486-500, 1991.

Davis, W.J., S. Lacroix, C. Gariépy, and N. Machado, Geochronology and radiogenic isotopes geochemistry of plutonic rocks from the central Abitibi subprovince: significance to the internal subdivision and plutono-tectonic evolution of the Abitibi belt, *Can. J. Earth Sci.*, 37, 117-133, 2000.

Davis, W.J., A.G. Jones, W. Bleeker, and H. Grütter, Lithosphere development in the Slave craton: a linked crustal and mantle perspective, *Lithos*, 71, 575-589, 2003.

Duchkov, A.D., Review of Siberian heat flow data. in *Terrestrial heat flow and the lithosphere structure*, edited by V. Cermak and L. Rybach. Springer-Verlag. 426-443, 1991.

Durham, W.B., V.V. Mirkovich, and H.C. Heard, Thermal diffusivity of igneous rocks at elevated pressure and temperature, *J. Geophys. Res.*, 92, 11,615-11,634, 1987.

Eade K.E., and W.F. Fahrig, Geochemical evolutionary trends of continental plates, A preliminary study of the Canadian Shield. *Geol. Surv. Canada Bull.*, 179, 1-59, 1971.

Fountain D.M., M.H. Salisbury, and K.P. Furlong, Heat production and thermal conductivity of rocks from the Pikwitonei-Sachigo continental cross section, central Manitoba: Implications for the thermal structure of Archean crust. *Can. J. Earth Sci.*, 24, 1583-1594, 1987.

Galushkin, Y.I., R.I. Kutas, and Y.B. Smirnov, Heat flow and analysis of the thermal structure of the lithosphere in the European part of the USSR. in *Terrestrial heat flow and the lithosphere structure*, edited by V. Cermak and L. Rybach. Springer-Verlag. 206-237, 1991.

Griffin, W.L., S.Y. O'Reilly, N. Abe, S. Aulbach, R.M. Davies, N.J. Pearson, B.J. Doyle, and K. Kivi, The origin and evolution of Archean lithospheric mantle, *Precambrian Research*, 127, 19-41, doi:10.1016/S0301-9268(03)00180-3, 2003.

Heier K.S., and I.B. Lambert, A compilation of potassium, uranium and thorium abundances and heat production of Australian rocks. *Technical Report*. Research School of Earth Science, Australian National University, Canberra, 1978.

James, D.E., F. Niu, and J. Rokosky, Crustal structure of the Kaapvaal craton and its significance for early crustal evolution, *Lithos*, 71, 413-429, doi:10.1016/j.lithos.2003.07.009, 2003.

Jaupart C., J.C. Mareschal, L. Guillou-Frottier, and A. Davaille, Heat flow and thickness of the lithosphere in the Canadian Shield, *J. Geophys. Res.*, 15,269-15,286, 1998.

Jaupart, C., and J.C. Mareschal, The thermal structure and thickness of continental roots, *Lithos*, 48, 93-114, 1999.

Jaupart, C., and J.C. Mareschal, Constraints on heat production from heat flow data, *Treatise on Geochemistry. Volume 3: The crust*. Edited by R. Rudnick. pp. 65-84. Elsevier. 2004.

Jeffreys, H., *The Earth* (4th edition), Cambridge University Press, pp. 438, 1959.

Jessop, A.M., and T.J. Lewis, Heat flow and heat generation in the Superior Province of the Canadian Shield. *Tectonophys.*, 50, 55-77, 1978.

Jones, M.Q.W., Heat flow in the Witwatersrand Basin and environs and its significance for the South African shield geotherm and lithosphere thickness, *J. Geophys. Res.*, 3243-3260, 1988.

Jones, M.Q.W., Heat flow anomaly in Lesotho: implications for the southern boundary of the Kaapvaal craton, *Geophys. Res. Lett.*, 19, 2031-2034, 1992.

Jull M. and P.B. Kelemen, On the conditions for lower crustal convective instability. *J. Geophys. Res.*, 106, 6423-6446, 2001.

Kay, R.W., and S.M. Kay, Delamination and delamination magmatism, *Tectonophysics*, 219, 177-189, 1993.

Kukkonen, I.T., W.D. Gosnold, and J. Safanda, Anomalously low heat flow density in eastern Karelia: a possible paleoclimatic signature, *Tectonophys.*, 291, 235-249, 1998.

Kukkonen, I.T., and P. Peltonen, Xenolith-controlled geotherm for the central Fennoscandian Shield: implications for lithosphere-asthenosphere relations, *Tectonophys.*, 304, 301-315, 1999.

Kutas, R.I., Heat flow, radiogenic heat and crustal thickness in southwest U.S.S.R., *Tectonophys.*, 103, 167-174, 1984.

Mareschal, J.C., and C. Jaupart, Variations of surface heat flow and lithospheric thermal structure beneath the North American craton, *Earth Planet. Sci. Lett.*, 223, 65-77, doi:10.1016/j.epsl.2004.04.002, 2004.

Mareschal, J.C., A. Nyblade, H.K.C. Perry, C. Jaupart, and G. Bienfait, Heat flow and deep lithospheric thermal structure at Lac de Gras, Slave Province, Canada, *Geophys. Res. Lett.*, 31, L12611, doi:10.1029/2004GL020133, 2004.

Michaut C., and C. Jaupart, Nonequilibrium temperatures and cooling rates in thick continental lithosphere, *Geophys. Res. Lett.*, 31, L24602, doi:10.1029/2004GL021092, 2004.

Morgan, P., Crustal radiogenic heat production and the selective survival of ancient continental crust, *J. Geophys. Res.*, 90, Supplement, C561-C570, 1985.

Morgan, P., Evolution of Composition, Thermal Structure, and Resistance to Deformation of Continental Lithosphere Through Time: Why Archean Diamonds Are Probably Proterozoic in Age.

Eos Trans. AGU, 85(47), Fall Meet. Suppl., Abstract T31A-1281, 2004.

Moser, D.E., R.M. Flowers, and R.J. Hart, Birth of the Kaapvaal tectosphere 3.08 billion years ago. *Science*, 291, 465-468, 2001.

Nicolaysen L.O., R.J. Hart, and N.H. Gale N.H., The Vredefort Radioelement Profile extended to supracrustal strata at Carletonville, with implications for continental heat flow, *J. Geophys. Res.*, 86, 10,653-10,661, 1981.

Nyblade A.A., and H.N. Pollack, A global analysis of heat flow from Precambrian terrains: Implications for the thermal structure of Archean and Proterozoic lithosphere, *J. Geophys. Res.*, 98, 12,207-12,218, 1993.

O'Neill, C.J., and L. Moresi, How long can diamonds remain stable in the continental lithosphere?, *Earth Planet. Sci. Lett.*, 213, 43-52, doi: 10.1016/S0012-821X(03)00294-2, 2003.

Pinet, C., C. Jaupart, J-C. Mareschal, C. Gariépy, G. Bienfait and R. Lapointe, Heat flow and structure of the lithosphere in the eastern canadian shield, *J. Geophys. Res.*, 96, 19,941-19,963, 1991.

Polyak, B.G., and Ya B. Smirnov, Relationship between terrestrial heat flow and tectonics of the continents, *Geotectonics*, 4, 205-213, 1968.

Ranalli, G., *Rheology of the Earth*, Chapman Hall, 2nd Ed., London, 413 pp., 1995.

Richardson, S.H., J.J. Gurney, A.J. Erlank, J.W. Harris, Origin of diamonds in old enriched mantle, *Nature*, 310, 198-202, 1984.

Richardson, S.H., S.B. Shirey, J.W. Harris and R.W. Carlson, Archean subduction recorded by Re-Os isotopes in eclogitic sulfide inclusions in Kimberley diamonds. *Earth Planet. Sci. Let.*, 191, 257-266, 2001.

Rolandone, F., C. Jaupart, J.C. Mareschal, C. Gariépy, G. Bienfait, C. Carbonne, and R. Lapointe, Surface heat flow, crustal temperatures and mantle heat flow in the Proterozoic Trans-Hudson Orogen, Canadian Shield, *J. Geophys. Res.*, 107(B12), 2341, doi:10.1029/2001JB000698, 2002.

Rolandone, F., J.C. Mareschal, C. Jaupart, and C. Gosselin, Heat Flow in the Western Superior Province of the Canadian Shield. *J. Geophys. Res.*, 1637, doi: 10.1029/2003GL017386, 2003.

Roy, S., and R.U.M. Rao, Heat flow in the Indian Shield, *Geophys. Res. Lett.*, 1005, 25,587-25,604, 2000.

Rudnick, R.L., and A.A. Nyblade, The thickness and heat production of Archean lithosphere: constraints from xenolith thermobarometry and surface heat flow. In Y. Fei, C.M. Bertka and B.O. Mysen (eds) *Mantle Petrology: Field Observations and High Pressure Experimentation: A Tribute to Francis R. (Joe) Boyd*, pp. 3-12. The Geochemical Society, St-Louis, MO, 1999.

Sandiford, M., and M. Hand, Australian Proterozoic high-temperature, low-pressure metamorphism in the conductive limit, in *What drives metamorphism and metamorphic reactions?*, edited by P.J. Treloar and P.J. O'Brien, *Geol. Soc. London Spec. Publ.*, 138, pp. 109-120, 1998.

Schärmeli, G., Identification of radiative thermal conductivity in olivine up to 25 kbar and 1500 K, in *Proceedings of 6th AIRAPT Conference*, edited by K.D. Timmerhauf and M.S. Barber, pp. 60-74, Plenum, New York, 1979.

Schmitz, M.D., and S.A. Bowring, Ultrahigh temperature metamorphism in the lower crust during Neo-Archean Ventersdorp rifting and magmatism, Kaapvaal craton, southern Africa. *Geol. Soc. Am. Bull.*, 115, 533-548, 2003.

Senthil Kumar, P., and G.K. Reddy, Radioelements and heat production of an exposed Archean crustal cross-section, Dharwar craton, south India, *Earth Planet. Sci. Lett.*, 224, 309-324, 2004.

Shapiro, N.M. and M.H. Ritzwoller, Thermodynamic constraints on seismic inversions, *Geophys. J. Int.*, 157, 1175-1188, doi:10.1111/j.1365-246X.2004.02254.x, 2004.

Shaw D.M., J.J. Cramer, M.D. Higgins, and M..G. Truscott, Composition of the Canadian Precambrian Shield and the continental crust of the earth. In *Nature of the Lower Continental Crust* (eds. J.B. Dawson *et al.*), pp. 257-282. Geological Society of London, 1986.

Shimizu, N., and N.V. Sobolev, Young peridotitic diamonds from the Mir kimberlite pipe, *Nature*, 375, 308-311, 1995.

Thompson, P.H., A.S. Judge, and T.J. Lewis, Thermal parameters in rock units of the Winter Lake-Lac de Gras area, central Slave province, Northwest Territories- implications for diamond genesis, *Current Research 1995-E*, Geol. Surv. Canada, 125-135, 1995.

C. Jaupart, Institut de Physique du Globe, 4 Pl. Jussieu, Paris, 75252, France. (cj@ccr.jussieu.fr)

J.C. Mareschal, GEOTOP-UQAM-McGill, University of Québec at Montréal, P.O.B. 8888, sta. "downtown", Montréal, Québec H3C3P8, Canada. (jcm@olympus.geotop.uqam.ca)

Creation and Preservation of Cratonic Lithosphere: Seismic Constraints and Geodynamic Models

C.M. Cooper[1], A. Lenardic, and A. Levander

Department of Earth Science, Rice University, Houston, Texas

L. Moresi

School of Mathematical Sciences, Monash University, Victoria, Australia

Cratons are areas of continental lithosphere that exhibit long-term stability against deformation. Seismic evidence suggests that cratonic lithosphere may have formed via thrust stacking of proto-cratonic lithosphere. We conducted numerical simulations and scaling analysis to test this hypothesis, as well as to elucidate mechanisms for stabilization. We found that formation of cratonic lithosphere via thrust stacking is most viable for buoyant and viscous lithosphere that is thin and/or possesses low effective friction coefficients. These conditions lead to low integrated yield strength within proto-cratonic lithosphere that allows it to fail in response to convection-generated stresses. Specifically, formation via thrust stacking is viable for lithosphere with chemical to thermal buoyancy ratios of $B = 0.75$–1.5, viscosity contrasts between the lithosphere and convective mantle of $\Delta\eta > 10^2$, and friction coefficients of $\mu = 0.05$–0.1. Preservation depends on the balance between the chemical lithosphere's integrated yield and convection-generated stresses. The physical process of thrust stacking generates a thickened cratonic root. This provides a higher integrated yield stress within cratons, which is more conducive to stability subsequent to formation. Increased friction coefficient values, due to dehydration, can also provide higher integrated yield stresses within cratons. To provide long-term stability, integrated yield stresses must be great enough to offset future mantle convection-generated stresses, which can increase with time as the mantle viscosity increases due to cooling. Thin or rehydrated cratonic lithosphere may not provide stability against the increasing convective stresses, thus providing an explanation as to why some cratons are not long-lived.

INTRODUCTION

[1] Now at Carnegie Institute of Washington, Washington, DC.

Archean Geodynamics and Environments
Geophysical Monograph Series 164
Copyright 2006 by the American Geophysical Union
10.1029/164GM07

Cratons are areas of continental crust that have not significantly been deformed since their origin [*Bates and Jackson*, 1980]. They are typically Archean in age and associated with thick lithospheric roots [*Jordan*, 1978; *Revenaugh and Jordan*, 1991]. How cratons were formed and then remained preserved is a long-standing problem [*Jordan*, 1978, 1988; *Pollack*, 1986; *Abbott*, 1991; *de Wit et al.*, 1992; *Lenardic and Moresi*, 1999]. In this paper, we will review craton formation and preservation hypotheses, present seismic evidence for craton formation via thrust stacking, and summarize the conditions required for craton formation via thrust stacking as determined from numerical simulations and scaling analysis.

Cratons are hypothesized to be formed in response to either a convective mantle upwelling or downwelling. Proponents of a hot, upwelling origin conjecture that the cratonic lithosphere is a thick layer of melt residuum formed from large-scale mantle melting induced by mantle plumes [*Pollack*, 1986; *Boyd*, 1989]. In the alternative hypothesis, cratons formed from the coalescence of some proto-cratonic lithospheric material over a mantle downwelling [*Jordan*, 1978, 1981; *Bostock*, 1998]. Regardless of the dynamic conditions, the formation mechanism must be able to produce the main observations that are commonly associated with cratonic lithosphere: thick roots, net neutral buoyancy (i.e., the positive chemical buoyancy offsets the negative thermal buoyancy), and long-term stability against deformation. Distinguishing the two alternative mechanisms relies upon the use of specific geochemical signatures within xenoliths and on seismic imaging of the cratonic lithosphere.

For example, *Pollack* [1986] proposed that a high-temperature adiabat due to a plume could cross the solidus at depths of 40 km and produce the high degree of melting (>50%) required to create and devolatilize thick cratonic lithosphere in a single melting event. His plume-based origin for cratons uses devolatilization induced by melting to provide both the increase in net buoyancy and viscosity of the cratonic lithosphere. The increased viscosity and elevated solidus produced by devolatilization provides the cratonic lithosphere with the needed stability against further melting and deformation. Alternatively, *Hart et al.* [1997] argue using diamond inclusions that proto-cratonic lithosphere originated in a setting similar to present-day oceanic lithosphere and cratonic roots formed through progressive stacking of this material. For further discussion of the geochemical signatures and implications of cratonic lithosphere, see *Lee* [this volume].

The most compelling argument for a downwelling-related craton formation process comes from the seismic imaging of cratonic lithosphere. *Bostock* [1998] and *Musacchio et al.* [2004] interpret seismic images of dipping reflectors lying deep beneath cratons as remnant subduction zone features. *Bostock* [1999] used these images to conceptualize craton formation through shallow angle subduction and layering of proto-cratonic lithosphere, such as buoyant oceanic lithosphere. This interpretation differs with *Jordan*'s [1978, 1981] proposal, which while still invoking subduction related processes, requires that the thick chemical boundary layer associated with cratons is formed through the amalgamation of continental masses underlain by depleted mantle wedges.

If cratons do originate via thrust stacking, then the proto-cratonic lithosphere must be highly deformable during formation, but once formed must also remain undeformed for billions of years. Previous geodynamic models have studied the conditions required to stabilize pre-existing cratonic lithosphere against disruption due to thrusting [*Lenardic et al.*, 2003]. The models showed that unless cratonic lithosphere exceeded a critical thickness, thrusting could penetrate the lithosphere allowing "slices" of deep cratonic roots to be recycled into the mantle (Plate 1). Thus, cratonic material is able to deform unless it exceeds a critical thickness; but how does the lithosphere thicken past the point where it can resist further deformation?

We begin by investigating a formation mechanism in which a chemically buoyant lithosphere is thickened. Thickening via thrust stacking above a mantle downflow can occur if the layer is thin enough such that thrusting can penetrate it. The layer's relative chemical buoyancy resists recycling into the mantle, allowing a cratonic root to grow via continued thrust stacking. The existence of a critical thickness beyond which a cratonic root can be stabilized [*Lenardic et al.*, 2003] suggests that this mechanism can allow for deformation during craton formation and provide for subsequent stability. The remainder of this paper quantifies the physical viability of the proposed mechanism as well as presenting recent seismic evidence for thrust stacking beneath the Kaapvaal craton.

SEISMIC EVIDENCE FOR CRATON FORMATION VIA THRUST STACKING

A combination of active and passive source seismic data provide evidence for organized structure in the cratonic mantle. For example, *Bostock* [1999] summarizes a number of key findings, including high-frequency (>10 Hz) reflection images of obvious subduction structures frozen in upper mantle dated to the Proterozoic [*Cook et al.*, 1997] and Archean [*Calvert*, 1995]. Despite the upper mantle being largely unreflective in comparison to the crust, those reflections that are identified are frequently of significant lateral extent (~100s of kilometers) and have thicknesses and physical properties that suggest they were once oceanic crust [e.g., *Warner et al.*, 1996]. Long-range refraction profiles have shown that abrupt lateral boundaries in the lithosphere can persist over billions of years [*Henstock et al.*, 1998], and the upper 200 km of the cratonic mantle shows some vertical stratification in the form of a lateral heterogeneity at several levels [e.g., *Nielsen et al.*, 2003] and a low velocity zone recognized beneath the cratons [e.g., *Thybo and Perchuc*, 1997]. Broadband seismic records of teleseismic earthquakes made by stations located in the same terrains as deep reflection profiles exhibit PdS conversions (teleseismic P waves diffracted, or refracted, to S waves on the receiver side of the path) from

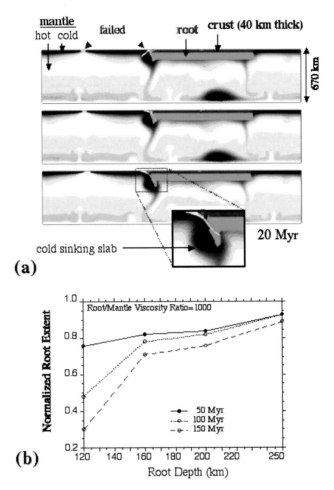

(a)

(b)

Plate 1. (a) Evolution of a model exploring craton stability and root longevity. The model assumes a cratonic root intrinsically more viscous than reference mantle. Chemically distinct crust and residuum (cratonic root material) are displayed as red and green tones, respectively, as labeled in the upper frame. Failure zones appear as narrow white zones. Brittle failure requires cool temperature conditions. Bulk mantle, crust, and cratonic root material all have a temperature dependent viscosity that allows for a five orders of magnitude variation from the hottest (dark gray) to the coldest (white) portions of the modeling domain. The viscosity of cratonic root material is three orders of magnitude greater than that of the mantle at equivalent temperatures while the viscosity of the crust is one order of magnitude lower at equivalent temperatures. (b) Normalized root extent versus root depth for models with a high root to mantle viscosity ratio. The normalized extent is defined as the lateral root extent at any time divided by the extent of the undeformed, initial extent. The normalized extent thus provides a measure of craton stability and deep root longevity.

reflective zones, suggesting both substantial velocity–density and anisotropy contrasts [e.g., *Bostock*, 1997]. Equally importantly, the PdS phases and the reflection data show that these structures are observable across a wide seismic bandwidth: roughly 0.03–0.3Hz for the teleseismically generated PdS phases, and 10–30 Hz from reflection surveys. *Bostock* [1999] and *Snyder* [2002] propose models of thrust stacking of subducted slabs to develop at least some parts of cratonic mantle. *Bostock* [1999] attributes the variability in depth of subduction-related mantle heterogeneity surfaces to the multiple and geographically widespread subduction events that form the cratons. This process would produce layering with less lateral coherence than either a pressure-sensitive phase transition (410 and 670) or a density-sensitive chemical boundary (Moho and core mantle boundary), both of which in comparison possess a regularity of depth.

In Plate 2, we show a pre-stack Kirchhoff depth migration of PdS signals showing the upper mantle beneath the Kaapvaal craton to 950 km depth [*Levander et al.*, 2005a]. The 410 and 660 discontinuities are well imaged. The lower two panels show a higher frequency migration of the same data showing the structures above the 410 discontinuity and beneath the crustal multiples that appear from about 100 to 225 km depth. A series of quasi-layered and dipping structures are apparent that extend for hundreds of kilometers (note scales). The amplitudes of the signals in this zone (~225-400 km) are appropriate for conversions from oceanic crust that has undergone transformation to eclogite within a depleted peridotite matrix [*Levander et al.*, 2005b]. The geodynamic modeling described in the next section shows that, rather than producing continuous slab-like features extending 1000 km or more, the thrust-stacking hypothesis allows for deformation to produce both moderate and steep dips, such as those seen in Plate 2.

MODEL SETUP

To dynamically test the viability of the thrust-stacking hypothesis, we first need to determine the conditions under which a chemically distinct layer, such as buoyant oceanic lithosphere or arc continental lithosphere, can thicken over a mantle downwelling in a localized manner. This can be achieved by using a simple model that incorporates a chemically distinct layer into the upper thermal boundary layer of a convecting mantle (top frame of Plate 3). Using this set-up in our simulations, the parameters affecting deformation can be systematically varied to identify the appropriate combination of parameters that controls the deformation response.

The deformation of the chemically distinct layer depends on its intrinsic properties (such as buoyancy, viscosity, and yield strength) and the deforming power provided by the convecting mantle. Therefore, we use numerical simulations that

solve the governing equations of mantle convection incorporating a viscoplastic flow law that applies in the cool part of the lithosphere. The mantle and the compositionally distinct chemical layer are given different material properties. The simulations solve the nondimensionalized conservation equations of mass, momentum, energy, and composition, respectively:

$$\nabla \cdot u = 0 \qquad (1)$$

$$\nabla \cdot (2\eta\dot{\varepsilon}) = \nabla p + Ra(T - BC)\hat{z} \qquad (2)$$

$$\frac{\partial T}{\partial t} = \kappa\nabla^2 T - u \cdot \nabla T \qquad (3)$$

$$\frac{\partial C}{\partial t} + u \cdot \nabla C = 0 \qquad (4)$$

where

$$Ra = \frac{\rho_m \alpha g \Delta T D^3}{\kappa\eta} \qquad (5)$$

$$B = \frac{\Delta C}{\alpha\Delta T} \qquad (6)$$

and u is the velocity, η is viscosity, $\varepsilon_{ij} = \partial u_i/\partial x_j + \partial u_j/\partial x_i$ is the strain rate tensor, p is pressure, C is a composition function, $\hat{z}$ is a unit vector pointing downwards, T is temperature, Ra is the Rayleigh number (which nondimensionally represents the convective vigor of the system), ρ_m is the mantle density, α is the thermal expansion coefficient, g is gravitational acceleration, ΔT is the temperature variation driving thermal convection in the mantle, D is the system thickness, κ is thermal diffusivity, B is the ratio of thermal to compositional buoyancy, and ΔC is the maximum compositional variation. The viscoplastic rheology operates such that, below a specified yield stress, deformation occurs in a Newtonian, viscous manner in response to the convective stresses. If the yield stress is exceeded, then deformation is governed by a strain-softening Von Mises plastic flow law that naturally produces localization into shear bands [*Moresi and Solomatov*, 1998; *Moresi et al.*, 2003]. The maximum yield stress of the chemically distinct layer, τ_b, increases linearly with depth, according to a continuum formulation of Byerlee's Law [*Byerlee*, 1968; *Moresi et al.*, 2003]:

$$\tau_b \sim d\mu \qquad (7)$$

where d is the thickness and μ is the friction coefficient of the chemically distinct layer. Temperature- and depth- dependent rheologies were not included within our simulations. However, we explored the effects of viscosity on deformation

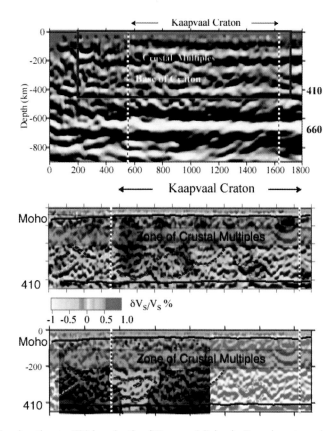

Plate 2. (Top) Pre-stack depth migration to 900 km depth of Kaapvaal Seismic Experiment receiver functions from 9 earthquakes [*Levander et al.*, 2005a]. The data were bandpass filtered to 0.033 < f < 0.167 Hz, so that the best resolution above the transition zone is ~20 km. The 410 and 660 discontinuities at the top and bottom of the transition zone are clearly imaged. The crustal multiple train appears at ~150–225 km depth. (Middle) Depth migration to 450 km depth of the same dataset, but with a bandpass of 0.033<f < 0.33 Hz, showing the base of the craton between the crustal multiple reflections and the 410-km discontinuity. Maximum resolution is ~10 km. The Moho is visible at the top of the section. The S tomography (cover overlay) of *James et al.* [2001] has been superimposed on the figure, and events corresponding to the basal region of the craton are noted as dashed red lines. The pipe of slow velocity reaching the surface at X~1200 km is associated with the Paleoproterozoic Bushveld complex. Negative polarity events, corresponding to a velocity decrease with depth, are indicated as red; positive polarity events, associated with a velocity increase with depth, are indicated as blue [*James et al.*, 2001]. Note that the events are 100s of kilometers in length. (Bottom) Seismic image superimposed on a scaled version of the layered cratonic root model described in this paper (see Plate 3). The convection model suggests that layering within the slab is more complicated than simply slabs stacked upon one another, in agreement with the complex events seen in the seismic data.

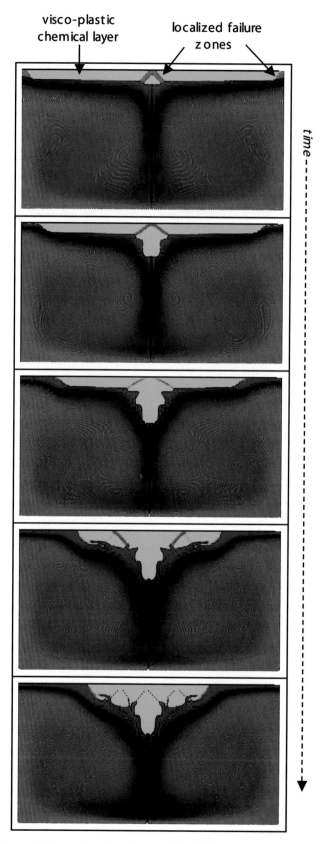

Plate 3. Numerical simulation setup and observed model thrust stacking behavior. Top frame: the initial setup of our models. The green layer is the chemically distinct layer that is emplaced within the upper thermal boundary layer (blue) of a 1 × 1 convecting mantle cell (red). The following frames show the time progression of the simulation. The chemically distinct layer thickens via thrust stacking with deformation occurring locally on the highlighted linear zones of failure (bright pink). The buoyancy ratio between the chemically distinct layer (in green) and the mantle (in red) is B = 1. In the final frame, the position of the formerly active faults are outlined in pink dotted lines.

by varying the viscosity contrast between the chemically distinct layer and an isoviscous mantle.

If the chemically distinct layer does deform, either in a localized or distributed manner, it will either recycle back into the mantle or thicken over the downwelling. The relative buoyancy between the chemically distinct layer and the mantle downwelling determines the specific response. The relative buoyancy depends on the buoyancy ratio, B, which is defined in its general form in equation (6), but can be defined more specifically as

$$B = \frac{\Delta \rho_{cbl}}{\rho_m \Delta T \alpha} \qquad (8)$$

where $\Delta \rho_{cbl}$ is the density difference between the chemically distinct layer and mantle at the same temperature.

Reflecting side boundary conditions were used to maintain unit-aspect ratio convection. The top and bottom boundaries were free-slip and isothermal. Initially, the chemically layer spanned the entire lateral extent of the system. The convective mantle flow is such that the chemical layer can thicken above a symmetric mantle downwelling. Although at present mantle downflow is asymmetric, during the time of craton formation it may have been more symmetric [Davies, 1990]. The strength of upwelling and downwelling in the system is determined by the convective vigor of the mantle, which is represented by the Rayleigh number. Convective vigor increases with increasing Rayleigh number.

MODEL RESULTS AND SCALING ANALYSIS

We mapped the parameter space controlling the deformation of the chemically distinct layer with a total of >700 simulations (parameter list in Table 1). We observed five deformation regimes: (1) stable or no deformation, (2) distributed recycling of the chemically distinct layer, (3) localized recycling of the chemically distinct layer, (4) distributed thickening of the chemically distinct layer, and (5) localized thickening of the chemically distinct layer. As we are

Table 1. Parameters Varied Within Study. All parameters listed are non-dimensionalized. Mantle convection cell is 1×1 and total system thickness, $D = 1$. Initial lateral extent of chemically distinct layer is also 1.

Parameter	Range
Rayleigh Number	10^5–10^7
Chemical Layer Thickness	3.6–7.5% of system
Viscosity Contrast	1–10^4
Buoyancy Ratio	0–2
Friction Coefficient	0.1–0.7

concerned with the formation of cratons via thrust stacking, we will focus on the localized thickening deformation regime.

Plate 3 illustrates the localized thickening response of the chemically distinct layer in our simulations. Localized deformation occurs as zones of localized plastic failure that are akin to ductile shear bands. These regions are highlighted as bright pink, linear features in the simulation images. As illustrated in Plate 3, the chemically distinct layer fails along these shear zones into fault blocks, which thrust under each other, forming a thickened cratonic root. The mode of thickening (either localized or distributed) depends on the viscosity contrast between the chemically distinct layer and the convecting mantle. If the viscosity contrast is relatively low, then deformation will occur in a distributed, ductile manner. However, a high-viscosity contrast coupled with mantle-imposed shear stresses can lead to high stresses within the chemically distinct layer. If these stresses reach the yield stress of the chemically distinct layer, then localized failure can occur. If the yield stress was not reached, then the high viscosity of the chemically distinct layer was able to provide stability against deformation. Within our simulations, localized deformation required viscosity contrasts of 10^2–10^3 between the chemically distinct layer and convecting mantle. Such high viscosity contrasts can be provided by dehydration of bound water in the lithosphere [Pollack, 1986; Hirth and Kohlstedt, 1998], which plays an important role in lithospheric ductile strength. Free water, which can enter pore space, can lubricate faults and can facilitate the generation of ductile shear zones via its effects on the plastic strength of the lithosphere [Bercovici, 1998]. These water-weakening processes can act independently. We will assume that formation of oceanic residuum removes bound water (due to melt processes), thus increasing ductile strength, but that the plastic strength (and effective friction coefficient) of proto-cratonic material can still be affected by free water; e.g., sediment subduction and dehydration of oceanic crust can introduce free water to mega-thrust zones.

In the simulations that exhibited localized deformation, localized shear zones are abandoned rapidly once stress levels drop below the yield stress. In other words, within the simulations, fault zones only exist as a plane of weakness when the yield stress is exceeded. Once stress levels drop below the yield stress, then the simulation faults or zone of weakness are no longer activated. Therefore, no long-lived zones of weakness exist within the chemically distinct layer. This allows for the progressive growth of the root through thrust stacking along peripheral faults while avoiding internal deformation associated with reactivation of pre-existing fault planes. Once the chemically distinct layer had thickened via thrust stacking, it resisted deformation for the duration of the simulation. All simulations were run for greater than five mantle overturn times (defined as the time it takes a particle to transverse the full mantle convection cell). Craton formation occurred after roughly one overturn time for simulations with parameters that allowed localized thickening.

It is important to note that the final craton thickness, in our simulations, is determined by the lateral extent and initial depth of the chemically distinct layer. However, in the Earth, the final thickness will depend on the time scale over which a stable mantle downwelling exists, as the residuum layer would be continually generated at mid-ocean ridges. Within a vigorously convecting mantle, as is likely for the Archean, the length of time any mantle downwelling is stable will be variable. This implies that the final thickness of cratonic lithosphere can also be variable, even if all cratonic roots form via similar physical means, i.e., via thrust stacking above regions of mantle downflow.

In order for the cratonic lithosphere to form via thrust stacking, three conditions must be met: (1) The viscosity contrast must be high enough to resist distributed deformation, (2) the buoyancy must be high enough to resist recycling, and (3) the yield stress must be exceeded by the convective stresses. Applying physical meaning to these conditions will allow us to gain a physical understanding of our modeling results, as well as extend them to parameter spaces beyond current computational power. Therefore, we have also developed scaling relationships that mark the transition between the recycling and thickening of the chemically distinct layer, as well as the transition between plastic yielding and nondeforming responses of the chemically distinct layer. The first condition follows empirically from the simulations in that, if the viscosity contrast between the chemically distinct layer and the mantle did not exceed 10^2–10^3, then localized deformation would not occur. However, though a distributed thickening of the chemically distinct layer could occur at lower viscosity contrasts, these simulations did not produce features that resisted deformation once formed.

For the second condition, the chemically distinct layer must resist recycling back into the mantle in order to thicken over the downwelling as opposed to being carried into the deep mantle by the downwelling. A chemically distinct layer will resist recycling if its positive chemical buoyancy exceeds the negative thermal buoyancy of the active upper thermal boundary layer of the convecting mantle. The chemical buoyancy, B_{cbl}, is defined by

$$B_{cbl} = \Delta\rho_{cbl}d. \qquad (9)$$

The thermal buoyancy of the mantle downwelling, B_{tbl}, is represented by

$$B_{tbl} = \rho_m \Delta T \alpha \delta \qquad (10)$$

where δ is the thermal boundary thickness, which scales in proportion to the Rayleigh number to the $-1/3$ power [*Turcotte and Oxburgh*, 1967]. Using this scaling and the buoyancy definitions in equations (8), (9), and (10), the criteria required for the chemically distinct layer to resist recycling are simplified to the relationship

$$Bd > \omega_1 Ra^{-\frac{1}{3}} \qquad (11)$$

where B is the buoyancy ratio as defined in equation (8) and ω_1 is a geometric scaling constant. If this condition is met, then the chemically distinct layer can thicken over a downwelling and remain stable from recycling back into the mantle, as we have confirmed via additional numerical simulations (the simulation constrained the scaling constant, ω_1, to a value of 15). While providing longevity from recycling, the chemical buoyancy alone does not enhance the stability of a thickened chemically distinct layer, as it does not control the layer's ability to yield or deform viscously.

The final condition that must exist for thrust stacking to occur is that the material must be able to deform via thrust faulting. Whether this will be the case depends on the material's plastic yield properties. More specifically, a chemically distinct layer will not yield in a plastic manner unless its inherent yield stress is reached by the stresses provided by the convecting mantle. The convective stresses then scale in proportion to

$$\tau_c \sim \eta_i u_0 D^{-1} \qquad (12)$$

where u_0 is average plate velocity. The convective stresses are related to the convective vigor of the mantle, as the average plate velocity and the interior mantle viscosity scales in proportion to the mantle Rayleigh number, Ra, to powers of $2/3$ and -1, respectively [*Turcotte and Oxburgh*, 1967]. Using those scaling relationships, equation (12) simplifies to

$$\tau_c \sim Ra^{-\frac{1}{3}} \qquad (13)$$

Applying the condition for stability, $\tau_b > \tau_c$, and using equation (7) leads to

$$d\mu > \omega_2 Ra^{-\frac{1}{3}} \qquad (14)$$

where ω_2 is a geometric scaling constant. This defines the regime transition between plastic yielding and stability.

Figure 1 shows the results from the simulations for variable friction coefficients, chemically distinct layer thicknesses, and Rayleigh numbers, as well as the scaling relationship defined in equation (14) plotted as the bold line. Simulations exhibiting the yielding behavior are plotted as solid circles,

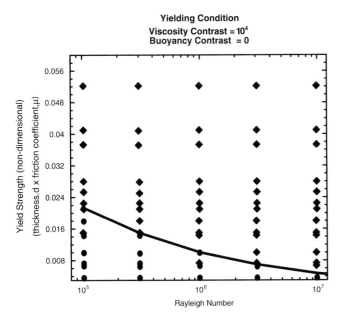

Figure 1. Simulation results and scaling analysis. Deformation responses of the chemically distinct layer for simulations with variable non-dimensional yield strength (thickness × friction coefficient) and Rayleigh numbers. Diamond symbols indicate simulations that showed no signs of deformation. Closed circles represent simulations that exhibited localized deformation. The solid line marks the transition from stable (above the line) to yielding responses derived from equation (6) with the best-fit geometric scaling constant of 1.

whereas stable simulations where no deformation occurs are plotted as solid diamonds. The theoretical scaling relationship marking the transition between the stable and deforming regimes coincides reasonably well with the transition mapped out by the numerical simulations with the best-fit geometric scaling constant, ω_2 equal to 1.

DISCUSSION

Using the scaling relationships derived in equations (11) and (14), we can make predictions about craton formation and preservation for parameter regions that are beyond our computational capabilities, but applicable to the early Earth. For example, Figure 2a shows the transition between the localized yielding and stable responses for variable lithospheric thicknesses, friction coefficients and Rayleigh numbers as calculated from equation (14). This schematic allows us to track the deformation response of the chemically distinct layer as it progresses in time from more to less vigorous convection. The diagram also provides the ability to track the deformation behavior of the chemically distinct layer in response to changes in thickness or friction coefficient.

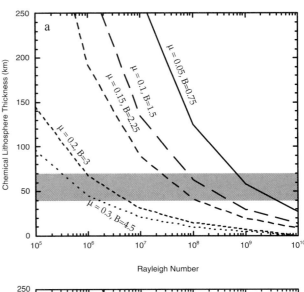

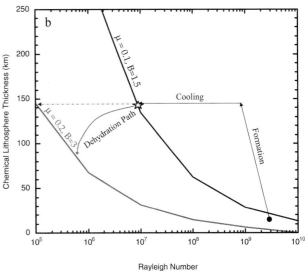

Figure 2. (a) Extended scaling analysis: Curves represent the transition from stable (above the line) to yielding responses derived from equation (11) for variable Rayleigh numbers, chemically distinct layer thicknesses and friction coefficient values. The best-fit geometric scaling constant for all curves is 1. The buoyancy ratios for each curve are determined from equation (14) by using a best-fit scaling constant of 15. The gray band represents range of present-day oceanic chemical lithosphere estimates [Lee et al., 2005] (b) Example of history path of a craton formed under low friction coefficient values ($\mu = 0.1$). The craton forms through progressive thickening until it reacs its maximum thickness as determined by the current dynamic conditions. After formation, the craton will cool and remain stable, maintaining its thickness until the point that it progresses past its stability field (star). At this point, the craton has entered an unstable regime and possesses the potential for deformation and "decratonization". If, however, the craton dehydrates during formation or cooling, then it will remain in a stable regime for a longer period of time (light gray path).

Also included is the buoyancy ratio required for the lithosphere to resist recycling back into the mantle, as calculated from equation (11). For the conditions above each curve, the scaling relationship predicts no deformation of the chemically distinct layer. Below each curve, localized deformation is predicted.

For cratons to form via thrust stacking, the chemical lithosphere must be able to thicken in a localized manner over a mantle downwelling. For higher mantle Rayleigh numbers, which may have been prevalent during the Archean, thrust stacking is only possible for chemical lithosphere that is either very thin or possesses low friction coefficient values (Figure 2a). Otherwise, the convective stresses cannot exceed the yield stress of the proto-cratonic lithosphere, and deformation is not possible. Present-day thickness estimates of the chemical boundary layer within the oceanic lithosphere range from 40 to 70 km, depending on the mantle's potential temperature [Lee et al., 2005]. The relative buoyancy of oceanic lithosphere may have been greater in the Archean due to the combined effects of increased melt production and thinner thermal boundary layers [Davies, 1992]. If buoyant oceanic lithosphere serves as proto-cratonic lithosphere, this range of estimated chemical boundary layer thicknesses limits craton formation to lithosphere with low effective friction coefficient values (Figure 2a). The alternative option, thin proto-cratonic lithosphere with high friction coefficient values, is harder to argue as higher temperatures in the past would have led to a thicker oceanic residuum layer. Low friction coefficient values are consistent with the hypothesis that cratons formed under hydrous conditions [de Wit et al., 1992] with pore water serving as a lubricant. In addition to allowing for thicker proto-cratonic lithosphere to thicken by thrust stacking, the low friction coefficient values also allow for lower, more realistic buoyancy ratios required to maintain thickening over the downwelling rather than recycling back into the mantle.

Although the high viscosity and increased buoyancy required for craton formation via thrust stacking aid in resistance to deformation, preservation is not achieved unless the total yield stress of the cratonic lithosphere exceeds convective stresses, which increase with time as the mantle viscosity increases and the Rayleigh number decreases. The low friction coefficient values required for yielding during formation via thrust stacking will not provide for long-term preservation unless the final thickness of the craton exceeds a critical value. Increasing the thickness of the cratonic lithosphere increases the total yield stress at the base. If thickened enough, the increased yield stress not only can offset the convective stresses present during formation, but also offset the future convective stresses that increase with decreasing mantle temperature. Such a mechanism can provide stability for a craton forming through the thickening of a proto-lithosphere material via subduction-related thrust stacking. If, however, cratons do not reach such a thickness, then they can be subject to future deformation and may become "decratonized" as the mantle cools and convective stresses increase. Note that the final thickness of cratons depends on the duration of the mantle downwelling and therefore can vary from craton to craton.

Preservation may also be achieved if the friction coefficient of the cratonic lithosphere increases due to loss of pore fluids. Increasing the friction coefficient also increases the total yield stress of the cratonic lithosphere, supplying it with greater potential to offset future increasing convecting stresses. As seen in Figure 2a, the curves with higher friction coefficients provide stability over larger parameter spaces. Following the history path illustrated in Figure 2b, a craton can form under low friction coefficient values and increase its preservation potential if, after formation, it becomes dehydrated of pore fluids. Note we invoke the effects of free water to explain the variations in yielding behavior, not the effects of bound water on the viscosity of cratonic lithosphere. However, inherent to our argument is that the proto-cratonic lithosphere has dehydrated of its bound water due to melting processes [Hirth and Kohlstedt, 1996; Pollack, 1986], such that it is able to resist distributed viscous deformation, allowing for localized yielding.

The dependence of craton preservation potential on its ability to offset increasing post-formation convective stresses provides a potential explanation as to why some cratons are not long-lived, such as the Sino-Korean craton [Griffin et al., 1998; Lebedev and Nolet, 2003]. Convective stresses increase as the mantle cools. If the cratonic lithosphere did not exceed a critical thickness or dehydrate of pore fluids, then its yield stress may not have been adequate to provide long-term stability from deformation, allowing it to become "decratonized". Furthermore, this may also imply that the current long-lasting cratons on the Earth's surface can lose their stability as the Earth cools or if they become rehydrated.

We have shown that craton formation and preservation is physically viable via thrust stacking, provided that faults are weak during formation yet strong enough to resist further post-formation deformation. One way to strengthen faults is to remove pore fluids as mentioned above. Alternatively, thrust faults may be strengthened if healed on a faster time scale than the life of the craton. Within our simulations, cratons formed in roughly one mantle overturn. At present this corresponds to 10^8 years but it would have been more rapid in the past because plate velocities scale with the Rayleigh number to the 2/3 power [Turcotte and Oxburgh, 1967]. Assuming a current Rayleigh number of 10^7–10^8 and a possible Archean value of 10^{10}, craton formation could have occurred on a time scale as rapid as the order of a Myr. If thrust faults remain weak for a much longer time scale, then cratonic lithosphere could unstack along these weak zones.

Finally, it is important to highlight that although craton formation via thrust stacking is physically possible within our numerical simulations and scaling analyses, it is viable for only a small range of parameters. Specifically, craton formation via thrust stacking is viable only for a chemical lithosphere with buoyancy ratios of $B = 0.75$–1.5, viscosity contrasts of $\Delta\eta > 10^2$, and friction coefficients of $\mu = 0.05$–0.1. Therefore, if cratons do form via thrust stacking, as suggested by seismic images, then the small size of the window in parameter space allowing for this behavior can provide specific predictions as to the proto-cratonic lithosphere's rheology and composition, which will depend on its formation environment (e.g., ocean ridges or back arcs). For example, if buoyant oceanic lithosphere is stacked to form cratons, then it must possess the required buoyancy ratio, viscosity contrast, and friction coefficient value. We have already suggested that the required viscosity contrast and friction coefficient values can be achieved due to the effects of bound and free water on the rheology of oceanic lithosphere. However, providing the necessary buoyancy ratio required to keep the oceanic lithosphere from recycling becomes more problematic.

Following the methods of *Lee* [2003] and *Lee et al.* [2005], we can determine the predicted range of Mg#s in the oceanic residuum that would be necessary to provide a buoyancy ratio range of $B = 0.75$–1.5, which is the range required by our models for craton formation via thrust stacking. The average Mg# of depleted oceanic residuum can be related to its average density through

$$\rho_r = -0.0144 Mg\# + 4.66, \qquad (15)$$

an empirically derived relationship [*Lee*, 2003].

To calculate the Mg# from the predicted buoyancy ratios, using equation (15), we must first make an assumption about the relative mass proportions of oceanic crust and mantle residuum within the chemical lithosphere column. This is because the buoyancy ratio definition, equation (11), provides only the average density over the entire chemical lithosphere, whereas equation (15) requires the oceanic residuum density to determine the Mg#. We do not assume a priori a mantle potential temperature and therefore we cannot calculate a melt fraction to then provide the percentage of oceanic crust and residuum thickness. Rather, we explored a range of percentages of oceanic crust and residuum thickness. We use the present-day case of the oceanic crust providing 10% of the entire chemical lithosphere thickness as the lower bound and an extreme case where the oceanic crust contributed 30% of the oceanic chemical lithosphere thickness as the upper bound. This extreme case could have been plausible in the past as higher mantle potential temperatures could have led to thicker and more buoyant chemical boundary layers within the oceanic lithosphere [*Davies*, 1992]. We assume a reference oceanic crust density of and a reference mantle density of $\rho_\infty = 2900\ kg/m^3$ and a reference mantle density of

$\rho_m = 3380\ kg/m^3$ We calculate the average chemical lithosphere density from equation (11) for both layered mantle convection and whole-mantle convection, as each convective style would provide different relative amounts of negative thermal buoyancy ($\Delta T = 1600°C$ for layered mantle convection and $\Delta T = 2900°C$ for whole-mantle convection).

For the present-day oceanic crust thickness percentage (10%), the required Mg#s for the oceanic residuum are >95 if layered mantle convection is assumed and become unrealistically high, e.g., Mg#s > 100, if whole-mantle convection is assumed. In either case, Mg#s are much higher than observed maximum values of 92–93 for cratonic lithosphere [*Jordan*, 1978; *Boyd*, 1989; *Griffin*, 1989; *Lee*, 2003; *Lee et al.*, 2005]. Therefore, present-day oceanic lithosphere is not a viable material for proto-cratonic lithosphere. In addition, under present conditions, craton formation via thrust stacking of oceanic lithosphere is not possible, as the buoyancy ratios required to resist deformation are not present (present-day buoyancy ratios range from $B = 0.2$–0.4, assuming a chemical oceanic lithosphere thickness column containing 10% of oceanic crust with a density of 2900 kg/m^3 and 90% of oceanic residuum layer with a density of 3360 kg/m^3 for either lower or whole-mantle convection of a mantle with reference density of 3380 kg/m^3). This is not surprising as we observe the oceanic lithosphere recycling into the mantle along subduction zones today.

However, in the past, higher mantle potential temperatures could lead to thicker and more buoyant chemical boundary layers within the oceanic lithosphere. For oceanic lithosphere with a larger crustal component (30%), the predicted Mg#s fall closer to the observed range with values of >87 for layered mantle convection. For an intermediate case (oceanic crust component = 20%) and layered mantle convection, the predicted Mg#s lower to values (>91) more typical to the observed range of cratonic lithosphere values. The predicted range of Mg#s (>97) for thicker oceanic crust within a whole-mantle convection setting, again, exceeds the observed average cratonic lithosphere values.

Therefore, if cratons do form via thrust stacking of oceanic lithosphere, it is more likely to occur for oceanic lithosphere with a greater percentage of oceanic crust, as this allows not only for increased chemical buoyancy but also for more realistic predicted Mg#s within the cratonic lithosphere. In addition, layered mantle convection, which may have been the prevalent convective style in the past [*Christensen and Yuen*, 1985], may be preferred, as it would provide a lower negative thermal buoyancy that the chemical lithosphere would need to offset in order to avoid early recycling into the mantle. However, though the oceanic lithosphere may be able to resist recycling during the formation period under layered mantle convection conditions, it may not be able to maintain that stability forever, as the thick oceanic crust would eventually encounter the eclogite transition when thrust to greater depths.

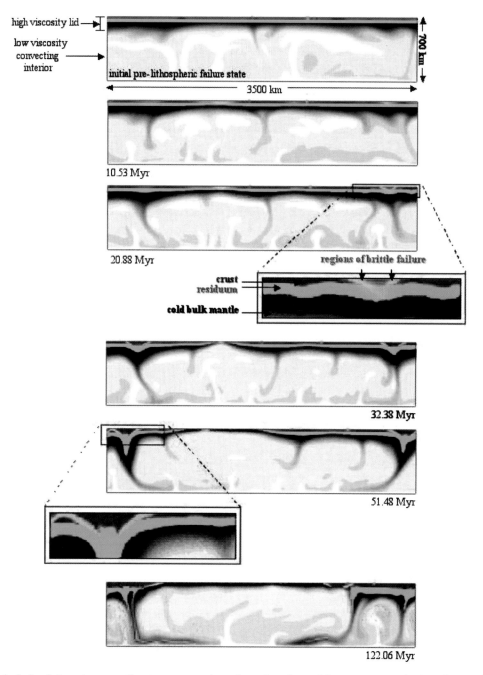

Plate 4. Numerical simulations incorporating temperature-dependent viscosity and larger aspect ratio. Top frame: the initial pre-lithospheric failure state. A chemical lithosphere was emplaced within the upper thermal boundary layer of the convecting mantle. This chemical lithosphere consists of a crustal (brown) and residuum layer (green) with thicknesses of 4% and 6% of the total system thickness and buoyancy ratios of B = 2 and 1, respectively. The temperature-dependence of viscosity allows for a five orders of magnitude viscosity variation from the hottest to the coldest portions of the model for all three components. The residuum layer maintains a higher viscosity than the convecting mantle as its prefactor determining the strength of the temperature dependence is 10× that of the mantle and the crust. The factor of 10× for residuum means its viscosity is 10× the mantle and crust at equivalent temperatures. Yielding occurs in the same manner as in the simulations presented earlier with the effective friction coefficient value of 0.1 for the crust, residuum, and mantle. Regions of failure in this model are highlighted in light blue. Wrap-around side boundary conditions are used and the top and bottom boundaries are free slip and isothermal. The bottom heating Rayleigh number was set at 2×10^7 and the ratio of internal to bottom heating Rayleigh numbers was set to 1. As the simulation progresses with time, the chemical lithosphere begins to thicken over a mantle downwelling along localized zones of failure until a craton-like feature forms.

The arguments presented above suggest that there are potential difficulties to the hypothesis that ancient ocean lithosphere formed cratonic roots. However, if cratons formed via thrust stacking of arc continental lithosphere, then the required buoyancy ratio to keep the chemically distinct layer from recycling back into the mantle is not hard to achieve. For example, assuming a continental lithosphere with a 40-km-thick crust with average density of 2800 kg/m^3 and 60-km-thick chemical lithosphere with average density of 3300 kg/m^3, the buoyancy ratios calculated from equation (11) are 1.53 for layered mantle convection and 0.89 for whole-mantle convection. Regardless of the convective style, the calculated buoyancy values exceed the required buoyancy values to keep the layer from recycling back into the mantle with the downwelling. Arc continental lithosphere could also provide the required viscosity contrast and low friction coefficient values if, as in the oceanic lithosphere case, it is dehydrated of bound water (to provide elevated viscosity) and its faults are lubricated. Whether arc continental lithosphere (either present-day or ancient) possesses Mg #s typical of average continental lithosphere is yet to be determined due to lack of data sampling.

FUTURE WORK

Though providing initial insight into the physical processes determining craton formation and preservation via thrust stacking, the simulations presented here are simplistic models. Initial investigations with larger aspect ratios and more complex rheologies have been explored as shown in Plate 4. Initial results confirm that, though thrust stacking is a physical viable mechanism for craton formation, it is limited to a narrow range of physical parameters. For robustness, 3D and spherical simulations should be conducted to account for potential geometric effects. Additionally, our proposed stabilizing mechanism depends on the assumption that thrust faults heal very rapidly in the models. Though faults have been known to heal rapidly within earthquakes cycles [*Tadokoro and Ando*, 2002], they may also be long-lasting zones of weakness available for reactivation [*Sykes*, 1978; *McConnell*, 1972]. As pre-existing zones of weakness could destabilize a cratonic region or deter initial stabilization, further studies should be conducted that incorporate a longer healing time for the thrust faults observed in our models.

CONCLUSIONS

As suggested from seismic data, cratons can form via thrust stacking of a chemical lithosphere. Our simulations and scaling analysis show that this formation mechanism is viable for only a small range of parameter values. Specifically, craton formation via thrust stacking is most favorable for proto-cratonic lithosphere that has buoyancy ratios of $B = 0.75$–1.5, viscosity contrasts of $\Delta\eta > 10^2$, and friction coefficients of $\mu = 0.05$–0.1. The required buoyancy ratios for craton formation can be provided with either continental arc lithosphere or buoyant oceanic lithosphere serving as the proto-cratonic lithosphere. However, craton formation via thrust stacking of buoyant oceanic lithosphere requires layered mantle convection during the time of formation; in addition, a thick oceanic crust is required to match Mg# constraints. This leads to the potential problem that the thick oceanic crust can transform to eclogite, which would offset the chemical buoyancy required to preserve cratonic lithosphere. Formation via thrust stacking of arc lithosphere can avoid this potential problem. Regardless of the origin of proto-cratonic material (i.e., oceanic vs. continental arc lithosphere), craton preservation is achieved if a craton exceeds a critical thickness during formation via thrust stacking. This critical thickness allows a craton's integrated yield stress to offset convective stresses, which increase as the mantle cools. The duration of stability depends on the thickness and friction coefficient of a craton with higher values of either parameter promoting longevity. Therefore, cratons possess the potential to "decratonize" if they do not achieve a critical thickness or are rehydrated.

Acknowledgments. The authors would like to thank Dallas Abbott and an anonymous reviewer for constructive criticism and helpful suggestions. In addition, we thank Cin-Ty Lee for very helpful discussions and geochemical insight. The work was supported by NSF grants EAR-0448871 and CMG EAR-0225670 (Levander) and NASA-MDAP grant NAG5-12166.

REFERENCES

Abbott, D., The case for accretion of the tectosphere by buoyant subduction, *Geophysical Research Letters,* 18 (4), 585-588, 1991.

Bates, R.L., and J.A. Jackson (Eds.), *Glossary of Geology*, Am. Geol. Inst., Falls Church, Va., 1980.

Bostock, M.G., Anistropic upper mantle stratigraphy and architecture of the Slave craton, *Nature,* 390, 392-395, 1997.

Bostock, M.G., Mantle stratigraphy and evolution of the Slave province, *Journal of Geophysical Research*, 103, 21,183-21,200, 1998.

Bostock, M.G., Seismic imaging of lithospheric discontinuities and continental evolution, *Lithos,* 48, 1-16, 1999.

Boyd, F.R., Compositional distinction between oceanic and cratonic lithosphere, *Earth and Planetary Science Letters,* 96, 15-26, 1989.

Byerlee, J. The brittle-ductile transition in rocks, *Journal of Geophysical Research,* 74, 4741-4750, 1968.

Calvert, A.J., E.W. Sawyer, W J. Davis and J.N. Ludden, Archaean subduction inferred from seismic images of a mantle suture in the Superior Province, *Nature,* 375, 670-674, 1995.

Christensen, U. and D. Yuen, Layered Convection Induced by Phase Transitions, *Journal of Geophysical Research,* 90, 10291-10300, 1985.

Cook, F.A., A.J. van der Velden, and K.W. Hall, Upper mantle reflectors beneath the SNORCLE transect-images of the base of the lithosphere? In Cook, F.A., and P. Erdmer (editors) Slave-Northern Cordillera Lithospheric Evolution (SNORCLE) Transect and Cordilleran Tectonics Workshop Meeting, Lithoprobe Report No. 56, University of Calgary, 58-62, 1997.

Davies, G.F., 1990, Heat and mass transport in the early Earth, *in* Newsom, H.E., and Jones, J.H., eds., Origin of the Earth: New York, Oxford University Press, p. 175-194.

Davies, G.F., 1992, On the emergence of plate tectonics. *Geology,* 20, 963-966.

de Wit, M.J., C. Roering, R.J. Hart, R.A. Armstrong, C.E.J. de Ronde, R.W.E. Green, M. Tredoux, E. Peberdy and R.A. Hart, Formation of an Archaean continent, *Nature,* 357, 553-562, 1992.

Griffin, W.L., Z. Andi, S.Y. O'Reilly, and C.G. Ryan, Phanerozoic evolution of the lithosphere beneath the Sino-Korean Craton, in *Mantle Dynamics and Plate Interactions in East Asia,* edited by M.F.J. Flower, S.-L. Chung, C.-H. Lo, and T.-Y. Lee, pp. 107-126, American Geophysical Union, Washington, D.C., 1998.

Hart, R.J., M. Tredoux, and M.J. de Wit, Refractory trace elements in diamond inclusions; further clues to the origins of the ancient cratons, *Geology,* 25 (12), 1143-1146, 1997.

Henstock, T.J., A. Levander, G.R. Keller, C.M. Snelson, K.C. Miller, S.H. Harder, A.R. Gorman, R.M. Clowes, M.J.A. Burianyk, and E.D. Humphreys, Probing the depths of western North America: contrasting Archean and Proterozoic lithosphere with controlled source seismology, *Geological Society of America Today* 8, 1-17, 1998.

Hirth, G. and D.L. Kohlstedt, Water in the oceanic upper mantle: implications for rheology, melt extraction and the evolution of the lithosphere, *Earth and Planetary Science Letters,* 144, 93-108, 1996.

Jordan, T.H., Composition and development of the continental tectosphere, *Nature,* 274, 544-548, 1978.

Jordan, T.H., Structure and formation of the continental tectosphere, *Journal of Petrology, 1988,* 11-37, 1988.

Lebedev, S. and G. Nolet, Upper mantle beneath Southeast Asia from S velocity tomography, *Journal of Geophysical Research,* 108, 26 pp., 2003.

Lee, C-T., Compositional variation of density and seismic velocities in natural peridoties at STP conditions: Implications for seismic imaging of compositional heterogeneities in the upper mantle, *Journal of Geophysical Research,* 108, 6-1–6-19, 2003.

Lee, C.-T., A. Lenardic, C. Cooper, F. Niu, and A. Levander, The role of chemical boundary layers in regulating the thickness of continental and oceanic thermal boundary layers, *Earth and Planetary Science Letters,* 230, 379-395, 2005.

Lenardic, A., and L.-N. Moresi, Some thoughts on the stability of cratonic Lithosphere: Effects of buoyancy and viscosity, *Journal of Geophysical Research,* 104, 12,747-12,758, 1999.

Lenardic, A., L.-N. Moresi, and H. Muhlhaus, Longevity and stability of cratonic lithosphere: insights from numerical simulations of coupled mantle convection and continental tectonic. *Journal of Geophysical Research,* 108, 2003.

Levander, A., F. Niu, and W.W. Symes, Imaging teleseismic P to S scattered waves using the Kirchhoff integral, in A. Levander and

G. Nolet, *Seismic Earth: Array Analysis of Broadband Seismograms,* American Geophysical Union Monograph 157, Washington, D.C., in press, 2005a.

Levander, A., F. Niu, C.-T.A. Lee, and Xin Cheng, *Imag(in)ing the Continental Lithosphere, Physics of the Earth and Planetary Interiors,* in press, 2005b.

McConnell, R.B., Geological Development of the Rift System of Eastern Africa, *Geological Society of America Bulletin,* 83(9), pp. 2549-2572, 1972.

Moresi, L.-N., and V.S. Solomatov, Mantle convection with a brittle lithosphere: Thoughts on the global styles of the Earth and Venus, *Geophysical Journal International,* 133, 669-682, 1998.

Moresi, L.-N., F. Dufour and H.B. Muhlhaus, A Lagrangian integration point finite element method for large deformation modeling of viscoelastic geomaterials, *Journal of Computational Physics,* 184 (2), 476-497, 2003.

Musacchio, G., White, D.J., Asudeh, I., Thomson, C.J., Lithospheric structure and composition of the Archean western Superior Province from seismic refraction/wide-angle reflection and gravity modeling, *Journal of Geophysical Research,* 109 (3-2), B03304, 2004.

Nelsen, L., H. Thybo, A. Levander, and L.N. Solodilov, Origin of upper-mantle scattering – evidence from Russian peaceful nuclear explosion data, *Geophysical Journal International,* 154, 196-204, 2003.

Pollack, H.N., Cratonization and thermal evolution of the mantle, *Earth and Planetary Science Letters,* 80, 175-182, 1986.

Revenaugh, J., and T.H. Jordan, Mantle Layering from ScS reverberations: 3. The upper mantle. *Journal of Geophysical Research,* 96, 19781-19810, 1991.

Snyder, D.B., Lithospheric growth at margins of cratons, *Tectonophysics,* 355, 7-22, 2002.

Sykes, L.R., Intraplate seismicity, reactivation of preexisting zones of weakness, alkaline magmatism, and other tectonism postdating continental fragmentation, *Reviews of Geophysics and Space Physics,* 16(4), pp.621-688, 1978.

Tadokoro, K. and M. Ando, Evidence for rapid fault healing derived from temporal changes in S wave splitting, *Geophysical Research Letters,* 29(4), pp. 6-1–6-4, 2002.

Thybo, H., and E. Perchuc, The seismic 8° discontinuity and partial melting in continental mantle, *Science,* 275, 1626-1629, 1997.

Turcotte, D.L., and E.R. Oxburgh, Finite amplitude convective cells and continental drift, *Journal of Fluid Mechanics,* 28, 1967.

Warner, M., J. Morgan, P. Barton, C. Price, and K. Jones, Seismic reflections from the mantle represent relict subduction zones within the continental lithosphere, *Geology,* 24, 39-42, 1996.

C.M. Cooper, A. Lenardic, and A. Levander, Department of Earth Science, 6100 Main Street MS-126, Rice University, Houston, Texas 77005, USA. (cooper@dtm.ciw.edu)

L. Moresi, School of Mathematical Sciences, Building 28, Monash University, Victoria 3800, Australia.

Geochemical/Petrologic Constraints on the Origin of Cratonic Mantle

Cin-Ty Aeolus Lee

Department of Earth Science, Rice University, Houston, Texas

Cratons are underlain by thick, cold, and highly melt-depleted mantle roots, the latter imposing a chemical buoyancy that roughly offsets the craton's negative thermal buoyancy associated with its cooler thermal state. Petrologic/geochemical predictions of three endmember scenarios for the origin of cratonic mantle are discussed: (1) high-degree melting in a very hot plume head with a potential temperature >1650°C, (2) accretion of oceanic lithosphere, and (3) accretion of arc lithosphere. The hot plume scenario predicts that cratonic peridotites were formed by high degrees of melting at very high pressures (≥7 GPa), whereas the two accretion scenarios predict an origin by melting on average at lower pressures (<~4 GPa) followed by subsequent transport of these residual peridotites to the greater depths (3–7 GPa) from which they presently derive. Major-element and mildly incompatible trace-element compositions of cratonic peridotite xenoliths suggest a low pressure origin, favoring the two accretion scenarios. The two accretion scenarios are difficult to distinguish geochemically, but one difference is that garnet pyroxenite xenoliths might be more clinopyroxene- and Si-rich in arc environments than in oceanic environments, which would be more olivine-rich (and Si-poor) due to lower pressures of crystallization in oceanic settings compared to arc settings. High-MgO cratonic garnet pyroxenite xenoliths in fact have major-element systematics similar to high-MgO garnet pyroxenites from Phanerozoic continental arcs. Taken at face value, this suggests that at least some component of cratons may have formed by accretion of arc lithosphere. Low-MgO garnet pyroxenite xenoliths from cratons represent either subducted oceanic crust or arc basalts. Cratonic mantle may be formed by a combination of arc and oceanic lithospheric mantle accretion.

1. INTRODUCTION

The term craton, as originally defined, describes that part of a continent that has been more or less tectonically quiescent for billion year timescales [*Stille*, 1936; *Hoffman*, 1988; *Hoffman*, 1989]. Almost by definition, this means that cratons represent the stable Archean and/or Proterozoic cores of continents. A number of somewhat unique characteristics are now attributed to cratons. For example, most cratons are characterized by low surface heat flows [*Pollack and Chapman*, 1977; *Pollack*, 1986; *Pollack et al.*, 1993] and underlain by thick and cold mantle keels composed of highly melt-depleted peridotitic mantle [*Boyd and McAllister*, 1976; *Jordan*, 1978, 1988; *Finnerty and Boyd*, 1987; *Canil*, 1992; *Herzberg*, 1993, 1999; *Grand*, 1994; *Griffin et al.*, 1999; *Pearson et al.*, 2003; *Carlson et al.*, 2005]. In addition, a large portion of cratonic mantle appears to have remained convectively isolated from the asthenosphere over billion year timescales [*Richardson et al.*, 1984; *Walker et al.*, 1989; *Pearson et al.*, 1995a; *Carlson et al.*, 1999]. From these observations has emerged the view that the long-term stability

Archean Geodynamics and Environments
Geophysical Monograph Series 164
Copyright 2006 by the American Geophysical Union
10.1029/164GM08

of cratons is, in some manner, related to the unique composition and thermal structure of their underlying mantle "keels".

Exactly how these variables conspire to ensure craton longevity is not immediately obvious. However, Jordan proposed an elegant and highly influential hypothesis ("isopycnic hypothesis"), which was successful in relating all of these variables [*Jordan*, 1975, 1978, 1988]. The motivation for this hypothesis was a simple paradox. Cratons are cold and hence should be negatively buoyant, yet they not only remain convectively stable but there appears to be no significant gravity excess over cold continents. To resolve this paradox, Jordan hypothesized that the negative thermal buoyancy (e.g., densification associated with thermal contraction) imposed by the craton's cooler thermal state is exactly canceled at every depth by intrinsic chemical buoyancy imparted by the low densities of melt-depleted peridotites (relative to fertile peridotites) making up cratonic keels [*Jordan*, 1988]. In essence, Jordan's isopycnic hypothesis states that cratons are in isostatic equilibrium due to a perfect balance between compositional buoyancy and thermal buoyancy at every depth within the thermal boundary layer. It has now been over two decades since the formulation of the isopycnic hypothesis, and while perturbations and slight modifications to this model have been proposed, overall the basic idea of a *present* balance between compositional and thermal buoyancy beneath cratons still holds.

One might interpret the long lifespan of Jordan's hypothesis to imply that there is not much left to do in cratonic studies. However, the isopycnic hypothesis, while solving one paradox, generates more paradoxes. If continents are presently isopycnic, that is, the thermal and chemical buoyancies are perfectly balanced, the question that arises is whether continents have always been isopycnic. If we allow for the possibility that the average temperature of the continental lithosphere has cooled with time relative to the ambient mantle (and this could have been the case because the peridotite that makes up cratonic mantle melted at some point), then it is not likely that continents have always been isopycnic. It then follows that if, at any point in time, continents were not isopycnic, destabilizing buoyancy forces would have existed. For instance, if the continents started off hotter then they are now, they would have tended to pancake out by gravitational collapse (e.g., in the spirit of *Bird* [1991]), leaving us without a thick keel. The way out of this is to invoke a cratonic mantle that is more viscous from the outset, perhaps as a consequence of its unique chemical composition, which we will discuss later. But if cratonic mantle is so intrinsically viscous that the pancaking effect is suppressed at higher temperatures, we again return to the conclusion that cratons could not always have been in isostatic balance, unless (1) the thermal states of the cratonic mantle and ambient mantle have remained constant through time,

(2) the elevation of the craton has undergone large adjustments in response to the changing thermal state of the craton, and/or (3) the internal density structure of the craton is continually reorganized in such a manner as to maintain isostasy during thermal evolution of the craton. These are only some of the outstanding questions and apparent paradoxes regarding the origin and evolution of cratons.

These paradoxes stem from the fact that, by itself, the isopycnic hypothesis, which is based on isosatic principles, provides no direct insight into craton origin and evolution, both of which inherently involve the dynamic, e.g., nonstatic, history of cratons. The origin and evolution of cratons thus form the theme of this manuscript. Clearly, we will not be able to answer all of the above questions. We can, however, critically examine some of the endmember hypotheses for craton formation and highlight possible observation-based tests of these hypotheses. Craton formation is discussed in the context of three endmember scenarios (Figure 1): formation from the residue of melting in a single hot plume head, formation by accretion/stacking of oceanic lithosphere, and formation by thickening of arc lithosphere. The hope, of course, is that with a better understanding of how cratons form (and evolve), we may come to a better understanding of why cratons are the way they appear now. This paper focuses primarily on the perspective from the geochemistry and petrology of mantle xenoliths, which represent the only direct samples of cratonic mantle.

2. SALIENT FEATURES OF CRATONIC MANTLE

Comprehensive reviews on the thermal and compositional structure of cratonic mantle are already available [*Griffin et al.*, 1999, 2003; *O'Reilly et al.*, 2001; *Pearson et al.*, 2003; *Carlson et al.*, 2005] and readers are urged to refer to these papers for more details. Our goal here, however, is to briefly review those features of cratonic mantle that are relevant to the foregoing theme of this manuscript. We also outline the assumptions and uncertainties in these thermal and compositional constraints.

2.1. Defining the Thickness of the Cratonic Mantle

The thickness of cratonic mantle is probably the one parameter that leads to the most confusion in scientific discussions, particularly across disciplines. This confusion stems from different definitions and approaches in estimating the thickness of cratonic mantle. Similar confusion arises from the term "lithosphere", which strictly speaking should only be used to describe the strong mechanical boundary layer at the surface of the Earth but has also been used to describe chemical boundary layers and thermal boundary layers without proper specification. For these reasons, the choice has

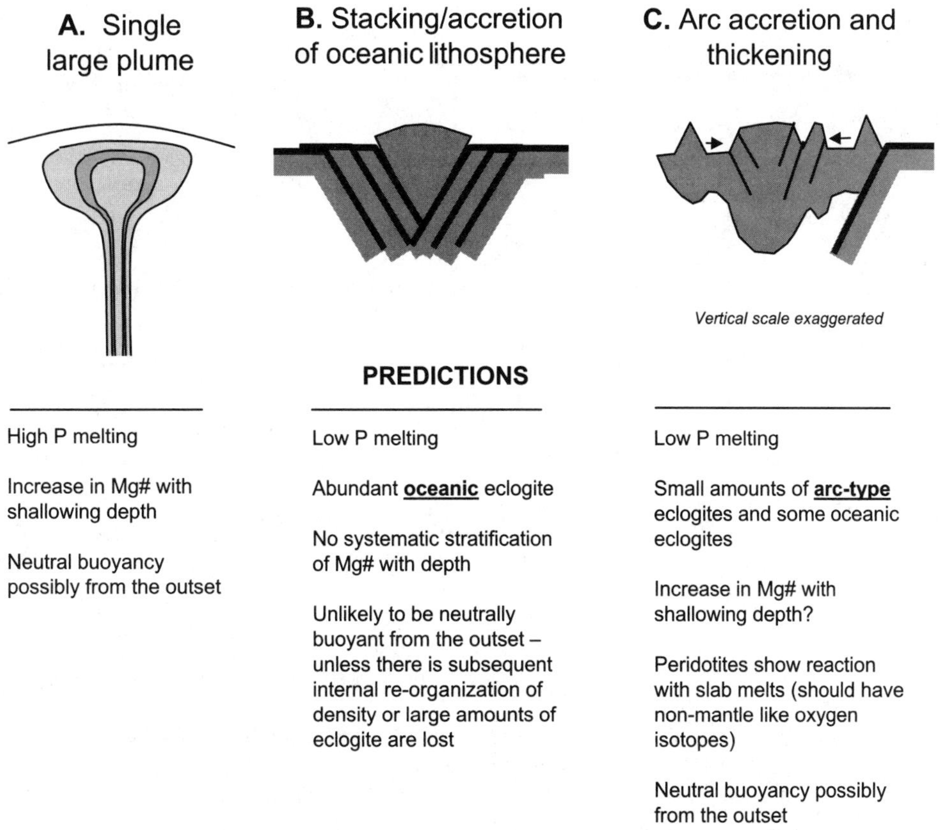

A. Single
large plume

B. Stacking/accretion
of oceanic lithosphere

C. Arc accretion and
thickening

Vertical scale exaggerated

PREDICTIONS

High P melting

Increase in Mg# with
shallowing depth

Neutral buoyancy
possibly from the outset

Low P melting

Abundant **oceanic** eclogite

No systematic stratification
of Mg# with depth

Unlikely to be neutrally
buoyant from the outset –
unless there is subsequent
internal re-organization of
density or large amounts of
eclogite are lost

Low P melting

Small amounts of **arc-type**
eclogites and some oceanic
eclogites

Increase in Mg# with
shallowing depth?

Peridotites show reaction
with slab melts (should have
non-mantle like oxygen
isotopes)

Neutral buoyancy possibly
from the outset

Figure 1. Endmember cartoons for craton formation scenarios and their associated predictions. (A) Plume head, (B) oceanic litho-sphere accretion, (C) arc accretion/thickening. The scenarios are not mutually exclusive. Vertical direction is greatly exaggerated.

been made to use the terms, "mechanical", "chemical" and "thermal boundary layers", for those instances where we wish to be very specific as to what definition of cratonic mantle thickness is of interest. When speaking in very general terms, "lithosphere" is used as a catch-all phrase for these three boundary layers.

We begin with the most unambiguous definition of the thickness of cratonic mantle, that is, to equate the thickness to that of the thermal boundary layer underlying cratons. The thermal boundary layer at the surface of the Earth is defined to be that part of the uppermost mantle across which the mode of vertical heat transfer transitions from thermal buoyancy-driven advection to conduction. The base of the thermal boundary layer can be uniquely defined with respect to the average potential temperature of the Earth's mantle by defining the base of the thermal boundary layer to correspond to the depth at which the temperature reaches some pre-defined fraction of the ambient mantle temperature, e.g., 0.9 or 0.95 times that of adiabatically upwelling mantle (Figure 2). Thus, if the temperature of the convecting mantle is known at depth (inferred from extrapolating the mantle potential temperature

to increasing depths along a solid adiabatic gradient), the thickness of the thermal boundary layer is known.

Because of the exponential increase in viscosity with decreasing temperature, the upper part of the thermal boundary layer must be highly viscous and hence strong. If cold enough, the viscosity will be high enough that this part of the thermal boundary layer will behave as a very viscous, and hence strong, layer. This layer will be effectively isolated from secondary convection and corresponds to a viscously defined mechanical boundary layer. The viscous mechanical boundary layer will be thinner than the thermal boundary layer. It is important to note that a continent's "elastic" thickness, i.e., the effective thickness over which the flexure of a continent under a surface load can be modeled as an elastic plate, does not necessarily equate with the viscously defined mechanical boundary layer, which will, in general, be thicker.

In the next subsection, we will learn that cratonic mantle is compositionally distinct from that of the underlying convecting mantle (asthenosphere). This compositionally distinct layer is referred to as the chemical boundary layer. An important

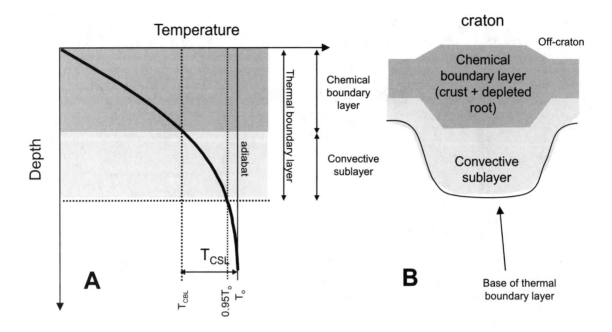

Figure 2. Schematic diagrams illustrating the relationships between the chemical boundary layer, thermal boundary layer, and convective sublayer (CSL) [*Cooper et al.*, 2004]). (A) Depth–temperature diagram showing that the base of the thermal boundary layer is defined by the point at which the temperature reaches 0.95 that of the ambient mantle temperature, T_0 (represented by the mantle adiabat). (B) Schematic cartoon corresponding to the diagram in (A); not to scale.

feature of the chemical boundary layer beneath cratons is that it is characterized by chemically buoyant peridotite as a consequence of high degrees of melt extraction. In the perfectly isopycnic condition proposed by Jordan, the negative thermal buoyancy associated with the craton's cooler thermal state relative to the ambient mantle, is exactly offset by an intrinsic compositional buoyancy at every depth. In this case, the base of the chemical boundary layer will coincide exactly with that of the thermal boundary layer. This equating of the two boundary layers is Jordan's "tectosphere".

If the cratonic mantle is not perfectly isopycnic, the base of the chemical boundary layer need not coincide precisely with the base of the thermal boundary layer. In particular, a possible by-product of extensive melt extraction is that the peridotite residues that make up the chemical boundary layer should have, at least initially, been dehydrated [*Pollack*, 1986; *Hirth and Kohlstedt*, 1996; *Aubaud et al.*, 2004]. Dehydration would potentially increase the intrinsic viscosity of the mantle residuum by two orders of magnitude [*Pollack*, 1986; *Hirth and Kohlstedt*, 1996]. The increase in viscosity coupled with the compositional buoyancy could help isolate this pre-existing chemical boundary layer from secondary convection [*Lenardic et al.*, 2003; *Cooper et al.*, 2004; *Dixon et al.*, 2004; *Lee et al.*, 2005b]. In this case, the chemical

boundary layer would equate with the viscously defined mechanical boundary layer (although in general the mechanical and chemical boundary layers do not necessarily have to coincide), not the thermal boundary layer, which is thicker. Underlying this strong chemical boundary layer would reside a convectively active sublayer ("convective sublayer") that makes up the base of the thermal boundary layer [*Cooper et al.*, 2004; *Lee et al.*, 2005b].

It is now clear that different approaches in determining the thickness of cratons will lead to different results because each approach is sensitive to different physical properties. For example, gravity and flexural studies of cratons provide constraints on elastic thickness. Seismic tomography will be most sensitive to the thermal boundary layer due to the high sensitivity of seismic velocities to temperature variations. In fact, if we define the thickness of the thermal boundary layer to be 0.95 times that of the ambient convecting mantle, seismic tomography will overestimate the thickness of the thermal boundary layer because it will still be sensitive to the remaining 5% allowable variation in temperature. Likewise, heat flow measurements and the assumption of steady-state conductive heat transfer provide information primarily on the thickness of the thermal boundary layer. Thermobarometric studies of mantle xenoliths may provide the most direct estimates of chemical and thermal boundary layer thicknesses, provided chemical equilibrium was maintained up until the time these xenoliths were erupted.

Below, we describe different approaches in estimating the thicknesses of cratons. What is important is that although each of these approaches comes with its own uncertainties, a consistent picture of craton thickness is emerging from the collective.

2.1.1. Thickness constraints from heat flow. The heat flow through cratons is inferred by measuring the temperature gradient down a borehole or well and multiplying it by the thermal conductivity of the rock. In general, cratons are characterized by surface heat flows between 30 and 50 mW/m^2, which is lower than the 50–80 mW/m^2 typically seen in Phanerozoic regions [*Pollack and Chapman*, 1977; *Nyblade and Pollack*, 1990; *Pollack et al.*, 1993; *Jaupart and Mareschal*, 1999; *Nyblade*, 1999]. Assuming one-dimensional steady-state and a purely conductive lid, surface heat flow can be used to construct a conductive geotherm for cratonic mantle, provided one has some knowledge (which in many cases is poorly constrained) of the thermal conductivity and concentration of heat-producing elements (U, Th, and K) in the crust and lithospheric mantle [*Pollack and Chapman*, 1977]. Examples of extrapolating a "skin" measurement (surface heat flow) to depth are shown in Figure 3, where heat production in cratonic mantle has been estimated from the average concentrations of K, U, and Th in mantle xenoliths. It can be seen that the low surface heat flow out of cratons implies that cratons are probably underlain by a thick and cold mantle keel (roughly 200 km), whereas that beneath Phanerozoic regions are much thinner and warmer.

Geotherm calculations, however, should be interpreted carefully because of the great number of assumptions inherent in such calculations. One of the largest uncertainties in extrapolating geotherms is the concentration of heat-producing elements at depth, particularly in the lithospheric mantle. Attempts have been made to use mantle xenoliths as constraints [*Rudnick et al.*, 1998], but this approach is not without uncertainty as the abundances of U, Th, and K in mantle xenoliths vary over an order of magnitude due to melt depletion and reenrichment processes. Additional uncertainties include the validity of assuming steady-state one-dimensional heat transport and purely conductive heat transfer [*Jaupart and Mareschal*, 1999]. For example, there is no doubt that cratons drift through the mantle, and because there are significant lateral variations in the temperature of the mantle (e.g., upwellings and downwellings), the transfer of heat through cratons, strictly speaking, cannot be in steady-state. Moreover, the timescales for radioactive decay of U, Th, and K are also on the same order of magnitude as thermal diffusion timescales, such that if there are nonnegligible amounts

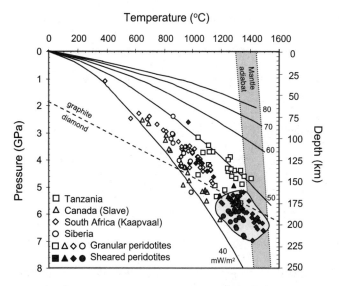

Figure 3. Thermobarometry data of xenoliths (using the *Brey and Kohler* [1990] Al in orthopyroxene coexisting with garnet barometer and the two-pyroxene thermometer) from different cratons combined with conductive geotherms inferred from surface heat flow (heat productions assumed for upper continental crust, middle continental crust, lower continental crust, and cratonic mantle are 1.1, 0.5, 0.1, and 0.02 μW/m^3, respectively). Open symbols represent low-temperature granular peridotites; colored symbols represent high-temperature sheared peridotites. The field of high-temperature sheared peridotites has been roughly outlined. Shaded vertical region represents range of mantle adiabatic temperatures at present.

of heat-producing elements in the lithospheric mantle (note that in the earlier *Pollack and Chapman* [1977] family of geotherms, heat production in the lithospheric mantle was assumed to be zero), the current thermal state in cratons cannot be in steady-state [*Michaut and Jaupart*, 2004]. Additionally, if there are any topographic variations at the base of a craton, heat transfer cannot be one-dimensional [*Lenardic and Moresi*, 2000]. Finally, the base of a craton is likely to be involved in secondary convection, and as a consequence, the assumption of a steady-state conductive geotherm must break down at great depths [*Lenardic et al.*, 2000].

2.1.2. Thickness constraints from xenolith thermobarometry. The thermal state of cratonic mantle can also be constrained from xenolith thermobarometry, which is based on the principle that the compositions of different mineral phases in a xenolith reflect a frozen-in equilibrium state, generally assumed to be the pressure and temperature at which the xenolith resided in the mantle just prior to eruption of the host kimberlite (although this does not always have to be the case). Thus, if it can be assumed that xenoliths are in thermodynamic equilibrium during their residence in the mantle, experimentally calibrated relationships between the activities of specific components within mineral phases can be used to estimate temperature and pressure [*Ellis and Green*, 1979; *Harley and Green*, 1982; *Finnerty and Boyd*, 1987; *Griffin et al.*, 1989; *Brey and Kohler*, 1990]. The results of such calculations on cratonic peridotites are well-known but are reproduced here for completeness. Note that there are no suitable barometers for spinel peridotites. Illustrated in Figure 3 are thermobarometric data for mantle xenoliths from the Tanzanian [*Rudnick et al.*, 1994; *Lee and Rudnick*, 1999], Slave (Canada; [*Kopylova et al.*, 1999a, b; *Kopylova and Russell*, 2000]), Kaapvaal (South Africa; [*Boyd*, 1987; *Finnerty and Boyd*, 1987; *Boyd et al.*, 1993; *Rudnick et al.*, 1998]), and Siberian cratons [*Boyd et al.*, 1997], all of which are Archean in age. The majority of cratonic mantle xenoliths derive from pressures between 3 and 7 GPa (corresponding to depths between ~90 and 210 km) and temperatures between 600°C and 1500°C. Cratonic peridotites thus define a thermal state which is well below the mantle adiabat and consistent with that determined from cratonic surface heat flow (Figure 3). Tanzanian xenoliths record slightly higher temperatures for a given depth, which may be due to the fact that the Tanzanian xenoliths sample the edge of the Tanzanian craton, which is bound by a Proterozoic mobile belt and is currently being affected by the southward-propagating East African Rift [*Nyblade and Brazier*, 2002]. The coincidence of the xenolith thermobarometry data with the heat flow-constrained geotherms is remarkable, given the potential uncertainties associated with both approaches.

Xenolith thermobarometric data can be used to reconstruct the thermal state of the cratonic mantle and hence provide information on both the thicknesses of the chemical and thermal boundary layers. This assumes, of course, that the last equilibrium states recorded by the xenoliths were set at the same time. In general, cratonic mantle xenoliths can be subdivided into roughly two textural groups [*Boyd*, 1987]. One group consists of granular peridotites. These typically have equilibration pressures and temperatures lower than 5.5 GPa and 1200°C. A second group consists of sheared or porphyroblastic textures, and these typically have equilibration pressures and temperatures exceeding 5.5 GPa and 1200°C. The low temperature granular peridotites further differ from the high temperature sheared peridotites in that the former are highly melt-depleted whereas the latter tend to be more fertile, some approaching the fertility of asthenospheric mantle. The transition (150–175 km) between the two groups most likely defines the base of a depleted mantle root, that is, the chemical boundary layer. Because the high-temperature sheared peridotites approach the temperature (~1500°C) of the mantle adiabat at depths of ~200 km, it must be concluded that the thermal boundary layer beneath cratons does not greatly exceed ~200 km (Figure 3), as previously pointed

out by *Rudnick et al.* [1998]. Thus, cratons are underlain by a ~200-km-thick thermal boundary layer, composed of a 150–175-km-thick chemical boundary layer beneath which lies a thin convective sublayer that makes up the base of the thermal boundary layer (Figure 3).

If, on the other hand, the high-temperature sheared peridotites and low-temperature granular peridotites equilibrated at different times, it is not exactly clear whether the thermobarometric constraints of these two xenolith groups can be pieced together as one continuous geotherm. It has been suggested that the high-temperature sheared peridotites may in fact reflect transient conditions associated with the kimberlite eruption or its precursor. This is based on the fact that many of these xenoliths actually show local textural and chemical disequilibria [*Smith and Boyd*, 1992; *Griffin et al.*, 1996, 1999]. If the high-temperature sheared peridotites have been affected by very recent heating, their estimated temperatures may overestimate the temperature at the base of the craton prior to heating. This would result in a slight underestimate of the thickness of the thermal boundary layer.

2.1.3. Thickness constraints from seismology. Seismic studies represent a powerful tool for constraining the current structure of cratonic mantle. Because seismic velocities are very sensitive to temperature (e.g., the sensitivity of V_P and V_S to temperature is ~0.7%/100°C and ~1%/100°C, respectively; [*Lee*, 2003; *James et al.*, 2004]), velocity anomalies are dominated by thermal variations. Seismic tomographic studies of cratons often show the presence of positive anomalies ("blue") extending to depths of 300–400 km beneath cratons [*Ritsema et al.*, 1998; *Simons et al.*, 1999; *James et al.*, 2001]. If these positive velocity anomalies correspond to the depth extent of cratonic mantle, there is a discrepancy with the depth extent of cratons as determined by xenolith thermobarometry.

This discrepancy can be reconciled in two ways. One way is to call upon disequilibrium and transient effects in some xenoliths as discussed above. Another way is to call upon the possibility of smearing effects in seismic tomography and the high sensitivity of seismic tomography to temperature variations. For example, seismic tomography studies of the Kaapvaal craton show "blue" P-wave velocity anomalies extending down to depths between 300 and 400 km [*James et al.*, 2001; *Shirey et al.*, 2002]. But at these depths, the velocity perturbations are only 0.5%. Using our above definition of thermal boundary layer thickness, that is, 0.95 that of the ambient mantle temperature, a temperature difference of roughly 75°C still persists below the defined base of the thermal boundary layer, assuming an ambient mantle temperature at 200 km depth of ~1500°C. This temperature difference corresponds to a ~0.5% variation in P-wave velocity. Without even considering smearing effects, the foregoing suggests that the "blue" regions in tomographic studies overestimate the true thickness of cratonic mantle.

More recent seismic approaches for estimating craton thicknesses are now appearing to be more consistent with the xenolith constraints. For example, S–P conversions can be used to map out the topography of the transition zone: Beneath Kaapvaal, there appears to be no measurable perturbation to the 410-km seismic discontinuity as would be expected if these cold keels penetrated the transition zone [*Niu et al.*, 2004]. In fact, when this observation is considered in the context of geodynamic constraints, the lack of an elevation of the 410-km discontinuity suggests that, at least for Kaapvaal, an upper bound to the thickness of the thermal boundary layer is ~300 km [*Cooper et al.*, 2004; *Niu et al.*, 2004]. Finally, recent surface-wave anisotropy studies of cratons also seem to suggest thinner keels (~200 km) to cratons [*Gung et al.*, 2003]. Thus, despite the uncertainties and assumptions that go with each of these different approaches in estimating craton thicknesses (heat flow, xenolith

thermobarometry, seismology), a consistent picture seems to be emerging.

2.2. Composition of the Cratonic Mantle and Chemical Buoyancy

One of the unique features of cratonic mantle is that it is dominated by high Mg# (molar Mg/(Mg + Fe) ×100) harzburgitic peridotite (91–94), e.g., olivine- and orthopyroxene-dominated lithologies [*Boyd and Mertzman*, 1987; *Boyd*, 1989; *Winterburn et al.*, 1989; *Boyd et al.*, 1993, 1997; *Rudnick et al.*, 1994; *Bernstein et al.*, 1998; *Griffin et al.*, 1999; *Lee and Rudnick*, 1999; *Pearson et al.*, 2003]. This can be seen in Figures 4–6. Cratonic peridotites are impoverished in clinopyroxene, Ca, Fe, and Al compared to fertile asthenospheric mantle and Phanerozoic peridotites, which are lherzolitic in composition (e.g., clinopyroxene 10–20%), have lower Mg#s (88–89 for fertile asthenosphere and 88–91 for Phanerozoic peridotites), and have higher Ca, Fe, and Al

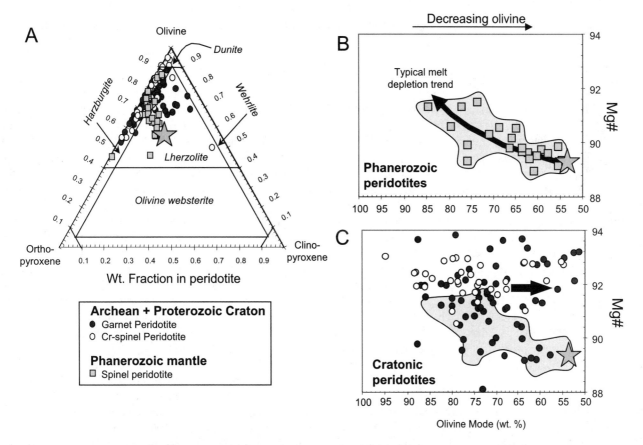

Figure 4. (A) Classification of peridotite lithologies based on weight fraction of olivine, orthopyroxene, and clinopyroxene. Star represents estimated composition of fertile asthenospheric mantle. (B) Mg# (molar Mg/(Mg + Fe) ×100) versus olivine mode in weight percent (note reverse axis) for Phanerozoic peridotites. Arrow denotes the expected trajectory of partial melting residues. (C) Same plot as in B but cratonic peridotites have been superimposed. A number of cratonic peridotites show anomalously high levels of orthopyroxene (low olivine mode) due to Si enrichment (large arrow pointing right).

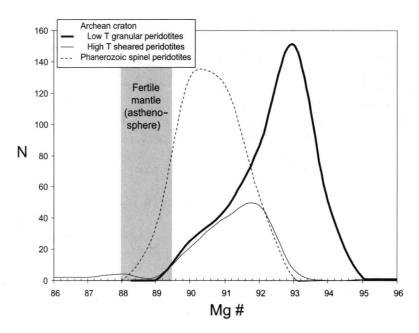

Figure 5. Mg# histogram of cratonic peridotites (low-temperature granular and high-temperature sheared), Phanerozoic mantle (continental lithospheric mantle xenoliths and obducted ophiolite massifs), and estimated fertile asthenospheric mantle (shaded vertical region).

(Figures 4–6). The depletion in clinopyroxene, Ca, Fe, and Al in cratonic peridotites relative to the asthenospheric mantle is most likely due to large degrees of melt extraction. Assuming all cratonic peridotites were originally derived from a more or less homogeneous convecting mantle having a primitive mantle-like composition (at least in terms of major elements and moderately incompatible trace elements), comparisons with partial melting experiments [*Walter*, 1998, 1999; *Herzberg*, 1999, 2004] suggest that most cratonic mantle peridotites have had 30–50% melt extracted (Figure 6). Only those cratonic peridotites that derive from the very base of the thermal boundary layer, e.g., the "high-temperature sheared peridotites" have fertile (lherzolitic) compositions (Mg# = 88–91; Figure 5). The higher fertility of the high-temperature sheared peridotites is due either to refertilization at the base of the thermal boundary layer [*Boyd*, 1987; *Smith and Boyd*, 1992; *Pearson et al.*, 1995b] or to the possibility that such peridotites have not undergone extensive partial melting.

One subset of cratonic peridotites (Figure 4c) is made up by those that appear to be anomalously Si-rich for their high Mg# and low Ca, Fe, and Al contents [*Boyd*, 1989; *Kelemen et al.*, 1998]. Melt depletion of a primitive mantle-like starting composition results in a subtle decrease in SiO₂ [*Walter*, 1998; *Herzberg*, 1999, 2004]; therefore, the high SiO₂ contents (or equivalently high orthopyroxene mode) for these highly melt-depleted harzburgites suggest that either silica

has been added or their initial starting compositions were Si-rich to begin with. It is largely assumed that the latter is not the case and, as such, most debates have centered around finding a mechanism for the Si-enrichment [*Canil*, 1992; *Herzberg*, 1993; *Rudnick et al.*, 1994; *Kelemen et al.*, 1998; *Herzberg*, 1999; *Smith et al.*, 1999]. One possibility is that harzburgitic residues reacted with silicic melts or fluids, perhaps from partial melting of a subducting slab. This process converts olivine to orthopyroxene, increasing the bulk SiO₂ content [*Kelemen et al.*, 1998; *Smith et al.*, 1999]; Si-enrichment has in fact been observed in a few subduction zone peridotites [*Ertan and Leeman*, 1996; *Smith et al.*, 1999; *McInnes et al.*, 2001; *Arai et al.*, 2004], possibly providing additional support for this hypothesis. Alternatively, the excess SiO₂ is derived by cumulate addition of orthopyroxene at depth, perhaps associated with melting in a large plume head [*Herzberg*, 1993, 1999; *Lee et al.*, 2003].

Regardless of how these Si-rich cratonic peridotites formed, an important conclusion is that the highly melt-depleted peridotites (including those with excess SiO₂) have lower intrinsic densities than fertile convecting mantle [*Boyd and McAllister*, 1976; *Lee*, 2003]. This is due to the decrease in clinopyroxene and garnet mode as well as the decreasing Fe content in olivine and orthopyroxene. Figure 7a shows how standard temperature and pressure (STP; 25°C and 1 atm) densities calculated for real cratonic peridotites decrease with increasing Mg#. Figure 7b shows the chemical

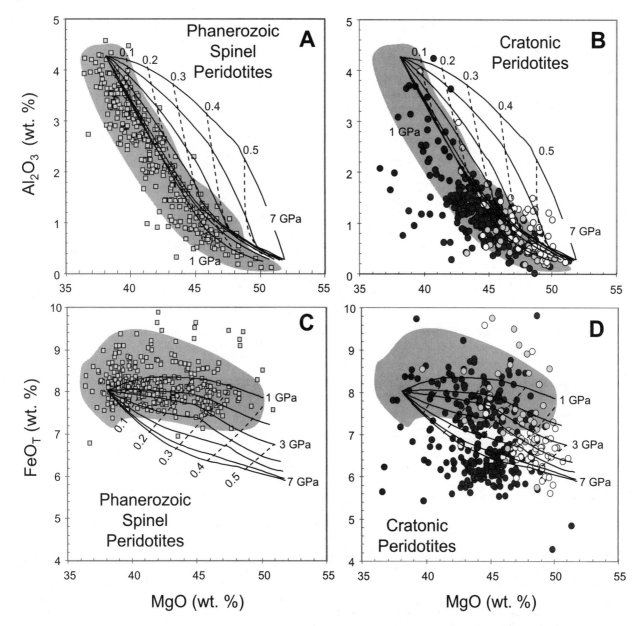

Figure 6. (A) Whole-rock Al_2O_3 versus MgO in Phanerozoic spinel peridotites. Solid curves represent the results of equilibrium melting experiments of a fertile lherzolite starting composition (KR-4003) from 1 to 7 GPa at 1-GPa intervals [*Walter*, 1998, 1999; *Herzberg and O'Hara*, 2002; *Herzberg*, 2004]. Dashed lines represent melt fraction contours at increments of 0.1. (B) Same plot as in (A), but cratonic peridotites have been superimposed. Shaded region represents Phanerozoic spinel peridotites from (A). Dark circles represent garnet-bearing cratonic peridotites. Open symbols represent Cr-spinel–bearing peridotites from the Greenland and Tanzanian cratons. (C) Phanerozoic spinel peridotites plotted in terms of FeO_T (total Fe as FeO) along with equilibrium melting experiments shown in (A). (D) Same as (C) but cratonic peridotites are superimposed.

buoyancy (density deficit) of cratonic mantle that is necessary to exactly offset its negative thermal buoyancy associated with its cooler thermal state (e.g., the isopycnic line). This chemical buoyancy curve can be converted into equivalent Mg# (Figure 7c) by using the empirical relationship shown in

Figure 7a. It can then be seen from Figsures 7c and 8 that the average Mg# of cratonic mantle between 3 and 7 GPa (equilibration pressure range over which most granular xenoliths are sampled) falls within the error range for neutral buoyancy (though perfect isopycnicity is not matched). Phanerozoic

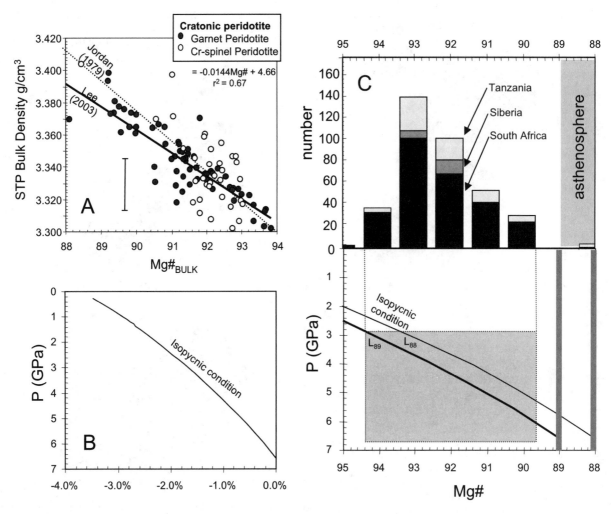

Figure 7. A. Normative (standard temperature and pressure) densities for cratonic peridotites versus Mg# [*Lee*, 2003]. Vertical bar represents largest estimated error. Line labeled "Lee (2003)" is a regression through the data. Line labeled "Jordan (1979)" represents Jordan's earlier parameterization based on estimated (rather than measured) mineral compositions [*Jordan*, 1979]. (B) Compositional buoyancy required to maintain neutral buoyancy at every depth for a craton characterized by a thermal state corresponding roughly to the xenolith geotherm shown in Figure 3 (roughly corresponding to the 45 mW/m² geotherm). (C) Top panel shows Mg# histogram of cratonic peridotites. Bottom panel shows the necessary Mg# required to achieve neutral buoyancy for the xenolith geotherm shown in Figure 3 and fertile asthenospheric mantle having Mg#s of 88 and 89 (L88 and L89). Shaded region represents the range of Mg#s and equilibration pressures seen in cratonic peridotites.

oceanic mantle, by contrast, becomes negatively buoyant within 20 Ma due to its less depleted nature (Figure 8).

2.3. Age and Longevity of Cratonic Mantle

One of the most important discoveries was that cratonic mantle is long-lived. The longevity of continental mantle was originally suggested on the basis of highly evolved Sr and Nd isotopic compositions of alkalic basalts and lamproiites derived from the continental lithospheric mantle [*Menzies*, 1989; *Hawkesworth et al.*, 1990]. These observations suggested that

continental lithospheric mantle resided in an incompatible-element–enriched environment isolated from the convecting mantle for long periods of time. Subsequently, similar isotopic studies of mantle xenoliths from cratons [*Menzies*, 1983; *Cohen et al.*, 1984; *Walker et al.*, 1989], as well as diamond inclusions [*Richardson et al.*, 1984], also suggested long-term isolation from the convecting mantle. The problem, however, is that Sr and Nd (and Pb) isotopes are based on incompatible element systems. Their isotopic compositions reflect the cumulative effects of metasomatic processes and, in general, do not yield information about the original partial melting

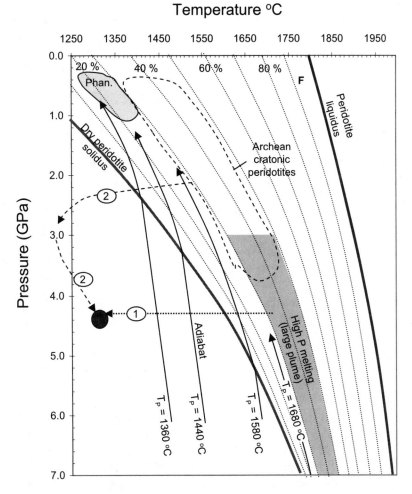

Figure 8. Schematic pressure–temperature diagram showing the dry peridotite solidus and peridotite liquidus along with constant melting degree F contours (in %) shown as dashed lines [*Katz et al.*, 2003]. Arrowed solid curves represent the P–T path of adiabatic decompression melting [*McKenzie*, 1984; *McKenzie and Bickle*, 1988]: The subsolidus leg represents the solid-mantle adiabat and the remaining part represents the melting adiabat, assuming the melt stays with the residue. Potential temperatures for each melting adiabat are denoted. Shaded region between 3 and 7 GPa refers to the range in pressures and melting degree F represented by cratonic peridotites. Shaded region at lower pressures (Phan) represents the region characterized by Phanerozoic spinel peridotites from various tectonic settings. Dashed open region for Archean cratonic peridotites reflects the preferred interpretation that cratonic peridotites formed by melting at pressures less than ~4 GPa. Dashed arrowed lines show hypothetical pressure–temperature paths: (1) isobaric cooling of plume head mantle residue, (2) low-pressure melt residues cooled isobarically and then tectonically transported to greater depths. Dark shaded circle at the endpoints of curves 1 and 2 is a *schematic* representation of present-day equilibration temperatures and pressures of cratonic peridotites.

event. Thus, these isotopes systems, although they suggest long-term isolation, cannot pin down exactly how long cratonic mantle has been convectively isolated.

Re-Os isotopic studies of mantle xenoliths have arguably provided the best (but still crude) constraint on the timing of cratonic mantle formation [*Walker et al.*, 1989; *Carlson and Irving*, 1994; *Pearson et al.*, 1995a, b, c; *Reisberg and Lorand*, 1995; *Handler et al.*, 1997; *Shirey and Walker*, 1998;

Carlson et al., 1999, 2005; *Chesley et al.*, 1999; *Lee et al.*, 2001b; *Shirey et al.*, 2002]. In contrast to Sr, Nd, and Pb, osmium is a highly compatible element and is therefore less easily overprinted by metasomatic processes (although Re can be overprinted). This makes Re-Os isotopic systematics a better recorder of initial melting conditions. Re-Os isotopic studies have shown that cratonic peridotites underwent melt extraction billions of years ago. Moreover, in many cases, the melt extraction ages (Re–Os model ages) fall within error of the age of the overlying crust, confirming that the present cratons are underlain by long-lived keels dating back to the time of craton formation.

2.4. Garnet Pyroxenites and Eclogites

Although peridotites typically dominate cratonic xenolith suites, garnet pyroxenite xenoliths (a term used to describe rocks composed of garnet and pyroxenes) are occasionally found [*MacGregor and Carter*, 1970; *Helmstaedt and Doig*, 1975; *Neal et al.*, 1990; *Jacob et al.*, 1994; *Fung and Haggerty*, 1995; *Beard et al.*, 1996; *Pyle and Haggerty*, 1998; *Jacob and Foley*, 1999; *Barth et al.*, 2001, 2002; *Taylor et al.*, 2003]. In South Africa, garnet pyroxenite xenoliths make up <1% of the xenolith population, peridotites making up the rest [*Schulze*, 1989]. However, in some rare instances, garnet pyroxenites dominate, such as in the Koidu kimberlite xenolith suite in Sierra Leone [*Fung and Haggerty*, 1995]. Whether xenolith demographics ever reflect the true distribution of lithologies in the mantle is unclear as it is impossible to prove that kimberlites randomly sample the underlying mantle. However, the high densities of garnet pyroxenites compared to peridotites make it unlikely (at least in the author's opinion) that such lithologies dominate cratonic mantle; if they did, cratons would be negatively buoyant.

Garnet pyroxenite xenoliths in cratons can, in general, be subdivided into high- and low-MgO groups [*Fung and Haggerty*, 1995; *Barth et al.*, 2001, 2002]. The low-MgO suites are dominated by true eclogites, that is, garnet pyroxenites containing only garnet and omphacitic clinopyroxene and no orthopyroxene. These low-MgO garnet pyroxenites are interpreted to represent fragments of subducted oceanic crust and oceanic gabbroic cumulates. This is based on the following: (1) except for slightly low SiO_2 contents, their major element compositions are very similar to that of mid-ocean ridge basalts [*Rollinson*, 1997], and (2) they have oxygen isotopic compositions suggestive of hydrothermal alteration [*Neal et al.*, 1990; *Jacob and Foley*, 1999; *Barth et al.*, 2001], which presumably could only have occurred at or near the surface of the Earth. In addition, many diamonds from kimberlites not only contain mineral inclusions of low-MgO eclogite affinity but also have extremely light carbon isotopic compositions interpreted to have a biogenic origin [*Kirkley et al.*, 1991; *Shirey et al.*, 2002; *Pearson et al.*, 2003]. It has been argued that the low SiO_2 of some of the low-MgO eclogites is due to the fact that they represent the residues of slab melting [*Rollinson*, 1997; *Barth et al.*, 2001] or of the melting of over-thickened basaltic crust [*Foley et al.*, 2002].

The high-MgO suite, on the other hand, consists of eclog-ites and garnet websterites. They have MgO contents too high to represent "typical" basaltic oceanic crust. They have

instead been interpreted as primitive cumulates, such as olivine–gabbro cumulates in oceanic settings [*Barth et al.*, 2002; *Taylor et al.*, 2003]. These xenoliths do not have anomalous oxygen isotopic compositions [*Barth et al.*, 2002]. In a later section, we will present an alternative hypothesis that explains these high-MgO eclogites as mafic cumulates associated with basaltic arc volcanism.

2.5. Seismic Heterogeneities in Cratonic Mantle

Recent seismic studies have revealed the presence of internal structure within cratonic mantle. For example, active source seismic experiments in and around the Slave craton have shown the existence of a pronounced dipping reflector on the edge of the craton [*Cook et al.*, 1997; *Bostock*, 1998, 1999]. This has been interpreted to represent a fossil subduction zone or an accreted sliver of oceanic lithosphere. More recently, teleseismic studies using receiver functions to map out distinct S–P conversions have revealed hints of dipping or possibly irregular interfaces within the Kaapvaal cratonic mantle [*Levander et al.*, 2005]. These seismic discontinuities may turn out to be the remnants of accreted oceanic lithosphere.

2.6. Summary

In summary, cratons are underlain by a thick thermal boundary layer not exceeding 300 km and probably falling within the range of 200–250 km. This thermal boundary layer is composed of a thick chemical boundary layer (150–175 km thick), which is made chemically buoyant due to high degrees of melt depletion. Preservation of ancient isotopic signatures indicates that, except for continent drift, cratons must be largely isolated from the convecting mantle. This implies that the chemical boundary layer underlying cratons must also be inherently strong, and it is speculated that this is the result of dehydration accompanying high degrees of melting. Beneath this strong chemical boundary layer lies the convectively active portion of the craton's thermal boundary layer, across which vertical heat transfer transitions from convection-dominated to conduction-dominated.

3. ENDMEMBER SCENARIOS FOR CRATON FORMATION

Craton formation scenarios can be classified into three endmembers. These include the formation of thick cratonic lithosphere from the mantle residuum of melting in a large mantle plume head (Figure 1a), by the accretion/stacking of oceanic lithosphere (Figure 1b) and/or by collision of arc lithosphere segment (Figure 1c). These endmember scenarios are not necessarily mutually exclusive. Below, each of these scenarios is described in more detail, emphasizing their specific geochemical and petrologic predictions.

3.1. Formation of Cratonic Mantle in a Single Hot Plume Head

The generation of a thick, highly melt-depleted mantle root requires high degrees of melting. At present, the high degrees of melt extraction characteristic of cratonic mantle (>30%) are represented only by localized hotspot magmatism or by the shallowest portions of the melting column associated with mid-ocean ridge basalt magmatism (Figure 8). Thus, a straightforward hypothesis for the formation of cratonic mantle is to call upon either a hotter and more vigorously convecting Archean mantle or the presence of larger and hotter thermal upwellings, e.g., plumes, in the Archean (Figure 1a). Both scenarios would lead to a thicker residual mantle column in addition to higher average degrees of melting (Figure 8). In such scenarios, the entire craton can be formed in one plume-melting event; that is, cratonic peridotites were formed in situ. The attractive aspect of the plume hypothesis is that a thick, strong, and chemically buoyant mantle keel can be generated by one seemingly simple and elegant process. The posterity of cratonic mantle would be ensured from its inception.

The geochemical/petrologic predictions of this hypothesis are as follows. If cratonic mantle represents the undisturbed residual mantle of a large plume head, that is, the last equilibration pressures of cratonic mantle xenoliths reflect the average pressure at which melt was last extracted, then a mantle with a very high potential temperature is required (Figure 8). For example, to achieve 30–50% melting at pressures of 3–7 GPa (the range in pressures from which cratonic mantle derives), a potential temperature significantly greater than 1650°C is required. This is hotter than the potential temperature of the hottest modern plumes (1400–1600°C) and at least 200°C higher than modern mantle beneath ridges (1250–1450°C) [*Herzberg*, 2004; *Putirka*, 2005]. Moreover, at these potential temperatures, partial melting would have actually started at pressures greater than 7 GPa, possibly up to 10 GPa (Figure 8). This is because the dT/dP of the lherzolite solidus decreases with increasing pressure such that at high enough pressures (>7 GPa), the melting adiabat roughly parallels that of the solidus (see Figure 8) and hence melting will occur over a very large depth interval. When we add on the fact that garnet is stable on the solidus over a large temperature range at high pressures [*Herzberg and Zhang*, 1996; *Walter*, 1998, 1999; *Herzberg and O'Hara*, 2002], the hot plume hypothesis predicts that cratonic peridotites should bear a distinct garnet trace-element signature if they indeed derive from a very hot plume head.

A second prediction of the plume hypothesis is that the residual mantle column should be stratified in terms of bulk composition as a consequence of decompression melting. The shallowest portions of the residual column will be the most depleted (e.g., highest Mg#) as this is where the temperature difference between the melting adiabat and the solidus is greatest [*McKenzie*, 1984; *Langmuir et al.*, 1992]. If the high Mg#s (91–94) of cratonic mantle peridotites were imparted by partial melting at their current equilibration pressures (3–7 GPa), one might also expect to find spinel peridotites (e.g., deriving from <3 GPa) with even higher degrees of melt extracted and accordingly higher Mg#s (>94); these are not seen, but their absence could be due to biased sampling. The high degrees of melt extraction also require the existence of large amounts of a complementary ultramafic magma, such as a komatiite [*Takahashi*, 1990]. Indeed, komatiites primarily occur in the Archean [*Nisbet et al.*, 1993; *Arndt*, 2003] but are not as abundant as one might expect (or require) from the plume scenario. However, at high enough pressures (~7–10 GPa), it is possible that komatiitic magmas may be neutrally buoyant with respect to olivine [*Stolper et al.*, 1981; *Agee and Walker*, 1993] and, as a consequence, may conveniently sink and dispose of themselves, obviating the need to explain the scarcity of komatiites in the geologic record.

3.2. Formation of Cratonic Mantle by Accretion/Stacking of Oceanic Lithosphere

A second way [*Helmstaedt and Schulze*, 1989] of generating cratonic mantle is by the accretion and stacking of oceanic lithosphere segments (Figure 1b). The motivation for this hypothesis was largely due to the presence of eclogite (a garnet + omphacite lithology) xenoliths in kimberlites, which has been interpreted to have hydrothermally altered oceanic crust protoliths (see section 2.5). This hypothesis has gained additional momentum in light of seismic studies done in the Slave craton, which showed the presence of a dipping seismic reflector on the edge of the craton [*Bostock*, 1998]. More recently, hints of dipping S–P conversions have been seen in the Kaapvaal craton [*Levander et al.*, 2005].

Unlike the plume-head model, the generation of a thick cratonic mantle keel by accretion of oceanic lithosphere is a sequential process. Thus, one might expect to see variation with depth in the timing of melt extraction. Attempts have been made using Re-Os isotopic systematics to determine if cratonic mantle is age-stratified [*Pearson et al.*, 1995c]. In the Kaapvaal craton, there is no systematic variation of Re-Os model ages with depth, at least to within 0.5 Gy. Originally, this was argued to provide evidence supporting the plume hypothesis [*Pearson et al.*, 1995c], but it does not rule out the accretion hypothesis if sufficient accretion and stacking of oceanic lithosphere can occur within 0.5 Gy. This time interval is within the noise range of Re-Os model ages. It may also be a reasonable amount of time to form a continent by accretion [*Şengör and Natal'in*, 1996].

Another distinction between the accretion and plume hypothesis is that cratonic mantle formed by accretion would be predicted to be composed of peridotites that had partially melted at fairly low pressures (<~4–5 GPa, depending on the potential temperature of the mantle) and subsequently transported to the greater pressures (3–7 GPa) from which they presently derive (Figure 8). If so, many cratonic mantle peridotites should have formed by melting in the spinel stability field. The garnets that are now present in cratonic peridotites would then have to represent subsolidus exsolution products from higher temperature (and lower pressure) orthopyroxenes, which are high in aluminum [*Cox et al.*, 1987; *Saltzer et al.*, 2001; *Simon et al.*, 2003]. Thus, in contrast to the plume hypothesis, we would not expect to see trace-element signatures of garnet even though garnet is now present in these xenoliths. The transposition of shallowly melted peridotites to greater depths also relaxes the requirement for extremely hot temperatures to generate high degrees of melting. This is because high degrees of melting can be generated by adiabatically decompressing a parcel of mantle to shallow pressures without calling upon extremely high temperatures (Figure 8). As a consequence, the large amounts of complementary komatiitic magmas required by the plume hypothesis are not required by the oceanic lithosphere accretion hypothesis.

Finally, if cratons are assembled by piecing together wholesale fragments of oceanic lithosphere, a range in fertility of the cratonic mantle might be expected. This is because the fertility of oceanic lithosphere is probably stratified. Although an entire section of oceanic lithosphere has never been sampled (typically, ophiolites sample only the uppermost part of the lithospheric mantle), the stratified nature of oceanic lithosphere is what would be predicted if oceanic lithosphere represents a residual melt column formed by passive upwelling and decompression melting [*Langmuir et al.*, 1992]. In such a scenario, Mg#s would be predicted to vary systematically from ~88 at the base of the melting column to ~91 at the top [*Langmuir et al.*, 1992; *Lee et al.*, 2005b]. Indeed, abyssal peridotites and ophiolitic peridotites, all of which undoubtedly sample the uppermost part of the residual melt column in oceanic lithosphere, are characterized by high degrees of melting and high Mg#s [*Dick and Bullen*, 1984; *Johnson et al.*, 1990]. Cratonic mantle peridotites, on the other hand, appear to be dominated by highly melt-depleted lithologies (Mg#s >91; Figure 5); only at the base of the cratonic mantle are fertile peridotites seen. Thus, if the oceanic lithosphere accretion hypothesis is correct, the deeper and more fertile portions of the oceanic lithosphere must be removed during or before the accretion process, so that the resulting cratonic mantle can be biased towards lower fertility as required by observation.

The main problem with the oceanic accretion hypothesis is that presently, oceanic lithosphere becomes negatively buoyant as it cools and hence becomes subductable as early as within 20 Ma of its inception (Figure 9). Thus, it seems unlikely that cratonic mantle could be formed by accretion of oceanic lithosphere segments having a compositional structure similar to today's oceanic lithosphere. On the other hand, if mantle potential temperature was just slightly higher in the Archean (>200°C), a thicker and hence more buoyant oceanic crust might have formed. A hotter mantle might also imply a more vigorously convecting Earth [*Davies*, 1992]. If so, mantle overturn times might have been shorter and the average age of oceanic lithosphere upon reaching trenches would be younger [*Davies*, 1992]. Collectively, such buoyancy may permit accretion and stacking of oceanic lithosphere, but there may be only a narrow range in parameter space for this to be a viable process [*Cooper et al.*, 2005].

3.3. Accretion/Thickening of Sub-Arc Mantle

Another scenario proposed for creating a thickened and highly melt-depleted mantle root is by thickening of sub-arc lithospheric mantle, perhaps through accretion or thickening of island arcs (Figure 1c). This was *Jordan's* [1988] preferred hypothesis for the formation of tectosphere (see also *Herzberg* [1999]). Dynamically, accretion of sub-arc lithosphere may differ fundamentally from accretion of oceanic lithospheric mantle. For example, arc lithosphere may already be buoyant, unlike oceanic lithosphere, which typically subducts. On the other hand, the petrologic predictions may appear superficially similar to the oceanic lithosphere accretion hypothesis because both accretion scenarios involve constituents formed by melting at shallower pressures and lower temperatures than implied by the hot plume scenario (Figure 8).

There are, however, some subtle differences between the two accretion scenarios. First, if arc lithospheric mantle was originally extracted from the asthenospheric mantle wedge in a subduction zone, one might expect partial melting of the mantle to be greatly influenced by fluids derived from dehydration and/or melting of the subducting slab. Interaction with extensive fluids is unlikely to be involved during the formation of oceanic lithospheric mantle. Of course, subsequent to accretion, it is possible that accreted segments of oceanic lithosphere could interact with later subducting fluids, but the fundamental difference is that such fluids would be interacting with already formed and presumably cold lithosphere whereas in the arc environment, the fluids would be interacting with hot asthenospheric mantle wedge and hence have

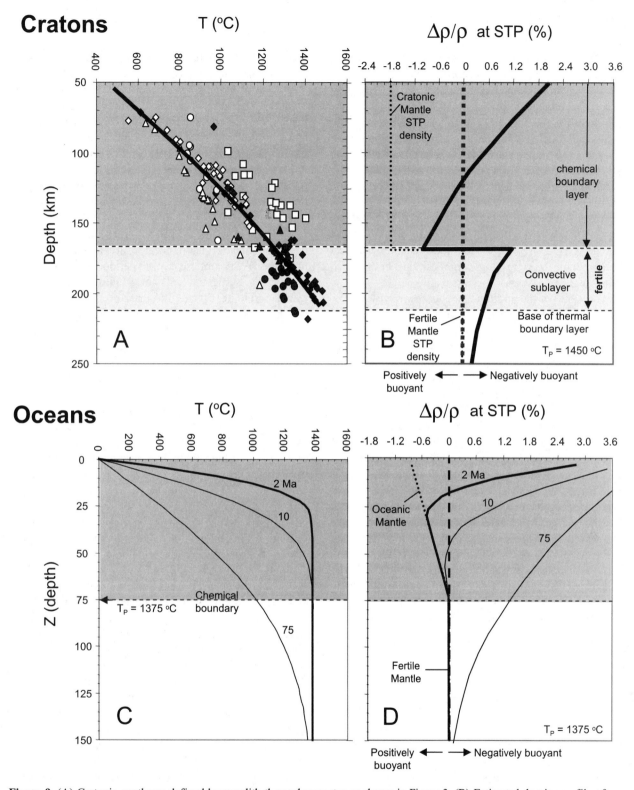

Figure 9. (A) Cratonic geotherm defined by xenolith thermobarometry as shown in Figure 3. (B) Estimated density profile of cratonic thermal boundary layer at roughly steady-state thickness, taking into account the thermal state in (A) and the compositional buoyancy of the chemical boundary layer and the convective sublayer (modified from *Lee et al.* [2005b]). Densities are given in terms of relative deviations from fertile peridotite (all calculations assuming STP conditions). (C) Transient geotherms of cooling oceanic lithosphere based on conductive cooling of an infinite half-space. Age of transient geotherms denoted. (D) Density profile of oceanic lithosphere, taking into account transient thermal states shown in (C) and a predicted compositional buoyancy for oceanic lithosphere. Slanted dashed line represents the standard temperature and pressure density profile of oceanic lithosphere associated with a gradual increase in Mg# with shallowing depth. Note oceanic lithosphere becomes negatively buoyant by 10 Ma.

a direct influence on the melting process. Thus, one useful prediction is that peridotites, originally derived from the mantle wedge, might show reaction with fluids/melts rising from the subducting slab. For example, interaction with silicic slab melts has the propensity to increase the SiO_2 content of highly melt-depleted harzburgites without changing the harzburgite's Mg# significantly [*Kelemen et al.*, 1998]. The anomalously Si-rich harzburgites seen in cratons may in fact have such an origin, although such harzburgites could also be the products of mixing residual peridotites with cumulate orthopyroxenes associated with melting in a large plume head [*Herzberg*, 1993, 1999]. The test of whether these Si-rich harzburgites represent melt–rock reaction products might lie in their oxygen isotopic compositions, but no such study has yet been systematically done.

Other possible petrologic differences between the two accretion hypotheses are the predicted abundances and compositions of garnet pyroxenite xenoliths in cratonic mantle. The oceanic lithosphere accretion scenario predicts a greater volume proportion of garnet pyroxentie lithologies (specifically low MgO eclogites) in the cratonic mantle, although admittedly, this would be a very difficult prediction to test since it is never clear whether xenolith sampling is representative of the true distribution in the mantle. However, there may be subtle differences in the composition of garnet pyroxenite lithologies found in cratons. For example, eclogite protoliths in the oceanic lithosphere accretion hypothesis would have major element compositions indicative of formation at shallower pressures, e.g., basaltic oceanic crust and cumulate gabbros. In contrast, the garnet pyroxenites formed in arc environments might have formed as higher pressure cumulates, such as at the base of the arc crust itself, which in general is thicker than oceanic crust. In the next section, we discuss whether some cratonic mantle garnet pyroxenite protoliths are arc cumulates.

4. THE PETROLOGIC/GEOCHEMICAL TESTS

4.1. Constraining the Pressures at Which Cratonic Peridotites Formed

As discussed in section 2.1.2, the pressures at which cratonic peridotites last equilibrated are in the range of 3–7 GPa (~90–200 km depth), but the question is whether they melted at these depths or were transposed to these depths since the time of melting. We can estimate melting pressures by examining their bulk compositions and assuming that they originate from a primitive mantle-like or pyrolitic source. Unlike mineral–mineral equilibria, which record conditions of last thermodynamic equilibrium, bulk composition can potentially preserve a record of past partial melting conditions (open system) insofar as the bulk system has remained closed to secondary disturbances since the melting event. In this sense,

major elements may be a natural choice as their concentrations in a peridotite should not change appreciably unless there has been significant mass loss or addition to the system.

Major element systematics, however, give ambiguous results. On the one hand, the low-Fe contents of some cratonic peridotites (Figures 6c, d) suggest, at face value, melting by 30–40% at 7 GPa. This can be seen in Figures 6c and d, where Phanerozoic spinel peridotites and Archean cratonic peridotites are superimposed on isobaric equilibrium melting experiments of a fertile peridotite [*Walter*, 1998, 1999; *Herzberg and O'Hara*, 2002; *Herzberg*, 2004]. We have taken Phanerozoic spinel peridotites as examples of modern-day peridotites formed in oceanic, arc, and continental environments. We do not use abyssal peridotites because they are extensively serpentinized; nevertheless, included in our spinel peridotite list are Phanerozoic ophiolite peridotites, which presumably formed in mid-ocean ridge or back-arc environments (see *Lee et al.* [2003] for references). For simplicity, we have adopted equilibrium isobaric melting rather than polybaric fractional melting. Although more realistic, polybaric fractional melting is highly model-dependent. In any case, the overall conclusions are similar regardless of whether the simplistic or complex approach is taken (for a more comprehensive treatment of polybaric fractional melting, the reader should refer to *Herzberg* [2004]). This conclusion is that high average pressures of melting (or initial pressures of melting in the polybaric case) results in residues having low-FeO contents due to the greater retention of orthopyroxene and garnet on the solidus at high pressures. It can be seen that Phanerozoic spinel peridotites have FeO contents consistent with melt extraction between 1 and 4 GPa (Figure 6c), whereas Archean cratonic peridotites have FeO contents indicating melting between 1 and 7 GPa, with some suggesting melting at even greater pressures (some of the cratonic peridotites with FeO contents greater than 8 wt.% are represented by the high-temperature sheared peridotites, which show evidence for Fe refertilization; hence the low apparent pressures in some may even be an artifact). In contrast to FeO, a very different picture arises from Al_2O_3–MgO systematics. The extensive involvement of garnet and pyroxene during melting at high pressures results in more aluminous residues for a given MgO content. Interestingly, the Al_2O_3–MgO systematics of cratonic peridotites are largely indistinguishable from those of Phanerozoic spinel peridotites and are easily explained by melting between 1 and 4 GPa (Figures 6a, b). The low Al_2O_3 contents of Archean cratonic peridotites has been used to argue that they actually melted at pressures lower than their current equilibration depths [*Bernstein et al.*, 1998].

We can expand on this debate by assessing the systematics of trace elements that may be sensitive to the presence of garnet. The elements most sensitive to garnet control are V, Y,

Sc, and the heavy rare-earth elements, these elements all being compatible in garnet. A convenient feature of these elements is that their *bulk* partition coefficients in peridotites are such that during partial melting they are only moderately incompatible relative to the bulk rock. This means that they do not become significantly depleted during partial melting and they are only slightly enriched in melts. As a consequence, these elements often remain relatively undisturbed by metasomatic processes, contrasting with the relative ease by which highly incompatible elements are overprinted by metasomatic processes [*Canil*, 2004]. Such behavior is confirmed by the fact that V, Sc, Y, and the heavy rare-earth elements in peridotites almost always preserve original melting systematics, while highly incompatible elements do not [*Lee et al.*, 2003; *Canil*, 2004]. *Canil* [2004] used Sc systematics of cratonic peridotites to show that they could not have formed by high-pressure melting. This is seen in Figure 10, where Sc is plotted against MgO. Partial melting trends of Sc versus MgO are negatively correlated because Sc is incompatible during melting whereas Mg is compatible. Each figure shows the effects of isobaric fractional melting at 1.5, 3, and 7 GPa [*Lee et al.*, 2005a; *Li and Lee*, 2004]. The extensive involvement of garnet during melting at 7 GPa results in less rapid depletion in Sc then at 1.5 and 3 GPa, where garnet is not present or present only during the very early stages of melting, respectively (Figure 10). Cratonic mantle peridotites (Tanzania, South Africa, and Siberia) have Sc systematics nearly indistinguishable from Phanerozoic spinel peridotites.

There thus appears to be no hint from Sc that cratonic peridotites originally melted at 7 GPa. Similar conclusions can be made from V–MgO systematics after the redox-sensitive behavior of V is accounted for [*Canil*, 2002, 2004; *Lee et al.*, 2003].

It thus seems likely that the majority of cratonic peridotites melted on average anywhere between 1 and 4 GPa, that is, mainly in the spinel stability field (although initial pressure of melting may have been deeper). If so, these peridotites were subsequently transported to greater depths (~7 GPa), after which subsolidus re-equilibration resulted in exsolution of garnet (as well as small amounts of clinopyroxene). An exsolution origin for the small amounts of garnet and clinopyroxene in cratonic harzburgites has been previously suggested on the basis of textural studies [*Cox et al.*, 1987]. These conclusions, if correct, require an explanation for the low FeO contents of some cratonic peridotites (Figure 11), which at face value imply very high pressures of melting [*Pearson et al.*, 1995b]. *Herzberg* [2004] suggested that the low FeO contents are related to Si-enrichment processes associated with subsequent metasomatism and/or cumulate addition of orthopyroxene (Figure 4b). This can be seen by first identifying those cratonic peridotites that appear to be

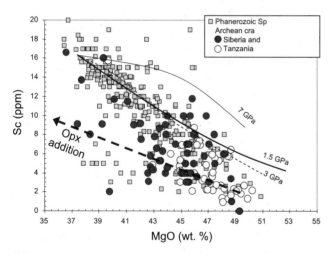

Figure 10. Plot of whole-rock Sc (ppm) versus MgO (wt.%) for Phanerozoic spinel peridotites and Archean cratonic peridotites (Siberia, Tanzania, South Africa). Melting curves for 1.5, 3, and 7 GPa are shown; Sc is elevated during 7-GPa melting due to the greater involvement of garnet on the solidus at high pressures and the high compatibility of Sc in garnet. Dashed arrowed line corresponds to the trend expected for the addition of orthopyroxene through melt-rock reaction or cumulate orthopyroxene addition [*Lee et al.*, 2003].

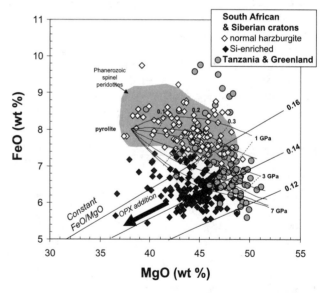

Figure 11. Plot of total FeO versus MgO for cratonic peridotites along with shaded field for Phanerozoic spinel peridotites. Curved lines represent 1–7 GPa equilibrium melting curves as in Figures 6 and 10. Dark-colored diamonds represent those samples identified as Si-enriched, based on their Mg/Si ratios. Arrow shows the direction of Si- and orthopyroxene-enrichment. Diagonal straight lines represent mixing with orthopyroxene having identical FeO/MgO with original peridotite (numbers correspond to FeO/MgO ratios). Note that addition of orthopyroxene can give results in low bulk FeO contents, giving an artifact of a high-pressure melting origin.

anomalously rich in Si (Figure 4b). For example, we can unambiguously identify Si-enriched peridotites as those having a Mg/Si ratio (an inverse proxy for Si-enrichment) less than that of Phanerozoic peridotites at a given Mg#. Using this criterion, we can see that Si-enriched peridotites systematically have lower FeO contents than "normal" cratonic peridotites (Figure 11). It also appears that the lower FeO contents of Si-enriched peridotites are also accompanied by a decrease in MgO content. Effectively, the low FeO and MgO contents of Si-enriched peridotites are consistent with the addition of a Si-rich component (cumulate or melt-rock reaction) having a FeO/MgO (or Mg#) ratio similar to that of the original peridotite. Mantle orthopyroxenes fit the composition of this endmember, consistent with the fact that the Si-enriched peridotites are anomalously rich in orthopyroxene [*Herzberg*, 2004].

If the Si-enriched peridotites are excluded, the consensus seems to be growing that few cratonic peridotites are residues of melting at average pressures as great as 7 GPa [*Bernstein et al.*, 1998; *Canil*, 2004; *Herzberg*, 2004]. The major and moderately incompatible trace element systematics of such cratonic peridotites are very similar to that for Phanerozoic spinel peridotites and can be modeled as 30–50% melt residues formed at pressures ~<4 GPa Thus, most cratonic peridotites probably melted at low pressures and have since been transported to the great depths (90–220 km) from which they currently derive. If this is the case, the potential temperature of the mantle in which these cratonic peridotites melted was likely to be between 1400°C and 1650°C (Figure 8). This straddles the upper end of accepted potential temperatures for modern mid-ocean ridge environments (1300–1450°C) [*Langmuir et al.*, 1992; *Asimow and Langmuir*, 2003; *Lee et al.*, 2005b; *Putirka*, 2005] and falls within the range of potential temperatures estimated for modern plume settings (1450–1650°C) [*Herzberg*, 2004; *Putirka*, 2005]. These conclusions, combined with the fact that there is no *gradual* decrease in fertility with increasing depth, suggest that cratonic mantle does not represent the residual mantle column of a single, large, and anomalously hot (potential temperature > 1650°C) plume head. Instead, cratonic peridotites could have formed at Archean mid-ocean ridges having a slightly higher potential temperature (by 100–200°C) than in the present. If the Archean mantle potential temperature was identical to the present, then cratonic peridotites must have formed in small thermal plumes, not unlike what is seen today. In either case, cratonic peridotites did not form in situ; thus they have since been transported to greater depths.

4.2. Constraining the Protoliths of Cratonic Eclogites

We now turn to assessing the two accretion hypotheses. To do so, we will focus on the composition of garnet pyroxenite

protoliths with particular emphasis on the high-MgO lithologies (see section 2 for discussion of garnet pyroxenite classification). Garnet pyroxenite protoliths in the oceanic lithosphere accretion hypothesis are likely to be dominated by basaltic oceanic crust and low-pressure gabbroic cumulates, the latter consisting of olivine–pyroxene lithologies (primitive cumulates) and/or pyroxene–plagioclase lithologies (evolved cumulates). The primitive olivine–pyroxene lithologies would give rise to MgO-rich, SiO_2-poor, and Al_2O_3-poor garnet pyroxenites upon subsolidus re-equilibration within the cratonic mantle.

Unlike ridge environments, arc magmas must traverse pre-existing lithosphere such that magmatic differentiation probably commences at greater pressures (e.g, >1 GPa) than beneath mid-ocean ridges (<1 GPa). At higher pressures, the olivine phase field diminishes significantly, so instead of precipitating olivine-dominated cumulates, as would occur in mid-ocean ridge settings, the most primitive cumulates in arc settings would be clinopyroxene-rich (this is corroborated by applying the pMELTS thermodynamic algorithm to the crystallization of a primary basaltic magma at 1 GPa). Thus, in the arc case, hig-MgO garnet pyroxenites would be represented by clinopyroxene-rich lithologies characterized by high MgO, high SiO_2, and low Al_2O_3 contents.

In Figure 12, we have plotted the bulk compositions of garnet pyroxenite xenoliths derived from the lower crust and lithospheric mantle underlying the Sierra Nevada batholith in California (western North America), an extinct Mesozoic continental arc [cf. *Ducea and Saleeby*, 1996, 1998; *Lee*, 2005; *Lee et al.*, 2006]. These pyroxenites can be subdivided into a high-MgO (>8 wt.%) and a low-MgO group [*Lee et al.*, 2006], the former being more primitive and the latter more evolved. Both groups are represented by garnet clinopyroxenites, but the high-MgO group has more than 50% modal clinopyroxene and the latter has more than 50% garnet. This is consistent with the fact that the high-MgO group has SiO_2 contents of ~50–52 wt.% (Figure 12A) and Al_2O_3 contents of 6–10 wt.%), while the low-MgO group has lower SiO_2 (Figure 12) and higher Al_2O_3 (Figure 12B). Also shown in Figure 12 are plutonic rocks from the Sierras, volcanic rocks from the Cascades volcanic arc (GEOROC database), and gabbroic cumulates (plagioclase and hornblende-rich) associated with the Sierra Nevada batholith [*Sisson et al.*, 1996]. The high SiO_2 contents of the high-MgO group eliminate an olivine-bearing protolith and are instead more consistent with a clinopyroxene-rich protolith. This clinopyroxene-rich protolith could also have had original cumulate garnet, but moderate enrichment in heavy rare earths (not shown) suggests that most of the garnet is subsolidus. The low-MgO group, on the other hand, is similar to the more evolved Sierran cumulates in terms of major elements. Thus, one possible interpretation is that the high-MgO Sierran garnet pyroxenites

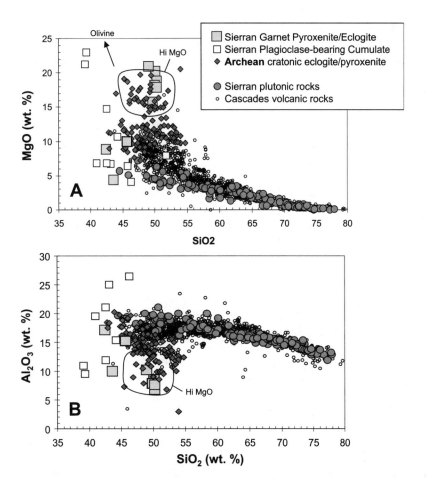

Figure 12. Whole-rock MgO (A) and Al₂O₃ (B) versus SiO₂, showing cratonic eclogites (Siberia, South Africa, West Africa) compared to Phanerozoic eclogites/garnet pyroxenites from an arc environment (Sierra Nevada batholith). References are given in text. Also shown are Cascades magmatic arc series and Sierra Nevada plutonic rocks. It can be seen that some of the high-MgO cratonic eclogites/pyroxenites have compositions similar to those of the high-MgO Sierran garnet pyroxenites.

represent primitive high-pressure cumulates, whereas the low-MgO pyroxenites represent more evolved cumulates. Indeed, initial crystallization of the high-MgO pyroxenites is consistent with the fact that the primitive Sierran plutons have anomalously low MgO and high Al₂O₃ contents for a given SiO₂. Subsequent crystallization of the more evolved low-MgO pyroxenites could then drive the initial increase in SiO₂ seen in the Sierran magmatic differentiation array.

In Figure 12, Archean cratonic garnet pyroxenite (mostly bimineralic eclogites) xenoliths from Siberia [*Jacob et al.*, 1994; *Beard et al.*, 1996; *Jacob and Foley*, 1999; *Taylor et al.*, 2003], South Africa [*Pyle and Haggerty*, 1998], and Sierra Leone in west Africa [*Fung and Haggerty*, 1995] have been superimposed. As discussed previously, the rough subdivision of the cratonic garnet pyroxenites into high- and low-MgO groups is apparent. The low-MgO group appears to have compositions similar to modern basaltic oceanic crust,

although Al₂O₃ and SiO₂ contents are slightly lower. The low-MgO group has been argued to have an oceanic crust protolith coupled with subsequent partial melting in the eclogite or amphibolite stability field [*Barth et al.*, 2001]. The high-MgO group has been interpreted to be either low-pressure olivine–plagioclase–pyroxene cumulates or high-pressure (2–3 GPa) garnet pyroxenite cumulates [*Barth et al.*, 2002]. However, as shown in Figure 12, the high-MgO Archean pyroxenites are remarkably similar to the high-MgO Sierran pyroxenites, interpreted here to be high-pressure arc cumulates.

If these compositional similarities are not a matter of coincidence, the easiest interpretation is that at least some of the high-MgO Archean garnet pyroxenites represent primitive cumulates associated with Sierran-like continental arc magmatism. If so, the high-MgO Archean garnet pyroxenites are evidence that at least some Archean cratons may have formed by accretion/thickening of arcs. The low-MgO Archean eclogites are generally interpreted to represent former subducted oceanic crust. Collectively, this suggests that craton formation may be a combination of arc and oceanic lithosphere accretion. Alternatively, the low-MgO Archean eclogites may represent arc basalts rather than oceanic crust basalts.

5. AFTERTHOUGHTS

In light of the foregoing discussions, it is worthwhile returning to the questions of why cratons are so long-lived and why are they presently neutrally buoyant? Any successful model for the origin of cratons must satisfy these conditions. In this context, the plume-head scenario seems highly attractive because it provides a thick chemically depleted and possibly dehydrated root that is strong from the outset, ensuring its longevity and approximate neutral buoyancy. This model, however, predicts that cratonic peridotites have a high pressure origin, which appears not to be the case in general.

The oceanic lithosphere accretion hypothesis is attractive in that it can explain the origin of low-MgO eclogite xenoliths in cratons as well as the presence of dipping reflectors in the margins of cratonic mantle. As pointed out above, however, for neutral buoyancy to be maintained, this hypothesis requires a very delicate balance between the amounts of eclogitized oceanic crust and depleted oceanic mantle: The more depleted (and hence chemically buoyant) the oceanic mantle, the more complementary eclogitized oceanic crust that must exist. Too much eclogite or garnet pyroxenite, regardless of how depleted the oceanic mantle is, will prevent the craton from ever being neutrally buoyant.

The arc accretion/thickening hypothesis is also attractive. In this hypothesis, highly depleted mantle would be expected throughout the cratonic mantle, unlike accretion/stacking of oceanic lithosphere, which would yield a periodic variation in fertility with depth. In addition, the amounts of eclogite/pyroxenite in the cratonic mantle would probably be less, especially if most arc-generated pyroxenites are delaminated or removed during or shortly after arc formation. By the time arc accretion occurs, only small amounts of garnet pyroxenite remain. Thus, the range of conditions, which permit neutral buoyancy in this hypothesis, seems (at face value) more easily achieved and maintained, but the dynamic feasibility of this process needs to be tested.

Clearly, craton formation is a complicated process and the true answer could very well be a mixture of formation scenarios. Identifying these formation processes undoubtedly will help us understand to what extent the dynamics of the Earth's mantle have changed through time. It has been shown that, on average, there may have been a rough secular change in the composition and structure of continental lithospheric mantle

[Griffin et al., 1999]; that is, the thickest and most melt-depleted lithospheric mantles are associated with Archean cratons, while the thinner and more fertile ones are Phanerozoic (Proterozoic lithospheres are intermediate). In light of the endmember formation scenarios, these secular changes could imply any of the following: the occurrence of larger and hotter plumes in the Archean, a hotter Archean mantle, or faster spreading rates in the Archean. Faster spreading rates might increase the probability of generating thickened arc lithosphere or stacked oceanic lithosphere sections.

In some cases, however, there appears to be no correlation between thickness and composition versus age. For example, late Archean to Paleo–Proterozoic lithospheric mantle underlies the Mojave block in southwestern United States [Lee et al., 2001b]. Although this mantle is fertile, the adjacent middle Proterozoic Colorado Plateau is underlain by a thick, highly melt-depleted cratonic root, not unlike "archetypal" Archean cratons. Another example is in Australia, where it has been shown seismically that some of the thickest cratons underlie Proterozoic regions rather than Archean regions [Simons et al., 1999]. These "exceptions to the rule" show that although there may very well be a secular change in the way or rate at which cratons are formed, one must conclude that at least some of the cratons seen today are simply a result of biased preservation. If so, this begs the question of how much crust and lithospheric mantle was generated in the past but just didn't have the necessary compositional, viscosity, and thermal structure to survive.

While we have focused primarily on the origin of cratons, it could be argued that an equally important perspective might come from studying the eventual fate of cratons and of continental lithosphere in general. Once a stable continental lithosphere is formed, does it last forever [O'Reilly et al., 2001]? If, as suggested here, cratonic mantle might have started out dehydrated due to the high degrees of melt extraction, cratonic mantle could be inherently strong and hence stable with respect to the convecting mantle. However, this outcome seems at face value inconsistent with the fact that many cratonic mantle xenoliths bear evidence for hydrous metasomatism (see Pearson et al. [2003] for a review). It has even been suggested that continental lithospheric mantle is weak, perhaps as a consequence of hydrous metasomatism [Maggi et al., 2000a, 2000b; Jackson, 2002]. If so, this challenges the view of a strong lithospheric mantle. The paradox, of course, is that rehydration might lead to weakening of the continental lithosphere, destroying all hope of a long life. Indeed, there are some lines of observational evidence that suggest continental lithospheric mantle, under some circumstances, may in fact be destabilized and recycled back into the convecting mantle [Kay and Kay, 1993; Ducea and Saleeby, 1996; Lee et al., 2000, 2001a; Gao et al., 2002]. Understanding the mechanisms of lithospheric destabilization will thus remain an area of open debate. A number of geodynamic studies have shown that lithospheric mantle can be destabilized during compressional orogenies [*Houseman et al.*, 1981; *Conrad and Molnar*, 1997; *Houseman and Molnar*, 1997; *Molnar et al.*, 1998; *Jull*, 2001], but none of these studies have incorporated the effects of dehydration (or rehydration); hence, the obvious question is what role does water play in the stabilization and potential destabilization of continental lithospheric mantle? The issue of water, that is, how it got there (or left), how it influences rheology, and how much is present, may ultimately be one of the most important foci for future research in the dynamics of continents.

6. CONCLUSIONS

The salient features of cratonic mantle were reviewed in this paper and discussed in the context of geochemical and petrologic predictions associated with three endmember formation scenarios: (1) formation by large degrees of melting in a single, large plume head, (2) formation by accretion/stacking of oceanic lithosphere, and (3) formation by accretion/thickening of arc lithosphere. The following conclusions were made:

1. Cratons are underlain by a thick thermal boundary layer composed of a highly melt-depleted and compositionally buoyant chemical boundary layer. This chemical boundary layer must be strong in order to ensure long-term preservation. The chemical boundary layer is underlain by a thin convectively active sublayer that makes up the base of the thermal boundary layer.
2. Cratonic peridotites appear to have melted on average at lower pressures (<4 GPa) than those from which they presently derive; hence, they have been transposed to greater depths after their formation.
3. The protoliths of Archean low-MgO eclogites are related to subducted basaltic crust of either oceanic or arc origin.
4. High-MgO Archean eclogites are similar to high-pressure cumulates found in Phanerozoic arc environments.

The low pressures from which cratonic peridotites originally melted suggest that cratonic mantle does not represent the residual melt column of a single, anomalously hot plume head. The low pressures are instead more consistent with formation by arc or oceanic lithosphere accretion. These latter scenarios are unfortunately difficult to distinguish petrologically. However, an arc component seems to be required in at least some cratons.

Acknowledgments. The following people are thanked for instigating and inspiring my interest in cratons, peridotites, and eclogites over the years: R. Rudnick, W. McDonough, P. Kelemen, D. Canil, C. Herzberg, G. Brimhall, and M. Barth. Special thanks go to my colleagues at Rice, A. Lenardic, W. Leeman, F. Niu, and A. Levander, for sharing these interests. Finally, I would like to thank the following graduate students for insightful, and sometimes humorous, craton discussions: C. Cooper, Z.-X. A. Li, S. Wood, X. Cheng, and E. Vanacore. This work also benefited from discussions and/or reviews by D. Anderson, D. Canil, C. Herzberg, P. Kelemen, K. Benn, and C. J. O'Neill. This work was supported by NSF grants EAR-0440033 and 0309121.

REFERENCES

Agee, C.B., and D. Walker, Olivine flotation in mantle melt, *Earth Planet. Sci. Lett.*, 114, 315-324, 1993.

Arai, S., S. Takada, K. Michibayahsi, and M. Kida, Petrology of peridotite xenoliths from Iraya volcano, Philippines, and its ipmilcation for dynamic mantle-wedge processes, *J. Petrol.*, 45, 369-389, 2004.

Arndt, N.T., Komatiites, kimberlites and boninites, *Journ. Geophys. Res.*, 108, 2293, 2003.

Asimow, P.D., and C.H. Langmuir, The importance of water to oceanic mantle melting regimes, *Nature*, 421, 815-820, 2003.

Aubaud, C., E. Hauri, and M.M. Hirschmann, Hydrogen partition coefficients between nominally anhydrous minerals and basaltic melts, *Geophys. Res. Lett.*, 31, doi:10.1029/2004GL021341, 2004.

Barth, M.G., R.L. Rudnick, I. Horn, W.F. McDonough, M.J. Spicuzza, J.W. Valley, and S.E. Haggerty, Geochemistry of xenolithic eclogites from West Africa; Part I, A link between low MgO eclogites and Archean crust formation, *Geochim. Cosmochim. Acta*, 65, 1499-1527, 2001.

Barth, M.G., R.L. Rudnick, I. Horn, W.F. McDonough, M.J. Spicuzza, J.W. Valley, and S.E. Haggerty, Geochemistry of xenolithic eclogites from West Africa, part 2: origins of the high MgO eclogites, *Geochim. Cosmochim. Acta*, 66, 4325-4345, 2002.

Beard, B.L., K.N. Fraracci, L.A. Taylor, G.A. Snyder, R.N. Clayton, T.K. Mayeda, and N.V. Sobolev, Petrography and geochemistry of eclogites from the Mir kimberlite, Yakutia, Russia, *Contrib. Mineral. Petrol.*, 125, 293-310, 1996.

Bernstein, S., P.B. Kelemen, and C.K. Brooks, Depleted spinel harzburgite xenoliths in Tertiary dykes from East Greenland: restites from high degree melting, *Earth and Planetary Science Letters*, 154, 221-235, 1998.

Bird, P., Lateral extrusion of lower crust from under high topography, in the isostatic limit, *J. Geophys. Res.*, 96, 10275-10286, 1991.

Bostock, M., Mantle stratigraphy and evolution of the Slave province, *J. Geophys. Res.*, 103, 21183-21200, 1998.

Bostock, M., Seismic imaging of lithospheric discontinuities and continental evolution, *Lithos*, 48, 1-16, 1999.

Boyd, F.R., High- and low-temperature garnet peridotite xenoliths and their possible relation to the lithosphere-asthenosphere boundary beneath southern Africa, in *Mantle xenoliths*, edited by P.H. Nixon, pp. 403-412, John Wiley & Sons Ltd, 1987.

Boyd, F.R., Compositional distinction between oceanic and cratonic lithosphere, *Earth Planet. Sci. Lett.*, 96, 15-26, 1989.

Boyd, F.R., and R.H. McAllister, Densities of fertile and sterile garnet peridotites, *Geophys. Res. Lett.*, 3, 509-512, 1976.

Boyd, F.R., and S.A. Mertzman, Composition and structure of the Kapvaal lithosphere, Southern Africa, in *Magmatic processes: physicochemical principles: a volume in honor of Hatten S. Yoder, Jr.*, edited by B.O. Mysen, pp. 13-24, Geochem. Soc. Spec. Pub., 1987.

Boyd, F.R., D.G. Pearson, P.H. Nixon, and S.A. Mertzman, Low-calcium garnet hazburgites from southern Africa: their relations to craton structure and diamond crystallisation, *Contributions to Mineralogy and Petrology*, 113, 352-366, 1993.

Boyd, F.R., N.P. Pokhilenko, D.G. Pearson, S.A. Mertzman, N.V. Sobolev, and L.W. Finger, Composition of the Siberian cratonic mantle: evidence from Udachnaya peridotite xenoliths, *Contrib. Mineral. Petrol.*, 128, 228-246, 1997.

Brey, G.P., and T. Kohler, Geothermobarometry in four-phase lherzolites II. New thermobarometers, and practical assessment of existing thermobarometers, *Journ. Petrol.*, 31, 1353-1378, 1990.

Canil, D., Orthopyroxene stability along the peridotite solidus and the origin of cratonic lithosphere beneath southern Africa, *Earth Planet. Sci. Lett.*, 111, 83-95, 1992.

Canil, D., Vanadium in peridotites, mantle redox and tectonic environments: Archean to present, *Earth. Planet. Sci. Lett.*, 195, 75-90, 2002.

Canil, D., Mildly incompatible elements in peridotites and the origins of mantle lithosphere, *Lithos*, 77, 375-393, 2004.

Carlson, R., and A.J. Irving, Depletion and enrichment history of subcontinental lithospheric mantle: an Os, Sr, Nd and Pb isotopic study of ultramafic xenoliths from the northwestern Wyoming Craton, *Earth Planet. Sci. Lett.*, 126, 457-472, 1994.

Carlson, R.W., D.G. Pearson, F.R. Boyd, S.B. Shirey, G. Irvine, A.H. Menzies, and J.J. Gurney, Re-Os systematics of lithospheric peridotites: implications for lithosphere formation and preservation, in *Proc. VIIth International Kimberlite Conference, B. J. Dawson Volume*, edited by J.J. Gurney, J.L. Gurney, M.D. Pascoe, and S.R. Richardson, pp. 99-108, 1999.

Carlson, R.W., D.G. Pearson, and D.E. James, Physical, chemical, and chronological characteristics of continental mantle, *Rev. Geophys.*, 43, RG1001, 2005.

Chesley, J.T., R.L. Rudnick, and C.-T. Lee, Re-Os systematics of mantle xenoliths from the East African Rift: age, structure, and history of the Tanzanian craton, *Geochim. Cosmochim. Acta*, 63, 1203-1217, 1999.

Cohen, R.S., R.K. O'Nions, and J.B. Dawson, Isotope geochemistry of xenoliths from East Africa: implications for development of mantle reservoirs and their interaction, *Earth and Planetary Science Letters*, 68, 209-220, 1984.

Conrad, C.P., and P. Molnar, The growth of Rayleigh-Taylor-type instabilities in the lithosphere for various rheological and density structures, *Geophys. Journ. Inter.*, 129, 95-112, 1997.

Cook, F.A., A.J. Van der Velden, and K.W. Hall, Upper mantle reflectors beneath the SNORCLE transect - images of the base of the lithosphere?, in *Slave-Northern Cordillera Lithospheric Evolution (SNORCLE) Transect and Cordilleran Tectonics Workshop Meeting; Lithoprobe Report*, edited by F.A. Cook, and P. Erdmer, pp. 58-62, University of Calgary, 1997.

Cooper, C.M., A. Lenardic, and L. Moresi, The thermal structure of stable continental lithosphere within a dynamic mantle, *Earth Planet. Sci. Lett.*, 222, 807-817, 2004.

Cooper, C.M., A. Lenardic, and L. Moresi, Thrust stacking and the creation and preservation of cratonic lithosphere, *Geology, submitted*, 2005.

Cox, K.G., M.R. Smith, and S. Beswetherick, Textural studies of garnet lherzolites: evidence of exsolution origin from high-temperature harzburgites, in *Mantle xenoliths*, edited by P.H. Nixon, pp. 537-550, John Wiley & Sons Ldt, 1987.

Davies, G.F., On the emergence of plate tectonics, *Geology*, 20, 963-966, 1992.

Dick, H.J.B., and T. Bullen, Chromian spinel as a petrogenetic indicator in abyssal and alpine-type peridotites and spacially associated lavas, *Contributions to Mineralogy and Petrology*, 86, 54-76, 1984.

Dixon, J.E., T.H. Dixon, D.R. Bell, and R. Malservisi, Lateral variation in upper mantle viscosity: role of water, *Earth Planet. Sci. Lett.*, 222, 451-467, 2004.

Ducea, M.N., and J.B. Saleeby, Buoyancy sources for a large, unrooted mountain range, the Sierra Nevada, California: evidence from xenolith thermobarometry, *Journ. Geophys. Res.*, 101, 8229-8244, 1996.

Ducea, M.N., and J.B. Saleeby, The age and origin of a thick mafic-ultramafic keel from beneath the Sierra Nevada batholith, *Contrib. Mineral. Petrol.*, 133, 169-185, 1998.

Ellis, D.J., and E.H. Green, An experimental study of the effect of Ca upon garnet-clinopyroxene Fe-Mg exchange equilibria, *Contributions to Mineralogy and Petrology*, 66, 13-22, 1979.

Ertan, I.E., and W.P. Leeman, Metasomatism of Cascades subarc mantle: evidence from a rare phlogopite orthopyroxenite xenolith, *Geology*, 24, 451-454, 1996.

Finnerty, A.A., and F.R. Boyd, Thermobarometry for garnet peridotites: basis for the determination of thermal and compositional structure of the upper mantle, in *Mantle Xenoliths*, edited by P.H. Nixon, pp. 844, John Wiley & Sons Ldt, 1987.

Foley, S.F., M. Tiepolo, and R. Vannucci, Growth of early continental crust controlled by melting of amphibolite in subduction zones, *Nature*, 417, 837-840, 2002.

Fung, A.T., and S.E. Haggerty, Petrography and mineral composition of eclogites from th eKoidu kimberlite complex, Sierra Leone, *J. Geophys. Res.*, 100, 20451-20473, 1995.

Gao, S., R.L. Rudnick, R.W. Carlson, W.F. McDonough, and Y. Liu, Re-Os evidence for replacement of ancient mantle lithosphere beneath the North China Craton, *Earth Planet. Sci. Lett.*, 198 (3-4), 307-322, 2002.

Grand, S.P., Mantle shear structure beneath the Americas and surrounding oceans, *Journ. Geophys. Res.*, 99, 11591-11621, 1994.

Griffin, W.L., D.R. Cousens, C.G. Ryan, and G.F. Suter, Ni in chrome garnet: a new geothermometer, *Contrib. Mineral. Petrol.*, 103, 199-202, 1989.

Griffin, W.L., S.Y. O'Reilly, N. Abe, S. Aulbach, R.M. Davies, N.J. Pearson, B.J. Doyle, and K. Kivi, The origin and evolution of Archean lithospheric mantle, *Precambrian Research*, 127, 19-41, 2003.

Griffin, W.L., S.Y. O'Reilly, and C.G. Ryan, The composition and origin of sub-continental lithospheric mantle, in *Mantle Petrology: Field observations and high pressure experimentation: a tribute to R. (Joe) Boyd*, pp. 13-45, Geochemical Society, 1999.

Griffin, W.L., D. Smith, C.G. Ryan, S.Y. O'Reilly, and T.T. Win, Trace-element zoning in mantle minerals: metasomatism and thermal events in the upper mantle, *Canadian Mineralogist*, 34, 1179-1193, 1996.

Gung, Y., M. Panning, and B. Romanowicz, Global anisotropy and the thickness of continents, *Nature*, 422, 707-711, 2003.

Handler, M.R., V.C. Bennett, and T.M. Esat, The persistence of off-cratonic lithospheric mantle: Os isotopic systematics of variably metasomatised southeast Australian xenoliths, *Earth Planet. Sci. Lett.*, 151, 61-75, 1997.

Harley, S.L., and D.H. Green, Garnet-orthopyroxene barometry for granulites and peridotites, *Nature*, 300, 697-701, 1982.

Hawkesworth, C.J., P.D. Kempton, N.W. Rogers, R.M. Ellam, and P.W. van Calsteren, Continental mantle lithosphere, and shallow level enrichment processes in the Earth's mantle, *Earth Planet. Sci. Lett.*, 96, 256-268, 1990.

Helmstaedt, H., and R. Doig, Eclogite nodules from kimberlite pipes of the Colorado Plateau - samples of Franciscan-type oceanic lithosphere, *Phys. Chem. Earth*, 9, 91-111, 1975.

Helmstaedt, H., and D.J. Schulze, Southern African kimberlites and their mantle sample; implication for Archean tectonics and lithosphere evolution, *Geol. Soc. Australia Special Publication*, 14, 358-368, 1989.

Herzberg, C., Phase equilibrium constraints on the formation of cratonic mantle, in *Mantle petrology, field observations and high pressure experimentation, a tribute to Francis R. (Joe) Boyd*, edited by Y. Fei, C.M. Bertka, and B.O. Mysen, pp. 241-257, Geochem. Soc. Spec. Pub., 1999.

Herzberg, C., Geodynamic information in peridotite petrology, *J. Petrol.*, 45, 2507-2530, 2004.

Herzberg, C., and M.J. O'Hara, Plume-associated ultramafic magmas of Phanerozoic age, *J. Petrol.*, 43, 1857-1883, 2002.

Herzberg, C., and J. Zhang, Melting experiments on anhydrous peridotite KLB-1: compositions of magmas in the upper mantle and transition zone, *J. Geophys. Res.*, 101, 8271-8295, 1996.

Herzberg, C.T., Lithosphere peridotites of the Kaapvaal craton, *Earth and Planetary Science Letters*, 120, 13-29, 1993.

Hirth, G., and D.L. Kohlstedt, Water in the oceanic upper mantle; implications for rheology, melt extraction and the evolution of the lithosphere, *Earth Planet. Sci. Lett.*, 144 (1-2), 93-108, 1996.

Hoffman, P.F., United plates of America, the birth of a craton: early Proterozoic assembly and growth of Laurentia, *Ann. Rev. Earth Planet. Sci.*, 16, 543-603, 1988.

Hoffman, P.F., Precambrian geology and tectonic history of North America, in *The geology of North America-An overview*, edited by B.A. W., and A.R. Palmer, pp. 447-511, Geological Society of America, Boulder, CO, 1989.

Houseman, G.A., D.P. McKenzie, and P. Molnar, Convective instability of a thickened boundary layer and its relevance for the thermal evolution of continental convergent belts, *Journ. Geophys. Res.*, 86, 6115-6132, 1981.

Houseman, G.A., and P.A. Molnar, Gravitational (Rayleigh-Taylor) instability of a layer with non-linear viscosity and convective thinning of continental lithosphere, *Geophys. Journ. Inter.*, 128, 125-150, 1997.

Jackson, J., Strength of continental lithosphere: time to abandon the jelly sandwich?, *Geol. Soc. Am. Today*, 12, 4-10, 2002.

Jacob, D.E., and S.F. Foley, Evidence for Archean ocean crust with low high field strength element signature from diamondiferous eclogite xenoliths, *Lithos*, 48, 317-336, 1999.

Jacob, D.E., E. Jagoutz, D. Lowry, D. Mattey, and G. Kudrjavtseva, Diamondiferous eclogites from Siberia: remnants of Archean oceanic crust, *Geochim. Cosmochim. Acta*, 58, 5191-5207, 1994.

James, D., F.R. Boyd, D. Schutt, D.R. Bell, and R.W. Carlson, Xenolith constraints on seismic velocities in the upper mantle beneath southern Africa, *Geochem. Geophys. Geosys.*, 5, Q01002, 2004.

James, D.E., M.J. Fouch, J.C. VanDecar, and S. van der Lee, Tectospheric structure beneath Southern Africa, *Geophys. Res. Lett.*, 28, 2485-2488, 2001.

Jaupart, C., and J.C. Mareschal, The thermal structure and thickness of continental roots, *Lithos*, 48, 93-114, 1999.

Johnson, K.T.M., H.J.B. Dick, and N. Shimizu, Melting in the oceanic upper mantle: an ion microprobe study of diopsides in abyssal peridotites, *Journ. Geophys. Res.*, 95, 2661-2678, 1990.

Jordan, T.H., The continental tectosphere, *Geophys. Space Phys.*, 13, 1-12, 1975.

Jordan, T.H., Composition and development of the continental tectosphere, *Nature*, 274, 544-548, 1978.

Jordan, T.H., Mineralogies, densities and seismic velocities of garnet lherzolites and their geophysical implications, in *The mantle sample: inclusions in kimberlites and other volcanics*, edited by F.R. Boyd, and H.O.A. Meyer, pp. 1-14, American Geophysical Union, Washington, D. C., 1979.

Jordan, T.H., Structure and formation of the continental tectosphere, *Journ. Petrol.*, 1988, 11-37, 1988.

Jull, M., Kelemen, P., On the conditions for lower crustal convective instability, *Journ. Geophys. Res.*, 106, 6423-6446, 2001.

Katz, R.F., M. Spiegelman, and C.H. Langmuir, A new parameterization of hydrous mantle melting, *Geochem. Geophys. Geosys.*, 4, doi:10.1029/2002GC000433, 2003.

Kay, R.W., and S.M. Kay, Delamination and delamination magmatism, *Tectonophys.*, 219, 177-189, 1993.

Kelemen, P.B., S.R. Hart, and S. Bernstein, Silica enrichment in the continental upper mantle via melt/rock reaction, *Earth. Planet. Sci. Lett.*, 164, 387-406, 1998.

Kirkley, M.B., J.J. Gurney, and A.A. Levinson, Age, origin, and emplacement of diamonds; scientific advances in the last decade, *Gems and Gemology*, 27, 2-25, 1991.

Kopylova, M.G., and J.K. Russell, Chemical stratification of cratonic lithosphere; constraints from the northern Slave Craton, Canada, *Earth Planet. Sci. Lett.*, 181 (1-2), 71-87, 2000.

Kopylova, M.G., J.K. Russell, and H. Cookenboo, Petrology of peridotite and pyroxenite xenoliths from Jericho Kimberlite; implications for the thermal state of the mantle beneath the Slave Craton, northern Canada, *J. Petrol.*, 40 (1), 79-104, 1999a.

Kopylova, M.G., J.K. Russell, and H.O. Cookenboo, Mapping the lithospere beneath the north central Slave Craton, in *Proceedings of the 7th International Kimberlite Conference*, pp. 468-479, 1999b.

Langmuir, C., E.M. Klein, and T. Plank, Petrological systematics of mid-ocean ridge basalts: constraints on melt generation beneath ocean ridges, in *Geophys. Monograph*, pp. 183-280, American Geophysical Union, 1992.

Lee, C.-T., and R.L. Rudnick, Compositionally stratified cratonic lithosphere: petrology and geochemistry of peridotite xenoliths from the Labait Volcano, Tanzania, in *Proc. VIIth International Kimberlite Conference, B. J. Dawson volume*, edited by J.J. Gurney, J.L. Gurney, M.D. Pascoe, and S.R. Richardson, pp. 503-521, 1999.

Lee, C.-T., R.L. Rudnick, and G.H. Brimhall, Deep lithospheric dynamics beneath the Sierra Nevada during the Mesozoic and Cenozoic as inferred from xenolith petrology, *Geochem. Geophys. Geosys.*, 2, 2001GC000152, 2001a.

Lee, C.-T., Q. Yin, R.L. Rudnick, and S.B. Jacobsen, Preservation of ancient and fertile lithospheric mantle beneath the southwestern United States, *Nature*, 411, 69-73, 2001b.

Lee, C.-T., Q.-Z. Yin, R.L. Rudnick, J.T. Chesley, and S.B. Jacobsen, Osmium isotopic evidence for Mesozoic removal of lithospheric mantle beneath the Sierra Nevada, California, *Science*, 289, 1912-1916, 2000.

Lee, C.-T.A., Compositional variation of density and seismic velocities in natural peridotites at STP conditions: implications for seismic imaging of compositional heterogeneities in the upper mantle, *Journ. Geophys. Res.*, 108, 2441, doi:10.1029/2003JB002413, 2003.

Lee, C.-T.A., X. Cheng, and U. Horodyskyj, The development and refinement of continental arcs by primary basaltic magmatism, garnet pyroxenite accumulation, basaltic recharge and delamination: insights from the Sierra Nevada, California, *Contrib. Mineral. Petrol.*, 2006, in press.

Lee, C.-T.A., A.D. Brandon, and M.D. Norman, Vanadium in peridotites as a proxy for paleo-fO$_2$ during partial melting: Prospects, limitations, and implications, *Geochim. Cosmochim. Acta*, 67 (16), 3045-3064, 2003.

Lee, C.-T.A., W.P. Leeman, D. Canil, and Z.-X.A. Li, Similar V/Sc systematics in MORB and arc basalts: implications for the oxygen fugacities of their mantle source regions, *Journal of Petrology*, 46, 2313-2336, 2005a.

Lee, C.-T.A., A. Lenardic, C.M. Cooper, F. Niu, and A. Levander, The role of chemical boundary layers in regulating the thickness of continental and oceanic thermal boundary layers, *Earth Planet. Sci. Lett.*, 230, 379-395, 2005b.

Lenardic, A., L. Guillou-Frottier, J.-C. Mareschal, C. Jaupart, L.-N. Moresi, and W.M. Kaula, What the mantle sees: the effects of continents on mantle heat flow, in *The history and dynamics of global plate motions*, American Geophysical Union, 2000.

Lenardic, A., and L. Moresi, A new class of equilibrium geotherms in the deep thermal lithosphere of continents, *Earth and Planetary Science Letters*, 176 (3-4), 331-338, 2000.

Lenardic, A., L.-N. Moresi, and H.-B. Muhlhaus, Longevity and stability of cratonic lithosphere: insights from numerical simulations of coupled mantle convection and continental tectonics, *Journ. Geophys. Res.*, 108, 10.1029/2002JB001859, 2003.

Levander A., F. Niu, and W.W. Symes, Imaging teleseismic P to S scattered waves using the Kirchhoff integral, in A. Levander and G. Nolet, editors, *Seismic Earth: Array Analysis of Broadband Seismograms*, Geophysical Monograph Series 157, American Geophysical Union, Washington, DC, 149-169, doi:10.1029/156GM10, 2005.

Li, Z.-X.A., and C.-T.A. Lee, The constancy of upper mantle fO$_2$ through time inferred from V/Sc ratios in basalts, *Earth Planet. Sci. Lett.*, 228, 483-493, 2004.

MacGregor, I.D., and J.L. Carter, The chemistry of clinopyroxene and garnets of eclogite and periodtite xenoliths from the Roberts Victor mine, South Africa, *Phys. Earth Planet. Inter.*, 3, 391-397, 1970.

Maggi, A., J.A. Jackson, D. McKenzie, and K. Priestley, Earthquake focal depths, effective elastic thickness, and the strength of the continental lithosphere, *Geology*, 28 (6), 495-498, 2000a.

Maggi, A., J.A. Jackson, K. Priestley, and P.A. Baker, A reassessment of focal depth distributions in southern Iran, the Tien Shan and northern India: do earthquakes really occur in the continental mantle?, *Geophys. Journ. Int.*, 143, 629-661, 2000b.

McInnes, B.I.A., M. Gregoire, R.A. Binns, P.M. Herzig, and M.D. Hannington, Hydrous metasomatism of oceanic sub-arc mantle, Lihir, Papua New Guinea: petrology and geochemistry of fluid-metasomatised mantle wedge xenoliths, *Earth. Planet. Sci. Lett.*, 188, 169-183, 2001.

McKenzie, D., The generation and compaction of partially molten rock, *Journal of Petrology*, 25 (3), 713-765, 1984.

McKenzie, D., and M.J. Bickle, The volume and composition of melt generated by extension of the lithosphere, *Journ. Petrol.*, 29, 625-679, 1988.

Menzies, M., Mantle Ultramafic Xenoliths in Alkaline Magmas: Evidence for Mantle Heterogeneity Modified by Magmatic Activity, in *Continental Basalts and Mantle Xenoliths*, edited by C.J. Hawkesworth, and M.J. Norry, pp. 272, Shiva Publishing Limited, Cambridge, 1983.

Menzies, M.A., Cratonic, circumcratonic and oceanic mantle domains beneath the western United States, *Journ. Geophys. Res.*, 94, 7899-7915, 1989.

Michaut, C., and C. Jaupart, Nonequilibrium temperatures and cooling rates in thick continental lithosphere, *Geophys. Res. Lett.*, 31, doi:10.1029/2004GL021092, 2004.

Molnar, P., G.A. Houseman, and C.P. Conrad, Rayleigh-Taylor instability and convective thinning of mechanically thickened lithosphere: effects of non-linear viscosity decreasing exponentially with depth and of horizontal shortening of the layer, *Geophys. Journ. Inter.*, 133, 568-584, 1998.

Neal, C.R., L.A. Taylor, J.P. Davidson, P. Holden, A.N. Halliday, P.H. Nixon, J.B. Paces, R.N. Clayton, and T. Mayeda, Eclogites with oceanic crustal and mantle signatures from the Bellsbank kimberlite, South Africa, Part 2. Sr, Nd, and O isotope geochemistry, *Earth Planet. Sci. Lett.*, 99, 362-379, 1990.

Nisbet, E.G., M.J. Cheadle, N.T. Arndt, and M.J. Bickle, Constraining the potential temperature of the Archean mantle: a review of the evidence from komatiites, *Lithos*, 30, 291-307, 1993.

Niu, F., A. Levander, C.M. Cooper, C.-T.A. Lee, A. Lenardic, and D.E. James, Seismic constraints on the depth and composition of the mantle keel beneath the Kaapvaal craton, *Earth Planet. Sci. Lett.*, 224, 337-346, 2004.

Nyblade, A., Heat flow and the structure of the Precambrian lithosphere, *Lithos*, 48, 81-91, 1999.

Nyblade, A.A., and R.A. Brazier, Precambrian lithospheric controls on the development of the East African rift system, *Geology*, 30, 755-758, 2002.

Nyblade, A.A., and H.N. Pollack, Terrestrial heat flow in East and Southern Africa, *Journ. of Geophys. Res.*, 95, 17371-17384, 1990.

O'Reilly, S.Y., W.L. Griffin, Y.H. Djomani, and P. Morgan, Are lithospheres forever? Tracking changes in subcontinental lithospheric mantle through time, *GSA Today*, 11 (4), 4-10, 2001.

Pearson, D.G., D. Canil, and S. Shirey, Mantle samples included in volcanic rocks: xenoliths and diamonds, in *Treatise of Geochemistry*, edited by H.D. Holland, and K.K. Turekian, pp. 171-275, Elsevier, Oxford, 2003.

Pearson, D.G., R.W. Carlson, S.B. Shirey, F.R. Boyd, and P.H. Nixon, Stabilisation of Archaean lithospheric mantle: a Re-Os isotope study of peridotite xenoliths from the Kaapvaal craton, *Earth Planet. Sci. Lett.*, 134, 341-357, 1995a.

Pearson, D.G., R.W. Carlson, S.B. Shirey, F.R. Boyd, and P.H. Nixon, Stabilization of Archean lithospheric mantle: a Re-Os isotope study of peridotite xenoliths from the Kaapvaal craton, *Earth and Planetary Science Letters*, 134, 341-357, 1995b.

Pearson, D.G., S.B. Shirey, R.W. Carlson, F.R. Boyd, N.P. Pokhilenko, and N. Shimizu, Re-Os, Sm-Nd, and Rb-Sr isotope evidence for thick Archaean lithospheric mantle beneath the Siberian craton modified by multistage metasomatism, *Geoch. Cosmochim. Acta*, 59, 959-977, 1995c.

Pollack, H.N., Cratonization and thermal evolution of the mantle, *Earth Planet. Sci. Lett.*, 80, 175-182, 1986.

Pollack, H.N., and D.S. Chapman, On the regional variation of heat flow, geotherms, and lithospheric thickness, *Tectonophys.*, 38, 279-296, 1977.

Pollack, H.N., S.J. Hurter, and J.R. Johnson, Heat flow from the Earth's interior; analysis of the global data set, *Rev. Geophys.*, 31, 267-280, 1993.

Putirka, K.D., Mantle potential temperatures at Hawaii, Iceland, and the mid-ocean ridge system, as inferred from olivine phenocrysts: evidence for thermally driven mantle plumes, *Geochem. Geophys. Geosys.*, 6, doi:10.1029/2005GC000915, 2005.

Pyle, J.M., and S.E. Haggerty, Eclogites and the metasomatism of eclogites from the Jagersfontein kimberlite: punctuated transport and implications for alkali magmatism, *Geochim. Cosmochim. Acta*, 62, 1207-1231, 1998.

Reisberg, L., and J.-P. Lorand, Longevity of sub-continental mantle lithosphere from osmum isotope systematics in orogenic peridotite massifs, *Nature*, 376, 159-162, 1995.

Richardson, S.H., J.J. Gurney, A.J. Erlank, and J.W. Harris, Origin of diamonds in old enriched mantle, *Nature*, 310, 198-202, 1984.

Ritsema, J., A.A. Nyblade, T.J. Owens, C.A. Langston, and J.C. VanDecar, Upper mantle seismic velocity structure beneath Tanzania, East Africa; implications for the stability of cratonic lithosphere, *J. Geophys. Res.*, 103 (9), 21,201-21,213, 1998.

Rollinson, H.R., Eclogite xenoliths in West African kimberlites as residues from Archaean granitoid crust formation, *Nature*, 389, 173-176, 1997.

Rudnick, R.L., W.F. McDonough, and R.J. O'Connell, Thermal structure, thickness and composition of continental lithosphere, *Chem. Geol.*, 145, 395-411, 1998.

Rudnick, R.L., W.F. McDonough, and A. Orpin, Northern Tanzanian peridotite xenoliths: a comparison with Kaapvaal peridotites and inferences of metasomatic reactions, in *Kimberlites, related rocks and mantle xenoliths*, edited by H.O.A. Meyer, and O.H. Leonardos, pp. 336-353, CPRM Special Publication, 1994.

Saltzer, R.L., N. Chatterjee, and T.L. Grove, The spatial distribution of garnets and pyroxenes in mantle peridotites: pressure-temperature history of peridotites from the Kaapvaal craton, *J. Petrol.*, 42, 2215-2229, 2001.

Schulze, D.J., Constraints on the abundance of eclogite in the upper mantle, *J. Geophys. Res.*, 94, 4205-4212, 1989.

Şengör, A.M.C., and B.A. Natal'in, Paleotectonics of Asia: fragments of a synthesis, in *The tectonic evolution of Asia*, edited by A. Yin, and T.M. Harrison, pp. 486-640, Cambridge University Press, Cambridge, UK, 1996.

Shirey, S.B., J.W. Harris, S.R. Richardson, M.J. Fouch, D.E. James, P. Cartigny, P. Deines, and F. Viljoen, Diamond genesis, seismic structure, and evolution of the Kaapvaal-Zimbabwe craton, *Science*, 297, 1683-1686, 2002.

Shirey, S.B., and R.J. Walker, The Re-Os isotope system in cosmochemistry and high-temperature geochemistry, *Annu. Rev. Earth Planet. Sci.*, 26, 423-500, 1998.

Simon, N.S.C., G.J. Irvine, G.R. Davies, D.G. Pearson, and R.W. Carlson, The origin of garnet and clinopyroxene in "depleted" Kaapvaal peridotites, *Lithos*, 71, 289-322, 2003.

Simons, F.J., A. Zielhuis, and R.D. van der Hilst, The deep structure of the Australian continent from surface wave tomography, *Lithos*, 48 (1-4), 17-43, 1999.

Sisson, T.W., T.L. Grove, and R.G. Coleman, Hornblende gabbro sill complex at Onion valley, California, and a mixing origin for the Sierra Nevada batholith, *Contrib. Mineral. Petrol.*, 126, 81-108, 1996.

Smith, D., and F.R. Boyd, Compositional zonation in garnets in peridotite xenoliths, *Contribution to Mineralogy and Petrology*, 112, 134-147, 1992.

Smith, D., J.C. Riter, and S.A. Mertzman, Water-rock interactions, orthopyroxene growth, and Si-enrichment in the mantle: evidence in xenoliths from the Colorado Plateau, southwestern United States, *Earth Planet. Sci. Lett.*, 165, 45-54, 1999.

Stille, H.W., The present tectonic state of the Earth, *Am. Assoc. Petroleum Geologists Bull.*, 20, 849-880, 1936.

Stolper, E., D. Walker, B.H. Hager, and J.F. Hays, Melt segregation from partially molten source regions: the importance of melt density and source region size, *J. Geophys. Res.*, 86, 6261-6271, 1981.

Takahashi, E., Speculations on the Archean mantle - missing link between komatiite and depleted garnet peridotite, *J. Geophys. Res.*, 95, 15941-15954, 1990.

Taylor, L.A., G.A. Snyder, R. Keller, D.A. Remley, M. Anand, R. Wiesli, J.W. Valley, and A.V. Sobolev, Petrogenesis of group A eclogites and websterites: evidence from the Obnazhennaya kimberlite, Yakutia, *Contrib. Mineral. Petrol.*, 145, 424-443, 2003.

Walker, R.J., R.W. Carlson, S.B. Shirey, and F.R. Boyd, Os, Sr, Nd, and Pb isotope systematics of southern African peridotite xenoliths: implications for the chemical evolution of subcontinental mantle, *Geochim. Cosmochim.*, 53, 1583-1595, 1989.

Walter, M.J., Melting of garnet peridotite and the origin of komatiite and depleted lithosphere, *J. Petrol.*, 39, 29-60, 1998.

Walter, M.J., Melting residues of fertile peridotite and the origin of cratonic lithosphere, in *Mantle petrology: field observations and high pressure experimentation: a tribute to Francis R. (Joe) Boyd,* edited by Y. Fei, C.M. Bertka, and B.O. Mysen, pp. 225-239, Geochemical Society Special Publication, 1999.

Winterburn, P.A., B. Harte, and J.J. Gurney, Peridotite xenoliths from the Jagersfontein kimberlite pipe I, primary and primary-metasomatic mineralogy, *Geochim. Cosmochim. Acta*, 54, 329-341, 1989.

C.-T. A. Lee, Department of Earth Science, MS-126, Rice University, 6100 Main Street, Houston, Texas 77005, USA. (ctlee@rice.edu)

Crustal Ductility and Early Continental Tectonophysics

R.C. Bailey

Departments of Geology and Physics, University of Toronto, Toronto, Canada

Observed effective viscosities of hot continental crust are low enough to mechanically decouple the upper brittle crust from the deeper lithosphere, permitting tectonic mobilization by gravity or plate boundary stresses. By extrapolating modern heat flow data back in time, these viscosity values can be used to assess decoupling during the Archean and Proterozoic eras in crust petrologically similar to a calibrated modern region (here "NG" or northwest German crust), assuming that secular changes in ductility are predominantly associated with changes in thermal activation. Such predictions suggest that NG-type crust, given cratonic thermal inputs (heat productivity and basal heat input) would have been pervasively decoupled during the early Archean, but became coupled before the end of the Archean. NG-type crust with thermal inputs characteristic of the Appalachians, or of cratonic crust which had not lost significant upper crustal heat productivity by erosion, would probably have remained decoupled through much or all of the Proterozoic. This limited sample suggests that during the early Archean, even crust with cratonic (low) thermal inputs may have been hot enough to have its elevation limited to below sea level by continuous extensional free-boundary gravitational collapse. A corollary conclusion is that the depths of Archean oceans were probably similar to today's. Once above sea level, decoupled continental crust would have switched to extrusion collapse as exhibited today by the Tibetan Plateau via the Himalayas. For thermally non-cratonic continental crust (that is, heat production higher than implied for cratonic crust), such extrusion collapse could have been endogenous throughout much or all of the Proterozoic, that is, generating mobile belts at continental boundaries which were not plate boundaries.

1. THE ARCHEAN CONTEXT

The much larger generation of heat by radioactive decay during the Archean era assures us that we should be considering continental tectonic processes which are associated with hotter crust. How much hotter is not precisely known (*Lenardic* [1998]; *Bickle* [1978]; *England and Bickle* [1984]). However, there is no doubt that, far enough back in

time, for example during the Hadean when a magma ocean existed (*Tonks and Melosh* [1993]; *Ohtani* [1985]; *Matsui and Abe* [1986]; *Abe* [1997]), the library of important tectonic processes must have been very different. There is no *a priori* reason to suppose that earliest date from which rocks have survived to the present day (by definition the beginning of the Archean) was also the date back to which a uniformitarian extension of modern tectonic processes is valid.

Archean tectonics is likely to have differed in many ways. A much debated point is how far back in time one can reliably use modern plate tectonics as a guide, in the uniformitarian tradition (*e.g. Burke et al.* [1976], *Hoffman and Ranalli* [1988], *Davies* [1992], *De Wit* [1998],

Archean Geodynamics and Environments
Geophysical Monograph Series 164
10.1029/164GM09

Hamilton [1998]). The initial successes of plate tectonics were in explaining the global behavior of horizontal motions as that of rigid plates, an approach which was particularly successful with oceanic crust and its interaction with continental boundaries. In this first approximation, continents could be regarded as passive markers riding on larger plates. In fact, this quality of continents (that they are generally passive and internally immobile except in the presence of abnormally high heat flow) is the aspect of continental tectonics which is most likely to have been different in the Archean. What is abnormal today (as exemplified by ongoing processes in western North America and the Tibetan Plateau) is likely to have been the norm then. In contrast, the main tectonic characteristics of oceanic crust (mobility and short lifetime) today are likely to have been similar during the Archean (although detailed processes would have differed). Since no undisturbed Archean oceanic crust remains today to be examined, all evidence of Archean tectonics has been filtered through Archean continental tectonics, and a clear picture of the unique features of Archean continental tectonics is essential for making progress.

It may well be that Archean crust was no hotter than modern orogenic crust, simply that such temperatures were the regional norm during the Archean rather than the local exception they are today. A primary characteristic of hot crust that distinguishes it tectonophysically from cold crust is the existence of a ductile layer in the middle or deep crust (*Wernicke* [1990], *McKenzie et al.* [2000], *McKenzie and Jackson* [2002]), a layer which can flow significantly over geological time, a layer which does not have much resistance to shear forces, which can mechanically decouple the upper crust from the deeper lithosphere. Such decoupling over small regions (as often occurs today) may have little effect in many cases; upper crustal plates unconnected mechanically to underlying lithospheric motion can still be tightly coupled via laterally adjoining cold crust. The possibility of entire continents with decoupled upper crust, however, raises interesting possibilities for continental tectonics.

But how will we recognize them? Deciphering the record of Earth evolution during the Archean era is a considerable challenge. Of the little evidence for Archean tectonic processes that is preserved in the geological record, most has been severely modified by subsequent events. In the absence of evidence to the contrary, the temptation is to apply a strictly uniformitarian approach: to use known Phanerozoic processes, drawn from plate tectonics, to explain as much as possible of what we see. Yet if applied too stringently, this approach can itself lead to serious errors: if a known process is even remotely plausible as an explanation for some of the evidence, the poor and meagre Archean geological record may be unable to refute it even if it is wrong. The most spectacular example of such an error was two centuries ago in Abraham Werner's (*Werner* [1774]; *Werner* [1786]), contraction of the

"respectable" repertoire of significant geological processes to sedimentary (and thus observable) ones (the plutonic processes invoked by James Hutton (*Hutton* [1788], *Hutton* [1795]) had never been observed in action by geologists, nor by their nature can they ever be; volcanism in Werner's view was a minor late effect of burning coal beds). Ironically, this uniformitarian restriction of process led Werner to the erroneous catastrophist theory of Neptunism, in order to explain observed geology. To avoid a similar error in interpreting the Archean record, we should ask not how few new processes we can possibly incorporate into Archean tectonics without conflicting with the evidence. Instead we should simply ask (about those which do not contradict the evidence) the question: which ones are the most plausible?

But what constitutes plausibility? Tectonophysical plausibility has three different levels: geometric, kinematic, and dynamic. A hypothesis is geometrically plausible if the final disposition of components in space is consistent with it. Kinematic plausibility requires more, specifically a history of tectonic velocity **v** consistent with the impenetrability of matter ($\nabla \cdot \mathbf{v} \simeq 0$). A modern example illustrating kinematic plausibility is given by lithospheric slabs undergoing "flat subduction": the arrangement of such slabs is clearly consistent kinematically with subduction at a very shallow angle. Yet flat subduction has yet to be proved dynamically possible: can the stresses of plate tectonics drive a slab horizontally for significant distances under continental crust without the assistance of gravity? An equally possible kinematic explanation of flat subduction geometry is continental free-boundary collapse as described later in this article. Where two physical processes are equally possible on geometric and kinematic grounds, dynamic plausibility then becomes an important tool for deciding between them. However, we then face the problem that although geometry and kinematics can be argued from field evidence, it is much more difficult to argue dynamic plausibility from field evidence. Even if we could measure the relevant stresses and forces throughout the lithosphere accurately enough, we can only do so accurately at the present instant. Dynamic plausibility in geology therefore has to be argued from model experiments or from theory, especially for Archean processes where preservation of unambiguous geometric and kinematic evidence is poor.

It is not the intent of this article to examine the implications of geometric or kinematic field evidence (nor the chemical or petrologic evidence) for the Archean; many studies (*e.g. Bleeker* [2002]; *James and Fouch* [2002]) have addressed this evidence. Instead, the intent of this article is to look at the likelihood of pervasive upper crustal decoupling during the early history of the Earth, given what we can infer about the dependence of such decoupling on the geothermal gradient. Such inferences suggest that, as a result of secular cooling, continents may have evolved through three distinct styles of dominant response to the thermal, mechanical and magmatic inputs of mantle convection: continental elevation limiting by extensional gravitational collapse, marginal extrusion (Himalayan) tectonics, and stable crust.

2. CRUSTAL DECOUPLING

2.1. Horizontal and Vertical Decoupling

Ductile materials do two things that rigid materials do not. They flow, and they lubricate. The first property, mobilization, can account for ductile material migrating in ways one might not expect of a rigid material; a classical modern example is salt dome diapirism. Such processes are endogenic ones made possible by ductility. The second process, lubrication, or more generally, decoupling, accounts for migration of attached material such as the upper crust, by decoupling it from resisting substrates; again, salt formations have been known to be important in sedimentary thrusts and nappe formation. These are exogenic processes which external forces drive with the assistance of lubrication. This classification is too simple to be anything other than a conceptual one, but is useful for analysis. It suggests that important parameters for Archean investigations are the thermal gradients or heat fluxes required to produce either decoupling or mobilization. Here I will discuss tectonic decoupling, which facilitates the response of crust to forces such as boundary forces or large scale topographic gradients such as might be imposed by global tectonics, whatever their form. There are interesting smaller scale endogenic processes which higher Archean crustal temperatures might facilitate, such as diapiric formation of granite-greenstone terrains, but these are beyond the scope of this article.

Both vertical and horizontal decoupling can occur in response to high geothermal gradients such as might have been more common in the Archean. By vertical decoupling in crustal tectonics is meant here the ability of the upper crust to move vertically without forcing coupled vertical motions of the Moho or deeper lithosphere. The way that the upper crust responds to plume driven uplift or to glacially driven loading depends on the degree of this vertical decoupling, and is a function of the time and length scale of the loading. Because such vertical decoupling is made possible by channel flow (Poiseuille flow) in the crustal ductile layer, vertical decoupling is tectonically synonymous with easy mid-crustal channel flow. The Tibetan plateau offers a clear modern example of vertical decoupling associated with easy mid-crustal channel flow (*Clark and Royden* [2000]) (although mid-crustal flow also has other spectacular effects here, such as exporting high elevations eastwards in a manner predicted by *Bird* [1991]).

Horizontal decoupling, on the other hand, occurs when upper crustal plates or blocks can easily move horizontally with respect to deeper crust and lithosphere. Horizontal movement of upper crust with respect to lower crust involves simple shear flow (Couette flow) rather than channel flow in the ductile zone. There are two ways in which this can occur: the upper brittle crustal plate can move as a rigid unit, or it can itself be deforming in extension or compression. In either case, the term "horizontal decoupling" will be used here to describe the situation in which the basal shear drag forces associated with displacing upper-crustal plates or blocks are negligible in comparison with the other horizontal forces such as boundary forces or gravitational potential energy gradients. The most important manifestation of horizontal decoupling is likely to be "free-boundary collapse" (*Rey et al.* [2001]) by extension of the high-standing crust over adjacent lower regions, specifically by boundary thrusting accompanied by extensional tectonics in the interior. This is probably occurring in western North America today (*e.g. Coney and Harms* [1984]; *Sonder et al.* [1987]), although complicated by transform-fault tectonics at the boundary.

2.2. Vertical Decoupling and Channel Flow

Vertical crustal motions over a ductile layer are both consequence and cause of channel flow in the ductile layer. The coupling of the vertical motions of the upper crust and the upper mantle through a ductile lower or middle crust has been analyzed by *Kuznir and Matthews* [1988] and *Kaufman and Royden* [1994]. Their analysis includes the flexural effects of an elastic upper crust, but not the effects of vertical motion of the Moho. *McKenzie and Jackson* [2002] provide an analysis which includes vertical motion of the Moho but has no elastic strength in the upper crust. For motions distributed over long horizontal wavelengths, much longer than the characteristic flexing length (the flexural parameter: *Turcotte and Schubert* [1982]) of the upper crust, the upper crustal elastic strength has negligible influence on the deformation. I use here the result of *McKenzie and Jackson* [2002] (although, as they point out, uncertainties in crustal viscosities are large enough to make the consequences of this choice unimportant). Here I recapitulate the results of their analyses. In the short term response to an applied load, the vertical motions of the upper crust and mantle are coupled; compensation is at mantle depths by depression of the entire lithosphere, because of the rapidity of whole mantle isostatic adjustment flow relative to mid-crustal channel flow (note that long wavelength post-glacial rebound times are of order 10^4 years: *Walcott* [1972]; *Walcott* [1973]). This short term response is described by the relaxation time τ_a of *McKenzie and Jackson* [2002]. There is also a significant long term response, however, if the deep crust is ductile. Lateral outward channel flow in the ductile layer permits the upper crust to sink further, as the Moho simultaneously rises. In the long term, if the lower crust stays ductile, the Moho will show no deflection due to the upper crustal load. Easy mid-crustal channel flow is synonymous with vertical decoupling. The time constant for this long-wavelength channel-flow response to loading is *McKenzie and Jackson* [2002]

$$\tau_b = \frac{3\lambda^2 \eta}{H^3 \rho_c g} \frac{\rho_m}{\pi^2 (\rho_m - \rho_c)} \quad (1)$$

where η is the effective viscosity of the mid-crustal ductile channel, H is the thickness of the channel, λ is the wavelength of the load and its response (for an isolated load, this would be of order twice the load length-scale), g is the acceleration of gravity, and ρ_m and ρ_c are mantle and crustal density respectively.

2.3. Horizontal Decoupling and Crustal Slip

A comprehensive modelling study of horizontal decoupling in the context of convergent margin tectonics has been given by *Royden* [1996]. Upper crustal horizontal slip in response to lateral forces proceeds by simple shear flow in the ductile layer (*Turcotte and Schubert* [1982]). For a uniform ductile layer of viscosity η and thickness H, this velocity is

$$v_s = \frac{H}{\eta} \sigma_h \quad (2)$$

where σ_h is the horizontal shear stress applied by the upper crust to the top of the ductile channel. If a total tectonic force F_t per unit strike length (striking perpendicular to the motion) is available for a crustal plate of horizontal extent L, this becomes

$$v_s = \frac{H F_t}{\eta L} \quad (3)$$

As will be argued later, an important Archean expression of horizontal decoupling is likely to be regional or continental gravitational collapse, such as may be occurring today in western North America, most visibly in the Basin and Range Province. It is therefore useful to shift the context of horizontal decoupling from Royden's localized orogenic context to one in which the stress gradients are related to gravitational potential energy rather than prescribed by orogenic backstops (*Jamieson et al.* [2002]). If gravitational energy is stored by uncompensated excess topography e, the lateral force (per unit strike length) available is (*Bailey* [2000])

$$F_t = \frac{\rho_c g e^2 \rho_m}{2(\rho_m - \rho_c)} \quad (4)$$

For excess topography of order 5 km (for example the height of continents above adjacent abyssal ocean floor, or the height of the Tibetan plateau), this is of order 10^{12} Newtons/meter, typical of modern plate boundary forces. Substitution of (4) into equation (3) gives

$$v_s = \frac{H \rho_c g e^2}{2 \eta L} \frac{\rho_m}{(\rho_m - \rho_c)} \quad (5)$$

A characteristic time for horizontal decoupling can now be defined as the time that this gravitational energy will take to slide this length L of upper brittle crust through a distance L (*i.e.* completely off its foundation), if no other energy sinks are present, as

$$\tau_s = \frac{L}{v_s} = \frac{2 \eta L^2}{H \rho_c g e^2} \frac{(\rho_m - \rho_c)}{\rho_m} \quad (6)$$

This can be related to the characteristic time τ_b for vertical decoupling over the same length scale by taking the ratio of equations (1) and (6) (by setting L in equation (6) equal to the wavelength λ in equation(1)). This gives

$$\frac{\tau_s}{\tau_b} = \frac{2}{3} \left[\frac{\pi H}{e} \left(1 - \frac{\rho_c}{\rho_m} \right) \right]^2 \quad (7)$$

Since ductile channel thicknesses and tectonically generated topography are both of the order of a few km, the right hand side of (7) is of order unity. We can therefore regard the timescales for vertical and horizontal decoupling as comparable in tectonically active regions.

3. THERMAL DEPENDENCE OF UPPER-CRUSTAL DECOUPLING

3.1. Crustal Viscosities

To relate the likelihood of upper-crustal decoupling during the Archean (or any other era) to the expected crustal thermal inputs (as described by two variables: the basal mantle heat input into the crust, and the internal crustal radiogenic heating) is a perilous undertaking, since factors other than temperature can influence crustal ductility, such as lithology, grain sizes, strain history and water content (*McKenzie and Jackson* [2002]).

Ductile behavior, manifested as deviatoric straining at rate $\dot{\varepsilon}$ driven by deviatoric stress σ, is often modelled by thermally activated power-law creep laws (*Ranalli* [1995]) like

$$\dot{\varepsilon} = A \sigma^n \exp\left(-\frac{Q}{RT} \right) \quad (8)$$

where Q is the molar activation energy of the process, R is the gas constant, and T is the absolute temperature, and the coefficient A incorporates all other effects such as grain-size dependence. This corresponds to an effective viscosity

$$\eta = \frac{\sigma}{2\dot{\eta}} = A^{-1}\sigma^{(1-n)}\exp\left(\frac{Q}{RT}\right) = \eta_0 \exp\left(\frac{Q}{RT}\right) \quad (9)$$

which is stress-dependent (non-Newtonian) unless $n = 1$.

Creep in the real Earth, of course, involves a multiplicity of mechanisms (*Ranalli* (1995)) (*e.g.* vacancy diffusion, dislocation creep, grain boundary creep, pressure solution creep) each potentially described by a different mechanism like (8), in a multiplicity of minerals occurring in a multiplicity of grain sizes. It is possible to measure values for A, n, and Q for rocks and minerals in the laboratory. It is virtually impossible to use such measurements to strongly constrain the viscosities that might be representative of deep hot continental crustal rocks (the situation is easier for the better defined composition of the mantle), for two reasons. First, power-law dependence of viscosity on stress implies that a creep mechanism with a high power n can dominate laboratory results even if it is unimportant at geologically realistic strain rates. Secondly, the variability of the activation energy with lithology implies a large uncertainty if crustal composition is not known well. All one can safely argue about the influence of temperature (or homologous temperature, as *McKenzie and Jackson* [2002] have done) is that effective crustal viscosities can range from geologically negligible at high temperatures (over about 400 °C) to values so high at low temperatures as to preclude creep on any reasonable geological time scale. Even such apparently better constrained aspects of equation (8) as the temperature range over which this transition from ductile to brittle occurs are not well determined, because the probable increase in Q with depth (as the lithology becomes more mafic) can combat the effects of increasing temperature with depth.

An apparent additional source of uncertainty is the thickness H of the ductile layer, which might potentially be strongly temperature dependent, since the depth of the brittle-ductile transition (base of the seismogenic zone) moves upward as the crustal geothermal gradient steepens. The *effective thickness* of the ductile layer is not the thickness in which ductile flow is *possible* (as determined, for example by the requirement that the Maxwell relaxation time be geologically short); it is the thickness in which the bulk of the flow or slip *actually takes place*, a crucial distinction noted by *Kaufman and Royden* [1994]. Because of the strong thermal dependence expected for a thermally activated creep mechanism, the flow is expected to be strongly concentrated at the bottom of the ductile layer. Somewhat surprisingly, their computed results for the effective thickness show a relatively weak dependence of H on activation energy Q. They show that the expected power-law dependence of stress on strain rate thickens the actively shearing layer relative to that expected for Newtonian viscosity. Their effective thickness is of the order of 5 km for a granitic rheology, and not very different for other rheologies. No great error, therefore, will be committed by assuming H to be 5 km for continental crust, even if one is ignorant of the lithological details. That this is much smaller than continental crustal thicknesses means that it is meaningful to talk about an effective depth, temperature and viscosity of the ductile layer.

3.2. Field-Based Estimates of Deep Crustal Viscosity

Field observations of crustal behavior are likely to provide more trustworthy constraints on deep crustal viscosity. Observations of vertical displacement histories in the Basin and Range province of the western United States have yielded viscosity estimates in the range 10^{17} to 10^{19} Pa.s, assuming a ductile channel thickness of 15 km (*Kruse et al.* [1991]; *Wdowinski and Axen* [1992]; *Kaufman and Royden* [1994]). For a tectonically active region of Germany, *Klein et al.* [1997] obtained a particularly well constrained estimate of crustal topographic diffusivity by using the recorded topographic response to a century of recorded crustal unloading by coal mining and groundwater pumping. Their model values for ductile thickness and viscosity are 15 km and $6 \cdot 10^{17}$ Pa.s *Clark and Royden* [2000] modelled the orogenic uplift of the Tibetan plateau as driven by lower crustal channel flow, assuming a 15-km-thick ductile channel. They estimated a viscosity for the lower crust of 10^{18} Pa.s beneath the low-gradient plateau margins, 10^{21} Pa.s beneath the steep plateau margins (where presumably hot lower crust abuts cold cratonic lower crust), and an upper bound of 10^{16} Pa.s beneath the plateau itself. In all of these estimates, there is a trade-off between ductile channel thickness and viscosity, but this does not produce the huge uncertainties that temperature and lithology do. Although these estimates are all for η_{15} (the viscosity needed to reproduce experimental results if the ductile channel is 15 km thick), it is possible that the true effective viscosities are smaller by an order of magnitude if ductile channels are actually of order 5 km in thickness as suggested above.

Heat flows have been estimated for some of these regions for which mid-crustal viscosities have been estimated. Figure (1) plots, as centers of ellipses, the ($\log_{10}$) viscosity and heat flow values for the three regions described above. The ellipses are used to circumscribe the errors. I have set the viscosity to plus or minus an order of magnitude (± 1 on a $\log_{10}$ plot) for the viscosity estimates so as to encompass the typical ranges of values published for each region. The Basin and Range values are 18 ± 1 $\log_{10}$ Pa.s as cited in *Clark*

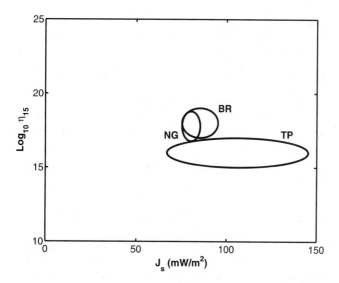

Figure 1. Viscosity versus surface heat flow. Ellipses represent experimentally derived values and their uncertainties for Basin and Range (BR), the Tibetan Plateau (TP), and northwest Germany (NG). See text for details.

and Royden [2000] and 85 ± 5 mW/m^2 from *Blackwell* [1983]). The Tibetan Plateau viscosity is 16 ± 1 log$_{10}$Pa.s, based on *Clark and Royden* [2000]; the heat flow used is 106 ± 40 mW/m^2 and is taken as the median and associated standard deviation of 13 measurements summarized in Shen [1993]. The values of 80 ± 10 mW/m^2 and $(17 \log_{10}6) \pm 1$ log$_{10}$ Pa.s for northwestern Germany are based on *Haenel* [1983] and *Klein et al.* [1997] respectively. In all cases, the viscosities are η_{15}, those which were derived using a nominal ductile channel thickness of 15 km.

3.3. Predicting Ancient Ductile Layer Viscosities

Can these estimates in Figure (1) be used to derive a relationship between surface heat flow and decoupling which can be applied elsewhere, and in particular to the Archean era? The answer to that depends on how consistent one believes the non-thermal factors influencing ductility to be, such as composition and water content. Clearly, a very dry granulite-grade cratonic crust is unlikely to mobilize in the same way that wet subduction-influenced non-cratonic crust does. However, since Archean crust was by definition largely non-cratonic prior to the cratonization that defines the end of the Archean, this may not be a serious impediment to deriving a very approximate relationship for "mobilizable" or non-cratonic continental crust, whose compositional factors may not differ too much from one instance to another. In any case, I attempt here a less ambitious prediction: that of estimating

the degree of ancient decoupling that would have existed in crust like the northwest German crust ("NG-type" crust) whose present-day viscosity has been estimated by Klein et al. [1997] (KJS). I choose the KJS site because it has arguably the best controlled estimate, based on a century of measurements of uplift resulting from (measured) historical coal mining and associated groundwater pumping, and also because the present day heat flow is reasonably well defined (*Haenel* [1983]).

Clearly the KJS viscosity estimate ($\eta_{15} \simeq 6 \cdot 10^{17}$ Pa.s) imposes a constraint on any viscosity-heat flow relationship, but does not determine its temperature dependence. However, it can be used to calibrate extrapolations to Archean thermal conditions for "similar" or NG-type crust. By similar is meant crust with similar physical properties to that of NG-type crust. "Similar" is not meant here to imply similar thermal inputs from crustal heat productivity and basal heat input from the mantle, since these will vary with time even if the crust in question is not tectonically disturbed. If the ductile zone is at an effective depth z_d, then η_{15} can be estimated for different thermal conditions by multiplying a known calibration value $\eta^{(C)}{}_{15}$ (such as obtained from the field results shown in Figure (1)) by the ratio of thermal activation factors for the past and calibration temperatures $T(z_d)$ and $T^{(C)}(z)$ at the depth z of the ductile layer. That is,

$$\frac{\eta_{15}}{\eta_{15}^{(C)}} = \frac{\exp\left(\frac{Q}{RT(z_d)}\right)}{\exp\left(\frac{Q}{RT^{(C)}(z_d)}\right)} \quad (10)$$

can be used to predict past values of η_{15} from past values of $T(z_d)$, which in turn can be inferred from estimates of past values of thermal inputs to the crust. There is here an implicit assumption that the effective depth to the ductile zone, if present, depends primarily on lithology and is not strongly temperature dependent, a reasonable assumption because of the results of *Kaufman and Royden* [1994] cited above.

Using equation (10) to predict the viscosity of NG-type crust at some time in the past requires estimates of past ($T(z_d)$) and present ($T^{(C)}(z_d)$) effective temperatures of the ductile layer. To do this we need a model for past and present geotherms. A plausible form for continental geotherms in regions at or near thermal equilibrium is (*Chapman* [1986]; *Lachenbruch* [1970])

$$T(z) = T_s + \left(\frac{J_s}{k} - \frac{A_0 h}{k}\right)z + (1 - \exp(-z/h))\frac{A_0 h^2}{k} \quad (11)$$

$T(z)$ is the geotherm (temperature as a function of depth z) for an Earth with thermal conductivity k, surface heat flow J_s and an exponentially downward decaying heat productivity $A_0 \exp(-z/h)$. This is more useful as

$$T(z) = T_s + \frac{J_m z}{k} + (1 - \exp(-z/h))\frac{J_c h}{k} \qquad (12)$$

where the surface heat flow J_s is separated into the sum of a crustal (J_c) and a mantle contribution (J_m). Ideally, specifying appropriate values for J_c and J_m in Archean crust (or any previous crust), would predict the corresponding Archean geotherm with equation (12) and thus the corresponding Archean ductile layer viscosity using equation (10).

Unfortunately, we require, but do not know, the effective depth z_d for NG crust to the ductile layer. We therefore cannot directly use the viscosity prediction equation (10) in a deterministic sense. Instead, a statistical approach is necessary, as follows: guess a plausible pair of values for z_d and T_d. Guess also a plausible pair of values for the properties k and h. With these and a specified surface temperature and surface heat flow, only one geotherm of type (12) is possible. This fixes the crustal and mantle contributions J_c and J_m separately. If they are implausible (*i.e.* exceed the surface heat flow, are negative, or less than some reasonable lower limit, here taken as 10 mW/m^2), this geotherm is discarded. Otherwise, this geotherm is taken as a "legal" candidate geotherm. (The set of "legal" present-day geotherms generated in this study for the calibration region NG is shown in Figure (2)). If a plausible activation energy Q is assumed, the factor η_0 in the viscosity formula (9) is then determined.

Using the assumed values for k and h, an ancient NG-type geotherm can now be computed for thermal inputs J_{ca} and J_{ma} appropriate to the ancient thermal environment (the

ancient heat fluxes J_{ca} and J_{ma} have higher values, since average radiogenic heat generation was predictably higher in the past). The ancient temperature T_{da} at the presumed ductile zone depth z_d is calculated, and used in the now-calibrated viscosity formula (9) to give an ancient effective viscosity η_{15a}. With this a decoupling time constant τ_b can be calculated using (1). (For present-day NG crust, τ_b is about 0.4 Ma using the parameters defined below.)

This single estimate of τ_b assumes a particular set of parameters k, h, Q, T_d and z_d for the calibration site. It would be dangerous to base conclusions on a single set of parameters, since they are not well constrained. Instead, all possible combinations using a range of plausible values for each of these five parameters are used, and the set of decoupling time constants τ_b is examined. In this way, the degree of constraint imposed by the calibration site and the other uncertainties are more fairly assessed, by yielding an ensemble of estimates for τ_b based on plausible ensembles of values for k, h, Q, T_d and z_d.

The following ranges of values were used: for k, 7 values approximately normally distributed with mean 2.7 W/°m and standard deviation 0.7 (based on values in Clauser and *Huenges* [1995]); for h, 6 values uniformly distributed over the range 5 to 25 km (a generous extension of the range of values found in *Jaupart and Mareschal* [1999]); for z_d, 7 values uniformly distributed over the range 13 to 25 km (based on adding a ductile layer half-thickness of ~2.5 km to the depths of the tops of ductile zones suggested by seismic reflectivity (*Klemperer* [1987])); for Q, 4 values uniformly distributed over the range 125,000 to 275,000 kJ/mole ("granitic" to "feldspathic" rheology, after *Ranalli* [1995]); for T_d, 7 values uniformly distributed over the range 400 to 700 °C (based on the discussion of *McKenzie and Jackson* [2002]). This yielded 13720 cases, of which 3672 yielded "legal" geotherms on which an analysis could be based. Final conversion of the implied viscosities to decoupling relaxation times was done using equation (1) with crustal and mantle densities of 2800 and 3300 kg/m^3 respectively, a ductile channel thickness of 5 km, and a horizontal spatial scale or wavelength λ of 1000 km. (Any errors in these parameters used for conversion are insignificant compared to the many orders of magnitude spanned by the viscosity values.)

Why does this statistical approach work better than simply making a best guess at an Archean geotherm and using a theoretical viscosity function? The advantage of the statistical approach is that the unknown or poorly determined factors controlling viscosity cancel out to first order, by multiplying an experimental value (which is known by observation) by a thermal activation factor. This thermal activation factor is the only influence on viscosity of which we have a predictively useful understanding. If, for example, the ductile zone is deeper than we believe, the effect of this on our temperature estimate will be similar for both the calibration region ductile

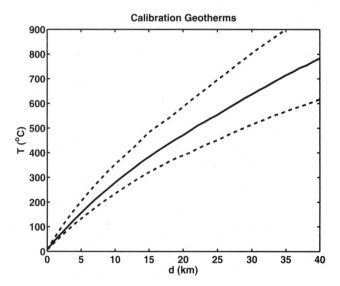

Figure 2. Ensemble of legal geotherms produced for the calibration site (NG). The solid line is the median at each depth of the suite of geotherms; dashed lines together enclose ± one standard deviation (68% of the values).

zone and that of the same crust extrapolated back to the Archean. Clearly this is not a general method. At best it predicts the past decoupling of crust like that of the calibration region, at different times and with different thermal inputs but the same physical properties. It is only as representative as the calibration region. However, since the viscosities of the calibration region are known only within an order of magnitude, the additional errors introduced by "non-representativeness" of the calibration region are unlikely to dominate.

A possible criticism is that viscosity dominated by partial melt contributions will not thermally track the viscosity formula (9). This would clearly be a concern if the Tibetan Plateau results had been used for calibration, since seismic and magnetotelluric studies image what may be laterally extensive partial melt (*Nelson et al.* [1996]), and the Tibetan Plateau viscosity is nearly two orders of magnitude lower than the KJS German result. However, the set of legal geotherms that resulted from the above process, shown in Figure (2), are for the most part, below temperatures at which crustal melting might be expected for lithologies characteristic of the depths used.

4. THE ARCHEAN THERMAL REGIME

4.1. Archean Geotherms

How decoupled was "normal" Archean crust? Given the above analysis, this is equivalent to asking how much hotter Archean crust was. Global radioactive heat generation must have been two to three times greater during the Archean eon than at present, yet preserved Archean rocks indicate, through their petrology, continental geothermal gradients which are only about 50% greater (*Burke et al.* [1976]) than current values (the "Archean thermal paradox": *Lenardic* [1998]). On the basis of geothermometric and geobarometric analyses of Archean high-grade rocks, Bickle [1978] proposed the two curves shown in Figure (3) as representative Archean continental geotherms.

The uppermost crustal thermal gradients of these are of the order of 30°/km. He assumed a crustal thermal conductivity of 3 W/m°, for which this corresponds to a surface heat flow of 90 mw/m². Bickle also noted that, because otherwise cool rocks can suffer transient heating by upwardly migrating magma, the petrologic data (shown as circles in Fig. 1) should be regarded as upper bounds on the average geotherm. *England and Bickle* [1984] subsequently also noted a second and opposite caveat: that transient crustal thickening could produce rocks that underestimate the geothermal gradient if those rocks do not stay deep long enough to re-equilibrate with the steady-state geotherm. Indeed, some recent models (*Vlaar et al.* [1994]; *Zegers and van Keken* [2001]) of Archean tectonics do involve short-duration crust-thickening

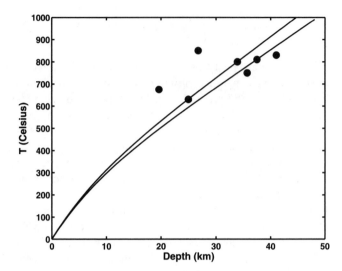

Figure 3. Archean metamorphic geotherms: black points are Archean high-grade rocks tabulated by Bickle [1978], (pressure-depth conversions are based on an Archean crustal density of 2860 kg/m³ derived from CRUST 2.0 (Bassin et al. [2000]), instead of Bickle's original 2750 kg/m³); solid curves are Bickle's geotherms (amphibolite: upper curve; granulite: lower curve).

events associated with lower crustal delamination, suggesting that the England and Bickle geotherms may underestimate the Archean case. In the absence of better information, 30°/km is a plausible Archean geothermal gradient for orogenic conditions. However, since the intent of this article is to discuss conditions which might have been pervasive in the Archean, and these preserved metamorphic geotherms necessarily indicate hotter than normal conditions and are not directly useful for assessing normal crustal conditions.

Instead, as indicated above, I extrapolate present day observed or inferred crustal and mantle heat flows back to Archean times for inactive continental crust, using the present day values of *Jaupart and Mareschal* [1999]. Since the objective is to assess the stability of "normal" crust not subject to orogenic reworking, the assumption that crustal radiogenic element concentrations have not been significantly changed by tectonic reworking is reasonable. Crustal radiogenic contribution J_c at a time t ago can therefore be calculated by multiplying present day values by $\exp(t/\tau_r)$, where τ_r is the effective decay time of a typical mix of radiogenic elements. Jaupart and Mareschal [1999] observe that for such a mixture, crustal radiogenic heat has decayed by a factor of about 3 over 3 Ga, giving $\tau_r \simeq 2.7$ Ga. The variation of mantle heat flux is less certain (*Lenardic* [1998]), but fortunately much smaller. Mantle temperatures may have declined by as much as 40% (*Nisbet et al.* [1993]) or as little as little as 7% (*Stone et al.* [1997]; *Parman et al.* [1997]) since the Archean; I conservatively assume a small linear decrease of J_m by 10% from 3 Ga.

4.2. Thermal Histories

Jaupart and Mareschal [1999] give present day surface heat flow and mantle heat flow for the Archean Superior province of the Canadian Shield as 40 and 10 to 15 mW/m² respectively, from which the crustal contribution can be inferred as $J_c = J_s - J_m = 40 - 12.5 = 27.5 = 27.5$ mW/m². Extrapolation backwards in time, for the range of physical property values given in Section (3.3) gives the suite of curves shown in Figure (4). A more useful representation of these results are provided by Figure (5) which shows the median and ± one standard deviation (68% of the values) of the values shown in Figure (4). A time constant of 1 Ma can be taken to divide coupled upper crust from decoupled upper crust. Figure (5)) indicates that the upper brittle layer of quiescent continental crust with cratonic heat productivity and mantle input was probably mechanically decoupled from the deeper crust and lithosphere during the early Archean (before 3.8 Ga in Figure (5)), and probably coupled from the early Proterozoic onwards (after 2.3 Ga in Figure (5)), with a late Archean median estimate (about 2.8 Ga on Figure (5)). This early coupling, and associated stability of cratonic quiescent crust, is directly the consequence of the low thermal input from the mantle, presumably the result of insulation by a thick cratonic root.

To illustrate this point, Figure (6) shows the same calculation with thermal inputs appropriate to the Appalachian region of North America (*Jaupart and Mareschal* [1999]). The Figure suggests that crust of this type would have, as a

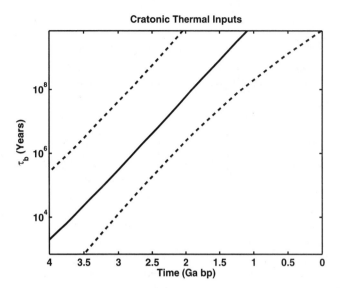

Figure 5. Envelope of possible time variation of decoupling time constant τ_b for northwest German crust given cratonic thermal inputs ($J_c = 0.027$ W/m² and $J_m = 0.013$ W/m², based on Jaupart and Mareschal [1999]), showing upper crustal vertical decoupling time constant τ_b for a 1000 km wide micro-continent, plotted against time. The solid line is the median of the suite of curves in Figure (4); dashed lines together enclose ± one standard deviation (68% of the values).

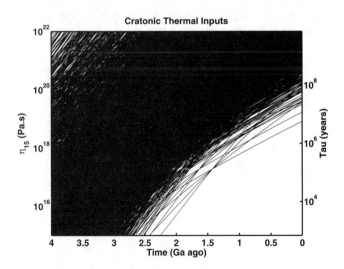

Figure 4. Suite of possible histories of decoupling time constant τ_b for northwest German crust given cratonic thermal inputs ($J_c = 0.027$ W/m² and $J_m = 0.013$ W/m², based on Jaupart and Mareschal [1999]), showing the ductile layer viscosity η_{15} and corresponding vertical decoupling time constant τ_b for a 1000 km wide micro-continent.

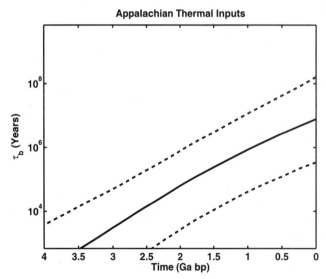

Figure 6. Envelope of possible time variation of decoupling time constant τ_b for northwest German crust given Appalachian thermal inputs ($J_c = 0.037$ W/m² and $J_m = 0.022$ W/m², based on Jaupart and Mareschal [1999]), showing upper crustal vertical decoupling time constant τ_b for a 1000 km wide micro-continent, plotted against time. The solid line is the median of the suite of individual test cases; dashed lines together enclose ± one standard deviation (68% of the values).

normal state, a decoupled upper brittle layer until late in the Proterozoic, as a direct result of the higher heat productivity and mantle input. This strongly supports the hypothesis of *Morgan* [1985] attributing the selective survival of Archean cratons to their low thermal inputs, rather than a special history. Crust which was not so favored (such as the "Appalachian" crust of Figure (6)) would not endure long the mobilizing processes (to be discussed below) associated with crustal decoupling.

Note that the evidence for early coupling of cratonic crust in Figure (5) depends strongly on the assumption that erosion to present levels for this crust was achieved early in the Archean. If this were not the case, added crustal thickness above current erosional levels would have added to the crustal heat production, and raised crustal temperatures. As an illustration, Figure (7) shows the same calculation as in Figure (5) except that the crustal component heat production J_c has been increased by 50% (mantle heat input is left unaltered). Decoupling now persists to somewhere between the late Archean and early Proterozoic.

It is important to remember that all of the geothermal arguments made here have been for "normal" crust in thermal equilibrium with "normal" mantle input and its own heat production. Where decoupling has been predicted in this analysis, it is therefore pervasive continental decoupling, not just that associated with special heat inputs from plate collisions or subduction-generated magmatism.

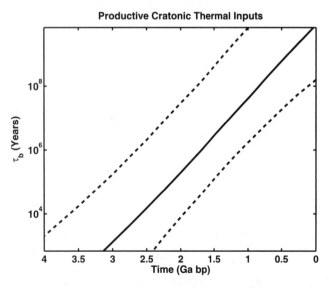

Productive Cratonic Thermal Inputs

Figure 7. Envelope of possible time variation of decoupling time constant τ_b for northwest German crust given a crustal productivity 50% higher relative to that of Figure (5). Shown is upper crustal vertical decoupling time constant τ_b for a 1000 km wide microcontinent, plotted against time. The solid line is the median of the suite of individual test cases; dashed lines together enclose ± one standard deviation (68% of the values).

4.3. Crustal Survival

A concrete example of *Morgan's* [1985] survival argument is obtained by considering England and Wales. Ignoring for the moment their actual geological ages, would it be plausible on thermal grounds for England and Wales to be preserved early Proterozoic crustal fragments? Of their 70 mW/m² average surface heat flow, the crustal contribution (based on the excess over the reduced heat flow) is about 45 (*Jessop* [1990]). Even in quiescent conditions (no orogenesis or basal plume input), this would have been about twice this 2 Ga ago, and the mantle contribution a little higher. The total surface heat flow would have been over 115 mW/m². This is well in excess of Basin and Range values (*Blackwell* [1983]); it would be surprising if England and Wales had not undergone immediate gravitational collapse into the nearest ocean basin, as discussed in the next section. On thermal grounds alone, England and Wales are unlikely to be surviving Proterozoic remnants.

By making this argument more quantitative, it is possible to predict, on this basis, the maximum modern day surface heat flux that would be consistent with the survival of a continental fragment of NG-type crust from any particular time in the past until the present day. Here, survival is assumed to occur if decoupling does not occur. A simple-minded way of making this prediction is to require that crust which has been stable for a time a (its tectonophysical age) have had a decoupling time constant at that time that was at least as large as a (otherwise its chances of mobilization between then and now would be high). That is, $a/\tau_b = 1$. Since τ_b is linearly proportional to ductile zone viscosity (equation (1)) and viscosity is thermally activated (equation (9)), then

$$\frac{a}{\tau_{b0}} = \exp\left(\frac{Q}{RT_a} - \frac{Q}{RT_0}\right) \quad (13)$$

where τ_{b0} is the decoupling time constant for modern NG crust (about 0.4 Ma as noted above, if a continental size of 1000 km is assumed), and T_{d0} and T_{da} are the current and ancient ductile zone temperatures. After some manipulation,

$$T_{da} = \frac{T_{d0}}{1 + \frac{RT_{d0}}{Q}\log\left(\frac{a}{\tau_{b0}}\right)} \quad (14)$$

The ancient ductile zone temperature T_{da} can be evaluated as before, using the geotherm given by equation (12), to give

$$T_s + \frac{J_{ca}z_d}{k} + (1 - \exp(-z_d/h))\frac{J_{ma}h}{k} = \frac{T_{d0}}{1 + \frac{RT_{d0}}{Q}\log\left(\frac{a}{\tau_{b0}}\right)} \quad (15)$$

where z_d is the depth of the ductile channel, J_{ca} and J_{ma} are the ancient crustal and mantle contributions to the heat flow. The ancient crustal contribution J_{ca}, as before, is related to the present day value J_{c0} by $J_{ca} = J_{c0}\exp(a/\tau_r)$, where τ_r^{-1} is the mean crustal radiogenic decay rate. (The minor enhancement in the mantle heat contribution will be neglected here.) Equation (15) can thus be solved to give J_{c0} and the present day mantle heat flow J_{m0} added to it to give the total present day heat flow J_{s0} as

$$J_{m0} + \frac{\exp(-a/\tau_r)}{1-\exp(-z_d/h)}\left[\frac{T_{d0}}{1+\frac{RT_{d0}}{Q}\log\left(\frac{a}{\tau_{b0}}\right)} - T_s - \frac{J_m z_d}{k}\right]$$

This expression estimates the maximum present-day heat flow associated with crust which has survived stably for a time a. It should correlate with the observed heat flows of continental crust as a function of tectonophysical age. Figure (8) shows the fit of this equation to experimental crustal survival data compiled by *Jessop* [1990]. It should not be surprising that a good fit can be obtained; although surface temperature (7 °C), continental size (1000 km) and mantle heat flow (30 mW/m²) were constrained, equation (15) has several relatively unconstrained parameters (K, h, z_d, T_d) which can be used to tune the fit. The best fitting solutions slightly preferred feldspathic activation energies rather than granitic ones, ductile channel depths of order 20 to 25 km, and relatively deep distributions of radiogenic elements.

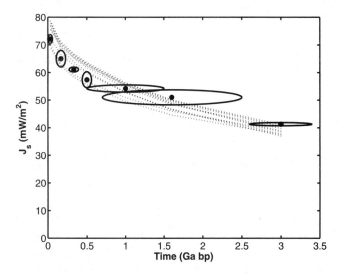

Figure 8. Present-day continental surface heat flow as a function of tectonic age in Ga. Ellipses surrounding points show the values and standard errors (for heat flow) or ranges (for ages) of observations compiled by Jessop [1990]. The dotted lines show a sample of curves predicted by equation (15).

However, a good fit here is sufficiently easy to obtain that it really says nothing about what plausible values of these adjustable parameters ought to be, except that a wide variety of plausible combinations worked (chosen from the ranges described in Section 3.3), of which ten are shown. Nevertheless, it is satisfying to know that the decoupling stability calculations are consistent with what is known about crustal survival.

5. THE FATE OF DECOUPLED CRUST

5.1. Styles of Gravitational Collapse

Continental crust which is decoupled from the underlying lithosphere has many possible fates; both gravity and tectonic boundary forces on the upper crustal plate can mobilize it. By good luck in avoiding collisions, continental crust can escape mobilization by tectonic boundary forces. It cannot, however, avoid the ubiquitous influence of gravity. The probably pervasive decoupling of early Archean upper crust (with cratonic thermal inputs or otherwise) from the rest of the lithosphere raises the possibility of continental-scale gravitational collapse, not so much as a catastrophic event but as a continuous process which limited early Archean continental elevations. *Dewey* [1988] pointed out that modern continental plateaus are observed to build to heights of only 4 to 5 kilometers above adjacent topography before gravitational collapse has prevented further growth. This limitation is associated with the decrease in the viscosity of lower crustal rocks with increasing depth and elevated temperature. Although elevations larger than 5 km would be stable for cold continental crust today, mountain building is normally associated with magmatism and elevated crustal temperatures. *Rey et al.* [2001] have described the possible modes of gravitational collapse. For excess topographic elevation generated by convergent tectonics, they describe two distinct end-member collapse styles. One is ductile extrusion by channel flow from under high-standing crust as observed in the Himalayas today (*e.g. Burchfiel et al.* [1992]; *Hodges et al. Beaumont et al.* [2001]; *Beaumont and Quinlan* [1994]). The second is "free-boundary collapse" by extension of the high-standing crust over an adjacent regions, by boundary thrusting accompanied by interior extensional tectonics in the interior (as is probably occurring in western North America today (*e.g. Coney and Harms* [1984]; *Sonder et al.* [1987]), although complicated by transform-fault tectonics at the boundary).

5.2. Prerequisites of Gravitational Collapse

Under what conditions will each type of gravitational collapse occur? Vertical decoupling of the upper crust is a prerequisite for extrusion collapse. During collapse mediated by channel flow and extrusion at the continental boundary, the brittle upper crust moves vertically, and only the viscous shear forces of the channel flow need be driven; by definition these are negligible for vertically decoupled crust. A second, and essential requirement for extrusional collapse, as seen in the Himalayan extrusion from under the Tibetan plateau, is that erosion is needed to sustain it (*e.g. Jamieson and Beaumont* [1989]; *Beaumont et al.* [1992]; *Koons* [1995]; *Koons et al.* [2002]). Without erosion, extruded material will reach the same height as the adjacent overthickened plateau and then prevent further extrusion by back pressure. (Both these arguments are consistent with the apparent free-boundary collapse of the western U.S., which has neither a confining cratonic boundary to the west, nor the strong erosional effects of monsoon rains.) Tectonically significant erosion is only possible in a sub-aerial environment. A third assisting factor, by virtue of inhibiting free-boundary collapse, is mechanical enclosure of the decoupled and overthickened region of crust by, for example, cold cratonic crust.

During free-boundary gravitational collapse, the upper crust must extend and this requires additional energy. What are the required conditions? Vertical and/or horizontal decoupling are not enough in themselves to guarantee free-boundary gravitational collapse. For free-boundary collapse, continental elevations must be high enough that the ratio (Argand ratio) of tensional stresses (that are gravitationally generated by the elevation) to the strength of the extending crust or lithosphere must exceed unity (*Rey et al.* [2001]). For free-boundary extension of a brittle crust above a decoupling ductile zone, it is only the brittle upper crust that is relevant to the strength, even though the relevant gravitational energy is contributed over the entire depth of the crust.

An alternate framework for viewing the same physics is that the energy liberated by collapse exceeds the energy consumed in dissipative processes such as shear flow in the ductile zone and normal faulting in the overlying brittle crust. A detailed consideration (*Bailey* [2000]) of the energy budget for this process shows that the dominant dissipative term is friction on upper crustal normal faults. The Argand ratio is therefore essentially independent of actual ductile zone viscosities; in order to get the thickness of the brittle crust low enough for collapse to occur, the required geothermal gradient happens to be so steep as to render the ductile channel hot and effectively inviscid. More specifically, collapse occurs when the brittle upper crust thickness b reduces to about 2.5 times the continental elevation e over adjacent ocean basins (*Bailey* [1999]; *Bailey* [2000]). For ocean basins of depth nearly 5 km, adjacent continental brittle upper crust cannot be thinner than about 12 km or the continent will undergo free-boundary gravitational collapse into the adjacent ocean basin.

6. ARCHEAN CONTINENTAL EMERGENCE

Of the suite of possible isotherms for thermally-cratonic crust used to obtain Figure (5), most reach, during the very early Archean, a plausible (*McKenzie and Fairhead* [1997]) brittle-ductile transition temperature of 350°C at depths shallower than 12 km (the critical value derived above for free-boundary collapse into an adjacent ocean basin of depth 5 km). This implies an upper crustal geothermal gradient of order 30°/km. However, only a minority of possible isotherms have such a thin upper brittle crust by the start of the Proterozoic (Figure (9)). This suggests that free-boundary gravitational collapse limited early Archean continental elevations to below sea level (a conclusion which would be even stronger for "productive" uneroded thermally-cratonic crust) but that they emerged some time during the Archean. However, this conclusion presupposes that Archean oceans had depths similar to today's, so this Archean emergence scenario needs independent support. Such support is available from petrologic observations. Unlike the Phanerozoic, submarine basalts showing crustal contamination are commonly found in Archean cratons, suggesting that large parts of the Archean continental crust were below sea level (Arndt [1999]).

But why is gravitational limiting the most plausible explanation of this shortfall, rather than a deeper Archean ocean,

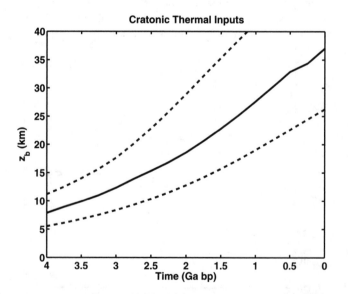

Figure 9. Envelope of possible time variation of the brittle-ductile transition depth z_b (presumed to be at 350°C, after McKenzie and Fairhead [1997]) for northwest German crust given cratonic thermal inputs. A value less than about 12 km would be expected to permit free-boundary gravitational collapse. The solid line is the median of the suite of individual test cases; dashed lines together enclose ± one standard deviation (68% of the values).

for example? The elevation of continents at any point in Earth's history is determined by the balance between continent-building and continent-reducing processes. Since the processes which build continental elevations obtain their energy from mantle convection, they can hardly have been less vigorous during the Archean. Thus one may infer that the reason for widespread Archean continental submergence must lie in the increased vigour of elevation-reducing processes. Since the most effective modern elevation-reducing process, erosion, cannot take elevations below sea level, some other process must have been responsible for Archean submergence. *Arndt* [1999] suggested that submergence was transient and resulted from sea level rise associated with enhanced midoceanic ridge volume during unusually strong episodes of global tectonic activity, thus being correlated with the majority of preserved submarine volcanics. However, it is equally plausible that enhanced tectonic activity would have increased continental elevation through increased continental compression and widespread orogeny. In any case, gravitational limiting of continental elevations must still have operated; the association of Archean continental emergence with secular reduction of the geothermal gradient to about 30°C/km heat flow requires Archean oceanic depths to coincide with Archean limiting continental elevations of about 5 km.

Thus, Arndt's emergence evidence (*Arndt* [1999]), Bickle's Archean geotherms, Bailey's free-boundary continental collapse criterion (*Bailey* [2000]), and a brittle-ductile transition temperature of 350°C, together require that the mean depths of Archean oceans were comparable with modern ones. This is an interesting conclusion in its own right. It has been suggested (*Murakami et al.* [2002]) that large amounts of water (~ five hydrospheres) may reside in the mantle. This water may be available for recycling through the oceans. If so, the correspondence of modern and Archean ocean depths is surprising; a very small imbalance in the subduction of water into the mantle at trenches and its recovery through volcanism could produce a strong secular evolution of ocean depth with time. Constancy of ocean depths since the Archean would then require a strong feedback mechanism, the nature of which remains unclear.

7. SECULAR EVOLUTION OF CONTINENTAL TECTONIC STYLE

What secular evolution of the "typical" tectonic style of continental crust do the above arguments suggest if they are valid? In answering such a question, one must remember that the thermal arguments above are about averages; individual crustal segments, with different thermal properties, need not always comply. Nevertheless, the thermal arguments do suggest that we can say something about the evolution of

typical continental behavior from the Archean through to the present.

Mantle convection must have driven global tectonics and associated continental thickening during the Archean, just as it does today. Whether plate tectonics, plume tectonics, flake tectonics (*Hoffman and Ranalli* [1988]), something different, or all four were responsible, such thickening would have been opposed by free-boundary gravitational collapse. This opposition would have kept continents on average below sea level, until mean geothermal gradients in thermally-cratonic crust cooled to about 30°/km some time during the Archean, and upper brittle crustal thicknesses exceeded about 12 km. For non-thermally-cratonic crust, with higher productivity and basal heat flows, lifetimes before remobilization would have been short.

When continents were then able to build above sea level by virtue of cooling and the suppression of free-boundary gravitational collapse, continental tectonics would have changed radically. In particular, extrusional (Himalayan-style) gravitational collapse mediated by erosion at continental margins must have have replaced free-boundary gravitational collapse as the normal elevation limiting process. Comparison of Figures (6) and (10) for thermally-Appalachian crust, and (5) and (9) for thermally-cratonic crust, suggests that vertical decoupling would persist longer than the possibility of free-boundary collapse; free-boundary collapse subsequent to

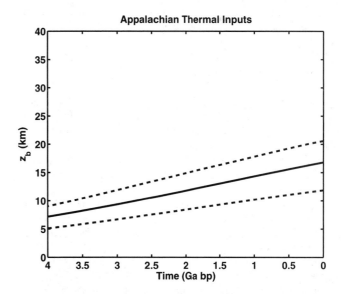

Figure 10. Envelope of possible time variation of the brittle-ductile transition depth z_b (presumed to be at 350°C, after McKenzie and Fairhead [1997]) for northwest German crust given Appalachian thermal inputs. A value less than about 12 km would be expected to permit free-boundary gravitational collapse. The solid line is the median of the suite of individual test cases; dashed lines together enclose ± one standard deviation (68% of the values).

mean continental emergence would only occur as a result of unusual sub-crustal heating events. This would apply largely to non-cratonic crust; stable thermally-cratonic crust would no longer be decoupled from the end of the Archean. As long as the conditions for vertical decoupling (easy channel flow) were still satisfied, thermally non-cratonic continental crust would be able to "leak" excess crustal elevation, wherever created, through eroding mobile belts on margins, whether or not the margins were plate boundaries. Such "orogenic export" from the Tibetan Plateau over distances of a thousand or more kilometers is visible in south and southeast Asia today. This mobile-belt tectonic style would be expected to continue for thermally non-cratonic crust through the Proterozoic. Subsequently, continental margins would then have become passive unless involved in plate-boundary tectonics.

REFERENCES

Abe,Y., Thermal and chemical evolution of the terrestrial magma ocean. *Phys. Earth Planet. Inter.*, 100: 27-39, 1997.

Arndt, N., Why was flood volcanism on submerged continental platforms so common in the Precambrian? *Precambrian Research*, 97: 155-164, 1999.

Bailey, R.C., Gravity-driven continental overflow and Archaean tectonics. *Nature*, 398: 413-415, 1999.

Bailey, R.C., Dynamical analysis of continental overflow. *Journal of Geodynamics*, 31: 293-310, 2000.

Bassin, C., G. Laske, and G. Masters. The current limits of resolution for surface wave tomography in North America. *EOS Trans AGU*, 81: F897, 2000.

Beaumont, C. and G. Quinlan. A geodynamic framework for interpreting crustal-scale seismic-reflectivity patterns in compresional orogens. *Geophys. J. Int.*, 116:754-783, 1994.

Beaumont, C., P. Fullsack, and J. Hamilton. Erosional control of active compressional orogens. In K. R. McClay, editor, *Thrust Tectonics*, pages 1-18, New York, 1992. Chapman and Hall.

Beaumont, C., R.A. Jamieson, M.H. Nguyen, and B. Lee. Himalayan tectonics explained by extrusion of a low-viscosity crustal channel coupled to focused surface denudation. *Nature*, 414:738-742, 2001.

Bickle, M.J., Heat loss from the earth: a constraint on Archaean tectonics from the relation between geothermal gradients and the rate of plate production. *Earth and Planetary Science Letters*, 40: 301-315, 1978.

Bird, P., Lateral extrusion of lower crust from under high topography. *J. geophys. Res.*, 96B:10,275-10,286, 1991.

Blackwell, D.D., Heat flow in the northern Basin and Range province. In Geothermal Resources Council, editor, *The Role of Heat in the Development of Energy and Mineral Resources in the Northern Basin and Range Province, Special Report 13*, page 81 93, 1983.

Bleeker, W., *The Early Earth: Physical, Chemical and Biological Development*, volume 199, chapter Archaean tectonics – a review,

with illustrations from the Slave craton, pages 151-181. Geological Society of London Special Publication, 2002.

Burchfiel, B.C., Z. Chen, K.V. Hodges, Y. Liu, L.H. Royden, C. Deng, and J. Xu. *The South Tibetan detachment system, Himalayan orogen: Extension contemporaneous with and parallel to shortening in collisional mountain belt*, volume 269. Geological Society of America Special Paper 269, Boulder, Colorado, 1992.

Burke, K., J.F. Delaney, and W.S.F. Kidd. Dominance of horizontal movements and microcontinental collisions during the later permobile regime. In B.F. Windley, editor, *The Early History of the Earth*, pages 113-130, London, 1976. Wiley.

Chapman, D.S., Thermal gradients in the continental crust. In J.B. Dawson, D.A. Carswell, J. Hall, and H.K.H. Wedepo, editors, *The Nature of the Lower Continental Crust*, volume 24 of *Spec. Publ. Geol. Society*, pages 63-70, London, 1986.

Clark, M.K., and L.H. Royden. Topographic ooze: Building the eastern margin of Tibet by lower crustal flow. *Geology*, 28: 703-706, 2000.

Clauser, C., and E. Huenges. *A Handbook of Physical Constants: Rock Physics and Phase Relations (Vol.3), AGU Reference Shelf Series*, volume 3, chapter Thermal Conductivity of Rocks and Minerals, page 270. 1995.

Coney, P.J., and T.A Harms. Cordilleran metamorphic core complexes, Cenozoic extensional relics of Mesozoic compression. *Geology*, 12:550-554, 1984.

Davies, G.F., On the emergence of plate tectonics. *Geology*, 20: 963-966, 1992.

Dewey, J.F., Extensional collapse of oroge ns. *Tectonics*, 7: 1123-1139, 1988.

De Wit, M.J., On Archean granites, greenstones, cratons and tectonics: Does the evidence demand a verdict? *Precambrian Research*, 91:181-226, 1998.

England, P., and M.J. Bickle. Continental thermal and tectonic regimes during the Archaean. *Journal of Geology*, 92:353-367, 1984.

Haenel, R., Geothermal investigations in the Rhenish Shield. In K. et al Fuchs, editor, *Plateau Uplift: The Rhenish Shield, A Case History*, pages 228-247, Berlin, 1983. Springer-Verlag.

Hamilton, W.B., Archaean magmatism and deformation were not products of plate tectonics. *Precambrian Research*, 91:143-179, 1998.

Hodges, K.V., R.R. Parrish, T.B. Housh, D.R. Lux, B.C. Burchfiel, L.H. Royden, and Z. Chen. Simultaneous Miocene extension and shortening in the Himalayan Orogen. *Science*, 258:1466-1470, 1992.

Hoffman, P.F., and G. Ranalli. Archean flake tectonics. *Geophys. Res. Lett.*, 150(10):1077-1080, 1988.

Hutton, J., Theory of the earth; or an investigation of the laws observable in the composition, dissolution, and restoration of land upon the globe. *Trans. Roy. Soc. Edinburgh*, 1:209-304, 1788.

Hutton, J., *Theory of the Earth, with proofs and illustrations*. Edinburgh, 1795.

James, D.E., and M.J. Fouch. Formation and evolution of Archaean cratons—insights from southern africa. In *Early Earth: Physical,*

Chemical and Biological Development, volume 199, pages 1-26. Geological Society of London Special Publication, 2002.

Jamieson, R.A., and C. Beaumont. Deformation and metamorphism in convergent orogens: A model for uplift and exhumation of metamorphic terrains. In J.S. Daly, R.A. Cliff, and B.W.D. Yardley, editors, *Evolution of Metamorphic Belts*, volume 43, pages 117-129. Geol. Soc. Spec. Publ., 1989.

Jamieson, R.A., C. Beaumont, M.H. Nguyen, and B. Lee. Interaction of metamorphism, deformation and exhumation in large convergent orogens. *Journal of Metamorphic Geology*, 200 (1): 9-24, 2002.

Jaupart, C., and J.C. Mareschal. The thermal structure and thickness of continental roots. *LITHOS*, 480(1-4):93-114, SEP 1999.

Jessop, A.M., Thermal geophysics. *Developments in Solid Earth Geophysics, Elsevier*, 17:277, 1990.

Kaufman, P.S., and L.H. Royden. Lower crustal flow in an extensional setting: Constraints from the Halloran Hills region, eastern Mojave Desert, California. *Journal of Geophysical Research*, 99:15723-15739, 1994.

Klein, A., W. Jacoby, and P. Smilde. Mining induced crustal deformation in northwest Germany: Modeling the rheological structure of the lithosphere. *Earth planet. Sci. Lett.*, 147:107-123, 1997.

Klemperer, S.L., A relation between continental heat flow and the seismic reflectivity of the lower crust. *J. Geophys.*, 61:1-11, 1987.

Koons, P.O., Modeling the topographic evolution of collisional belts. *Ann. Rev. Earth Planet. Sci.*, 23:375-408, 1995.

Koons, P.O., P.K. Zeitler, C.P. Chamberlain, D. Craw, and A.S. Meltzer. Mechanical links between erosion and metamorphism in Nanga Parbat, Pakistan Himalaya. *Am. J. Sci.*, 302: 749-773, 2002.

Kruse, S.E., M.K. McNutt, J. Phipps-Morgan, L. Royden, and B.P. Wernicke. Lithospheric extension near Lake Mead, Nevada: A model for ductile flow in the lower crust. *Journal of Geophysical Research*, 96:4435-4456, 1991.

Kuznir, N.J., and D.H. Matthews. Deep seismic reflections and the deformational mechanics of the continental lithosphere. In *Journal of Petrology Special Lithosphere Issue*, pages 63-87. 1988.

Lachenbruch, A.H., Crustal temperature and heat production: implications of the linear heat-flow relation. *J. Geophys. Res.*, 75:3291-3300, 1970.

Lenardic, A., On the partitioning of mantle heat loss below oceans and continents over time and its relationship to the Archaean paradox. *Geophysical Journal International*, 134:706-720, 1998.

Matsui, T., and Y. Abe. Formation of a "magma ocean" on the terrestrial planets due to the blanketing effect of an impact induced atmosphere. *Earth Moon Planets*, 34:223-230, 1986.

McKenzie, D., and D. Fairhead. Estimates of the effective elastic thickness of the continental lithosphere from Bouger and free air gravity anomalies. *J. Geophys. Res.*, 102B:27523-27552, 1997.

McKenzie, D., and J. Jackson. Conditions for flow in the continental crust. *Tectonics*, 210(6):1055–1061, 2002.

McKenzie, D., F. Nimmo, J.A. Jackson, P.B. Gans, and E.L. Miller. Characteristics and consequences of flow in the lower crust. *J. Geophys. Res.*, 105:11,029 11,046, 2000.

Morgan, P., Crustal radiogenic heat-production and the selective survival of ancient continental-crust. *J. Geophys. Res.*, 90:C561-C570 Suppl. S, 1985.

Murakami, M., K. Hirose, H. Yurimoto, Nakashima S., and N. Takafuji. Water in the earth's lower mantle. *Science*, 295:1885-1887, 2002.

Nelson, K.D., W. Zhao, L.D. Brown, J. Kuo, J. and Che, X. Liu, Y. Klemperer, S.L. and Makovsky, R. Meissner, J. Mechie, R. Kind, F. Wenzel, J. Ni, J. Nabelek, H. Chen, L. and Tan, W. Wei, A.G. Jones, J. Booker, M. Unsworth, W.S.F. Kidd, M. Hauck, D. Alsdorf, A. Ross, M. Cogan, C. Wu, E. Sandvol, and M. Edwards. Partially molten middle crust beneath southern tibet: Synthesis of project indepth results. *Science*, 2740(5293):1684-1688, December 1996.

Nisbet, E.G., M.J. Cheadle, N.T. Arndt, and M.J. Bickle. Constraining the potential temperature of the Archaean mantle: A review of the evidence from komatiites. *Lithos*, 30:291-307, 1993.

Ohtani, E., The primordial terrestrial magma ocean and its implication for stratification of the mantle. *Phys. Earth Planet. Inter.*, 38:7080, 1985.

Parman, S.W., J.C. Dann, T.L. Grove, and M.J. De Wit. Emplacement conditions of komatiite magmas from the 3.49 ga Komati formation, Barberton Greenstone Belt, South Africa. *Earth Plan. Sci. Let.*, 150:303-323, 1997.

Ranalli, G., *Rheology of the Earth*. Chapman & Hall., London, 1995.

Rey, P., O. Vanderhaeghe, and C. Teyssier. Gravitational collapse of the continental crust: Definition, regimes, and modes. *Tectonophysics*, 342:435-449, 2001.

Royden, L., Coupling and decoupling of crust and mantle in convergent orogens: Implications for strain partitioning in the crust. *Journal of Geophysical Research*, 101:17679-17705, 1996.

Shen, X., Kinematics and tectonothermal modeling - interpretation of heat flow observed on the Tibetan plateau. *Tectonophysics*, 225:91-106, 1993.

Sonder, L., P. England, B. Wernicke, and R. Christianson. A physical model for Cenozoic extension of western North America. In et al. Coward, M., editor, *Continental extensional tectonics: Geological Society of London Special Publication*, volume 28, pages 187-201, 1987.

Stone, W.E., E. Deloule, M.S. Larson, and C.M. Lesher. Evidence for hydrous high-mgo melts in the precambrian. *Geology*, 250(2):143-146, 1997.

Tonks, W.B., and H.J. Melosh. Magma ocean formation due to giant impacts. *J. Geophys. Res.*, 98:5319-5333, 1993.

Turcotte, D., and G. Schubert. *Geodynamics*. John Wiley and Sons, New York, 1982.

Vlaar, N.J., P.E. van Keken, and A.P. van den Berg. Cooling of the earth in the Archaean: Consequences of pressure-release melting in a hotter mantle. *Earth and Planetary Science Letters*, 121: 1-18, 1994.

Walcott, R.I., Late Quaternary vertical movements in eastern North America. *Rev. Geophys. Space Phys.*, 10:849-884, 1972.

Walcott, R.I., Structure of the earth from glacio-isostatic rebound. *Annu. Rev. Earth Planet. Sci.*, 1:115-37, 1973.

Wdowinski, S., and G.J. Axen. Isostatic rebound due to tectonic denudation: A viscous flow model of a layered lithosphere. *Tectonics*, 11:303-315, 1992.

Werner, A.G., *Von den usserlichen Kennzeichen der Fossilien.* Leipzig, 1774.

Werner, A.G., Kurze klassifikation der verschiedenen gebirgsarten. In *Abh. B´hm. Gesellsch. d. Wiss.*, pages S. 272-297, Prague, 1786.

Wernicke, B., The fluid crustal layer and its implications for continent dynamics. In M. Salisbury and D. Fountain, editors, *Exposed cross-sections of the continental crust*, pages 509-544, Boston, 1990. Kluwer Academic.

Zegers, T.E., and P.E. van Keken. Middle Archaean continent formation by crustal delamination. *Geology*, 29, 2001.

Bailey, R.C., Departments of Geology and Physics, University of Toronto, 60 St. George St., Toronto, Ontario M5A 1A7, Canada. (bailey@physics.utoronto.ca)

Thermal and Mechanical Controls on the Evolution of Archean Crustal Deformation: Examples From Western Australia

Simon Bodorkos and Mike Sandiford

School of Earth Sciences, University of Melbourne, Victoria, Australia

Dome-and-keel formation in Archean granite–greenstone terrains was uniquely favored by high abundances of heat-producing elements (HPEs) in the crust, and the ubiquity of pre-doming "greenstone-over-granite" stratification, which constituted large-scale density inversions and facilitated long-term steepening of the geotherm via burial of felsic basement-hosted HPEs. This contribution investigates the influence of these factors on the thermo-mechanical stability of the crust, with reference to two contrasting Archean granite–greenstone terrains in Western Australia. In the Neoarchean Eastern Goldfields Province, the rapid (2715–2665 Ma) accumulation of a thin (8-km) greenstone succession atop HPE-poor felsic crust (radiogenic heat flow contribution $q_c = 45$ mW m^{-2}) raised mid-crustal temperatures from 230 °C to 370 °C prior to 2650 Ma dome-and-keel formation. In such cold, strong crust, the development of broad, granite-cored antiforms and narrow, greenstone-cored synforms with little strike variation is likely to directly reflect far-field, horizontal tectonic forces, oriented perpendicular to the regional structural grain. In contrast, the Mesoarchean East Pilbara Granite–Greenstone Terrane underwent slow (3515–3325 Ma) accumulation of thick (14-km) greenstones atop HPE-rich ($q_c = 70$ mW m^{-2}) basement, which raised mid-lower crustal temperatures from 400 °C to 750–800 °C, prior to 3300 Ma dome-and-keel formation. The effective viscosity of this hot, weak crust was further reduced by partial melting of the felsic basement; the resulting "classical" architecture (featuring granite domes flanked by greenstone keels with a variety of strike orientations) may therefore reflect partial convective overturn and vertical reorganization of the crust during thermo-mechanical stabilization.

INTRODUCTION

"Dome-and-keel" structure is a distinctive feature of Archean granite–greenstone terrains worldwide. In general, variably deformed and metamorphosed granitoids define broad, high-amplitude "domes" that are separated from each other by narrow, synformal "keels" consisting of steeply dipping supracrustal greenstone successions [e.g., *Macgregor*,

Archean Geodynamics and Environments
Geophysical Monograph Series 164
Copyright 2006 by the American Geophysical Union
10.1029/164GM10

1951; *Anhaeusser et al.*, 1969; *Hickman*, 1983, 1984; *Marshak et al.*, 1992, 1999; *Choukroune et al.*, 1995, 1997; *Chardon et al.*, 1998; *Collins et al.*, 1998; *Van Kranendonk et al.*, 2004]. This crustal architecture reflects the prevalence of a peculiarly Archean phenomenon: the widespread early development of a gross tectono-stratigraphy comprising one or more relatively dense supracrustal successions (dominated by mafic and ultramafic volcanic rocks) overlying less-dense felsic crust of tonalite–trondhjemite–granodiorite (TTG) composition [e.g., *de Wit and Ashwal*, 1997; *Bleeker*, 2002; *Sandiford et al.*, 2004; and references therein]. In detail, Archean dome-and-keel provinces may be divided into two major groups, based primarily on the first-order structural geometry of the upper and middle crust [e.g., *Peschler et al.*, 2004], but also encapsulating important differences in age and tectono-magmatic relationships.

The first group comprises predominantly Mesoarchean (3.5–3.0 Ga) terrains featuring "classical" architecture. Curviplanar greenstone keels with short strike lengths and a wide range of strike orientations typically flank and separate several relatively large domal granitoid complexes (50–90-km diameter in plan view). In most cases, episodic but voluminous felsic magmatism (in the form of granite plutonism in dome cores and coeval, thick volcanic intercalations within overlying supracrustal successions) played an active role in the formation and amplification (or inflation) of dome-and-keel structure over an extended period (>500 My). Examples include the Zimbabwe Craton [e.g., *Macgregor*, 1951; *Anhaeusser et al.*, 1969; *Jelsma et al.*, 1993], the eastern Pilbara Craton (Figure 1b) of Western Australia [e.g., *Hickman*, 1983, 1984; *Van Kranendonk et al.*, 2002], and the western Dharwar Craton of southern India [e.g., *Choukroune et al.*, 1995, 1997; *Chardon et al.*, 1996, 2002].

In contrast, the second group is dominated by Neoarchean terrains (post 3.0 Ga, and mostly 2.75–2.65 Ga) and are typified by granitoid domes of widely varying shape and size, though domes with well-developed outlines tend to be pseudo-elliptical in plan view. These granitoids occur as antiformal culminations enveloped by laterally extensive greenstone belts, with long strike lengths and little variation in strike orientation. Felsic magmatism was relatively short-lived (<100 My), and mostly predated large-scale dome-and-keel amplification. The resulting map patterns resemble (at least superficially) those of post-Archean orogenic belts, and examples include the Eastern Goldfields Province (Figure 1c) of the Yilgarn Craton, Western Australia [e.g., *Griffin*, 1990; *Swager et al.*, 1997; *Krapez et al.*, 2000; *Blewett et al.*, 2004a; *Goleby et al.*, 2004], and the western Superior Province of Canada [e.g., *Card*, 1990; *Henry et al.*, 2000; *Thurston*, 2002; *Tomlinson et al.*, 2003; *White et al.*, 2003; *Percival et al.*, 2004].

Notwithstanding the existence of some notable exceptions to the above generalizations, the potential reasons for these apparent secular contrasts in the timing, duration, and nature of the interplay between large-scale upper-crustal deformation and felsic magmatic activity remain poorly understood [e.g., *Choukroune et al.*, 1997; *Bleeker*, 2002]. Do the observed patterns reflect a secular Archean evolution of tectonic processes [e.g., *Hamilton*, 1998], an evolving response of the Archean crust and upper mantle lithosphere to essentially time-invariant tectonic processes [e.g., *de Wit*, 1998], or some combination of the two? Could the relationship between deformation pattern and age reflect some systematic bias in the preserved record of Archean granite–greenstone

terrains [*Morgan*, 1985]? It is likely that the answers to these questions are linked and reflect an important feedback between the physical properties of the Archean lithosphere, the nature of tectonic processes that shape it, and the tendency of those processes to stabilize (and therefore preserve) cratonic nuclei.

The primary aims of this paper are to explore the potential link between syn-tectonic thermal regimes and the styles of crustal-scale deformation in Archean granite–greenstone terrains and to examine how this relationship may have changed through time. We commence with a brief review of the factors exerting long-term influence on the thermal evolution of a typical Archean greenstone-over-granite crust, with emphasis on the abundance (q_c) and depth distribution (z_c) of crustal heat production, and the deep mantle heat flux (q_m; see next section). This is followed by a geological overview of two West Australian (Figure 1a) granite–greenstone terrains [the Mesoarchean East Pilbara Granite–Greenstone Terrane (Figure 1b) and the Neoarchean Eastern Goldfields Province (Figure 1c)] that are characterized by broadly comparable histories of greenstone accumulation atop ancient felsic basement, but disparate dome-and-keel morphology. In each terrain, the available stratigraphic, structural, geophysical, geochemical, and geochronological constraints are used to estimate syn-tectonic q_c, z_c, and q_m values and to formulate simplified (one-dimensional) greenstone accumulation histories. These data are used as input for thermal models of each terrain, aimed at quantifying the potential extent of mid-crustal heating (attributable to the burial of relatively HPE-rich felsic crust beneath insulating greenstones) and the potential influence of such heating on the effective viscosity (and therefore mechanical behavior) of the mid-crust in each case. Finally, we consider the feedback relations affecting the thermo-mechanical behavior of the Archean crust and upper lithosphere in terms of the potential for secular change in both the processes governing crustal deformation and the record of those processes preserved by granite–greenstone terrains.

THERMAL CONSEQUENCES OF A REVERSE-STRATIFIED ARCHEAN CRUST

Most Archean granite–greenstone terrains were characterized by two fundamental crustal properties that incontrovertibly favored dome-and-keel formation. Firstly, granitic rocks comprised 50–60% of the crust, and their concentrations of the radioactive heat-producing elements (HPEs) K, Th, and U were at least two to three times higher than at present [e.g., *Pollack*, 1997]. Secondly, the early stages of crustal growth and active tectonism typically resulted in the widespread development of greenstone-over-granite tectono-stratigraphy [e.g., *Bleeker*, 2002]. This combination of properties had two

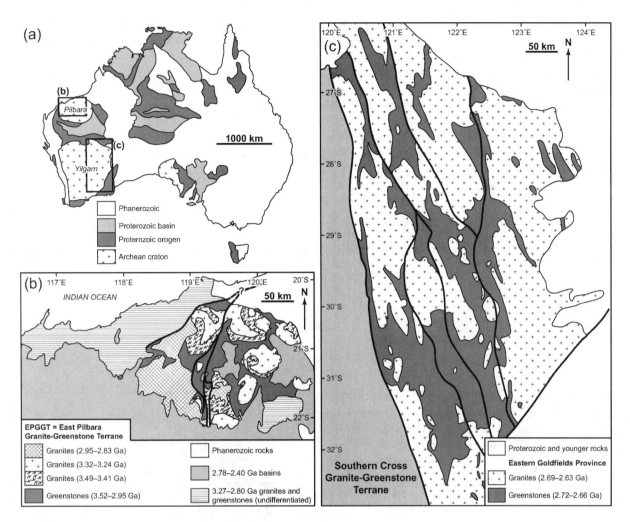

Figure 1. (a) Precambrian tectonic elements of Australia (modified after *Myers et al.* [1996]), showing the areas of the Archean cratons that form the focus of this study. Note that (b) and (c) have a common scale. (b) The Mesoarchean East Pilbara Granite–Greenstone Terrane (modified after *Van Kranendonk et al.* [2002] and *Hickman and Van Kranendonk* [2004]). Granite domes are typically pseudo-circular in plan view and separated from each other by greenstone keels with a variety of strike orientations. (c) The Neoarchean Eastern Goldfields Province (modified after *Myers* [1997] and *Blewett et al.* [2004a]). In plan view, granite domes are elongated parallel to the NNW–SSE structural grain, and there is little strike variation within the greenstone keels.

important thermo-mechanical consequences: (1) a substantial gravitational instability was generated by the density inversion within the reverse-stratified crust [e.g., *Mareschal and West*, 1980; *Collins et al.*, 1998; *de Bremond d'Ars et al.*, 1999], and (2) significant long-term steepening of the geotherm (termed "conductive incubation" by *Sandiford et al.* [2004]) occurred as a direct result of the burial of HPE-rich felsic crust beneath a thick greenstone edifice [see also *West and Mareschal*, 1979; *Ridley and Kramers*, 1990; *Sandiford and McLaren,* 2002; *Pysklywec and Beaumont*, 2004].

In the absence of significant advection, one-dimensional (vertical) heat flow at the Earth's surface (q_s) may be approximated by the sum of two independent components: "deep"

conductive heating driven by asthenospheric convection (q_m), and "shallow" radiogenic heating controlled by the abundance (q_c) of predominantly crust-hosted HPEs:

$$q_s = q_m + q_c \qquad (1)$$

The one-dimensional geotherm, $T(z)$, where z is depth, is governed by a similar relationship:

$$T(z) = T_m(z) + T_c(z) \qquad (2)$$

For the syn-tectonic greenstone-over-granite tectonostratigraphy characteristic of most Archean granite–greenstone terrains, the vertical distribution of HPEs may be approximated (in one dimension) by a three-layer configuration. This typically consists of a greenstone layer (of thickness z_1 and negligible HPE content), overlying a TTG layer (of thickness $z_2 - z_1$ and depth-independent volumetric heat production H), which in turn overlies a lower crust and conductive lithosphere with zero HPE content. Assuming a constant deep mantle heat flux q_m at the base of the conductive lithosphere and a uniform, temperature-independent thermal conductivity (k) within the lithosphere, the steady-state geotherm is given by Sandiford et al. [2004]:

$$T(z) = \frac{q_m}{k}z + \left[\frac{H(z_2 - z_1)}{k}\right]z \quad 0 < z < z_1 \quad (3a)$$

$$T(z) = \frac{q_m}{k}z + \left[\frac{H(2z_2 - z)}{2k}\right]z - \frac{Hz_1^2}{2k} \quad z_1 < z < z_2 \quad (3b)$$

$$T(z) = \frac{q_m}{k}z + \frac{H(z_2^2 - z_1^2)}{2k} \quad z > z_2 \quad (3c)$$

Beneath the base of the heat-producing crust (i.e., for $z > z_2$), the maximum crustal contribution to the geotherm (i.e., the second term of equation (3c)) is independent of z [Sandiford et al., 2002], and is denoted T_c'. Sandiford and McLaren [2002] showed that T_c' may be re-expressed in terms of the depth-integrated heat production (equivalent to q_c) and the characteristic depth of the heat production distribution (z_c). In the scenario described above, q_c is given by:

$$q_c = \int_{z_1}^{z_2} H\,dz = H(z_2 - z_1) \quad (4a)$$

and z_c corresponds to the depth-midpoint of the TTG layer:

$$z_c = z_1 + \frac{z_2 - z_1}{2} = \frac{z_1 + z_2}{2} \quad (4b)$$

Substituting equations (4a) and (4b) into the expression for T_c' within equation (3c) gives:

$$T_c' = \frac{q_c z_c}{k} \quad (5)$$

The sensitivity of deep crustal temperatures to q_c and q_m is well understood, but equation (5) highlights the additional importance of z_c. Its influence is illustrated by considering the long-term thermal evolution of the material point corresponding to the base of the TTG layer, following an instantaneous increase (Δz) in the thickness of the

supracrustal greenstone layer (without altering $z_2 - z_1$, H, or q_m). The initial temperature at $z = z_2$ is given by equation (3c):

$$T(z_2) = \frac{q_m z_2 + q_c z_c}{k} \quad (6a)$$

Following greenstone emplacement (and subsequent thermal equilibration), the temperature at the base of the TTG layer is given by:

$$T(z_2 + \Delta z) = \frac{q_m(z_2 + \Delta z) + q_c(z_c + \Delta z)}{k} \quad (6b)$$

Subtracting (6b) from (6a), the long-term temperature increase (ΔT) is:

$$\Delta T = \frac{\Delta z(q_m + q_c)}{k} \quad (7)$$

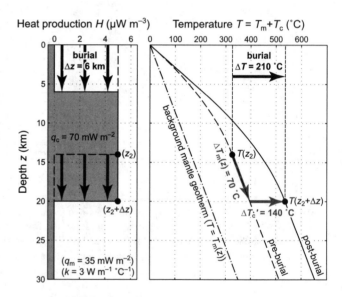

Figure 2. Steepening of the ambient geotherm as a consequence of the burial of HPE-rich felsic crust beneath a greenstone edifice. In the left panel, burial associated with greenstone accumulation is indicated by the thick black arrows; the unfilled dashed outline and the gray-filled solid outline correspond to the pre- and post-burial depth distributions, respectively, of HPEs. The right panel shows the time-invariant "mantle" component of the geotherm ($T_m(z)$), the pre-burial geotherm, and the post-burial geotherm, assuming that thermal equilibrium has been attained. The black circle (corresponding to the base of the heat-producing layer) has undergone a temperature increase of 210 °C during burial, only one-third of which is due to migration down the mantle geotherm ($\Delta T_m(z)$); the remainder ($\Delta T_c'$) is independent of z and is due to conductive incubation resulting from deepening of HPEs [e.g., Sandiford et al., 2002].

The significance of equation (7) is realized with the appreciation that the magnitude of q_c may have been as much as twice that of q_m in the Archean Earth, due to the abundance of HPEs [*Sandiford et al.*, 2004]. This is best illustrated by example (Figure 2): The emplacement of a 6-km-thick greenstone succession atop an Archean terrain with TTG-hosted $q_c = 70$ mW m^{-2} and $q_m = 35$ mW m^{-2} translates to a long-term temperature *increase* $\Delta T = 210$ °C at the base of the heat-producing crust (Figure 2). Two-thirds of this increase is directly attributable to conductive incubation via the burial of HPEs ($\Delta T_c' = 140$ °C; Figure 2); the remainder is due to migration of the base of the heat-producing crust down the mantle geotherm ($\Delta T_m(z) = 70$ °C; Figure 2), defined by the first term in each of equations (3a)–(3c).

However, the rate of deep crustal heating is dictated by the thermal response time of the lithosphere, so the magnitude of ΔT also depends upon the extent of thermal equilibration within the greenstone-over-granite stratified crust, prior to significant external modification via erosion, deformation, and/or magmatism. In the next section, we summarize the geology of the Mesoarchean East Pilbara Granite–Greenstone Terrane and the Neoarchean Eastern Goldfields Province, with emphasis on the stratigraphic and geochronological constraints on the nature, timing, and extent of greenstone accumulation. This provides a basis for quantitative assessment of the changes in the thermal structure of the crust prior to dome-and-keel formation.

GEOLOGICAL OVERVIEW

The East Pilbara Granite–Greenstone Terrane and the Eastern Goldfields Province (hereafter referred to as the Pilbara and the Goldfields, respectively) display several lithologic and tectono-stratigraphic similarities, despite significant differences in the timing and style of crustal growth and deformation. Both are characterized by the early development of regional greenstone-over-granite stratification [*Swager*, 1997; *Krapez et al.*, 2000; *Van Kranendonk et al.*, 2002] and an upper crust comprising felsic-dominated plutonic rocks and mafic-dominated volcanic rocks in roughly equal proportions by volume [*Wellman*, 2000; *Goleby et al.*, 2004]. In addition, both preserve isotopic evidence for the widespread existence of felsic crust that is considerably older than the exposed granites: *Smithies et al.* [2003] obtained 48 pre-3450 -Ma Nd model ages from 50 Pilbara granites with magmatic (U–Pb zircon) ages in the range 3450–2850 Ma, and *Champion and Sheraton* [1997] demonstrated that Nd model ages are at least 100 My older than magmatic crystallization ages for granites throughout the Goldfields.

Pilbara

The oldest exposed greenstone succession is the 3515–3495 Ma Coonterunah Group [*Buick et al.*, 1995],

which locally reaches 6 km in thickness [*Green et al.*, 2000] and is overlain by the regionally extensive 3490–3425 Ma lower Warrawoona Group, which is up to 10–12 km thick [*Hickman*, 1983; *Van Kranendonk et al.*, 2002]. Plutonic rocks of TTG composition were episodically emplaced into the lower parts of the both successions over the interval 3490–3410 Ma, and granite–greenstone contacts are intrusive and/or sheared. After an extended (70–80 My) hiatus, the 3350–3325 Ma upper Warrawoona Group, which is up to 8 km thick [*Van Kranendonk et al.*, 2002; *Bagas et al.*, 2004], was deposited atop the pre-3425 Ma crust. It is therefore likely that the regional stratigraphic thickness of the greenstone edifice was approximately 11–17 km, prior to the first major episode of dome-and-keel formation at ca. 3300 Ma [*Sandiford et al.*, 2004].

Dome-and-keel formation was accompanied by voluminous granitic magmatism in the easternmost Pilbara [e.g., *Williams and Collins*, 1990; *Barley and Pickard*, 1999], although the means of dome-and-keel formation remains the subject of considerable debate. Granite ascent through the overlying greenstones was initially attributed to solid-state diapirism [*Hickman*, 1983, 1984], but more recent interpretations of the granite domes include (1) magmatic diapirism and gravity-driven partial convective overturn of the crust [e.g., *Collins*, 1989; *Collins et al.*, 1998; *Van Kranendonk et al.*, 2002, 2004], (2) steepened metamorphic core complexes unroofed via extension of the overlying greenstones [e.g., *Zegers et al.*, 1996; *Kloppenburg et al.*, 2001], and (3) granite-cored antiformal culminations produced by complex cross-folding [e.g., *Blewett*, 2002; *Blewett et al.*, 2004b].

Goldfields

Pre-doming greenstones in the Goldfields are characterized by a relatively restricted range of ages. With the exception of the southwesternmost Goldfields (where a package of ca. 2900 Ma mafic volcanic rocks up to 1–2 km in thickness is locally exposed [*Nelson*, 1997; *Swager*, 1997]), the volcano-sedimentary successions were deposited over the 50 My interval 2715–2665 Ma [*Brown et al.*, 2001, and references therein]. These were intruded by 2690–2655 Ma high-Ca granites [*Champion and Sheraton*, 1997], though granite–greenstone contacts are commonly sheared. The maximum stratigraphic thickness of the greenstone succession is approximately 4–5 km [*Krapez et al.*, 2000], though significant structural repetition during regional pre-2670 Ma deformation has been recognized in the Kalgoorlie area [*Archibald et al.*, 1978; *Swager and Griffin*, 1990; *Blewett et al.*, 2004a]. However, it is unlikely that the regional thickness of the supracrustal complex exceeded 8 km when greenstone accumulation ceased at ca. 2665 Ma.

Dome-and-keel formation in the Goldfields at ca. 2650 Ma is widely attributed to regional WSW–ENE shortening, which resulted in the exposure of granites as domal complexes in antiformal culminations [*Swager*, 1997; *Krapez et al.*, 2000]. *Blewett et al.* [2004a] attributed vertical amplification of the structures to relatively rapid switches in tectonic mode (coaxial horizontal shortening–extension–shortening over the interval 2670–2650 Ma). *Weinberg et al.* [2003] suggested that antiform nucleation was focused by pre-tectonic felsic plutons, with subsequent deformation at least partly attributable to solid-state, buoyancy-driven ascent of granites under conditions of slow compression.

THERMAL PARAMETERS

We now turn our attention to geophysical and geochemical data with the potential to constrain the syn-tectonic magnitudes of q_m, q_c, and z_c in the Pilbara and the Goldfields. The estimates obtained will be used (in combination with z_c–time evolutions synthesized from the relevant stratigraphic and geochronological data) in the next section, as input for thermal models simulating the greenstone accumulation history appropriate to each terrain, prior to dome-and-keel formation.

Archean q_c

Present-day surface heat flow (q_s) data provide important constraints on the magnitudes of q_c and q_m. Heat flow measurements in the Australian Archean are sparse [*Cull*, 1982], but typical q_s values are 40–50 mW m^{-2} in the Pilbara and 35–45 mW m^{-2} in the Goldfields [*Howard and Sass*, 1964; *Jaeger*, 1970; *Sass et al.*, 1976; *Cull and Denham*, 1979; *Cull*, 1991]. The overall tectonic quiescence of cratonic Western Australia during the Phanerozoic means that transient contributions to the measured surface heat flow are likely to be negligible, so equation (1) may be applied to the Archean terrains, in order to approximate q_s as the one-dimensional (vertical) sum of its independent components q_m and q_c. A first-order upper bound on present-day q_m is then obtained by using Fourier's Law:

$$q_m = k \frac{\Delta T_L}{\Delta z_L} \qquad (8)$$

where ΔT_L is the temperature difference between the Earth's surface and the base of the conductive lithosphere, and Δz_L is the lithosphere thickness. Seismic tomography data suggest that the West Australian Archean lithosphere is 200–300 km thick [*Simons et al.*, 1999] and thermal modeling implies a depth of 170–230 km for the 1300 °C isotherm [*Artemieva and Mooney*, 2001]. Assuming $\Delta z_L = 200$–250 km, $\Delta T_L = 1280$–1300 °C, and uniform thermal conductivity $k = 3$ W m^{-1} °C^{-1} for the lithosphere, equation (8) yields present-day q_m values in the range 15–20 mW m^{-2} for

Western Australia. Such values are not significantly different from those obtained for a variety of Archean cratons worldwide [e.g., *Jones*, 1988; *Gupta et al.*, 1991; *Jaupart and Mareschal*, 1999; *Russell et al.*, 2001]. In fact, q_m must be lower than this, perhaps by as much as 5 mW m^{-2}, because of the contribution of heat production in the crust [e.g., *Jaupart and Mareschal*, 2003]. Of course, the present-day mantle heat flow has little bearing on the magnitude of the heat flow during Archean tectonism, which was likely to have been much higher (see later discussion).

By difference, equation (1) yields a lower bound (and therefore conservative) estimate for present-day q_c values of 20–35 mW m^{-2} for the Pilbara [*Bodorkos et al.*, 2004] and 15–30 mW m^{-2} for the Goldfields. Assuming these q_c values represent the respective present-day contributions of crust-hosted HPEs to the surface heat flux, we can calculate q_c values (corrected for the secular decline in radionuclide abundances) for each terrain at the time of Archean dome-and-keel formation. We used the present-day bulk-crustal element ratios K/U = 1.01×10^4 and Th/U = 3.9 [e.g., *McLennan and Taylor*, 1996] and obtained $q_c \sim 50$–90 mW m^{-2} for the Pilbara at 3300 Ma, and $q_c \sim 30$–60 mW m^{-2} for the Goldfields at 2650 Ma.

Initial (Pre-Greenstone) and Final (Post-Greenstone, Pre-Doming) Archean z_c Values

One-dimensional HPE depth distributions simulating greenstone-over-granite stratification were derived for each terrain by using two simple first-order assumptions. First, the Archean HPE content of the supracrustal greenstones is considered negligible. Second, we assume that the greenstone edifice overlies a geochemically undifferentiated felsic layer of unknown thickness, with an overall HPE budget dictated by its syn-tectonic q_c value and a uniform depth distribution of HPEs.

For the Pilbara, the geochemistry of this ancient felsic substrate was approximated directly by the 3490 Ma North Shaw granite suite, which has average K$_2$O = 2.5 wt%, Th = 11 ppm, and U = 2.5 ppm [*Bickle et al.*, 1989, 1993]. At a mean density $\rho = 2800$ kg m^{-3}, the volumetric heat production of this average composition was 4.10 µW m^{-3} at 3300 Ma. Our q_c estimate (50–90 mW m^{-2}) then corresponds to a North Shaw granitic layer thickness of approximately 12–22 km, prior to dome-and-keel formation at 3300 Ma. Unfortunately, analogous calculations in the Goldfields are complicated by the absence of felsic plutonic rocks that demonstrably predate accumulation of the greenstone succession. We therefore used the arithmetic mean of three geochemical averages from the high-Ca suite (K$_2$O = 2.9 wt%, Th = 12 ppm, and U = 3 ppm [*Champion and Sheraton*, 1997]), which is the oldest granite suite in the Goldfields, and accounts for over 60% of the granite exposure. For an average density $\rho = 2800$ kg m^{-3}, the volumetric heat production of the mean composition was 3.72 μW m^{-3} at 2650 Ma. Our q_c estimate (30–60 mW m^{-2}) then corresponds to a high-Ca granitic layer thickness of approximately 8–16 km, prior to dome-and-keel formation at 2650 Ma.

These layer-thickness data were then used to estimate the initial (pre-greenstone accumulation) and final (post-greenstone, pre-doming) characteristic depth of the heat production distribution (z_c) in each terrain [e.g., *Sandiford and Hand*, 1998; *Sandiford et al.*, 2002], which corresponds in each case to the depth midpoint of the HPE-bearing granitic layer (Figure 2). Initial z_c values for each terrain were obtained by simply halving the inferred thickness of the heat-producing layer (17 ± 5 km in the Pilbara, 12 ± 4 km in the Goldfields). Final z_c values were obtained by adding the initial z_c value to the pre-doming greenstone thickness (14 ± 3 km in the Pilbara, 6 ± 2 km in the Goldfields) in each case, yielding $z_c = 22.5 ± 4$ km in the Pilbara at 3300 Ma and $z_c = 12 ± 3$ km in the Goldfields at 2650 Ma [see also *Sandiford et al.*, 2004]. Equation (5) may then be used to estimate the *maximum* radiogenic contribution to the geotherm at and below the base of the heat-producing crust (i.e., T_c' at $z > z_2$; Figure 3), assuming that thermal equilibrium was attained in the interval separating the conclusion of green-tone accumulation and the initiation of dome-and-keel formation. The ranges of equilibrium T_c' values for the Goldfields (200 ± 100 °C) and the Pilbara (550 ± 250 °C) differ significantly, despite the partial overlap of the inferred

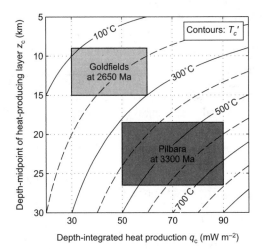

Figure 3. Diagram showing temperature at the base of the heat-producing crust with the ambient mantle component subtracted (T_c'), as a function of the abundance (q_c) and the depth distribution (z_c) of HPEs [see also *Sandiford and McLaren*, 2002]. The sensitivity of T_c' to z_c is highlighted by the large difference between the Pilbara ($T_c' = 300$–800 °C) and the Goldfields ($T_c' = 100$–300 °C) values, despite the partial overlap in their q_c ranges.

q_c ranges (Figure 3), emphasizing the dependence of T_c' on z_c.

Archean q_m

It is widely believed that Archean mantle heat flow was significantly higher than the q_m values inferred for present-day Archean cratons, because the young Earth held more of its primordial heat [*McKenzie and Weiss*, 1975; *Thompson*, 1984; *Richter*, 1985; see also *Marshak*, 1999]. In addition, it is unlikely that the 200–300-km-thick lithospheric roots currently underlying almost all Archean cratons [e.g., *Artemieva and Mooney*, 2001] were present in their entirety during major tectono-magmatic activity in the overlying crust (although the thicknesses of such roots may have increased rapidly during the final stages of cratonization [e.g., *Pearson et al.*, 2002]). Secondly, geochemical data from Archean and modern mantle-derived mafic rocks suggest a decrease in primitive upper-mantle liquidus temperatures of the order of 150–200 °C since Mesoarchean times [e.g., *Abbott et al.*, 1994], and the results of parametrized mantle convection models imply a decrease in the mean mantle temperature of similar magnitude.

With respect to equation (8), these two considerations imply lower Δz_L and higher ΔT_L values, respectively, during the Archean, which act in concert to yield higher q_m values during active tectonism. Our simplified calculations assume a generalized Archean ΔT_L value of 1450 °C [*Abbott et al.*, 1994], and a syn-tectonic Δz_L value of 125 km, which is approximately half the seismic thickness of the lithospheric root currently underlying the Archean cratons of Western Australia. Combining these parameters with a time- and temperature-independent $k = 3$ W m^{-1} °C^{-1}, equation (8) yields an Archean q_m value of approximately 35 mW m^{-2}. We have adopted this q_m value for both terrains, for simplicity and in order to showcase the influence of q_c and z_c on the results of our thermal models.

THERMAL MODELING RESULTS

Now that we have derived Archean values for q_c, z_c and q_m in the Pilbara and the Goldfields, the thermal evolutions of crustal columns corresponding to the relevant greenstone deposition histories (prior to dome-and-keel formation) may be assessed. We solve the one-dimensional, time-dependent heat flow equation in a moving medium, which has the general form:

$$\rho c \left(\frac{\partial T}{\partial t} + U \frac{\partial T}{\partial z} \right) = k \frac{\partial^2 T}{\partial z^2} + H(z) \qquad (9)$$

where ρ is density, c is heat capacity, U is the vertical velocity (positive downward) of the reference frame (i.e., $U = \partial z/\partial t$), and $H(z)$ is the depth distribution of radiogenic heat production in the system. We adopted mid-range q_c

Table 1. Thermal Parameters and Geological Histories Modeled for the Pilbara and Goldfields[a].

Event/parameter	Pilbara	Goldfields
Initial configuration	3515 Ma	2715 Ma
Depth-integrated heat production (q_c)	70 mW m^{-2}	45 mW m^{-2}
Depth midpoint of uniform heat-producing layer (z_c)	8.5 km	6 km
First greenstone package		
Overall thickness (Δz_1)	8 km	6 km
Accumulation interval (Δt_1)	3515–3425 Ma	2715–2665 Ma
Time-averaged vertical velocity U_1 (=$\Delta z_1/\Delta t_1$)	89 m My^{-1}	120 m My^{-1}
New z_c (=$z_c + \Delta z_1$)	16.5 km at 3425 Ma	12 km at 2665 Ma
First period of quiescence ($U = 0$)	3425–3350 Ma (75 My)	2665–2650 Ma (15 My)
Second greenstone package		
Overall thickness (Δz_2)	6 km	–
Accumulation interval (Δt_2)	3350–3325 Ma	–
Time-averaged vertical velocity U_2 (=$\Delta z_2/\Delta t_2$)	240 m My^{-1}	–
New z_c (=$z_c + \Delta z_1 + \Delta z_2$)	22.5 km at 3325 Ma	–
Second period of quiescence ($U = 0$)	3325–3300 Ma (25 My)	–
Dome-and-keel formation	3300 Ma	2650 Ma

[a] The following time-invariant parameters are common to both models: $\rho = 3000$ kg m^{-3}, $c = 1$ kJ kg^{-1} °C^{-1}, $k = 3$ W m^{-1} °C^{-1}, $q_m = 35$ mW m^{-2}.

and z_c values for the initial configuration in each terrain (Table 1; see also Figure 3), and modeled the greenstone accumulation history by using the stratigraphic and geochronological data presented previously (Table 1). A basal boundary condition of constant heat flux ($q_m = 35$ mW m^{-2}) is applied throughout the modeled time, simply so we can clearly illustrate the thermal effects of changing heat production distribution.

Figure 4 shows the results in the form of a series of geotherms, corresponding to critical time-steps in each evolution. The results are most easily interpreted by considering the thermal evolution of the material point at the *base* of the heat-producing layer (indicated by a circle on each geotherm in Figure 4) in each model.

Goldfields

The initial (2715 Ma) temperature at the base of the 12-km-thick HPE-bearing layer in the Goldfields model was 230 °C, increasing to 350 °C (at a depth of 18 km) during the 2715–2665 Ma accumulation of the overlying 6-km-thick greenstone succession (Figure 4a). Nearly two-thirds of this temperature increase is due to the migration of the point down the mantle geotherm; the remainder is due to conductive incubation (i.e., steepening of the geotherm in response to the burial of HPEs). The heating rate is dictated by the thermal response time of the lithosphere, and in this case, significant thermal equilibration is achieved *during* greenstone accumulation, due to the relatively slow time-averaged burial rate (120 m My^{-1}; Table 1). Thermal equilibration and conductive incubation continued beyond the cessation of

greenstone accumulation but were limited by the initiation of dome-and-keel formation at 2650 Ma, only 15 My after greenstone deposition was terminated. The conductive incubation-related temperature increase at 18 km depth over this period was only 20 °C (to $T = 370$ °C at 2650 Ma), representing approximately one-third of the temperature increase corresponding to complete thermal equilibration (depicted in Figure 4a by a theoretical 2550 Ma geotherm: $T = 405$ °C at 18 km depth after 115 My).

However, it is clear that conductive incubation in isolation will not result in significant heating within the greenstone-over-granite configuration corresponding to the Goldfields q_c, z_c, and q_m values, irrespective of equilibration time. Our model results suggest that at the time dome-and-keel formation was initiated at 2650 Ma, the effects of conductive incubation had not raised crustal temperatures ($z \leq 40$ km) above 600 °C, and that the HPE-bearing layer remained below the wet tonalite solidus throughout. Consequently, the 2650–2620 Ma granite magmatism observed in the Goldfields during and after dome-and-keel formation [e.g., *Smithies and Champion*, 1999] requires a significant additional source of heat.

Pilbara

The corresponding model for the Pilbara displays some important differences. Due to the combined effect of greater HPE abundances and a greater thickness, the initial temperature at the base of the heat-producing layer was significantly higher ($T = 396$ °C at 3515 Ma), and its increase during the first phase (3515–3425 Ma) of greenstone emplacement was much greater ($T = 588$ °C at 3425 Ma). This is partly due to

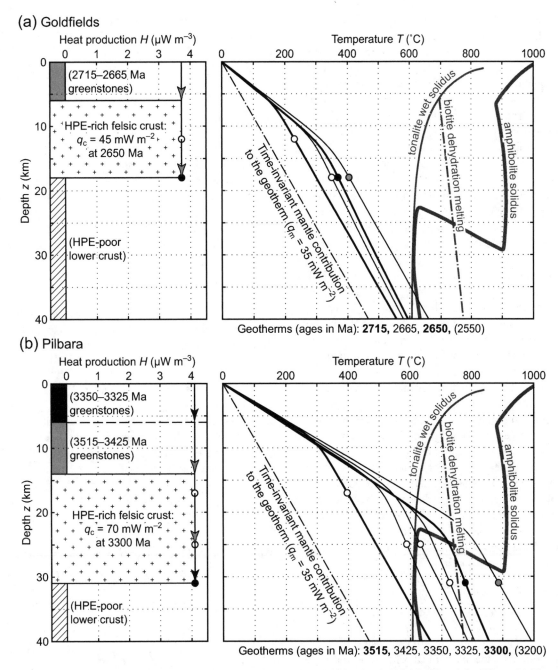

(a) Goldfields

Geotherms (ages in Ma): **2715**, 2665, **2650**, (2550)

(b) Pilbara

Geotherms (ages in Ma): **3515**, 3425, 3350, 3325, **3300**, (3200)

Figure 4. Input parameters and results of the one-dimensional thermal models for the Goldfields (a) and the Pilbara (b). In each case, the left panel shows the thickness and depth-independent heat production rate of the model TTG layer, with arrows showing the greenstone accumulation history (Table 1). The right panel shows the model results as a series of geotherms at critical times, with filled circles corresponding to the base of the HPE-bearing layer throughout. The heavy lines (with white and black circles, respectively) represent the initial and final (pre-doming) configurations; the gray circle represents a theoretical "final" geotherm, assuming thermal equilibrium was attained prior to dome-and-keel formation. The positions of the wet tonalite solidus, the amphibolite solidus, and the biotite dehydration melting reaction are those of *Zegers* [2004], following *Clemens and Wall* [1981], *Wolf and Wyllie* [1993], *Rapp* [1997], and *Wyllie et al.* [1997].

a greater greenstone thickness (8 km), but also because substantial syn-depositional thermal equilibration was permitted by the slow time-averaged accumulation rate (89 m My^{-1}; Table 1), similar to that observed in the Goldfields model. Conductive incubation further increased the temperature at 25 km depth to 634 °C over the 75-My period of quiescence from 3425 Ma until the second greenstone deposition event commenced at 3350 Ma.

During accumulation of the 6-km-thick 3350–3325 Ma greenstone succession, the temperature at the base of the heat-producing layer increased by 92 °C (to 726 °C at 31 km depth), mostly due to migration down the mantle geotherm, because the relatively rapid burial rate did not permit significant thermal equilibration. However, during the 25-My period of quiescence following the 3325 Ma cessation of greenstone accumulation, conductive incubation was responsible for a further temperature increase of 51 °C (to 777 °C at 31 km depth), prior to the initiation of dome-and-keel formation at 3300 Ma. The extent of this final heating episode is similar to the Goldfields model, in that it represents approximately one-third of the temperature increase corresponding to complete thermal equilibration (depicted in Figure 4b by a theoretical 3200 Ma geotherm: $T = 887$ °C at 31 km depth after 125 My).

Our model results suggest that for the combination of q_c, z_c, q_m, and greenstone accumulation history appropriate to the Pilbara prior to dome-and-keel formation, the base of the heat-producing layer probably lay on the high-temperature side of the wet solidi of tonalite and amphibolite and possibly also the fluid-absent biotite dehydration melting reaction (Figure 4b). This implies that conductive incubation *in isolation* was capable of generating granites via partial melting of pre-existing felsic crust and that such melting was broadly synchronous with dome-and-keel formation [e.g., *Sandiford et al.*, 2004]. The potential link between the two processes is considered in the following section.

THERMAL CONTROLS ON THE MECHANICAL BEHAVIOR OF THE CRUST

We now examine the influence of increasing mid-lower crustal temperature on the mechanical behavior of our greenstone-over-granite configurations, with emphasis on the role (and interplay) between two critical factors. First, the effective viscosities of crustal rocks are stress-dependent and decrease with increasing temperature [e.g., *Paterson*, 1987]. Second, the accumulation of thick greenstone successions atop HPE-rich felsic basement may result in a temperature increase sufficient to trigger partial melting (as illustrated in the previous section), and the presence of pseudo-Newtonian granitic magma in significant volumes may influence the rheology of the lower crust. Below, we outline the theoretical basis for our one-dimensional treatment of the effective

(power-law) viscosity of the mid-crust as a function of temperature and melt fraction. This is followed by an assessment of the effective viscosity evolution at the base of the heat-producing layer in the Pilbara and the Goldfields.

Relating Effective Viscosity to Temperature and Partial Melting

The influence of temperature on the effective viscosity of our greenstone-over-granite crust is considered in simplified fashion, by considering the buoyant ascent of a spherical, low-density felsic dome into an overlying, higher-density greenstone layer [e.g., *Weinberg and Podladchikov*, 1994]. Under natural physical conditions, most crustal rocks approximate power-law fluids, and in the case of uniaxial stress (σ), the equation

$$\dot{\varepsilon} = Ae^{-E/RT}\sigma^n \qquad (10)$$

relates uniaxial strain rate ($\dot{\varepsilon}$), composition (where A, E, and n are material constants governing creep rates), and temperature (T), where R is the universal gas constant. The characteristic uniaxial stress is defined by

$$\sigma = \Delta\rho gr \qquad (11)$$

where $\Delta\rho$ is the greenstone–granite density contrast, g is gravitational acceleration, and r is the dome radius. In this scenario, the effective viscosity (η_{eff}) is given by *Weinberg and Podladchikov* [1994] as follows:

$$\eta_{\text{eff}} = \frac{\sigma}{\dot{\varepsilon}} = \frac{e^{E/RT}}{A(\Delta\rho gr)^{n-1}} \qquad (12)$$

We incorporated the influence of partial melting on effective viscosity by using a generalized model [following *Gerya and Yuen*, 2003] in which the melt fraction (M) within felsic crust at a specified pressure varies linearly over the temperature interval between the wet solidus (T_{wetsol}) and the dry liquidus (T_{dryliq}):

$$M = \frac{T - T_{\text{wetsol}}}{T_{\text{dryliq}} - T_{\text{wetsol}}} \qquad (13)$$

This obviously represents a significant simplification, as most experimental data show nonlinear relationships between melt fraction and viscosity. The classical notion [e.g., *Arzi*, 1978; *van der Molen and Paterson*, 1979] that catastrophic loss of strength of a partially molten rock occurs at a critical melt fraction on the order of 0.3–0.5 (corresponding to breakdown of the solid framework) has recently been challenged by *Rosenberg and Handy* [2004], who contended that

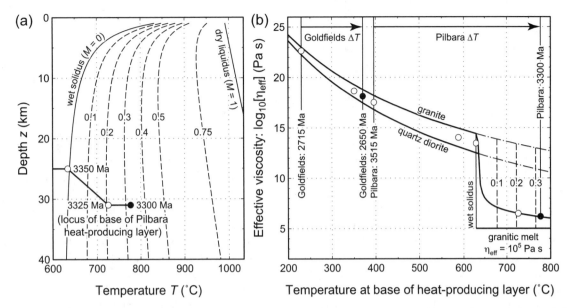

Figure 5. (a) Diagram summarizing the partial melting model used for felsic crust, following *Gerya and Yuen* [2003]. Melt fraction contours were constructed by using equation (13), and the *P–T* time path of the base of the Pilbara heat-producing layer is superimposed. (b) Diagram showing the temperature dependence of effective viscosity for a felsic dome with radius $r = 20$ km and density contrast $\Delta\rho = 250$ kg m^{-3} with respect to the overlying greenstones, over the temperature interval appropriate to the thermal evolutions modeled for the Pilbara and the Goldfields (circle fills are the same as Figure 4). The solid curves (at temperatures below the wet solidus) and their dot-dashed extensions (above the wet solidus) were constructed by using equation (12) and the following material parameters: "granite" $A = 2 \times 10^{-6}$ MPa^{-n} s^{-1}, $E = 187$ kJ mol^{-1}, $n = 3.3$ [*Ji et al.*, 2003]; "quartz diorite" $A = 3 \times 10^{-2}$ MPa^{-n} s^{-1}, $E = 212$ kJ mol^{-1}, $n = 2.4$ [*de Bremond d'Ars et al.*, 1999]. Above the wet solidus, the solid curve was constructed by using equations (13) and (14) with a temperature- and composition-independent η_{liq} value of 10^5 Pa s [e.g., *Clemens and Petford*, 1999]. The vertical dashed lines correspond to melt-fraction contours from (a).

the loss of strength corresponds to the onset of partial melting. Within the realm of low-volume partial melting appropriate to the thermal regimes considered in this study, our formulation corresponds most closely to the viscosity–melt fraction relationship proposed by *Rosenberg and Handy* [2004]. We used the pressure–temperature (*P–T*) loci for the wet solidus and dry liquidus (Figure 5a) fitted by *Gerya and Yuen* [2003] to the experimental data of *Johannes* [1985], *Schmidt and Poli* [1998], and *Poli and Schmidt* [2002], to estimate *M* as a function of temperature. The effective viscosity of the partially melted crust ($0 < M < 1$) was then estimated by using the equation derived by *Gay et al.* [1969] for high-concentration suspensions [see also *Pinkerton and Stevenson*, 1992]:

$$\eta_{eff} = \eta_{liq}\left\{\left[2.5 + \left(\frac{1-M}{M}\right)^{0.48}\right](1-M)\right\} \quad (14)$$

where η_{liq} is the effective viscosity of granitic liquid. We assumed a uniform η_{liq} value of 10^5 Pa s to be representative of felsic liquids over the pressure, temperature, and water

content ranges appropriate to the melting of TTG composition [see also *Scaillet et al.*, 1998; *Clemens and Petford*, 1999; *Holtz et al.*, 1999].

Crustal Mechanics in the West Australian Archean

We use the base of each heat-producing layer as a proxy for the Pilbara and Goldfield mid-crusts, which permits the effective viscosity evolutions directly corresponding to the thermal histories modeled in the previous section to be assessed. At subsolidus temperatures, equation (12) was used to estimate η_{eff}; in melt-bearing rocks, η_{eff} was approximated with equations (13) and (14). Figure 5b shows η_{eff} over the temperature interval spanning the Pilbara and Goldfields thermal histories for a felsic dome with radius $r = 20$ km, and density contrast $\Delta\rho = 250$ kg m^{-3} with respect to the overlying greenstones. The $\eta_{eff}(T)$ curves were constructed using equations (12) and (14), in combination with material parameters (*A*, *E*, and *n*) derived from the experimental data of *Hansen and Carter* [1982] for the end-members "granite" [*Ji et al.*, 2003] and "quartz diorite" [*de Bremond d'Ars et al.*, 1999].

No attempt has been made in either terrain to quantify the mechanical behavior of the upper crust in terms of its brittle yield strength. Consequently, the effective viscosity of the initial (2715 Ma) configuration in the Goldfields (from equation (12)) is very high ($\eta_{eff} = 10^{22}$–10^{23} Pa s; Figure 5b). Although it is likely that a power-law rheology is not appropriate at such low temperatures ($T = 230$ °C), the η_{eff} value obtained is a useful maximum against which to compare higher-T η_{eff} values. As documented in the previous section, the modeled 65-My history of burial and conductive incubation in the Goldfields produced a temperature increase $\Delta T = 140$ °C at the base of the heat-producing layer, corresponding to $\eta_{eff} \sim 10^{18}$ Pa s at 2650 Ma (Figure 5b). Conductive incubation may therefore have decreased the effective viscosity of the felsic layer by four to five orders of magnitude, prior to dome-and-keel formation. However, the low temperatures ($T < 400$ °C throughout; Figure 5b) precluded partial melting of the felsic substrate in the absence of a substantial external heat source. It is thus unlikely that burial and conductive incubation in isolation had the capacity to modify the rheological behavior of the Goldfields crust to the point where purely buoyancy-driven dome-and-keel formation was a viable process.

The contrast between the Goldfields and Pilbara models is best illustrated by the similarity between the *final* (2650 Ma) configuration of the former (prior to dome-and-keel formation) and the *initial* (3515 Ma) condition of the latter, in terms of temperature (and therefore η_{eff}) at the base of the felsic crust. For the same range of A, E, and n values, Pilbara η_{eff} values decreased by approximately four orders of magnitude (from 5×10^{17} Pa s at 3515 Ma to 5×10^{13} Pa s at 3350 Ma) *prior* to the initiation of partial melting (Figure 5). Once melt production commenced (at $T \sim 635$ °C; Figure 5a), η_{eff} declined sharply with increasing M (Figure 5b), until the effective viscosity of the crystal–melt suspension at the base of the heat-producing layer resembled that of a granitic liquid, prior to dome-and-keel formation ($\eta_{eff} \sim 10^6$ Pa s for $M > 0.3$ at 3300 Ma; Figure 5b). Although our $\eta_{eff}(T)$–M model neglects the obvious possibility of progressive, batch-wise melt removal [e.g., *Sawyer*, 1994; *Rutter*, 1997; *Bons et al.*, 2004] and the potential for associated "stiffening" of the residual heat-producing layer, it is very likely that the modeled thermal history resulted in a substantially decreased effective viscosity in the Pilbara mid-crust. Dome-and-keel formation may have been initiated by large-scale, buoyancy-driven ascent of the crystal–melt suspension, triggering partial convective overturn of the crust [*Collins et al.*, 1998; *Van Kranendonk et al.*, 2002] that was assisted and sustained by the gravitational instability imposed by the overlying 11–17-km-thick greenstone edifice.

IMPLICATIONS FOR THE EVOLUTION OF ARCHEAN DEFORMATION STYLE

Our model results highlight the relative inefficiency of conductive incubation as a mechanism for raising upper- to mid-crustal temperatures in the Goldfields at the time of 2650 Ma dome-and-keel formation. This is primarily due to low HPE abundances in the felsic basement ($q_c = 45 \pm 15$ mW m^{-2}), and the relatively short period (65 My) over which the relatively thin (6–10-km-thick) greenstone edifice accumulated. Consequently, temperatures remained below 400 °C throughout, and it is unlikely that the effective viscosity of the mid-crust was significantly affected by conductive incubation (Figure 5b). We therefore contend that the link between the deformation history preserved by macroscopic structures and the physical processes governing deformation is similar to that characteristic of post-Archean tectonism at terrain-scale: Most large-scale structures developed in direct response to large, horizontally directed body forces and were "locked in" by a mechanically strong upper- to mid-crust. Dome-and-keel structure in the Goldfields (Figure 1) is characterized by the preservation of broad, granite-cored D2 antiforms in preference to the intervening greenstone-dominated synforms [*Blewett et al.*, 2004a], which may reflect antiform nucleation controlled by buoyancy-driven ascent of the underlying granites in the solid state [*Weinberg et al.*, 2003].

The Pilbara model results represent a stark contrast, because the heat production rate of the Pilbara felsic basement was 50–100% higher ($q_c = 70 \pm 20$ mW m^{-2}), the greenstone carapace was 50–100% thicker (11–17 km), and the duration of accumulation and thermal equilibration (215 My) was 200% longer. As a result, initial temperatures of approximately 400 °C at the base of the Pilbara heat-producing layer were elevated to 750–800 °C by conductive incubation, potentially reducing the effective viscosity of the mid-crust by several orders of magnitude (Figure 5). Thermal and mechanical softening was exacerbated by partial melting of the mid- to lower-crustal felsic layer and was amplified by the inherent large-scale density inversion. This resulted in a patently unstable crustal configuration (characterized by high q_c and z_c values) for which there is no equivalent in the modern Earth.

This thermo-mechanical instability had two potentially important implications. First, in terms of buoyancy and gravitational effects, the internal instability of the reverse-stratified system rendered it susceptible to large-scale deformation by external body forces that need not have been large in comparison with those acting along the boundaries of converging plates in the modern Earth. Second, and equally importantly, the rock record of such a perturbation event may be obscured, or even obliterated, by subsequent large-scale deformation associated with the attainment of a more stable configuration.

In this context, the crustal scale at which "classical" dome-and-keel structure developed is important, because low mid-crustal effective viscosities in isolation are unlikely to trigger large-scale dome-and-keel formation without significant weakening and/or brittle rupture of the overlying greenstone-dominated upper crust at similar scale [e.g., *Bleeker*, 2002]. Early extension in the upper crust may therefore be critical to the successful development of dome-and-keel structure [e.g., *Martinez et al.*, 2001; *Bleeker*, 2002], but the record of such a process may be poorly preserved and/or difficult to recognize within a deformation framework dominated by large-scale partial convective overturn of the crust. Consequently, fundamental differences may exist between the hot, soft syn-tectonic crust typical of the Pilbara (and possibly other Mesoarchean granite–greenstone terrains featuring classical dome-and-keel architecture; *Choukroune et al.* [1995]) and the colder, stronger crust typical of younger terrains, with respect to the relationship(s) between the physical processes governing deformation and the structures developed and preserved.

Such thermal-mechanical feedbacks are likely to have important implications for crustal behavior and tectonic style in Archean granite–greenstone terrains. In thermal terms, large-scale diapiric ascent of granitic magmas derived from ancient felsic basement represented an extremely efficient mechanism for reconcentrating HPEs in the upper crust [*Sandiford et al.*, 2002; 2004]. In the Pilbara, the resulting re-establishment of a configuration comparable to the initial condition via granite doming inevitably led to significant long-term cooling (and therefore strengthening) of the lower crust. It is thus possible that the physical conditions prevailing during Archean crustal construction were not merely favorable with respect to dome-and-keel formation, but that such vertical reorganization of the crust was *necessary* (together with rapid development of thick lithospheric mantle roots; e.g., *Pollack* [1986]; *Pearson et al.* [2002]) for the preservation of HPE-rich crustal segments at craton-scale, without significant reworking [e.g., *Morgan*, 1985].

CONCLUSIONS

Conductive incubation is an important source of "internal" heat in Archean granite–greenstone terrains. In this study we have illustrated the sensitivity of mid-crustal temperature (and, by association, effective viscosity) to the first-order geological and tectono-stratigraphic features of the contrasting Pilbara and Goldfields terrains in Western Australia. Our one-dimensional thermal modeling of greenstone-over-granite crustal evolution has highlighted a large temperature difference at the base of the heat-producing layer in each terrain,

prior to dome-and-keel formation ($T = 370$ °C at 18 km depth in the Goldfields at 2650 Ma; $T = 777$ °C at 31 km depth in the Pilbara at 3300 Ma). This contrast primarily reflects the concerted influence of interterrain differences in the syn-tectonic crustal HPE budget (q_c), the thicknesses of the greenstone successions accumulated atop ancient felsic basement (z_c), and the timescales over which greenstone accumulation took place.

The prevalence of low temperatures (and relatively small temperature increases) in the Goldfields model implies a dominantly brittle upper-mid-crust rheology and probably precludes a significant role for conductive incubation in isolation with respect to reducing the effective viscosity of the Goldfields crust ($\eta_{eff} = 10^{18}$–10^{23} Pa s throughout). In contrast, higher initial temperatures (and much larger temperature increases) at the base of the heat-producing layer in the Pilbara model resulted in significant effective viscosity decreases, from $\eta_{eff} = 5 \times 10^{17}$ Pa s at 3515 Ma to $\eta_{eff} < 10^{10}$ Pa s prior to dome-and-keel formation at 3300 Ma. These low effective viscosities were largely due to in situ partial melting of the felsic crust during conductive incubation which, in combination with the density inversion inherent in the greenstone-over-granite configuration, uniquely favored large-scale, classical dome-and-keel formation in the Pilbara.

The differing crustal rheologies produced by the contrasting Pilbara and Goldfields thermal regimes may reflect secular differences between Mesoarchean and younger terrains, with respect to the physical processes responsible for first-order structural development of the Archean crust. This study advocates the development of classical dome-and-keel structure in the Pilbara as a predominantly thermal response to substantial greenstone accumulation, driven "internally" by the imperative of attaining (via large-scale reconcentration of HPEs in the upper crust) a thermo-mechanically sustainable configuration [e.g., *Sandiford et al.*, 2004]. This does not preclude the possibility that dome-and-keel formation was triggered or assisted by a horizontally directed "external" tectonic force (significant or otherwise), but the structural record of such an event may be difficult to recognize due to the pervasive structural overprinting associated with subsequent vertical reorganization of the crust.

In contrast, the rheology of the Goldfields crust was probably not significantly affected by conductive incubation, so the greenstone-over-granite configuration had greater intrinsic stability. Dome-and-keel formation was therefore probably driven by external horizontally directed plate tectonic forces [e.g., *Blewett et al.*, 2004a], with the shallowing of HPEs associated with granite ascent playing only a supporting role in stabilizing the crust. This colder, stronger crust is correspondingly more likely to lock in a direct record of such far-field forcing, analogous to the record typically preserved in modern orogenic belts.

Acknowledgments. We thank Roberto Weinberg for stimulating and informative discussions that refined these ideas, and Jean-Claude Mareschal, Olivier Vanderhaeghe, and an anonymous reviewer for formal reviews that resulted in important clarifications to an earlier version of the manuscript. The Australian Research Council supported this research, via grants DP0208176 (awarded to S. Bodorkos) and DP0209157 and F10020050 (awarded to M. Sandiford).

REFERENCES

Abbott, D., L. Burgess, J. Longhi, and W.H.F. Smith (1994), An empirical thermal history of the Earth's upper mantle, *J. Geophys. Res.*, 99, 13835-13850.

Anhaeusser, C.R., R. Mason, M.J. Viljoen, and R.P. Viljoen (1969), A reappraisal of some aspects of Precambrian shield geology, *Geol. Soc. Am. Bull.*, 80, 2175-2200.

Archibald, N.J., L.F. Bettenay, R.A. Binns, D.I. Groves, and R.J. Gunthorpe (1978), The evolution of Archaean granite-greenstone terrains, Eastern Goldfields Province, Western Australia, *Precambrian Res.*, 6, 103-131.

Artemieva, I.M., and W.D. Mooney (2001), Thermal thickness and evolution of Precambrian lithosphere: a global study, *J. Geophys. Res.*, 106, 16387-16414.

Arzi, A.A. (1978), Critical phenomena in the rheology of partially melted rocks. *Tectonophysics*, 44, 173-184.

Bagas, L., M.J. Van Kranendonk, and M. Pawley (2004), Geology of the Split Rock 1:100 000 sheet, Western Australia, *1:100 000 Geological Series Explanatory Notes*, 43 pp., Geol. Surv. West. Aust., Perth.

Barley, M.E., and A.L. Pickard (1999), An extensive, crustally-derived, 3325 to 3310 Ma silicic volcanoplutonic suite in the eastern Pilbara Craton: evidence from the Kelly Belt, McPhee Dome, and Corunna Downs Batholith, *Precambrian Res.*, 96, 41-62.

Bickle, M.J., L.F. Bettenay, H.J. Chapman, D.I. Groves, N.J. McNaughton, I.H. Campbell, and J.R. de Laeter (1989), The age and origin of younger granitic plutons of the Shaw Batholith in the Archaean Pilbara Block, Western Australia, *Contrib. Mineral. Petrol.*, 101, 361-376.

Bickle, M.J., L.F. Bettenay, H.J. Chapman, D.I. Groves, N.J. McNaughton, I.H. Campbell, and J.R. de Laeter (1993), Origin of the 3500–3300 Ma calc-alkaline rocks in the Pilbara Archaean: isotopic and geochemical constraints from the Shaw Batholith, *Precambrian Res.*, 60, 117-149.

Bleeker, W. (2002), Archaean tectonics: a review, with illustrations from the Slave Craton, in *The Early Earth: Physical, Chemical and Biological Development*, edited by C.M.R. Fowler, C.J. Ebinger, and C.J. Hawkesworth, *Geol. Soc. London Spec. Publ.* 199, pp. 151-181.

Blewett, R.S. (2002), Archaean tectonic processes: a case for horizontal shortening in the North Pilbara Granite-Greenstone Terrane, Western Australia, *Precambrian Res.*, 113, 87-120.

Blewett, R.S., K.F. Cassidy, D.C. Champion, P.A. Henson, B.S. Goleby, L. Jones, and P.B. Groenewald (2004a), The Wangkathaa Orogeny: an example of episodic regional 'D₂' in the late Archaean Eastern Goldfields Province, Western Australia, *Precambrian Res.*, 130, 139-159.

Blewett, R.S., S. Shevchenko, and B. Bell (2004b), The North Pole Dome: a non-diapiric dome in the Archaean Pilbara Craton, Western Australia, *Precambrian Res.*, 133, 105-120.

Bodorkos, S., M. Sandiford, B.R.S. Minty, and R.S. Blewett (2004), A high-resolution, calibrated airborne radiometric dataset applied to the estimation of crustal heat production in the Archaean northern Pilbara Craton, Western Australia, *Precambrian Res.*, 128, 57-82.

Bons, P.D., J. Arnold, M.A. Elburg, J. Kalda, A. Soesoo, and B. van Milligen (2004), Melt extraction and accumulation from partially molten rocks, *Lithos*, 78, 25-42.

Brown, S.J.A., B. Krapez, S.W. Beresford, K.F. Cassidy, D.C. Champion, M.E. Barley, and R.A.F. Cas (2001), Archaean volcanic and sedimentary environments of the Eastern Goldfields province, Western Australia – a field guide, *Geol. Surv. West. Aust. Record*, 2001/13, 66 pp.

Buick, R., J.R. Thornett, N.J. McNaughton, J.B. Smith, M.E. Barley, and M. Savage (1995), Record of emergent continental crust ~3.5 billion years ago in the Pilbara craton of Australia, *Nature*, 375, 574-577.

Card, K.D. (1990), A review of the Superior Province of the Canadian Shield: a product of Archaean accretion, *Precambrian Res.*, 48, 99-156.

Champion, D. C., and J. W. Sheraton (1997), Geochemistry and Nd isotope systematics of Archaean granites of the Eastern Goldfields, Yilgarn Craton, Australia: implications for crustal growth processes, *Precambrian Res.*, 83, 109-132.

Chardon, D., P. Choukroune, and M. Jayananda (1996), Strain patterns, décollement and incipient sagducted greenstone terrains in the Archaean Dharwar craton (south India), *J. Struct. Geol.*, 18, 991-1004.

Chardon, D., P. Choukroune, and M. Jayananda (1998), Sinking of the Dharwar Basin (South India): implications for Archaean tectonics, *Precambrian Res.*, 91, 15-39.

Chardon, D., J.J. Peucat, M. Jayananda, P. Choukroune, and C.M. Fanning (2002), Archean granite-greenstone tectonics at Kolar (South India): Interplay of diapirism and bulk inhomogeneous contraction during juvenile magmatic accretion, *Tectonics*, 21(3), DOI: 10.1029/2001TC901032.

Choukroune, P., H. Bouhallier, and N.T. Arndt (1995), Soft lithosphere during periods of Archaean crustal growth or crustal reworking, in *Early Precambrian Processes*, edited by M.P. Coward and A.C. Ries, *Geol. Soc. London Spec. Publ.*, 95, pp. 67-86.

Choukroune, P., J.N. Ludden, D. Chardon, A.J. Calvert, and H. Bouhallier (1997), Archaean crustal growth and tectonic processes: a comparison of the Superior Province, Canada and the Dharwar Craton, India, in *Orogeny Through Time*, edited by J.-P. Burg and M. Ford, *Geol. Soc. London Spec. Publ.*, 121, pp. 63-98.

Clemens, J.D., and N. Petford (1999), Granitic melt viscosity and silicic magma dynamics in contrasting tectonic settings, *J. Geol. Soc. London*, 156, 1057-1060.

Clemens, J.D., and V.J. Wall (1981), Origin and crystallization of some peraluminous (S-type) granitic magmas, *Can. Mineral.*, 19, 111-131.

Collins, W.J. (1989), Polydiapirism of the Archean Mount Edgar Batholith, Pilbara Block, Western Australia, *Precambrian Res.*, 43, 41-62.

Collins, W.J., M.J. Van Kranendonk, and C. Teyssier (1998), Partial convective overturn of Archaean crust in the east Pilbara Craton, Western Australia: driving mechanisms and tectonic implications, *J. Struct. Geol.*, 20, 1405-1424.

Cull, J.P. (1982), An appraisal of Australian heat-flow data, *BMR J. Aust. Geol. Geophys.*, 7, 11-21.

Cull, J.P. (1991), Heat flow and regional geophysics in Australia, in *Terrestrial Heat Flow and the Lithosphere Structure*, edited by V. Cermak and L. Rybach, pp. 486-500, Springer-Verlag, Berlin.

Cull, J.P., and D. Denham (1979), Regional variations in Australian heat flow, *BMR J. Aust. Geol. Geophys.*, 4, 1-13.

de Bremond d'Ars, J., C. Lécuyer, and B. Reynard (1999), Hydrothermalism and diapirism in the Archean: gravitational instability constraints, *Tectonophysics*, 304, 29-39.

de Wit, M.J. (1998), On Archean granites, greenstones, cratons and tectonics: does the evidence demand a verdict?, *Precambrian Res.*, 91, 181-226.

de Wit, M.J., and L.D. Ashwal (1997), Greenstone Belts, Oxford University Press, Oxford.

Gay, E.C., P.A. Nelson, and W.P. Armstrong (1969), Flow properties of suspensions with high solids concentrations, *Am. Inst. Chem. Eng. J.*, 15, 815-822.

Gerya, T.M., and D.A. Yuen (2003), Rayleigh-Taylor instabilities from hydration and melting propel 'cold plumes' at subduction zones, *Earth Planet. Sci. Lett.*, 212, 47-62.

Goleby, B.R., R.S. Blewett, R.J. Korsch, D.C. Champion, K.F. Cassidy, L.E.A. Jones, P.B. Groenewald, and P.A. Henson (2004), Deep seismic reflection profiling in the Archaean northeastern Yilgarn Craton, Western Australia: implications for crustal architecture and mineral potential, *Tectonophysics*, 388, 119-133.

Green, M.G., P.J. Sylvester, and R. Buick (2000), Growth and recycling of early Archaean continental crust: geochemical evidence from the Coonterunah and Warrawoona Groups, Pilbara Craton, Australia, *Tectonophysics*, 322, 69-88.

Griffin, T.J. (1990), Eastern Goldfields Province, in *Geology and Mineral Resources of Western Australia, Geol. Surv. West. Aust. Memoir*, 3, pp. 77-119.

Gupta, M.L., A. Sundar, and S.R. Sharma (1991), Heat flow and heat generation in the Archaean Dharwar cratons and implications for the southern Indian Shield geotherm and lithospheric thickness, *Tectonophysics*, 194, 107-122.

Hamilton, W.B. (1998), Archaean magmatism and deformation were not products of plate tectonics, *Precambrian Res.*, 91, 143-179.

Hansen, F.D., and N.L. Carter (1982), Creep of selected crustal rocks at 1000 MPa, *EOS Trans. AGU*, 63, 437.

Henry, P., R.K. Stevenson, Y. Larbi, and C. Gariepy (2000), Nd isotopic evidence for Early to Late Archaean (3.4–2.7 Ga) crustal growth in the Western Superior Province (Ontario, Canada), *Tectonophysics*, 322, 135-151.

Hickman, A.H. (1983), Geology of the Pilbara Block and its environs, *Geol. Surv. West. Aust. Bulletin*, 127, 268 pp.

Hickman, A.H. (1984), Archaean diapirism in the Pilbara Block, Western Australia, in *Precambrian Tectonics Illustrated*, edited by A. Kröner, and R. Greiling, pp. 113-127, E. Schweizerbart'sche Verlagsbuch-handlung, Stuttgart.

Hickman, A.H., and M.J. Van Kranendonk (2004), Diapiric processes in the formation of Archaean continental crust, East Pilbara Granite-Greenstone Terrane, Australia, in *The Precambrian Earth: Tempos and Events*, edited by P.G. Eriksson, W. Altermann, D.R. Nelson, W.U. Mueller, and O. Catuneanu, pp. 118-139, Elsevier, Amsterdam.

Holtz, F., J. Roux, S. Ohlhorst, H. Behrens, and F. Schulze (1999), The effects of silica and water on the viscosity of hydrous quartzofeldspathic melts, *Am. Mineral.*, 84, 27-36.

Howard, L.E., and J.H. Sass (1964), Terrestrial heat flow in Australia, *J. Geophys. Res.*, 69, 1617-1626.

Jaeger, J.C. (1970), Heat flow and radioactivity in Australia, *Earth Planet. Sci. Lett.*, 8, 285–292.

Jaupart, C., and J.C. Mareschal (1999), The thermal structure and thickness of continental roots, *Lithos*, 48, 93-114.

Jaupart, C., and J.C. Mareschal (2003), Constraints on crustal heat production from heat flow data, in *Treatise on Geochemistry: Volume 3: The Crust*, edited by R.L. Rudnick, pp. 65-84, Elsevier Pergamon, Oxford.

Jelsma, H.A., P.A. van der Beek, and M.L. Vinyu (1993), Tectonic evolution of the Bindura-Shamva greenstone belt (northern Zimbabwe) – progressive deformation around diapiric batholiths, *J. Struct. Geol.*, 15, 163-176.

Ji, S., P. Zhao, and B. Xia (2003), Flow laws of multiphase materials and rocks from end-member flow laws, *Tectonophysics*, 370, 129-145.

Johannes, W. (1985), The significance of experimental studies for the formation of migmatites, in *Migmatites*, edited by J.R. Ashworth, pp. 36-85, Blackie, Glasgow.

Jones, M.Q.W. (1988), Heat flow in the Witwatersrand Basin and environs and its significance for the South African shield geotherm and lithospheric thickness, *J. Geophys. Res.*, 93, 3243-3260.

Kloppenburg, A., S.H. White, and T.E. Zegers (2001), Structural evolution of the Warrawoona Greenstone Belt and adjoining granitoid complexes, Pilbara Craton, Australia: implications for Archaean tectonic processes, *Precambrian Res.*, 112, 107-147.

Krapez, B., S.J.A. Brown, J. Hand, M.E. Barley, and R.A.F. Cas (2000), Age constraints on recycled crustal and supracrustal sources of Archaean metasedimentary sequences, Eastern Goldfields Province, Western Australia: evidence from SHRIMP zircon dating, *Tectonophysics*, 322, 89-133.

Macgregor, A.M. (1951), Some milestones in the Precambrian of southern Rhodesia, *Proc. Geol. Soc. S. Afr.*, 54, 27-71.

Mareschal, J.-C., and G.F. West (1980), A model for Archean tectonism. Part 2. Numerical models of vertical tectonism in greenstone belts, *Can. J. Earth Sci.*, 17, 60-71.

Marshak, S. (1999), Deformation style way back when: thoughts on the contrast between Archean/Palaeoproterozoic and contemporary orogens, *J. Struct. Geol.*, 21, 1175-1182.

Marshak, S., F.F. Alkmim, and H. Jordt-Evangelista (1992), Proterozoic crustal extension and the generation of dome-and-keel structure in an Archean granite-greenstone terrane, *Nature*, 357, 491-493.

Martinez, F., A.M. Goodliffe, and B. Taylor (2001), Metamorphic core complex formation by density inversion, *Nature*, 411, 930-934.

McKenzie, D., and N. Weiss (1975), Speculations on the thermal and tectonic history of the Earth, *Geophys. J. R. Astronom. Soc.*, 42, 131-174.

McLennan, S.M., and S.R. Taylor (1996), Heat flow and the chemical composition of continental crust, *J. Geol.*, 104, 369-377.

Morgan, P. (1985), Crustal radiogenic heat production and the selective survival of ancient continental crust, *J. Geophys. Res.*, 90, C561-C570.

Myers, J.S. (1997), Preface: Archaean geology of the Eastern Goldfields of Western Australia – regional overview, *Precambrian Res.*, 83, 1-10.

Myers, J.S., R.D. Shaw, and I.M. Tyler (1996), Tectonic evolution of Proterozoic Australia, *Tectonics*, 15, 1431-1446.

Nelson, D.R. (1997), Evolution of the Archaean granite-greenstone terranes of the Eastern Goldfields, Western Australia: SHRIMP U-Pb constraints, *Precambrian Res.*, 83, 57-81.

Paterson, M.S. (1987), Problems in the extrapolation of laboratory rheological data, *Tectonophysics*, 133, 33-43.

Pearson, D.G., G.J. Irvine, R.W. Carlson, M.G. Kopylova, and D.A. Ionov (2002), The development of lithospheric keels beneath the earliest continents: time constraints using PGE and Re–Os isotope systematics, in *The Early Earth: Physical, Chemical and Biological Development*, edited by C.M.R. Fowler, C.J. Ebinger, and C. J. Hawkesworth, *Geol. Soc. London Spec. Publ.*, 199, pp. 65-90.

Percival, J.A., V. McNicoll, J.L. Brown, and J.B. Whalen (2004), Convergent margin tectonics, central Wabigoon subprovince, Superior Province, Canada, *Precambrian Res.*, 132, 213-244.

Peschler, A.B., K. Benn, and W.R. Roest (2004), Insights on Archean continental dynamics from gravity modelling of granite greenstone terranes, *J. Geodyn.*, 38, 185-207.

Pinkerton, H., and R.J. Stevenson (1992), Methods of determining the rheological properties of magmas at sub-liquidus temperatures, *J. Volcanol. Geotherm. Res.*, 53, 47-66.

Poli, S., and M.W. Schmidt (2002), Petrology of subducted slabs, *Ann. Rev. Earth Planet. Sci.*, 30, 207-235.

Pollack, H.N. (1986), Cratonization and thermal evolution of the mantle, *Earth Planet. Sci. Lett.*, 80, 175-182.

Pollack, H.N. (1997), Thermal characteristics of the Archaean, in *Greenstone Belts*, edited by M.J. de Wit and L.D. Ashwal, pp. 223-232, Clarendon Press, Oxford.

Pysklywec, R.N., and C. Beaumont (2004), Intraplate tectonics: feedback between radioactive thermal weakening and crustal deformation driven by mantle lithosphere instabilities, *Earth Planet. Sci. Lett.*, 221, 275-292.

Rapp, R.P. (1997), Heterogeneous source regions for Archaean granitoids: experimental and geochemical evidence, in *Greenstone Belts*, edited by M.J. de Wit and L.D. Ashwal, pp. 267-279, Clarendon Press, Oxford.

Richter, F.M. (1985), Models for the Archean thermal regime, *Earth Planet. Sci. Lett.*, 73, 350-360.

Ridley, J.R., and J.D. Kramers (1990), The evolution and tectonic consequences of a tonalitic magma layer within Archean continents, *Can. J. Earth Sci.*, 27, 219-228.

Rosenberg, C.L., and M.R. Handy (2004), Experimental deformation of partially melted granite revisited: implications for the continental crust, *J. Metamorph. Geol.*, 23, 19-28.

Russell, J.K., G.M. Dipple, and M.G. Kopylova (2001), Heat production and heat flow in the mantle lithosphere, Slave craton, Canada, *Phys. Earth Planet. Interiors*, 123, 27-44.

Rutter, E.H. (1997), The influence of deformation on the extraction of crustal melts: a consideration of the role of melt-assisted granular flow, in *Deformation-Enhanced Fluid Transport in the Earth's Crust and Mantle*, edited by M.B. Holness, pp. 82-110, Chapman and Hall, London.

Sandiford, M., and M. Hand (1998), Australian Proterozoic high-temperature, low-pressure metamorphism in the conductive limit, in *What Drives Metamorphism and Metamorphic Reactions?*, edited by P. J. Treloar and P.J. O'Brien, *Geol. Soc. London Spec. Publ.*, 138, pp. 109-120.

Sandiford, M., and S. McLaren (2002), Tectonic feedback and the ordering of heat producing elements within the continental lithosphere, *Earth Planet. Sci. Lett.*, 204, 133-150.

Sandiford, M., S. McLaren, and N. Neumann (2002), Long-term thermal consequences of the redistribution of heat-producing elements associated with large-scale granitic complexes, *J. Metamorph. Geol.*, 20, 87-98.

Sandiford, M., M.J. Van Kranendonk, and S. Bodorkos (2004), Conductive incubation and the origin of dome-and-keel structure in Archean granite-greenstone terrains: a model based on the eastern Pilbara Craton, Western Australia, *Tectonics*, 23, TC1009, DOI: 10.1029/2002TC001452.

Sass, J.H., J.C. Jaeger, and R.J. Munroe (1976), Heat flow and near surface radioactivity in the Australian continental crust, *U.S. Geol. Surv., Open File Report*, 76-250.

Sawyer, E.W. (1994), Melt segregation in the continental crust, *Geology*, 22, 1019-1022.

Scaillet, B., F. Holtz, and M. Pichavant (1998), Phase equilibrium constraints on the viscosity of silicic magmas–1. Volcano-plutonic association, *J. Geophys. Res.*, 103, 27257-27266.

Schmidt, M.W., and S. Poli (1998), Experimentally based water budgets for dehydrating slabs and consequences for arc magma generation, *Earth Planet. Sci. Lett.*, 163, 361-379.

Simons, F.J., A. Zielhuis, and R.D. van der Hilst (1999), The deep structure of the Australian continent from surface wave tomography, *Lithos*, 48, 17-43.

Smithies, R.H., and D.C. Champion (1999), Late Archaean felsic alkaline igneous rocks in the Eastern Goldfields, Yilgarn Craton, Western Australia: a result of lower crustal delamination?, *J. Geol. Soc. London*, 156, 561-576.

Smithies, R.H., D.C. Champion, and K.F. Cassidy (2003), Formation of Earth's early Archaean continental crust, *Precambrian Res.*, 127, 89-101.

Swager, C.P. (1997), Tectono-stratigraphy of late Archaean greenstone terranes in the southern Eastern Goldfields, Western Australia, *Precambrian Res.*, 83, 11-42.

Swager, C.P., B.R. Goleby, B.J. Drummond, M.S. Rattenbury, and P.R. Williams (1997), Crustal structure of granite-greenstone terranes in the Eastern Goldfields, Yilgarn Craton, as revealed by seismic reflection profiling, *Precambrian Res.*, 83, 43-56.

Swager, C.P., and T.J. Griffin (1990), An early thrust duplex in the Kalgoorlie-Kambalda greenstone belt, Eastern Goldfields Province, Western Australia, *Precambrian Res.*, 48, 63-73.

Thompson, A.B. (1984), Geothermal gradients through time, in *Patterns of Change in Earth Evolution*, edited by H.D. Holland and A. F. Trendall, pp. 345-355, Springer-Verlag, Berlin.

Thurston, P.C. (2002), Autochthonous development of Superior Province greenstone belts?, *Precambrian Res.*, 105, 11-36.

Tomlinson, K.Y., D.W. Davis, D. Stone, and T.R. Hart (2003), U-Pb age and Nd isotopic evidence for Archean terrane development and crustal recycling in the south-central Wabigoon subprovince, Canada, *Contrib. Mineral. Petrol.*, 144, 684-702.

van der Molen, I., and M.S. Paterson (1979), Experimental deformation of partially-melted granite, *Contrib. Mineral. Petrol.*, 70, 299-318.

Van Kranendonk, M.J., W.J. Collins, A.H. Hickman, and M.J. Pawley (2004), Critical tests of vertical vs. horizontal tectonic models for the Archaean East Pilbara Granite-Greenstone Terrane, Pilbara Craton, Western Australia, *Precambrian Res.*, 131, 173-211.

Van Kranendonk, M.J., A.H. Hickman, R.H. Smithies, D.R. Nelson, and G. Pike (2002), Geology and tectonic evolution of the Archean North Pilbara Terrain, Pilbara Craton, Western Australia, *Econ. Geol.*, 97, 695-732.

Weinberg, R.F., L. Moresi, and P. van der Borgh (2003), Timing of deformation in the Norseman-Wiluna Belt, Yilgarn Craton, Western Australia, *Precambrian Res.*, 120, 219-239.

Weinberg, R.F., and Y. Podladchikov (1994), Diapiric ascent of magmas through power law crust and mantle, *J. Geophys. Res.*, 99, 9543-9959.

Wellman, P. (2000), Upper crust of the Pilbara Craton, Australia: 3D geometry of a granite/greenstone terrain, *Precambrian Res.*, 104, 175-186.

West, G.F., and J.-C. Mareschal (1979), A model for Archean tectonism. Part 1. The thermal conditions, *Can. J. Earth Sci.*, 16, 1942-1950.

White, D.J., G. Musacchio, H.H. Helmstaedt, R.M. Harrap, P.C. Thurston, A. van der Velden, and K. Hall (2003), Images of a lower-crustal oceanic slab: direct evidence for tectonic accretion in the Archean western Superior Province, *Geology*, 31, 997–1000.

Williams, I.S., and W.J. Collins (1990), Granite-greenstone terranes in the Pilbara Block, Australia, as coeval volcano-plutonic complexes; evidence from U-Pb zircon dating of the Mount Edgar Batholith, *Earth Planet. Sci. Lett.*, 97, 41-53.

Wolf, M.B., and P.J. Wyllie (1993), Garnet growth during amphibolite anatexis: implications of a garnetiferous restite, *J. Geol.*, 101, 357-373.

Wyllie, P.J., M.B. Wolf, and S.R. van der Laan (1997), Conditions for formation of tonalites and trondhjemites: magmatic sources and products, in *Greenstone Belts*, edited by M.J. de Wit and L.D. Ashwal, pp. 256-266, Clarendon Press, Oxford.

Zegers, T.E. (2004), Granite formation and emplacement as indicators of Archaean tectonic processes, in *The Precambrian Earth: Tempos and Events*, edited by P.G. Eriksson, W. Altermann, D.R. Nelson, W. U. Mueller, and O. Catuneanu, pp. 103-118, Elsevier, Amsterdam.

Zegers, T.E., S.E. White, M. de Keijzer, and P. Dirks (1996), Extensional structures during deposition of the 3460 Ma Warrawoona Group in the eastern Pilbara Craton, Western Australia, *Precambrian Res.*, 80, 89-105.

S. Bodorkos, Geological Survey of Western Australia, 100 Plain Street, East Perth, Western Australia 6004, Australia. (simon.bodorkos@doir.wa.gov.au)

M. Sandiford, School of Earth Sciences, University of Melbourne, Melbourne, Victoria 3010, Australia. (mikes@unimelb.edu.au)

Experimental Constraints on TTG Petrogenesis: Implications for Archean Geodynamics

Jean-François Moyen and Gary Stevens

Department of Geology, University of Stellenbosch, Stellenbosch, South Africa

Archean TTGs (tonalite–trondhjemite–granodiorite) are sodic granitoids that represent the bulk of the Archean continental crust. They are formed by fluid-absent partial melting of amphibolites. A compilation of the published data on experimental melting of amphibolites allows a mineralogical model for amphibolite melting to be derived for three different starting lithologies. A major and trace element model for melt compositions is produced using the mineralogical model. This model suggests that TTGs formed at $P > 15$ kbar and T between 900°C and 1100°C, corresponding to low (15°C/km) geothermal gradients that are likely to be attained only in subduction zones. Furthermore, it appears that Nb/Ta, La/Yb, Eu/Eu*, Sr, and HREE contents are intercorrelated in TTGs and are indicators of the pressure of melting. TTGs were generated over a large range of depths, from at least 10 to 25 kbar, and this is reflected in TTG compositions.

1. INTRODUCTION

The TTG (tonalite–trondhjemite–granodiorite) series consists of sodic leuco-granitoids that appear to have formed the dominant granitoid type in the Archean, having been estimated to represent at least two-thirds of the Archean continental crust [*Condie*, 1981; *Jahn et al.*, 1984; *Martin*, 1994; *Windley*, 1995]. However, potential modern equivalents (the adakites) are a very rare rock type, found only in peculiar geodynamic settings that are characterized by hot subduction zones [e.g., *Defant and Drummond*, 1990; *Martin*, 1999; *Maury et al.*, 1996]. This suggests that understanding the petrogenesis of TTGs may be a key to understanding Archean geodynamic processes. There is a generally held petrological perspective that the typical sodic and HREE-depleted character of TTG granitoids arises through the partial melting of amphibolitic mafic sources within the garnet stability field, and that garnet remains in the residuum following efficient melt segregation and ascent. There is also a general understanding that garnet stability in high-temperature metamafic rocks requires

high-pressure conditions. Despite the obvious implication that melts with a typical TTG signature are likely to arise from melting at substantial depths, few studies have systematically examined the specific pressure constraints that a garnet-bearing residua place on depth of TTG melt genesis. This study aims to contribute to a better understanding of the geodynamic implications of TTG predominance in the Archean by using the extensive database that exists on the experimental partial melting of metamafic rocks to better constrain TTG petrogenesis; the scope of this work is therefore limited to investigating the melting processes from an experimental perspective. It should be kept in mind that other processes (such as fractional crystallization or interaction with pre-existing crust) can, and will, affect the composition of real TTG plutons. Nevertheless, we believe that the melting processes will strongly control the composition of the emplaced magmas. If this is so, melt (and pluton) chemistry may allow for strong constraints to be placed on the site of melt genesis (regardless of their subsequent evolution), and through this a better understanding of Archean geodynamic processes.

2. TTG DETAILS

Archean TTGs are dominated by plagioclase (typically oligoclase). Quartz is typically the second most abundant

Archean Geodynamics and Environments
Geophysical Monograph Series 164
Copyright 2006 by the American Geophysical Union
10.1029/164GM11

mineral. Ferro-magnesian phases are biotite (nearly always) and hornblende (commonly). Small amounts of K-feldspar are occasionally found. Accessory minerals are typically epidote, allanite, sphene, zircon, magnetite, and ilmenite. In terms of chemistry, TTGs are sodic granitoids, with molecular K/Na ratios below 0.4 and SiO_2 content between 65 and 75 wt%. Most are metaluminous [molecular Al/(Ca + Na + K), or A/CNK, < 1.1], with a small minority being slightly peraluminous, with A/CNK up to 1.2. Their Mg# is commonly between 30 and 40. On Na–Ca–K triangles, TTGs plot near the Na apex or along the Na–Ca side; rare suites evolve towards relatively potassic compositions, and some of these probably are "secondary TTGs", i.e., partial melts of pre-existing tonalitic crust [*Jébrak and Harnois*, 1991; *Johnston and Wyllie*, 1988; *Van der Laan and Wyllie*, 1992; *Winther and Newton*, 1991].

Apart from their sodic character, the most typical geochemical feature of TTGs is their rare earth elements (REE) patterns; REEs in TTGs are strongly fractionated (average $(La/Yb)_N$ ratio is 38.4 [*Martin*, 1994], but in some cases it is as high as 150), with a marked HREE and Y depletion (Yb_N = 2.6 on average, but with the 25th percentile at 1.7). A concave HREE pattern is frequently observed. Most TTGs show no Eu anomaly, or a slightly negative one; a small positive Eu anomaly is reported only occasionally, mostly in granulitic terranes [*Condie et al.*, 1985; *Rollinson and Windley*, 1980; *Weaver and Tarney*, 1980]. In some cases, especially for older suites, a fairly atypical REE pattern is observed, with a less pronounced Yb depletion and a small negative Eu anomaly. Again, this suggests that TTGs are more complex than a general approach to their petrogenesis may suggest and that several different processes might have been involved in their genesis.

On the basis of numerous experimental data (Table 1) and geochemistry [*Barker and Arth*, 1976; *Drummond and Defant*, 1990; *Martin*, 1987, 1994], most workers agree that TTG melts are generated by partial melting of metabasalt (amphibolites) in the garnet stability field. In contrast, the geodynamic setting of Archean TTG petrogenesis remains more controversial. Two main hypotheses persist: (1) TTGs were formed by partial melting of the subducting slab in relatively hot subduction zones; (2) TTGs were formed by partial melting of underplated hydrous basalt at the base of continental crust or overthickened oceanic crust (basaltic plateaus: *Albarède* [1998]; *Rudnick* [1995]). These two hypotheses differ in terms of the expected pressure–temperature (*P–T*) conditions of melting. In a subduction setting, slab melting would occur at 700–900°C and 15–25 kbar or even more, corresponding to geothermal gradients in the range 10–30°C/km; in contrast, melting at the base of thick plateaus would occur at 8–15 kbar and 700–1000°C, implying geothermal gradients reaching 30–50°C/km.

In recent years, numerous experimental studies (Table 1) have documented the partial melting of amphibolites. The published literature has been compiled for the present study, from which has been extracted information on the modal composition of experimental charges and major element compositions of the melts over a large range of *P–T* conditions and starting materials. This has allowed for the generation of generalized models of melt composition, both for major elements (from the published compositions) and trace elements (recalculated from modal compositions and partition coefficients) of the experimental melts, for different starting compositions and throughout the *P–T* space over which anatexis occurs in these rocks; comparison of the results of this modeling with TTG compositions allows additional constraints to be formulated on the locus and conditions of TTG magmas genesis and on the geodynamic context of TTG genesis.

3. REVIEW OF EXPERIMENTAL STUDIES

A substantial body of data exists on the experimental genesis of TTG-type liquids from a variety of sources. The early works (in the '70s) [*Allen and Boettcher*, 1975; *Allen and Boettcher*, 1978; *Allen and Boettcher*, 1983; *Green*, 1982; *Green and Ringwood*, 1968; *Lambert and Wyllie*, 1972] focused mostly on the determination of phase diagrams. Subsequent experimental work generally included more information on the nature of the starting materials and the products. These studies are the source of data for the present investigation and are summarized in Tables 1 and 2.

3.1. Starting Materials

The starting materials used in the experiments are of broadly basaltic composition (Table 2), ranging from basalts to basaltic andesites (Figure 1) and generally belonging to the tholeitic series, or being close to the calc-alkali/tholeitic boundary. Nevertheless, there are fairly significant differences between the different materials used, in terms of modal proportions of minerals, bulk rock chemistry and mineral chemistry:

3.1.1. Modal composition. Most of the starting materials were amphibolites, with the exception of *Skjerlie and Patiño-Douce* [2002], who used an eclogitic composition. Both synthetic [*Patiño-Douce and Beard*, 1995] and natural materials (other studies) have been used; in general, the starting materials were mineralogically simple. All the amphibolites consisted of amphibole (amp) and plagioclase (pl; plag), with amp/pl ratios that vary from 0.18 to > 4 (Table 2). Quartz was present in several starting materials (Table 2) but commonly occurs just in trace amounts; it can, however, represent up to 24%. Other

Table 1. List of Experimental Works on Partial Melting of Amphibolites. In each case, the reference is indicated together with the code by which this work will be referred to in the subsequent figures; starting material is stated. The number of experiments is given, together with the *P–T* and fluid-saturation conditions of these (see text for discussion of the four cases). Not listed in this table are works with no melt compositions published: *Green* [1982] (Grn), *Lambert and Wyllie* [1972] (LW72), and *Liu et al.* [1996] (L&al) give only phase diagrams; *Schmidt and Poli* [1998] (SP98) discuss the subsolidus stability field of hydrous minerals; *Wyllie and Wolf* [1993] discuss the "S-shaped" solidus (see text); and *Vielzeuf and Schmidt* [2001] discuss the solidus position as a function of water saturation and the role of hydrous minerals.

Reference	Code	Starting material	No. experiments	*P-T* range	Water saturation?
Rapp et al., 1991; *Rapp and Watson*, 1995	RWM91, RW95	No 1 (Josephine Ophiolite)	5+13	8-32 kbar, 1000-1150°C	fluid absent (b)
"	"	No 2 (low-K amphibolite)	9+16	"	fluid absent (b)
"	"	No 3 (migmatitic amphibolite, "Barker's")	8+14	"	fluid absent (b)
"	"	No 4 (greenstone amphibolite)	7+11	"	fluid absent (b)
Rushmer, 1991	Rus91	ABA (alkali basalt)	2	8 kbar, 950-1000°C	fluid absent (b)
"	"	IAT (island arc tholeite)	1	8 kbar, 950°C	fluid absent (b)
"	"	MMA (mechanically mixed amphibolite)	1	8 kbar, 975°C	fluid absent (b)
Winther and Newton, 1991; *Winther*, 1996	WN91	Synthetic Archean tholeite	9	5-20 kbar, 800-1000°C	fluid absent (b), fluid present (c), and water saturated (d). Only (d) melts are published.
Sen and Dunn, 1994	SD94	Amphibolite, British Columbia	13	15 and 20 kbar, 850-1150°C	fluid absent (b)
Wolf and Wyllie, 1994	WW94	Amphibolite, Sierra Nevada	15	10 kbar, 750-1000°C	fluid absent (b)
Zamora, 2000	Zam00	Ophiolite Bahia Barrientes	53	7-30 kbar, 975-1150°C	fluid present (c)
Beard and Lofgren, 1991	BL91	557 (low-K, calc-alkaline andesite)	8	1, 3, and 6.9 kbar, 800-1000°C	probably fluid absent(b)
"	"	"	8	"	water saturated(d)
"	"	555 (very low-K andesite)	9	"	probably fluid absent(b)
"	"	"	10	"	water saturated(d)
"	"	478 (low-K andesite)	11	"	probably fluid absent(b)
"	"	"	11	"	water saturated(d)
"	"	466 (low-K basalt)	7	"	probably fluid absent(b)
"	"	"	6	"	water saturated(d)
"	"	571 (low-K andesite)	8	"	probably fluid absent(b)
"	"	"	10	"	water saturated(d)

Table 1. Continued

Reference	Code	Starting material	No. experiments	*P-T* range	Water saturation?
Springer and Seck, 1997	SS97	S6 (granulitic metabasalt)	3	5-15 kbar; 700-1200°C	water saturated (d)
"	"	S35 (granulitic metabasalt)	2	"	fluid absent (b)
"	"	S37 (granulitic metagabbro)	4	"	water saturated (d)
Skjerlie and Patiño-Douce, 1995	SPD95	AGS11.1 (N. Idaho amphibolite)	3	10 kbar, 900-950°C	fluid absent (b)
Skjerlie and Patiño-Douce, 2002	SPD02	Verpenesset eclogite (Norway)	19	10-32 kbar, 900-1150°C	fluid present (c)
Lopez and Castro, 2001	LC01	Acebuches amphibolite (Spain)	7	4-14 kbar, 725-950°C	fluid absent (b)
Patiño-Douce and Beard, 1995	PDB95	SQA (Synthetic quartz-amphibolite)	20	3-15 kbar, 840-1000°C	fluid absent (b)
Yearron, 2003	Y03	AmX12-a (Barberton amphibolite)	3	16 kbar, 875-1000°C	fluid absent (b)

minerals (such as chlorite, greenschist facies amphiboles, or epidote) have occasionally been present in the starting material, but mostly only as traces, probably not common enough to play a significant role. Finally, Ti-oxides (ilmenite) and/or sphene is present in most of the natural rocks used as starting materials, suggesting that the compositions were commonly buffered by a Ti-rich phase and that the hornblendes were consequently Ti-saturated.

3.1.2. Bulk compositions used. SiO_2 values range from ca. 47% to ca. 60% (average, 51%). K_2O values range from 0.1% to 1.8% (generally below 1%); Na_2O values are between 1% and 4.3%. Total alkali values are between 1.1 and 5.2 wt%. There is no correlation between K_2O and SiO_2 values; Na_2O and the total alkali content are loosely positively correlated with SiO_2. TiO_2 values are between 0.4 and 2 and are correlated with neither SiO_2 nor the modal compositions of the starting materials. Finally, Mg# values are between 38 and 71. These values are not correlated with SiO_2 but are negatively correlated with total alkali content.

3.1.3. Mineral chemistry. The plagioclase compositions are not cited in all cases but appear to vary between An_{30} and An_{50}. In contrast, the amphibole mineral chemistry used does show some important variations, in terms of T-site occupancy, Mg# of the hornblende, and Ti per formula unit (p.f.u.) (Table 2). These three parameters are not independent; the more Mg-rich hornblendes are also more Si-poor, and Ti-rich.

These modal bulk chemical and mineral chemical parameters are, of course, not independent. In general, the starting materials define a large compositional range, from compositions intermediate to tholeitic and komatiitic basalts—which are quartz- and plagioclase-poor, alkali-poor, and relatively magnesian and contain a Mg-, Ti-, and Al(T)-rich hornblende —to arc tholeites to andesitic basalts, which conversely are quartz- and plagioclase-rich, less magnesian, and richer in alkalis and are made up of an iron-rich, Ti-poor siliceous hornblende.

3.2. Water Saturation

Vielzeuf and Schmidt [2001] described in detail how water saturation affects the solidus and the melting reactions of metabasalts. Several cases can be considered:

(a) H_2O-free melting: There is absolutely no water in the system, not even as constitutive water in hydrous minerals. In this case, only dry melting would be possible, a situation outside of the scope of the present study.

(b) Fluid-absent melting: All the water in the system is accommodated in hydrous minerals that are stable at least to the conditions of the relevant solidus (hornblende in all cases examined here). According to *Zamora* [2000], this implies a water content of less than ca. 1.8 wt% water for rocks of this type (effectively the maximum water content of hornblende). In this case, melting will occur through typically fluid-absent incongruent partial melting reactions involving hornblende breakdown. This is the most common mode of melting in the studies examined here (Table 1). One hundred ninety-two experiments in our dataset belong to this group.

Table 2. Nature of the Source Material Used for Amphibolite Experimental Melting. Shown are modal characteristics, major elements content, and amphibole mineral chemistry, if available. Codes as in Table 1.

Code	Source mode			Source chemistry										Amphibole mineral chemistry		
	Qz%	Amp %	Pl %	SiO_2	Al_2O_3	Fe_2O_3	MnO	MgO	CaO	Na_2O	K_2O	TiO_2	P_2O_5	Si(T)	Ti p.f.u.	Mg#
RWM91, RW95 (#1)	no	-	-	51.19	16.62	11.32	0.23	6.59	5.49	4.33	0.82	1.18	-	6.06	0.47	0.74
" (#2)	traces	-	-	48.60	17.03	10.69	0.21	6.07	9.66	3.30	0.21	2.06	-	6.02	0.42	0.63
" (#3)	yes[a]	-	-	48.30	15.30	10.70	0.19	8.40	12.60	2.27	0.08	0.72	-	5.85	0.34	0.80
" (#4)	no	-	-	47.60	14.18	13.77	0.19	6.86	10.99	2.56	0.19	1.19	-	5.96	0.30	0.68
Rus91 (ABA)	8	54	36	49.04	16.37	9.18	0.18	7.45	10.81	3.42	0.44	1.27	0.16	6.57	0.16	0.57
" (IAT)	17	44	32	51.69	16.31	8.70	0.11	7.51	8.90	3.09	0.26	1.00	0.10	7.10	0.14	0.77
" (MMA)[b]	10	54	36-46													
WN91	yes[a]	-	-	49.10	14.80	15.78	0.20	6.50	11.40	2.30	0.30	1.30	-	-	-	-
SD94	2	76	20	46.88	15.00	13.09	0.26	8.25	11.28	2.51	0.80	1.22	-	6.37	0.16	0.62
WW94	no	67	33	48.40	14.60	9.33	0.20	10.70	14.30	1.00	0.10	0.40	-	7.02	0.04	0.79
Zam00	no	ca. 45	ca. 45	52.20	15.05	7.92	0.14	6.13	7.70	5.15	0.56	1.40	-	-	-	-
BL91 (557)	12.3	29.3	47.5	57.02	15.39	8.01	0.17	5.52	9.20	2.54	0.44	0.60	0.18	-	-	-
" (555)	7.9	26	57	55.11	14.94	11.28	0.21	4.01	6.07	4.29	0.03	1.66	0.30	-	-	-
" (478)	11.7	36.2	43.4	52.47	15.29	11.79	0.22	5.29	9.21	2.55	0.16	1.74	0.29	ca. 7.3	-	? < 0.5
" (466)	8.3	34.7	48.6	49.48	17.76	12.49	0.26	4.74	10.90	1.96	0.15	1.18	0.30	-	-	-
" (571)	2.6	37.2	52.7	51.39	15.82	12.23	0.26	4.42	8.95	3.30	0.37	1.55	0.30	ca. 7.3	-	? < 0.5
SS97 (S6)	no	23.5	32	45.97	14.81	11.29	0.21	8.11	13.16	0.10	0.37	1.58	0.35	-	-	-
" (S35)	no	7	40	46.92	14.37	13.82	0.17	7.32	12.14	0.17	1.80	2.02	0.24	-	-	-
" (S37)	no	7.5	51.5	48.79	17.13	8.03	0.22	10.86	11.66	0.05	1.54	0.54	0.12	-	-	-
SPD95	16	54	13	54.60	13.70	12.20	0.20	5.30	10.00	1.10	0.60	0.93	0.10	6.42	0.09	0.55
SPD02	4	0	0	49.66	19.97	5.83	0.11	8.15	12.72	2.57	0.14	0.76	-	-	-	-
LC01	no	49	46	49.14	16.00	10.94	0.22	7.17	10.70	3.29	0.09	1.61	-	-	-	-
PDB95	24.5	53.9	19.6	60.40	11.30	7.90	0.20	6.70	7.60	1.90	0.70	1.70	-	6.72	0.209	0.69
Y03	no	70	30	52.52	10.57	10.91	0.22	10.83	10.81	1.74	0.83	1.31	0.14	6.68	0.104	0.69

[a] Amount unspecified.
[b] Source chemistry and amphibole mineral chemistry unspecified.

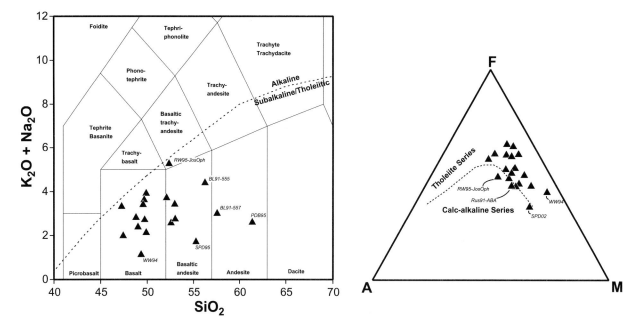

Figure 1. Compositions of the starting materials of various authors, in a total alkali vs. silica (TAS) diagram [*Le Bas et al.*, 1986] and in an Al_2O_3–Fe_2O_3–MgO (AFM) [*Irvine and Baragar*, 1971] diagram. The atypical starting materials are labeled with codes as in Table 1.

(c) Fluid-present melting but without a free fluid phase in the starting material: All the water in the system is accommodated in hydrous minerals, such as amphiboles and greenschist facies minerals (chlorite, epidote, etc.). However, such minerals are not stable in the upper amphibolite facies, and the system should become water-saturated before reaching the solidus because of the water release that accompanies the breakdown of these phases. In this case, the system commonly becomes fluid-absent as melting progresses, because the water content is restricted. Sixty-two experiments from our database correspond to this situation.

(d) Fluid-present melting with water-saturated starting compositions. In this case, the system is typically water-saturated throughout the melting range. A relatively uncommon situation in modern experiments, this was the general situation of the "early" studies of the 1970s and '80s. Fifty-two melt compositions corresponding to this case have been published, mostly from relatively low-pressure conditions.

Fluid-present melting (cases c and d above) seems to have little relevance for natural situations—although they could conceivably occur in some specific settings, such as water-fluxed melting in an active shear zone, or water-induced melting close to dehydrating sediments in an amphibolite–metasediment greenstone package. This may have some relevance in specific cases but clearly is not a general process accounting for the bulk of TTGs.

3.3. P–T Range of Investigations

Experiments have been conducted over a fairly large *P–T* range, from 1 to 35 kbar and from 750°C to 1200°C. Modern studies focusing on the genesis of TTG/adakites were mostly conducted in the garnet stability field, i.e., over ca. 10 kbar. Nevertheless, the *P–T* space has not been completely explored. While some studies aimed at describing the melting behavior of the system over the widest possible range, some others considered other parameters—for instance, time of equilibration [*Rushmer*, 1991; *Wolf and Wyllie*, 1994]—and were restricted to one single pressure. Furthermore, only *P–T* conditions thought to be relevant for TTG genesis were explored, more or less along a positively sloped "band" from 700°C to 900°C at *P* < 10 kbar to 1000–1100°C at *P* > 25 kbar. While this is certainly a sound approach, it means that the behavior of the system is not described in some "exotic" parts of *P–T* space. In particular, the high pressure/low temperature domain (above amphibole and plagioclase breakdown, in the eclogitic domain) has been the subject of very few studies. Theoretical considerations [*Wyllie and Wolf*, 1993; *Vielzeuf and Schmidt*, 2001] suggest that in this domain, melting at low temperature is possible (see below), but this has not been experimentally tested.

3.4. Solidus and Melt Fractions

Figure 2a summarizes the solidus positions, as determined in different experimental works.

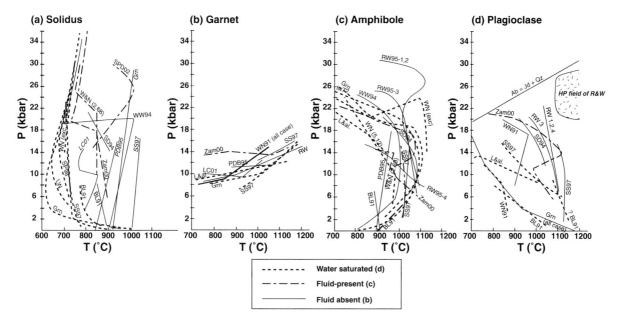

Figure 2. Position of the solidus (a) and mineral stability fields (b–d) in the *P–T* space for the different experiments (codes as in Table 1) and for different degrees of water saturation (exc, excess). Letters (b), (c) and (d) refer to the discussion on fluid saturation in text. Ab = Jd + Qz: position of the albite = jadeite + quartz reaction, after *Holland* [1980]. A high-pressure field of plagioclase stability, as evidenced by *Rapp and Watson* [1995], is shown in panel (d) (stippled).

3.4.1. Fluid-absent solidus and melt fractions. The fluid-absent solidus (case b, above) generally has a steep positive slope. Large variations in the solidus position are observed between different studies (Figure 2); melt fraction evolution suggests that the "true" solidus is at ca. 850–900°C; the apparently higher solidus temperature in some experiments is likely to be related to the low-melt fractions close to the solidus, leading to an "apparent" higher temperature solidus [*Sen and Dunn*, 1994]. At higher pressure, it has been suggested [*Vielzeuf and Schmidt*, 2001; *Wyllie and Wolf*, 1993] that the solidus backbends and is located at temperatures close to those of the water-saturated case. This is due to amphibole breakdown above 20–22 kbar to form phengite or paragonite [*Schmidt and Poli*, 1998], which are not able to accommodate as much water as amphibole does. Therefore, even fluid-absent systems (as defined above) may evolve to have fluid-present portions in parts of *P–T* space along the solidus at *P* > 20–22 kbar. This "S-shaped" solidus [*Wyllie and Wolf*, 1993] has not really been mapped; its existence is partially supported by experimental data of *Winther and Newton* [1991] ("dry" case). Furthermore, we point out that the S-shaped solidus could be largely an experimental arti-fact, not completely relevant to natural systems; in natural rocks, depending on the *P–T* loop, the water released by amphibole breakdown on the subsolidus, prograde path, is likely to escape the system and be unable to induce any

melting. Actually an S-shaped solidus will appear only if the amphibole-out curve is crossed at temperatures above the fluid-present solidus.

The evolution of melt fraction (F) as a function of temperature (Figure 3) is mostly independent of the source composition. At low pressures (below garnet-in), F values rise from 0 to ca. 50%, from ca. 850°C to 1000°C. At medium pressures (between garnet-in and amphibole-out), a similar evolution takes place between 900°C and 1100°C, with slightly higher maximum F values (50–60%). The rela-tively steep rise of F fractions between the solidus and the amphibole-out curve is consistent with the melting being primarily controlled by amphibole breakdown. In the eclogitic domain (above amphibole-out), the melt fractions reported increase from 20% at ca. 1000°C to 40–509% at 1100°C. In the high-pressure case, the low melt-fraction domain has not been explored. In *P–T* space, this defines positively sloped iso-melt curves with a "bulge" around 10 kbar. Quartz-rich sources tend to have slightly higher melt fractions for comparable conditions, relative to their quartz-poor and quartz-free counterparts.

3.4.2. Solidus position and melt fractions in water-saturated experiments with restricted water availability. For the water-saturated case with restricted water availability (case c), including fluid-present melting in systems where water was

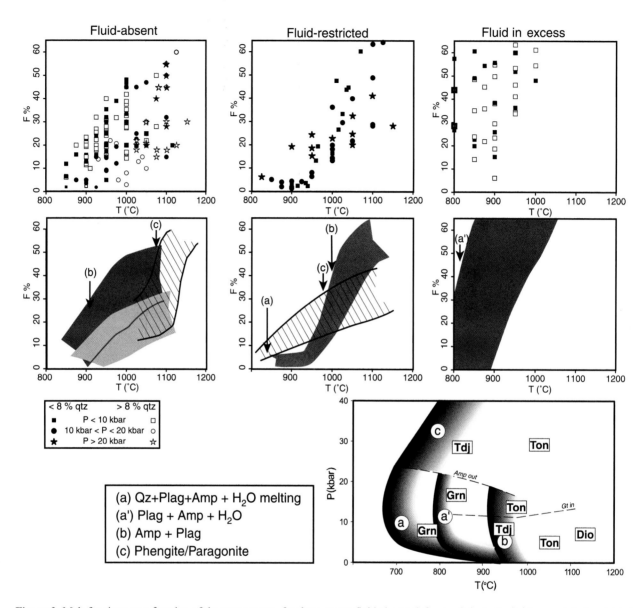

Figure 3. Melt fractions as a function of the temperature, for three cases: fluid-absent (left; case b in text), fluid present, but restricted (middle; case c), and fluid in excess (right; case d). Symbols are affected according to the pressure (shape) and the quartz content (color). For each case, the bottom panel shows abstracted F evolutions; the dark gray band corresponds to $P < 10$ kbar, the light gray to 10 kbar $< P < 20$ kbar, and the hatched band to $P > 20$ kbar. Capital letters denote the occurrence of the melting reactions summarized at the bottom right, with their emplacement in P–T space (amp: amphibole, gt: garnet, Tdj: trondhjemite, Ton: tonalite, Grn: granite, Dio: diorite).

added in the form of low-temperature metamorphic minerals, the solidus is at about 800°C at low pressure (Figure 2). Melting will occur at the relevant water-saturated solidus, i.e., arranged in order of increasing temperature, either $Pl + Q + Hbl + H_2O$, $Pl + Hbl + H_2O$, or $Hbl + H_2O$. If water is supplied by low-temperature minerals, melting will occur via an identical process where the low-temperature metamorphic minerals decompose before the solidus is attained. This may produce very small melt fractions. Consequently, the

fact that some studies record melting at a temperature well above the amphibolite fluid-present solidus probably corresponds to an "apparent" solidus [*Sen and Dunn*, 1994], with enough liquid to allow interconnection of melt pockets (with liquid content of more than a few %), rather than to a "true" solidus.

Melt fractions (Figure 3) are mostly known from *Zamora* [2000]. In this quartz-free system, melting starts at ca. 850°C at low and medium pressures (amphibolitic domains) and stays below 10% up to 950°C. Only at this temperature, similar to the temperature where F values rise in the fluid-absent case, do the melt fractions start to rise, up to similar values of 50–60% at 1100°C. This behavior can be related to a two-stage evolution:

(1) At low temperatures, melting occurs through a water-present reaction. Since this reaction yields granitoid (s.l.) melts, it is obviously incongruent for examples that do not have both quartz and plagioclase at the solidus; peritectic phases are Fe–Ti oxides, clinopyroxene, and garnet or ortho-pyroxene. The limited amount of water is quickly exhausted, and no further melting is possible at low temperature. The reaction is most likely of the type $Pl + Hbl_1 + H_2O = melt + Hbl_2$. The new generation of hornblende (Hbl_2) is likely to be less Si-rich and have a higher Mg# and higher Ti contents.

(2) At higher temperatures, melting occurs—just as it does in the water-absent case—by fluid-absent amphibole breakdown.

At higher pressures (eclogitic domain), melt fractions evolve smoothly, from 5% at 750–800°C to 50% at 1100–1150°C. This is consistent with the S-shaped solidus noted earlier. At temperatures above 1000°C, melt fraction evolution is also similar to what was observed in the fluid-absent case.

The presence of abundant quartz in the starting material is likely to have the effect of making the early, water-present melting more efficient (see below); the F vs. T curve, in this case, can be expected to have the same shape, shifted upwards.

3.4.3. Solidus position and melt fractions in water-saturated experiments. The fluid-present solidus (case d) is subhorizontal for low pressures, then progressively turns upwards and becomes subvertical, parallel to the fluid-absent solidus, above 10 kbar (Figure 2). The zone of minimum temperature is at 6–10 kbar (600–650°C), increasing to 700–800°C at higher pressures.

Melt fractions (Figure 3) are, unfortunately, mostly known from low-pressure (<7 kbar) experiments [*Beard and Lofgren*, 1991], all of which had quartz present. Melt fractions display a very steep rise at about 800°C, yielding high melt fractions (about 50–60%) at temperatures as low as 850–900°C. Melt fractions then stay at this level up to 1000°C. We suggest that this corresponds to a very efficient, low-temperature melting of $Qz + Pl + Amp + H_2O$; amphibole is quickly exhausted by this reaction (none is reported above 950–1000°C in Beard and Lofgren's experimental charges), such that, when reaching amphibole-breakdown temperatures, nothing more happens.

3.4.4. Summary. In summary, several melting reactions are possible during amphibolite melting.

At low pressure (where amphibole is stable), the reactions are as follows:

(1) The water-present melting of $Pl + Qz + Amp + H_2O$, at about 700–750°C (Figure 3). Because this reaction gives rise to very high melt fractions at relatively low temperature, amphibole may be quickly exhausted in this case, where quartz and plagioclase are abundant.

(2) The water-present incongruent melting of $Pl + Amp + H_2O$ (reaction a'), at about 850°C. This reaction seems to be dominant only in *Zamora* [2000]'s experiments, where it yields only 5–10% melt. However, in this work the total amount of available water during melting was limited by the fact that the water came from chlorite breakdown.

(3) The fluid-absent incongruent melting of amphibole, at ca. 900°C (reaction b). This reaction is not significantly different in the presence or absence of quartz, although the quartz-present version will be more congruent and will there-fore produce more abundant liquids.

At high pressure (eclogitic domain), the reactions are less well known. Melting reactions involve the breakdown of phengite and paragonite and probably involve a water phase at low temperature.

3.5. Mineral Stability Limits During Melting

Mineral stability limits are important parameters in constraining the melt chemistry, because the assemblage coexisting with melt determines the major and trace element composition of the melts. Therefore, one can model the melt trace element compositions only if consistent mineral stability behavior can be established. Most of the mineral stability limits relevant to this study are dependent on bulk composition and vary between the studies examined.

3.5.1. Garnet stability. Curves are fairly consistent in all the published works (Figure 2). No systematic differences are observed for different water contents or quartz absence/presence. This is relatively surprising, as garnet stability in other anatectic systems (e.g., pelitic system [*Stevens et al.*, 1997]) has been demonstrated to be greatly dependent on bulk rock Mg#. This parameter does play a role in garnet stability in mafic compositions, but the more grossular-rich compositions that result from incongruent melting in the studies examined here, as well as the restricted Mg# range of the starting materials, to some degree mask this effect. The only apparent compositional control has been discussed by *Zamora* [2000], who pointed out that high Na contents in the starting materials tend to shift the garnet-in curve upwards, which is consistent with the uncommonly high position of this curve in his work (ca. 14 kbar regardless of T, for a starting material above 5% Na_2O). Presumably, this works through stabilization of higher-pressure more

Na-rich amphibole, delaying garnet appearance with increasing pressure.

3.5.2. Amphibole stability. In contrast, the amphibole-out curve shows large variations between the studies, with no obvious control from the presence or absence of quartz or the water saturation of the system. In contrast to *Gardien et al.* [2000], who showed that water saturation helps in stabilizing the amphibole, the largest field of stability is observed in the dehydration melting experiments of Rapp and colleagues [*Rapp and Watson*, 1995; *Rapp et al.*, 1991]. In general, the dehydration melting experiments have amphibole-out boundaries that scatter far more widely than those of the water-saturated experiments, for which that limit is grouped between 1000°C and 1100°C.

In general, the amphibole-out boundary draws a positively slopped curve that progressively backbends to a flat segment at 20–22 kbar connecting to the subsolidus amphibole-out phase boundary of *Schmidt and Poli* [1998] (Figures 2 and 3). In fluid-absent melting experiments, this boundary is close to the solidus, indicating that the domain of coexistence of amphibole and melt is very limited, which has important consequences for TTG/adakite genesis [*Martin*, 1999; *Maury et al.*, 1996].

Several parameters seem to exert a limited control on the position of the fluid-absent amphibole-out boundary. There is a loose correlation between the initial amount of amphibole and the position of the amphibole-out boundary; the largest stability field is observed for starting material having large amphibole:plagioclase ratios, whereas the smaller stability fields are observed for experiments with lower amphibole: plagioclase ratios (and quartz in the starting material). This is consistent with the amphibole stability limit corresponding to the exhaustion of amphibole during dehydration melting reactions (see below), which is obviously dependent on the quantity of reactants. Furthermore, the presence of quartz causes the amphibole stability to decrease (900–1050°C with quartz, 950–1150°C without).

The bulk rock composition (bulk Mg#) seems to exert little or no direct control on the position of amphibole-out curve.

Finally, the composition of the initial amphibole (in the starting material) appears to exert some control on the position of the amphibole-out curve (Figure 4). However, no systematic study of these parameters has been conducted, and they are obviously not independent in the published works: Si-poor amphiboles tend to be also Ti- and Mg- rich; additionally, most studies were performed with natural rocks, and therefore the modal composition of the source is not independent from the amphibole mineral chemistry. Ti-rich (and also Si-poor, 5.8–6.4 Si(T) p.f.u., and Mg-rich) amphiboles collectively are stable at higher temperatures (1000–1100°C) than Ti-poor (< 1000°C) ones. Si-poor (magnesian, Ti-rich) amphiboles are stable to higher temperatures (1050–1100°C) than Si-rich ones and also appear to be stable at higher pressures (25–30 kbar). The associated amphibole-out boundary is approximately vertical at low pressures, before becoming nearly horizontal at higher pressures. At the other extremity of the "behavior spectrum", the Si-rich amphiboles define a negatively sloped boundary, extending no further than ca. 1000°C and not above ca. 15 kbar.

3.5.3. Plagioclase stability. There is relatively good agreement between the published work suggesting systematic

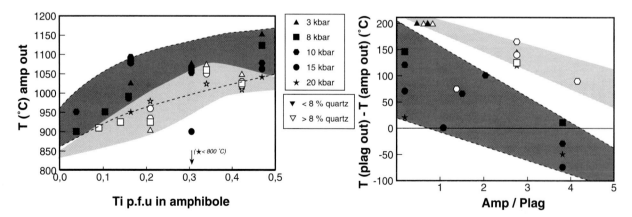

Figure 4. Mineral stability, as a function of the starting material composition. (Left) Temperature of amphibole ultimate stability ("amp out") (for diverse pressures) as a function of Ti content (atoms p.f.u., calculated on the basis of 23 oxygens) in amphibole. (Right) Difference between ultimate plagioclase ("plag out") and amphibole ("amp out") stabilities, as a function of the amphibole:plagioclase ratio (amp/plag) of the source. In both cases, two cases are distinguished as a function of the abundance of quartz; as in Figure 3, shading corresponds to the presence or absence of quartz and shape corresponds to the pressure (see caption). The dark band outlines the quartz-poor behavior, the light band the quartz-rich.

differences in the limits of plagioclase stability between fluid-absent melting and fluid-saturated experiments (Figures 2 and 4). In general, fluid-absent melting situations result in plagioclase being stable at higher temperature than in water-saturated cases, with the exception of *Zamora* [2000], who reported a theoretically water-present series of experiments wherein the plagioclase stability field was actually similar to that of typical fluid-absent melting cases.

In general, the plagioclase-out boundary of fluid-absent experiments is very close to the amphibole-out boundary, again emphasizing the role of reactions of the form amphibole + plagioclase = melt + peritectic phases. At pressures above garnet stability, both plagioclase-out and ampholite-out lines are fairly close to the solidus, suggesting that the melting reaction occurs over a narrow temperature range. A consequence of this is that the interval of plagioclase– melt coexistence is very restricted, which has profound consequences for TTG genesis [*Martin and Moyen*, 2002]. A special case has been described by *Rapp and Watson* [1995], who observed plagioclase at high temperature (1100–1200°C, 22–28 kbar). This plagioclase coexisted with a significant proportion of dioritic liquid and seems to correspond to a new cotectic phase formed during high-temperature reactions of the form garnet + clinopyroxene + $melt_1$ = plagioclase + $melt_2$.

Further investigations of the relative positions of the plagioclase-out and amphibole-out phase boundaries during vapor-absent melting (Figure 4) again suggest a control exerted by the source modal composition. In quartz-rich (also plagioclase-rich) lithologies, the plagioclase is stable up to temperatures that are actually above the amphibole-out curve; in contrast, quartz-poor (and plagioclase poor) lithologies show the reverse behavior, with plagioclase disappearing before amphibole.

3.5.4. Other mineral phases. The stability curves for other (minor) mineral phases are generally not given, and only general statements are possible.

Clinopyroxene is the dominant mineral in all experiments. It appears as a peritectic phase in the early melting reactions and is stable in all of the studied *P–T* range; the clinopyroxene-out boundary probably corresponds to the liquidus, at 1200°C and above [*Allen and Boettcher*, 1983; *Green*, 1982]. As could be expected, the jadeitic component in clinopyroxene is more abundant at high pressure [*Rapp and Watson*, 1995; *Zamora*, 2000].

Orthopyroxene [*Beard and Lofgren*, 1991; *Patiño-Douce and Beard*, 1995; *Rapp and Watson*, 1995; *Rushmer*, 1991; *Vielzeuf and Schmidt*, 2001] and/or olivine [*Rapp and Watson*, 1995; *Vielzeuf and Schmidt*, 2001] are often, but not always, observed as peritectic phases below the garnet-in curve. They are probably liquidus phases under low-pressure conditions.

Fe-Ti oxides are common throughout the *P–T* field. They are probably present in the starting compositions based on natural rocks and may also be formed as peritectic phases of some melting reactions (see below). *Zamora* [2000] systematically investigated the character of the oxide assemblage coexisting with melt. Ulvospinel was observed below the garnet-in curve; ilmenite from garnet-in to 16 kbar; and rutile above 18 kbar. Since rutile as a very high partitioning coefficient for high field strength elements (HFSE: Nb, Ta, etc.) [*Brenan et al.*, 1994; *Foley et al.*, 2000, 2002; *Kalfoun et al.*, 2002; *Ryerson and Watson*, 1987; *Schmidt et al.*, 2004], its stability limit has profound implications on the chemistry of the melts and on the genesis of the Nb-Ta anomaly [*Foley et al.*, 2002; *Green*, 1995; *Prouteau et al.*, 2000; *Rudnick et al.*, 2000; *Rudnick and Fontain*, 1995].

In experiments at high pressure, quartz or coesite [*Allen and Boettcher*, 1983] appears as a product of the "eclogitic" reaction albite = jadeite + quartz [*Holland*, 1980]. It is then progressively consumed and disappears at ca. 900–1000°C.

High-pressure hydrous phases (phengite and paragonite) are formed via subsolidus reactions, above the amphibolite-out curve [*Schmidt and Poli*, 1998]. Even though the high-pressure solidus must correspond to the breakdown of these phases, the melting reactions in the eclogite facies have been studied only partially [*Skjerlie and Patiño-Douce*, 2002; *Vielzeuf and Schmidt*, 2001].

Finally, nepheline has been reported [*Zamora*, 2000] at 975–1000°C and 11 kbar. This has since been confirmed by D. Vielzeuf (pers. comm.) in the same *P–T* range and with the same alkali-rich (> 5 wt% Na_2O) starting material.

3.5.5. A general mineralogical model for fluid-absent melting. The above discussions on the positions of mineral-out and solidus curves have demonstrated that the relative positions of these limits are controlled by the nature of the source. Three main cases can be distinguished, and these correlate with end-member behavior in the relevant amphibolite compositions (Table 3 and Figure 5).

Lithology 1 (KoB) is a intermediate between a komatiitic and a tholeitic basalt. Such a lithology has a high Mg#, little plagioclase, and little or no quartz. It is also alkali-poor and depleted in trace elements (see below). Samples No. 1 and No. 2 from *Rapp and Watson* [1995] and the starting material of *Sen and Dunn* [1994] are good representatives of this type of composition: The solidus is fairly high (940°C, subvertical), and the melt fraction rises rather quickly, with iso-melt curves close to vertical in *P–T* space. As stated above, the amphibole is stable to relatively high pressures and temperatures (1050°C and 25–30 kbar). This controls the position of the solidus backbend, which is pushed to high pressures. The plagioclase-out curve is close to the solidus, defining only a very restricted domain for coexistence of plagioclase and melt.

Lithology 2 is a tholeitic to quartz tholeitic basalt (ThB). It has a higher plagioclase:amphibole ratio; contains some quartz (1–5%); and has average alkali contents and moderately depleted trace element concentrations. Representative examples of this type of composition are ABA [*Rushmer*, 1991]; Sample No. 3 [*Rapp and Watson*, 1995]; and the composition of *Lopez and Castro* [2000]. In this case, the solidus is at ca. 880°C and backbends at 20–25 kbar; the melt

Table 3. Summary of the Characteristics of the Three Modeled Sources (see text), for Mineralogical and for Major and Trace Characteristics.

	Komatiitic basalt KoB	Tholeitic basalt ThB	Arc basalt AB
Modal proportions			
Quartz	0	1	10
Plagioclase	75	59	36
Amphibole	25	40	54
Amp:Pl	3:1	3:2	2:3
Major elements			
CaO	11.0	10.0	9.0
Na$_2$O	2.2	2.8	3.3
K$_2$O	0.1	0.5	1.0
TiO$_2$	1.2	2.1	0.8
Amphibole composition			
	Ti-rich high Mg# Si(T)-poor	Inter-mediate	Ti-poor low Mg# Si(T)-rich
Trace elements			
Rb	1	3.9	4.6
Sr	124	180	300
Ba	12	68	110
Hf	2.9	2.5	1
Zr	85	75	22
Y	29	22	16.8
Nb	2.4	8.1	0.7
Ta	0.15	0.5	0.06
La	2.4	6.3	12
Ce	8	15	20
Sm	2.5	3	3.5
Eu	1	0.9	1
Gd	3.6	3	3
Dy	4.5	3.6	3
Er	3.1	2.3	1.6
Yb	3.3	2.2	1.1
Nb/Ta	15.8	16.2	11.7
(La/Yb)$_N$	0.49	1.91	7.29

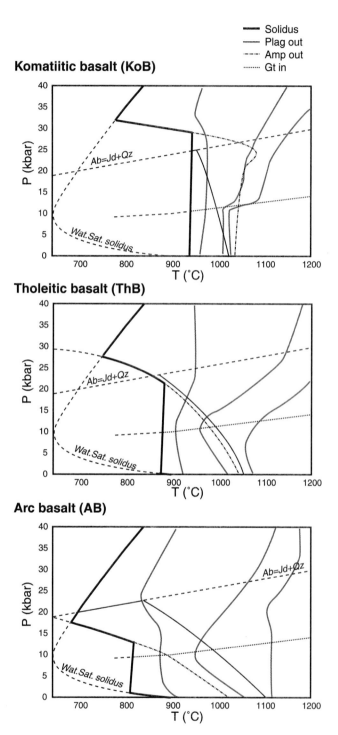

Figure 5. Generalized mineral stability diagrams for the three different sources used in the modeling below. In each diagram, the theoretical water saturated solidus ("wat sat solidus") is indicated and distinguished from the actual solidus (heavy line); plagioclase, amphibole and garnet stability fields are indicated ("plag out", "amp out" and "gt in"). The position of the albite = jadeite + quartz [*Holland*, 1980] is also shown. The gray lines correspond to approximate iso-melt lines of 10%, 30%, and 50%.

fraction increases more slowly, and iso-melt curves are positively sloped in P–T space. For constant temperatures, the garnet-in boundary corresponds to a marked decrease in melt fractions. The amphibole-out and plagioclase-out phase boundaries are approximately at the same position, from 20 kbar and 880°C to 1050°C at 0 kbar.

Finally, lithology 3 corresponds to an arc tholeite to andesitic basalt (AB). The other starting materials approximately match this composition. This lithology is quartz- and plagioclase-rich; it is also alkali-rich and has enriched trace element contents. Its amphibole is silicic and has a low-Mg#. In this case, the solidus is as low as 820°C and backbends only at 10–15 kbar; the melt fractions increase slowly below plagioclase-out, and the iso-melt fractions are positively sloped. Amphibole disappears well before plagioclase, from 820°C at 12 kbar to 1020°C at 0 kbar; plagioclase in contrast is stable at 850°C (at 22 kbar) to 1100°C (at 0 kbar).

Obviously, such different lithologies can also be expected to have contrasted trace element contents. KoB was given a depleted characteristic, with a negatively sloped REE pattern. ThB is nearly chondritic, with a flat pattern, and AB is already enriched, has a positively sloped REE pattern, and shows a Nb–Ta anomaly (Table 3).

4. MELT COMPOSITIONS

4.1. Major Elements

4.1.1. Analytical difficulties with Na. In the studies examined here, glass (quenched melt) compositions have been determined by either wavelength dispersive spectrometry (WDS) or energy dispersive spectrometry (EDS) in conjunction with either an electron microprobe or a scanning electron microscope. Analysis of sodium-bearing hydrous alumino-silicate glasses by electron micro-beam techniques is problematic because of the tendency for sodium counts to decay during the analytical counting period [*Stevens et al.*, 1997; *Vielzeuf and Holloway*, 1988]. Na loss during analysis is obviously critical for studies of TTG genesis, and the studies investigated here have attempted to deal with it in a number of ways; none of the studies reported here, however, used the one technique believed to result in reliable glass compositions over a wide compositional range and using a single standardization procedure: the use of a cryogenic stage to freeze the sample to liquid nitrogen temperatures prior to analysis [*Stevens et al.*, 1997; *Vielzeuf and Holloway*, 1988]. As a result, the glass analyses should be regarded as likely to underestimate sodium concentrations. This will be most acute at higher pressures and lower temperature, where melt water contents are higher.

This appears to be reflected in the data: In the >350 glass analysis from our database, about one-third have impossibly

high A/CNK ratios above 1.2, clearly incompatible with amphibolite melting: Such highly peraluminous melts are ordinarily produced only in equilibrium with highly aluminous residual assemblages. This suggests that the published Na_2O analysis are largely unreliable, which presents a problem because Na_2O contents are important for the definition of the tonalite field [*O'Connor*, 1965].

4.1.2. Potassium as a trace element. During partial melting of amphibolite, no potassic mineral phases are present; K is accommodated in the minerals only when substituted for Na. It is, therefore, not a major element but rather behaves as a trace element. Trace element composition in melts is described by the equation

$$\frac{C_l}{C_0} = \frac{1}{F + D.(1-F)}$$

where C_l is the composition in the liquid, C_0 is that in the source, D is the bulk repartition coefficient, and F is the melt fraction [*Rollinson*, 1993]. A plot of $K_2O_{melt}/K_2O_{source}$ vs. F (Figure 6) for the experimental melts indeed shows that compositions of the melts appear to plot along curves corresponding to D values between 0.05 and 0.1. Therefore, for given F values, the composition in K_2O of the melt depends mostly on the source composition. This has important implications on TTG chemistry; partial melts of amphibolites can be expected to have relatively high K_2O contents, should the source be itself relatively rich in potassium. In other words, the K values are not a function of the melting processes but rather of the source composition; thus, the low K values of TTG melts is not an intrinsic characteristic linked to their genesis but is a product of relatively low K sources. A recent experimental confirmation of this interpretation has been brought forward by *Sisson et al.* [2005], who showed that melting of high-K amphibolites at 0.7 GPa with 1.01–2.32 wt% K_2O yielded granitic melts (3–6% K_2O) for relatively high melt fractions (12–25%).

Consequently, plotting the melt compositions in the classical K–Na–Ca diagram is problematic. The Na values are likely to be flawed, and the K values will only reflect source differences. This diagram can hardly be used as a signature of the amphibolite melting processes.

4.1.3. Pressure effect? As suggested by *Rapp and Watson* [1995] and *Prouteau* [1999], but in contrast with *Zamora* [2000], no obvious difference in melt major element composition can be observed as a function of pressure; the diagrams in Figure 6 show that in most cases, the same evolution can be observed in terms of C_{melt}/C_{source}, regardless of the pressure. The "pressure effect" observed when comparing

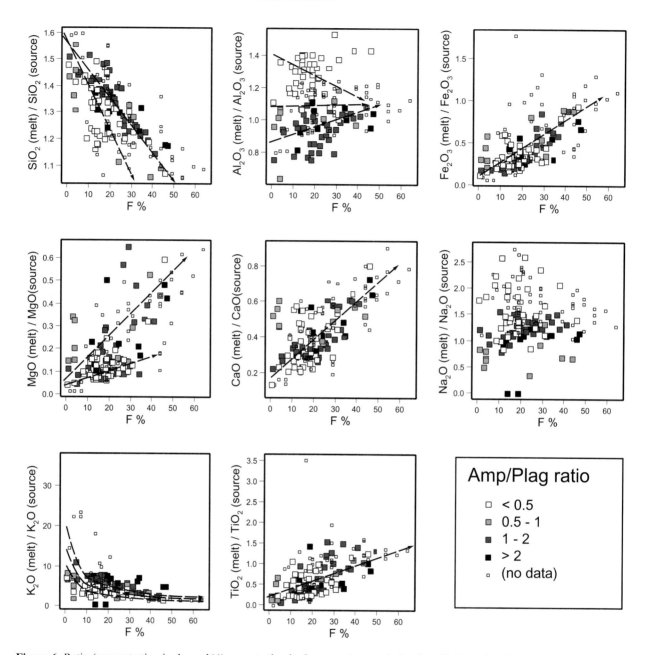

Figure 6. Ratio (concentration in the melt)/(concentration in the source) vs. melt fraction (F) for major elements. Dashed lines correspond to the modeled lines (see text and Table 4). K_2O is modeled as a trace element (hence the curved trends), and Na_2O models depend not on the source, but on the pressure. Symbols according to the type of source (Table 3).

low- and high-pressure melts at the same temperature is likely to actually result from the fact that, given the positive slope of the iso-melt curves in the P–T space, high-pressure melts correspond to lower melt fractions and therefore to more alkali-rich melts.

The striking lack of difference between the evolution of the melts as a function of F suggests that, for a given F, the melting residuum has a more or less constant composition in terms of major elements, whatever the pressure. In other words, the low-pressure clinopyroxene + orthopyroxene + Ca-rich plagioclase assemblage is chemically equivalent to the high-pressure garnet + (jadeitic) clinopyroxene assemblage, at least in terms of major elements.

The only exception to this is Na_2O; here, the high-pressure melts (coexisting with an eclogitic or garnet-amphibolite residue) tend to be more enriched in Na than the low-pressure melts are; this probably relates to the low-Na contents of the high-pressure assemblage, where the only Na-bearing phase is jadeite, which tends to become unstable at high temperature.

4.1.4. A model for major element melt composition during fluid-absent melting. Plotting individual elements vs. SiO_2 or versus F gives no obvious, interpretable trends, probably owing to the relatively large diversity of sources. However, this difficulty can be overcame by plotting, rather than one element's composition, the ratio C_{melt}/C_{source} for this element as a function of F. In this case (Figure 6), well-defined linear trends appear that are controlled by the nature of the source.

For each of our three "end-member" sources, we propose a set of equations describing the "enrichment" of the melt in a given element, relative to the source. This is expressed as

$$\frac{C_{melt}}{C_{source}} = a.F + b$$

where F is still the melt fraction; coefficients a and b for each major element, in all cases, are given in Table 4. These trends, drawn on Figure 6, correspond to the best fit in all cases, except for K and Na. K trends have been modeled in terms of a trace element, with a bulk repartition coefficient of 0.1 to 0.05 (depending on the case); for Na, owing to the analytical difficulties and the underestimation of Na contents in melts, we used the upper envelope of the data points, rather than the best trend. In addition, pressure control (rather then source control) has been incorporated in the model for Na_2O.

4.2. A Model for Trace Element Concentrations in the Experimental Melts

Trace element contents in melts can be predicted by applying the equation for equilibrium melting used above, where

$$D = \sum_i K_d^i X_i$$

(X_i: proportion of mineral i in the residue; K_d^i: partition coefficient of an element between melt and mineral i). Building such a model over the $P-T$ space therefore requires (1) a model of melt fractions; (2) a model of modal composition of the residue; (3) a set of K_d values; and (4) an hypothesis on the source composition.

First, a model of typical modal proportions for melt and of the residual mineral assemblage in the experimental charges as a function of P and T had to be established for each of the three cases outlined above (KoB, ThB, and AB). In each situation, iso-abundance curves for each mineral species (and melt) have been manually drawn on the $P-T$ diagram, smoothing the data scatter between different experimental works. Based on these curves, the mineral modal proportions have been extrapolated over the relevant parts of the $P-T$ space, using triangulated irregular network interpolation between the iso-abundance curves. For Ti-bearing minerals (ilmenite and rutile), which have very high K_d values for Nb and Ta and are therefore important to constrain precisely, we refined this approach by estimating the amount of Ti in the residue by mass balance and calculating the rutile/ilmenite

Table 4. Model Parameters for the Major Element Composition Model. Enrichment in an element C_{melt}/C_{source} is expressed by an equation of the form $C_{melt}/C_{source} = a\,F + b$, where F (in %) is the melt fraction. a and b parameters are given here for the three sources defined in Table 2.

	AB		ThB		KoB	
	a	b	a	b	a	b
SiO_2	−0.013	1.500	−0.011	1.525	−0.009	1.550
TiO_2	0.016	0.200	0.016	0.200	0.016	0.200
Al_2O_3	−0.006	1.400	0.003	1.080	0.005	0.850
Fe_2O_3	0.015	0.125	0.015	0.125	0.015	0.125
MgO	0.004	0.030	0.007	0.050	0.010	0.070
CaO	0.011	0.150	0.011	0.150	0.011	0.150
Pressure-dependent behaviour						
Na_2O		Low P (amphibolite domain): a = 0.003, b = 1.450				
		Medium P (garnet-amphibolite domain): a = −0.016, b = 2.500				
		High P (eclogite domain): a = −0.025, b = 3.250				
Trace element behaviour						
K_2O		$\dfrac{C_l}{C_0} = \dfrac{1}{F + D(1-F)}$				
	D = 0.1		D = 0.07		D = 0.05	

content from that. For each point in the *P–T* space, the modal composition has then been normalized to 100%. The modal composition model is depicted in Plate 1a.

For most elements, the K_d values used are from *Rollinson* [1993]; they do not change with melt composition—a reasonable approximation in general [*Rollinson*, 1993]. However, considering the recent discussion on Nb and Ta contents in melts, on the one hand [*Foley et al.*, 2002, 2004; *Rapp et al.*, 2003], and the recent work on partitioning of Nb and Ta in Ti-bearing phases, on the other hand (e.g., *Schmidt et al.* [2004]), we decided for a more precise model in that case, taking into account the melt composition. Following *Schmidt et al.* [2004], we use a $K_d^{melt/rutile}$ that varies with the melt SiO_2 content, from 25 (at 48% SiO_2) to 150 (at 70% SiO2) for Nb and from 50 to 200 (over the same SiO_2 range) for Ta.

Finally, the source composition for each lithology is as described above (Table 3). Plates 1 and 2 show the results of this modeling for selected trace elements and ratios.

5. FEATURES OF THE MODEL AND RELEVANCE FOR TTG GENESIS

5.1. Ca–Na–K Systematics

On a Na_2O–K_2O–CaO diagram, TTGs plot in a rounded area close to the Na_2O apex. As pointed out by *Martin* [1994], they do not define a single trend, and this suggests that TTGs actually derive from different sources and/or processes. Our model shows that partial melts from amphibolites start on or close to the Na–K side, in a position dependent on the K_2O content of the source, and evolve towards the Ca apex with increasing melt fractions. The lack of evidence for such evolution within the TTG series suggests that either (a) no low-melt fraction liquids are represented in the rocks, or (b) the source of the magmas was a very K-poor basalt.

A total alkali vs. silica or SiO_2 vs. K_2O/Na_2O diagram (Figure 7) shows that the felsic and K-rich source (AB) yields melts that plot significantly above most TTGs for a given SiO_2 content; therefore, we suggest that the low K values in TTGs reflect a K-depleted source, as, according to our model, a K-rich amphibolite (more than 1 wt% K_2O) could easily yield potassic, granitic, or even syenitic melts at low-melt fractions.

More generally, the melts show two different behavior patterns on a *P–T* diagram (Figure 8): Low-pressure melts evolve from rare granites (very low F, first melts close to the solidus) to granodiorites and tonalites, whereas at higher pressures, the melts are trondhjemitic close to the solidus and become tonalitic with increasing temperature. TTG melts (tonalitic or trondhjemitic, between 60% and 70% SiO_2, with low K/Na values) form for melt fractions between ca. 15%

and 40%, roughly corresponding to temperatures between 900°C and 1100°C (this is, of course, pressure-dependent).

5.2. REE Contents

REE contents of the melt are, unsurprisingly, strongly controlled by garnet abundance. It is, however, important to note that the typical TTG HREE and Y depletion occurs only when significant amounts of garnet are found in the residue; to achieve a twofold depletion of Yb contents (from a MORB-like source, $Yb_N = 10$ source, to a TTG-like melt, $Yb_N = 5$) requires a minimum of 20% of garnet in the residue. The degree of melting exerts relatively little influence, in that the melting reactions incongruently produce garnet; therefore, increasing melt fractions also result in increasing proportions of garnet in the residue. Even a source with half this amount in Yb ($Yb_N = 5$) still requires ca. 15% of garnet in the residue to achieve appropriate degrees of depletion in the melt. Therefore, the HREE depletion of TTGs not only requires garnet to be stable, it also requires garnet to occur in significant amounts in the residue. This, in turn, involves pressures well above the garnet-in curve, since garnet becomes a volumetrically significant mineral only well above this pressure. Consequently, the appropriate level of Yb depletion can be reached only for pressures above ca. 15 kbar (in the ThB case).

In addition to garnet abundance, the source composition plays some role in shaping the REE pattern produced in the melts. A depleted source (KOB in our lithology), with an initially negatively sloped REE pattern, is unlikely to ever yield melts with a positively sloped REE pattern (Plate 1, right panels). This suggests that the source was undepleted to moderately enriched.

5.3. Sr and Y

Sr content is linked to plagioclase stability, and Sr contents increase sharply above plagioclase-out; this has been demonstrated experimentally by *Zamora* [2000] and is discussed in *Martin and Moyen* [2002]. Y has the same behavior as Yb, and low Y concentrations in the melt are also correlated with significant amounts of garnet in the residuum. These two factors together produce the Sr/Y ratio that behaves in a very similar way to the La/Yb ratio during melting events that produce garnet and consume plagioclase. Consequently, Sr/Y appears also to be a good pressure indicator, low-pressure melts being Sr-poorer and Y-richer than high-pressure ones.

5.4. Nb and Ta

Nb and Ta contents are largely controlled by the presence of rutile. The extremely high K_d of rutile, for both elements,

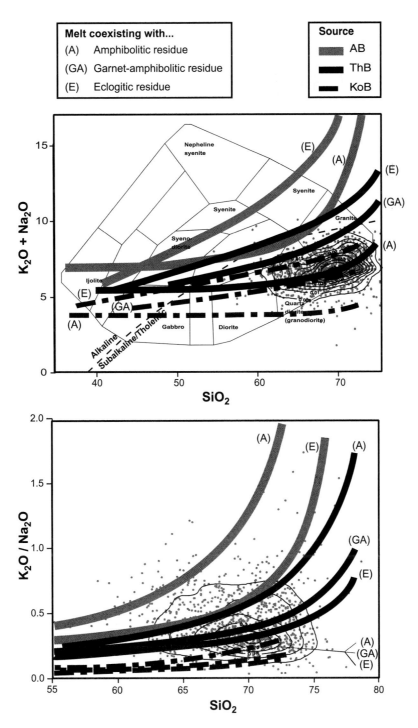

Figure 7. Total alkali vs. silica (TAS) [*Le Bas*, 1986] and SiO$_2$ vs. K$_2$O/Na$_2$O diagrams showing the composition of the TTGs (gray dots) (ca. 2,000 analysis, [*Martin and Moyen*, 2002]) compared with the predicted model compositions (lines). TTG compositions are contoured by data point density, estimated by kernel density ([*Venables and Ripley*, 2002], a built-in function in the statistical package "R": http://www.r-project.org/). The numbers of the contours correspond to the values of the density function. The three models (corresponding to the three sources) discussed above are represented: AB, "arc basalt" source (thick gray line); ThB, "Tholeiitic basalt" (black line); KoB, "Komatiitic basalt" (dashed black line). Capital letters denote the nature of the residuum coexisting with the melt (amphibole, amphibole + garnet, or garnet). In both cases, the AB source yields magmas typically too rich in potassium to adequately model TTGs, while the KoB source is generally too poor in potassium. Only the ThB source allows the model to reproduce the TTG compositions.

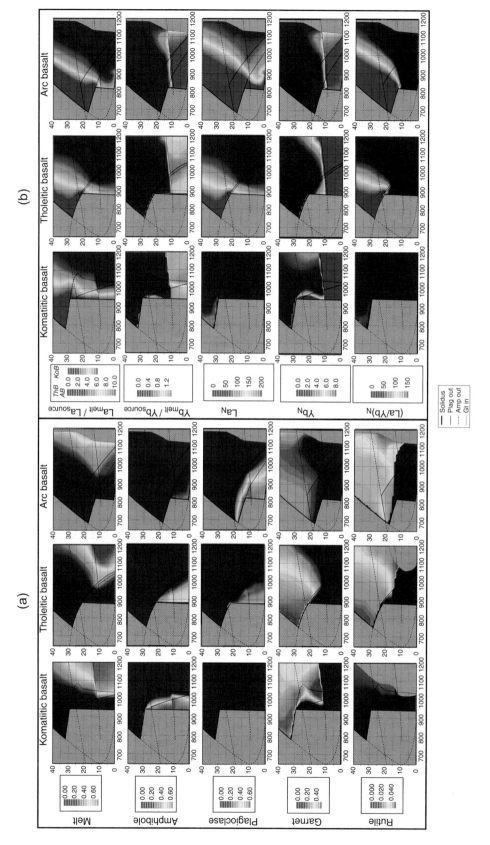

Plate 1. (a) *P–T* representation of the modal proportion model for each of the three modeled cases. Color codes correspond to melt fraction (first line) or to modal proportion in the residue (next four lines), legend adjacent to each diagram. Additionally, the mineral stability curves of Figure 6 are indicated. (b) Results of the modeling for REE, color-coded in the *P–T* space. Disposition same as in (a); the subscript "source" indicates the concentration in the source, and "melt" denotes the concentration in the melt (in other words, the C_{melt}/C_{source} values correspond to the enrichment of the studied element during melting); the subscript "N" indicates melt concentration, normalized to chondrites (normalization values of *Thompson* [1982]).

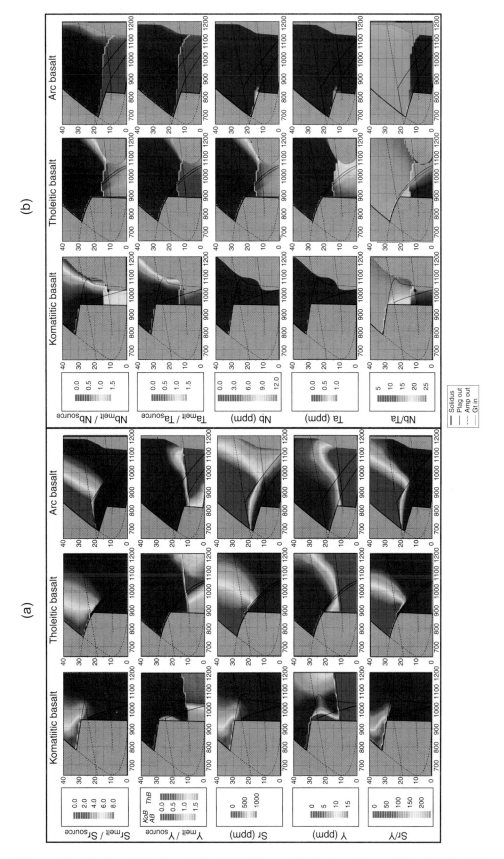

Plate 2. (a) Results of the modeling for Sr and Y, color-coded in the *P–T* space. (b) Results of the modeling for Nb and Ta, color-coded in the *P–T* space. Comments as in Plate 1.

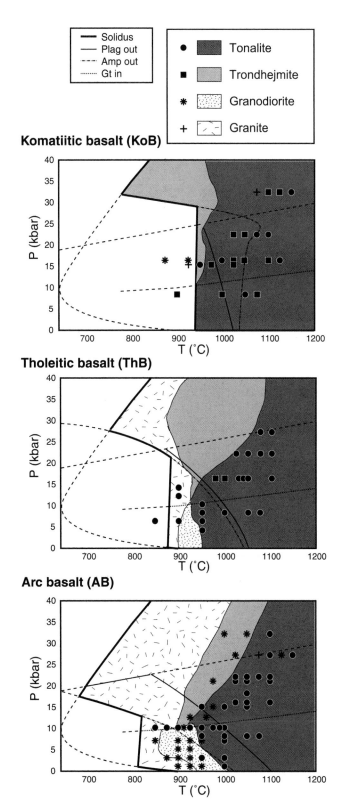

Figure 8. Nature of the modeled melts in *O'Connor* [1965] systematic, compared with the nature of actual experimental melts. Modeled compositions are indicated by the shadings, real experimental melt nature by symbols.

effectively regulates their concentration in the melt. *Foley et al.* [2002] and *Schmidt et al.* [2004], on this basis, showed that Nb/Ta is a very good pressure indicator, rutile-present melts having high Nb/Ta values and being overall depleted in these elements. Our work indeed confirms this; the "cutoff value", based on our modeling, seems to be around 15–20 for Nb/Ta values, slightly lower than the Nb/Ta value of ca. 25 considered by *Foley et al.* [2002] as a minimum for rutile-present melting.

However, neither our modeling nor the work by *Foley et al.* [2002] and *Schmidt et al.* [2004] is able to predict the very low Nb/Ta ratios (between 10 and 3) observed in some TTGs, irrespective of their degree of differentiation (SiO_2 content). The only plausible solution would be a source rock already having very low Nb/Ta ratios; our sources have initial Nb/Ta values around 15 and give melts not below 10. It appears that a Nb/Ta value of 5 can be attained only with a source having a Nb/Ta value not exceeding 7, strongly distinct from known basaltic compositions. This observation prompted some workers, e.g., *Kambers et al.* [2002], to suggest that TTGs are not the product of amphibolite melting, but rather of garnet-present fractionation of common wet mantle melts [*Alonso-Perez et al.*, 2003; *Grove et al.*, 2003; *Müntener et al.*, 2004]. This interpretation, however, seems to be at odds with a large corpus of experimental (summarized here) and field observations showing that partial melting of amphibolite is a viable mechanism to yield TTGs; amphibolite-derived melts indeed match TTG composition for nearly all elements (major or trace), apart from their Nb/Ta ratio.

We found, furthermore, that our natural TTG database (more than 2000 analyses, expanded after *Martin and Moyen* [2002]) encompasses rocks with Nb/Ta values (see Figure 9) up to at least 40 (omitting some very high, suspect values), which contrasts with the maximum Nb/Ta of ca. 20 reported by *Foley et al.* [2002] and *Kambers et al.* [2002].

The systematics of Nb and Ta in TTG rocks remains poorly understood, and the present models are unable to account for the low-Nb/Ta members of the TTG group.

5.5. Other Systems

To a lesser degree, other trace elements also appear to be pressure-sensitive. The Zr/Hf ratio, for instance, increases from typically ca. 20 to > 50 between 5 and 25 kbar, spanning the whole range of observed values in natural TTGs.

6. DISCUSSION AND CONCLUSIONS

6.1. Eclogite vs. Amphibolite Melting

The conditions of melting that are appropriate for the generation of TTGs has been the subject of several recent discussions. While the conventional view [*Martin*, 1987, 1994] has been that the source of TTGs is garnet-amphibolite,

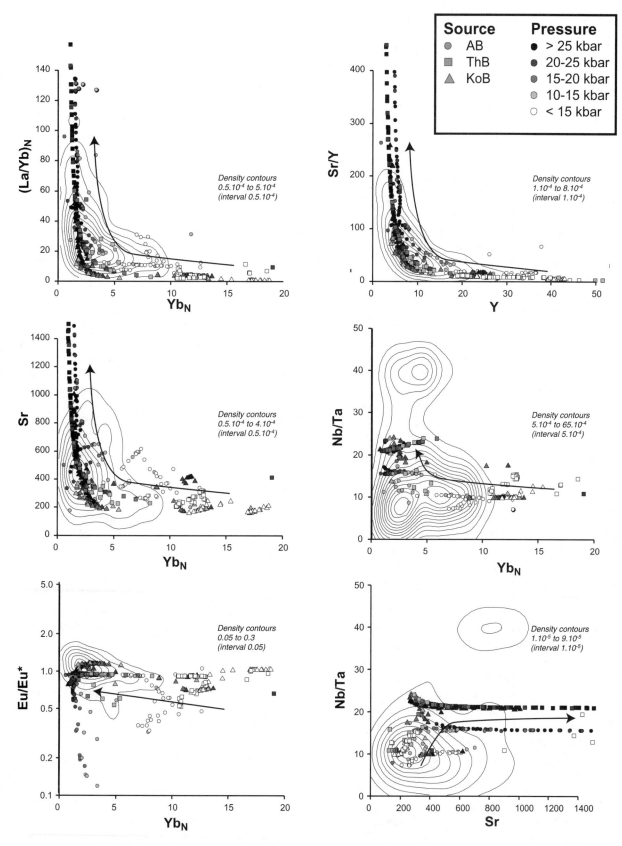

Figure 9. Melt compositions for trace elements as predicted by the model, compared with TTG compositions, contoured as in Figure 7 (only contours are shown). For the modeled compositions, the shape of the symbol corresponds to the source (see Figure 7), and the color (white to black) corresponds to pressure (high pressure is darker). In each diagram, arrow graphically shows the effect of increasing pressure on the modeled compositions.

Rapp et al. [2003] recently proposed, on the basis of experimental data, that eclogite melting is an equally plausible source. This view has been in turn challenged by *Foley et al.* [2002, 2004] and *Schmidt et al.* [2004], on the basis of Nb and Ta contents in TTGs; they concluded that a rutile-bearing eclogitic residue would yield melts with higher Nb/Ta values than those that typify TTGs.

At this point, however, we should note that the terms "amphibolite" or "eclogite" melting are somewhat misleading. Because amphibole disappears relatively quickly during melting, as discussed earlier, even melting starting at amphibolite-facies depths will very quickly yield melts in equilibrium with an amphibole-free assemblage (garnet + clinopyroxene, generally), which is actually eclogitic. The only relevant distinction, in terms of geochemistry, is whether rutile is present or not during melting—which roughly corresponds to pressures above or below 15 kbar, respectively.

Foley et al. [2002] and *Schmidt et al.* [2004] ruled out rutile as a common residual phase, on the grounds of the low Nb/Ta values in TTGs. However, as stated above, our database shows a far greater diversity in TTG compositions, suggesting that Nb/Ta values may not be a clear discriminant feature of TTG magmas but instead are quite variable within this group. This important variation in Nb/Ta values, combined with equally significant variations in other elements, leads us to conclude that melting did occur both in and out of the rutile stability field; TTGs with characteristics matching both possibilities are observed.

As a side note, in the database of *Martin et al.* [2005] on adakites, the acid adakites (considered to be more or less pure slab melts) have Nb/Ta values between 7 and 20. This suggests a far greater homogeneity of adakites, with less variation than among Archean TTGs. This also suggests that, although some TTG magmas formed in the presence of rutile, this is probably not the case for most adakites.

6.2. Pressure Proxys in TTG Geochemistry

Our modeling suggests that some elements are good indicators of the pressure of melting. Sr is linked to plagioclase stability; Eu/Eu* is also classically an indicator of plagioclase stability. Nb/Ta is a good indicator of rutile stability (despite the above-mentioned limitations); since rutile appears at ca. 15 kbar, regardless of temperature, Nb/Ta is very pressure-sensitive. Finally, Yb is a good indicator of garnet stability, or more correctly, of garnet abundance, and is also a pressure-sensitive parameter.

Figure 9 shows the interrelationships between these indicators, as well as their correlation with pressure. TTGs occur all along the curves in these diagrams, suggesting that TTG genesis occurs over a large pressure range, from ca. 10 kbar to ca. 20–25 kbar.

These indications should be used with some caution, though. As can be seen in the figures, they are slightly sensitive to the source composition. Worse still, later magmatic evolution will indeed affect the concentrations of these elements: The Nb/Ta ratio is unlikely to be strongly modified, but Yb and Sr contents will be modified during fractional crystallization (assuming fractionation of plagioclase + amphibole, after *Martin* [1987, 1994]). The direction of the Sr and Yb variations is not easily predictable, since amphibole has high K_d for Yb but low K_d for Sr; the situation is reversed for plagioclase. Therefore, the direction of the evolution will depend on the exact proportion of both minerals involved, and of course on the inclusion of accessory minerals such as apatite [*Martin*, 1987]. However, precisely because of the opposite effects of both principal minerals, the overall partition coefficient will be close to 1, resulting in variations of ca. 30% for 30% fractional crystallization—considerably less than the observed differences.

Thus, the data appear to support a continuum of melting conditions to produce TTG magmas. This range extends from ca. 10 kbar—where at low pressure TTGs form that are Sr-poor (< 400 ppm), relatively undepleted in Yb ($Yb_N = 5$–10), slightly negative for Eu anomaly, and low Nb/Ta (ca. 10)—to 20–25 kbar, where at high pressure, TTGs are produced that show the opposite characteristics (Sr = 500–1000 ppm, $Yb_N \approx 2$, Nb/Ta $\approx$ 20).

6.3. Geothermal Gradients for Making TTG

The previous discussion leads to the conclusion that TTGs with appropriate compositions, both in terms of major and trace elements, can be formed only in a relatively restricted part of the *P–T* diagram. Major element considerations indicate that tonalites and trondhjemites are formed between ca. 900°C and 1100°C (Figure 8). The trace element modeling suggest pressures varying between 10 and 25 kbar; garnet is present in sufficient amounts to produce a suitable HREE and Y depletion above about 15 kbar; this value is also the pressure above which *Yearron et al.* [2003] found garnet coexisting with trondhjemitic liquids (at 850°C).

Therefore, the *P–T* conditions for making TTGs are represented in Figure 10: *P* > 15 kbar and 900°C < *T* < 1100°C. This corresponds to relatively low geothermal gradients of ca 20°C/km (assuming a density of 3.0 for crustal rocks), lower than most recorded Archean gradients (which are up to 60°C/km [*Percival*, 1994]).

Occasionally, atypical low-pressure TTGs are found, with relatively unfractionated REE patterns having small Eu anomalies and fairly high Nb and low Sr values. The present study suggests that this group of rocks can form at lower pressures (10–12 kbar) but equivalent temperatures, corresponding to higher geothermal gradients of ca. 30°C/km.

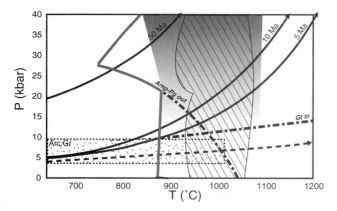

Figure 10. Summary diagram showing (1) the domain of tonalitic and trondhjemitic melts (dashed); (2) the domain of melts with TTG-like trace element signatures (shaded; darker shade corresponds to higher pressure melts, Yb- and Y- poorer sources and Sr-richer sources with higher Nb/Ta); (3) the field of Archean granulites (Arc. Gr.), after *Percival* [1994], and associated geotherm (dashed, with arrow); (4) Geothermal gradients after *Delong et al.* [1979] along a subducting slab of different ages (solid line, with arrow; age indicated). Only subduction of a young lithosphere seems to be able to give geothermal gradients along which *P–T* conditions adequate for TTG genesis are reached.

6.4. Geodynamic Implications

P–T conditions as discussed above correspond to fairly low geothermal gradients of ca. 20°C/km. They also imply melting at ca. 50 km or more. All this is in poor agreement with an intraplate (plateau melting) setting but is in better agreement with subduction zone processes. *Martin and Moyen* [2002], on the basis of interpretation of a compilation of TTG analyses and their secular evolution, arrived at the same conclusion; furthermore, an increasing number of examples of interactions between TTG-like melts and mantle peridotites are being described (the "sanukitoids"; e.g., *Smithies and Champion* [2000]; *Moyen et al.* [2003]; *Martin et al.* [2005]), which strongly suggests that TTG magmas were formed in a place where they could interact with the mantle—thus implying a subduction zone.

However, melting of the subducted slab is not a common process in modern subduction zones; it is actually restricted to some rare locations, but in such places the volcanic activity associated with subduction is "adakitic", i.e., lavas having chemical compositions similar to those of Archean TTGs [e.g., *Defant and Drummond*, 1990; *Maury et al.*, 1996; *Martin*, 1999]. On the other hand, the frequency of TTG occurrence in the Archean suggests that these conditions were far more commonly achieved during that time.

A possible explanation [*Martin*, 1994; *Martin and Moyen*, 2002] is that, due to higher overall radioactive heat production

during the Archean, the Earth as a whole was warmer; therefore, upper mantle temperatures were higher, resulting in hotter subductions. However, this model implicitly considers that the higher heat production resulted in a homogeneous increase of the temperatures of each part of the Earth; an alternative possibility is that the higher heat production was rather accommodated by more active or more numerous spreading centers, the rest of the upper mantle being at temperatures not significantly different from now [*Bickle*, 1978, 1986; *de Wit and Hart*, 1993]. In this case, more abundant ridges would result in faster and smaller plates, with a mean age of subducting lithosphere significantly younger that at present [*Lagabrielle et al.*, 1997]. *P–T* conditions in subduction zones have been studied by (e.g.) *Delong et al.* [1979]; according to their modeling, the *P–T* "window" identified in the present study for making TTGs can be achieved in three cases: (1) subduction of a young lithosphere (< 30 Ma when entering the subduction); (2) fast subduction (> 10 cm/yr); (3) young subduction (subduction processes having started less than 10 Ma ago).

This view has been challenged by *Parsons* [1982], *Galer* [1991], and *Kambers et al.* [2002], on the ground of thermal modeling of mantle potential temperature and oceanic plate formation. These workers consider that the rate of oceanic plate generation did not greatly change between the Archean and the present, and that the higher heat production (and higher upper mantle potential temperature) was instead accommodated by thicker—but not younger, on average—oceanic plates. Even in this case, though, the higher potential temperatures of the upper mantle would generate hotter subduction zones, whereas the thicker oceanic plates would allow large amounts of mafic materials to go into subductions, likewise allowing for the generation of large amounts of TTG melts in subduction zones.

Regardless of the model adopted for Archean oceanic plate formation, the geochemical features of TTG point to magma generation along a geothermal gradient of ca. 20°C/km, too low for intraplate settings, even if it is somewhat higher than expected in a modern subduction zone.

In contrast, the low-pressure TTGs do not necessarily need to be formed in a subduction setting; they can be formed at the base of a thick oceanic plateau, by melting of underplated basalts, or in a stack of oceanic crust "slices" [*de Wit and Hart*, 1993]. In the Barberton greenstone belt area, *Moyen and Stevens* [2004] showed that the low-pressure TTGs occurred early in the cratonic history (the 3.55 Ga Steynsdorp pluton [*Kröner et al.*, 1996]), whereas the "normal" high-pressure TTGs are more recent (the 3.45 and 3.22 Ga generations [*de Ronde and Kamo*, 2000; *Kamo and Davis*, 1994]). This suggests that the early stages of continental accretion consist of the formation of an intra-oceanic continental nucleus, which later focuses the subduction zones; the subsequent continent development is mostly related to subduction zone magmatism probably associated with arc accretion.

6.5. Nature of the Source

Our modeling suggests that the suitable range for source composition is actually relatively restricted. A felsic source (our AB lithology) is too rich in plagioclase to produce melts with appropriate compositions of major elements; only a mafic source, corresponding to our KoB or ThB sources, gives appropriate melts with low K/Na values even at high silica contents.

In terms of trace elements, an arc-style source, already bearing a Nb–Ta anomaly, is ruled out because the resulting melts would be extremely depleted in Nb and Ta, which is inconsistent with the TTG compositions. On the other hand, melting of an N-MORB source (our KoB lithology) does not replicate the trace element contents (La/Yb$_N$ in particular) of TTGs. Therefore, a slightly enriched (E-MORB or similar) source is needed to account for TTG characteristics. In our three models, the ThB source therefore seems to be the most appropriate to account for the genesis of most TTG. This would be consistent with the Archean upper mantle being less depleted than its modern counterpart (probably owing to a lesser amount of continental crust extracted from the mantle).

However, while these observations tell us that the bulk of the TTGs are probably derived from a slightly enriched, mafic, MORB-like source, they do not imply that melting of other sources is impossible. Actually, we suggest that some strange "TTG looking" or "near-TTG" plutonic rocks could be interpreted in terms of melting of an amphibolite with uncommon characteristics, e.g., an arc-related rock (during closure and subduction of a back-arc basin?), or komatiitic basalts (during regional metamorphism after accretion of a komatiitic greenstone sequence?).

6.6. Further Work

This review suggests several directions in which experimental work on partial melting of amphibolites could be extended:

(1) The role of epidote during partial melting, and the nature of the liquids formed by melting associated with epidote breakdown;

(2) The parameters actually controlling the position of the amphibole-out and plagioclase-out phase boundaries: modal composition of the source and mineral chemistry of the amphibole, most likely;

(3) A more correct determination of Na contents in the experimental glasses;

(4) The behavior of the system (and the melt compositions) in the eclogitic domain, close to the high-pressure solidus (> 20 kbar, 750–900°C);

(5) The development of a proper thermodynamic model for TTG melts, comparable to that exists for peraluminous melts [*Holland and Powell*, 2001; *White et al.*, 2001] or mantle melts (MELTS) [*Ghiorso et al.*, 2002; *Ghiorso and Sack*, 1995].

Finally, our study suggests that TTGs encompass at least some degree of geochemical diversity, part of which, we propose, can be related to differences in melting depth of similar sources. We suggest that more attention should be paid to subtle differences of rocks broadly belonging to the "TTG" group, to try to discuss with a sufficient degree of precision the conditions for their formation.

7. CONCLUSIONS AND SUMMARY

The following conclusions can be drawn from this review:

(1) Archean TTGs are the product of fluid-absent partial melting of metabasites, at pressures commonly above 15 kbar and temperatures between 900°C and 1100°C. This corresponds to relatively low Archean geothermal gradients (typically 20°C/km). These conditions are likely to have been achieved in subduction zones that were significantly hotter than their modern counterparts.

(2) The TTG source was basaltic (rather than andesitic) and relatively enriched. Neither an arc-style source nor a depleted MORB or komatiite can account for the compositions of major and trace elements in TTGs. Surprisingly, a large proportion of previous experimental studies have used lithologies that were too felsic as starting material and are therefore only partially adequate to address the question of TTG formation.

(3) Nb/Ta, Sr, and HREE contents in TTGs appear to be intercorrelated and represent an indicator of the pressure of melting. These indicators span a relatively wide range, suggesting that TTGs were generated over a large range of pressures, from 10 to 25 kbar at least.

Acknowledgments. JFM's postdoctoral stay at Stellenbosch University has been funded by NRF grant GUN 2053698 and a grant from the Department of Geology. Hervé Martin kindly made available his databases on TTGs, adakites, and experimental melts [*Martin et al.*, 2005]. David Zamora supplied a copy of his thesis, which was greatly inspirational, in addition to being an important source of data [*Zamora*, 2000]. Comments on an early draft by Hugh Rollinson also greatly improved the discussion. Most figures have been drawn by using GCDkit, a geochemical plotting software by V. Janousek (http://www.gla.ac.uk/gcdkit)

REFERENCES

Albarède, F., The growth of continental crust, *Tectonophysics*, 196, 1-14, 1998.

Allen, J.C., and A.L. Boettcher, Amphiboles in andesite and basalt: I. Stability as a function of P-T-fH$_2$O, *Am.Min.*, 60, 1069-1085, 1975.

Allen, J.C., and A.L. Boettcher, Amphiboles in andesite and basalt: II. Stability as a function of P-T-fH$_2$O-fO$_2$, *Am. Min.*, 63, 1074-1087, 1978.

Allen, J.C., and A.L. Boettcher, The stability of amphibole in andesites at high pressures, *Am. Min.*, 68, 307-314, 1983.

Alonso-Perez, R., P. Ulmer, O. Müntener, and A.B. Thompson, Role of garnet fractionation in H$_2$O undersaturated andesite liquids at high pressure, *Lithos*, 73 (1-2), S116, 2003.

Barker, F., and J.G. Arth, Generation of trondhjemite-tonalite liquids and Archean tondhjemite-basalt suites, *Geology*, 4, 596-600, 1976.

Beard, J.S., and G.E. Lofgren, Dehydration melting and water-saturated melting of basaltic and andesitic greenstones and amphibolites at 1, 3 and 6.9 kb, *J. Petrol.*, 32, 465-501, 1991.

Bickle, M.J., Heat loss from the Earth: a constraint on Archean tectonics from the relationships between geothermal gradients and the rate of plate production, *Earth.Plan.Sci.Lett.*, 40, 301-315, 1978.

Bickle, M.J., Implications of melting for stabilization of the lithosphere and heat loss in the Archean, *Earth. Plan. Sci. Lett.*, 80, 314-324, 1986.

Brenan, J.M., H.F. Shaw, D.L. Phinney, and F.J. Ryerson, Rutile-aqueous fluid partitioning of Nb, Ta, Hf, Zr, U and Th: implications for high field strength elements depletion in island-arc basalts, *Earth.Plan.Sci.Lett.*, 128, 327-339, 1994.

Condie, K.C., *Archean greenstone belts*, 434 pp., Elsevier, Amsterdam, 1981.

Condie, K.C., G.P. Bowling, and P. Allen, Missing Eu anomaly and Archean high-grade granites, *Geology*, 13, 633-636, 1985.

Defant, M.J., and M.S. Drummond, Derivation of some modern arc magmas by melting of young subducted. lithosphere, *Nature*, 367, 662-665, 1990.

Delong, S.E., W.M. Schwarz, and R.N. Anderson, Thermal effects of ridge subduction, *Earth. Plan. Sci. Lett.*, 44, 239-246, 1979.

de Ronde, C.E.J., and S.L. Kamo, An Archean arc-arc collisional event: a short-lived (ca 3 Myr) episode, Weltvreden area, Barberton greenstone belt, South Africa, *J. Afr. Earth. Sci.*, 30 (2), 219-248, 2000.

de Wit, M.J., and R.A. Hart, Earth's earliest continental lithosphere, hydrothermal flux and crustal recycling, *Lithos*, 30, 309-335, 1993.

Drummond, M.S., and M.J. Defant, A model from trondhjemite-tonalite-dacite genesis and crustal growth via slab melting: Archean to modern comparisons, *J. Geophys. Res.*, 95, 21503-21521, 1990.

Foley, S.F., M.G. Barth, and G.A. Jenner, Rutile/melt partition coefficients for trace elements and an assessment of the influence of rutile on the trace element characteristics of subduction zone magmas, *Geoch. Cosmoch. Acta*, 64, 933-938, 2000.

Foley, S.F., M. Tiepolo, and R. Vannucci, Growth of early continental crust controlled by melting of amphibolite in subduction zones, *Nature*, 417, 637-640, 2002.

Foley, S.F., R. Vannucci, M. Tiepolo, R. Oberti, and A. Zanetti, Recognition of melts of subducted slabs by high field strength element fractionation, in *EGU meeting, Nice, April* 2004., Nice, 2004.

Galer, S.J.G., Interrelationships between continental freeboard, tectonics and mantle temperature, *Earth.Plan.Sci.Lett.*, 105, 214-228, 1991.

Gardien, V., A.B. Thompson, and P. Ulmer, Melting of Biotite + Plagioclase + Quartz gneisses: the role of H$_2$O in the stability of amphibole, *J.Petrol.*, 41, 651-666, 2000.

Ghiorso, M.S., M.M. Hirschmann, P.W. Reiners, and V.C. Kress, The pMELTS: A revision of MELTS for improved calculation of phase relations and major element partitioning related to partial melting of the mantle to 3 GPa, *Geoch.Geophy.Geosystems*, 3, art. no.-1030, 2002, 2002.

Ghiorso, M.S., and R.O. Sack, Chemical Mass-Transfer in Magmatic Processes 4. A Revised and Internally Consistent Thermodynamic Model for the Interpolation and Extrapolation of Liquid-Solid Equilibria in Magmatic Systems at Elevated-Temperatures and Pressures, *Contrib. Mineral. Petrol.*, 1995.

Green, T.H., Anatexis of mafic crust and high pressure cristallisation of andesite, in *Andesites*, edited by R. Thorpe, pp. 465-486, J. Wiley and sons, New-York, 1982.

Green, T.H., Significance of Nb/Ta as an indicator of geochemical processes in crust-mantle system, *Chem.Geol.*, 120, 347-359, 1995.

Green, T.H., and A.E. Ringwood, Genesis of the calc-alkaline igneous rock suite, *Contrib.Mineral.Petrol.*, 18, 105-162, 1968.

Grove, T.L., L.T. Elkins, S.W. Parman, N. Chatterjee, O. Müntener, and G.A. Gaetani, Fractional crystallization and mantle-melting controls on calc-alkaline differentiation trends, *Contrib. Mineral. Petrol.*, 145, 515-533, 2003.

Holland, T.J.B., The reaction albite = jadeite + quartz determined experimentally in the range 600-1200°C, *Am. Min.*, 65, 129-134, 1980.

Holland, T.J.B., and R. Powell, Calculation of phase relations involving haplogranitic melts using an internally-consistent thermodynamic data set, *J. Petrol.*, 42, 673-683, 2001.

Irvine, T.N., and W.R.A. Baragar, A guide to the chemical classification of the common volcanic rocks, *Can. J. Earth. Sci*, 8, 523-548, 1971.

Jahn, B.M., P. Vidal, and A. Kröner, Multi-chronometric ages and origin of Archean tonalitic gneisses in finnish Lapland: a case for long crustal residence time, *Contrib.Mineral.Petrol.*, 86, 398-408, 1984.

Jébrak, M., and L. Harnois, Two-stage evolution in an Archean tonalite suite: the Taschereau stock, Abitibi, *Can. J. Earth. Sci*, 28, 172-183, 1991.

Johnston, A.D., and P.J. Wyllie, Constrains on the origin of Archean trondhjemites based on phase relationships of Nük gneisses with H$_2$O at 15 kbar, *Contrib. Mineral. Petrol.*, 100, 35-46, 1988.

Kalfoun, F., D. Ionov, and C. Merlet, HFSE residence and Nb/Ta ratios in metasomatized, rutile-bearing mantle peridotites, *Earth. Plan. Sci. Lett.*, 199, 49-65, 2002.

Kambers, B., A. Ewart, K.D. Collerson, M.C. Bruce, and G.D. McDonald, Fluid-mobile trace elements constraints on the role of slab melting and implications for Archean crustal growth models, *Contrib. Mineral. Petrol.*, 144, 38-56, 2002.

Kamo, S.L., and D.W. Davis, Reassessment of Archean crustal development in the Barberton mountain land, South Africa based on U-Pb dating, *Tectonophysics*, 13, 167-192, 1994.

Kröner, A., E. Hegner, J.I. Wendt, and G.R. Byerly, The oldest part of the Barberton granitoid-greenstone terrain, South Africa: evidence for crust formation between 3.5 and 3.7 Ga, *Precamb. Res.*, 78, 105-124, 1996.

Lagabrielle, Y., J. Goslin, H. Martin, J.-L. Thiriot, and J.-M. Auzende, Multiple active spreading centers in the hot North Fiji basin (SW Pacific): a possible model for Archean seafloor dynamics?, *Earth. Plan. Sci. Lett.*, 149, 1-13, 1997.

Lambert, I.B., and P.J. Wyllie, Melting of gabbro (quartz eclogite) with water excess to 35 kbar, with geological application, *J. Geol.*, 80, 693-703, 1972.

Le Bas, M.J., R.W. Le Maître, A. Streckeisen, and B. Zanettin, A chemical classification of volcanic rocks based on the total alkali-silica diagram, *J. Petrol.*, 27, 745-750, 1986.

Le Maître, R.W., *Igneous Rocks: A Classification and Glossary of Terms*, Cambridge University Press, Cambridge, 2002.

Liu, J., S.R. Bohlen, and W.G. Ernst, Stability of hydrous phases in subducting oceanic crust, *Earth. Plan. Sci. Lett.*, 143, 161-171, 1996.

Lopez, S., and A. Castro, Determination of the fluid-absent solidus and supersolidus phase relationships of MORB-derived amphibolites in the range 4-14 kbar, *Am. Min.*, 86, 1396-1403, 2000.

Martin, H., Petrogenesis of Archean trondhjemites, tonalites and granodiorites from eastern Finland; major and trace element geochemistry, *J. Petrol.*, 28 (5), 921-953, 1987.

Martin, H., The Archean grey gneisses and the genesis of the continental crust, in *Archean crustal evolution*, edited by K.C. Condie, pp. 205-259, Elsevier, Amsterdam, 1994.

Martin, H., The adakitic magmas: modern analogues of Archean granitoids, *Lithos*, 46 (3), 411-429, 1999.

Martin, H., and J.-F. Moyen, Secular changes in TTG composition as markers of the progressive cooling of the Earth, *Geology*, 30 (4), 319-322, 2002.

Martin, H., R.H. Smithies, R.P. Rapp, J.-F. Moyen, and D.C. Champion, An overview of adakite, tonalite-trondhjemite-granodiorite (TTG) and sanukitoid: relationships and some implications for crustal evolution, *Lithos*, 79 (1-2), 1-24, 2005.

Maury, R., F.G. Sajona, M. Pubellier, H. Bellon, and M.J. Defant, Fusion de la croûte océanique dans les zones de subduction/collision récentes: l'exemple de Mindanao (Philippines), *Bull. Soc. Geol. Fr.*, 167, 579-595, 1996.

Moyen, J.-F., H. Martin, M. Jayananda, and B. Auvray, Late Archean granites: a typology based on the Dharwar Craton (India), *Precamb. Res.*, 127 (1-3), 103-123, 2003.

Moyen, J.-F., and G. Stevens, Evolution of TTG geochemistry around the Barberton greenstone belt from 3.5 to 3.2 Ga: transition from shallow melting below under-plated crust to subduction-associated magmatism?, in *Geocongress*, Johannesbourg (South Africa), 2004.

Müntener, O., R. Alonso-Perez, and P. Ulmer, Phase Relations of Garnet, Amphibole and Plagioclase in H_2O undersaturated Andesite Liquids at high Pressure and Implications for the Genesis of Lower Arc Crust, in *EGU meeting*, Nice, 2004.

O'Connor, J.T., A classification for quartz-rich igneous rocks based on feldspar ratios, *U.S. G.S. Prof. Paper*, 525 (B), 79-84, 1965.

Parsons, B., Causes and consequences of the relation between area and age of the ocean floor, *J. Geophys. Res.*, 87, 289-302, 1982.

Patiño-Douce, A.E., and J.S. Beard, Dehydration-melting of Bt gneiss and Qtz amphibolite from 3 to 15 kB, *J. Petrol.*, 36, 707-738, 1995.

Percival, J.A., Archean high-grade metamorphism, in *Archean crustal evolution*, edited by K.C. Condie, pp. 315-355, Elsevier, Amsterdam, 1994.

Prouteau, G., Contribution des produits de la fusion de la croûte océanique subductée au magmatisme d'arc: exemple du Sud-Est asiatique et approche expérimentale, Thèse d'Université, Université de Bretagne Occidentale, Brest, 1999.

Prouteau, G., B. Scaillet, M. Pichavant, and R. Maury, Evidence for mantle metasomatism by hydrous silicic melts derived from subducted oceanic crust, *Nature*, 410, 197-200, 2000.

Rapp, R.P., N. Shimizu, and M.D. Norman, Growth of early continental crust by partial melting of eclogite, *Nature*, 425, 605-609, 2003.

Rapp, R.P., and E.B. Watson, Dehydratation melting of metabasalt at 8-32 kbar: implications for continental growth and crust-mantle recycling, *J. Petrol.*, 36 (4), 891-931, 1995.

Rapp, R.P., E.B. Watson, and C.F. Miller, Partial melting of amphibolite/eclogite and the origin of Archean trondhjemites and tonalites, *Precamb. Res.*, 51, 1-25, 1991.

Rollinson, H., *Using geochemical data*, 352 pp., Longman, London, 1993.

Rollinson, H., and B.F. Windley, An Archean granulite - grade tonalite - trondhjemite - granite suite from Scourie, NW Scotland: Geochemistry and origin, *Contrib. Mineral. Petrol.*, 72, 265-281, 1980.

Rudnick, R.L., Making continental crust, *Nature*, 378, 571-578, 1995.

Rudnick, R.L., M.G. Barth, I. Horn, and W.F. McDonough, Rutile-bearing refractory eclogites: missing link between continents and depleted mantle, *Science*, 287, 278-281, 2000.

Rudnick, R.L., and D.M. Fontain, Nature and composition of the continental crust: a lower crustal perspective, *Rev. Geophys.*, 33, 267-309, 1995.

Rushmer, T., Partial melting of two amphibolites: contrasting experimental results under fluid-absent conditions, *Contrib. Mineral. Petrol.*, 107, 41-59, 1991.

Ryerson, F.J., and E.B. Watson, Rutile saturation in magmas: implications for Ti-Nb-Ta depletion in island-arc basalts, *Earth. Plan. Sci. Lett.*, 86, 225-239, 1987.

Schmidt, M.W., A. Dardon, G. Chazot, and R. Vannucci, The dependence of Nb and Ta rutile-melt partitioning on melt composition and Nb/Ta fractionation during subduction processes, *Earth. Plan. Sci. Lett.*, 226, 415-432, 2004.

Schmidt, M.W., and S. Poli, Experimentally based water budgets for dehydrating slabs and consequences for arc magma generation, *Earth. Plan. Sci. Lett.*, 163, 361-379, 1998.

Sen, C., and T. Dunn, Dehydration melting of a basaltic composition amphibolite at 1.5 and 2.0 Gpa: implications for the origin of adakites, *Contrib. Mineral. Petrol.*, 117, 394-409, 1994.

Sisson, T.W., K. Ratajeski, W.B. Hankins, and A.F. Glazner, Voluminous granitic magmas from common basaltic sources, *Contrib. Mineral. Petrol.*, 148, 635-661, 2005.

Skjerlie, K., and A.E. Patiño-Douce, Anatexis of interlayered amphibolite and pelite at 10 kbar: effect of diffusion of major components on phase relations and melt fraction, *Contrib. Mineral. Petrol.*, 122, 62-78, 1995.

Skjerlie, K., and A.E. Patiño-Douce, The fluid-absent partial melting of a zoisite bearing quartz eclogite from 1.0 to 3.2 GPa: implications for melting of a thickened continental crust and for subduction-zone processes, *J. Petrol.*, 43, 291-314, 2002.

Smithies, R.H., and D.C. Champion, The Archean high-Mg diorite suite: Links to Tonalite-Trondhjemite-Granodiorite magmatism and implications for early Archean crustal growth, *J. Petrol.*, 41 (12), 1653-1671, 2000.

Spicer, E.M., G. Stevens, and J.-F. Moyen, The quantitative measurement of Na_2O in silicate melts by scanning electron microscope, fitted with an energy dispersive spectrometry system and cryo-stage, *Min. Mag.*, submitted.

Springer, W., and H.A. Seck, Partial fusion of granulites at 5 to 15 kbar: implications for the origin of TTG magmas, *Contrib. Mineral. Petrol.*, 127, 30-45, 1997.

Stevens, G., J.D. Clemens, and G.T.R. Droop, Melt production during granulite-facies anatexis: experimental data from "primitive" metasedimentary protoliths, *Contrib. Mineral. Petrol.*, 128, 352-370, 1997.

Thompson, R.N., British tertiary volcanic province, *Scott. J. Geol.*, 18, 49-107, 1982.

Van der Laan, S.R., and P.J. Wyllie, Constraints on Archean trondhjemite genesis from hydrous crystallization experiments on Nûk gneisses at 10-17 kbar, *J. Geol.*, 100, 57-68, 1992.

Venables, W.N., and B.D. Ripley, *Modern Applied Statistics with S.*, Springer, 2002.

Vielzeuf, D., and J.R. Holloway, Experimental determination of the fluid-absent melting reactions in the politic system: consequences for crustal differenciation, *Contrib. Mineral. Petrol.*, 98, 257-276, 1988.

Vielzeuf, D., and M.W. Schmidt, Melting reactions in hydrous systems revisited: application to metapelites, metagreywackes and metabasalts, *Contrib. Mineral. Petrol.*, 141, 251-267, 2001.

Weaver, B.L., and J. Tarney, Rare-Earth geochemistry of lewisian granulite facies gneisses, northwest Scotland: implications for the petrogenesis of the Archean lower continental crust, *Earth. Plan. Sci. Lett.*, 51, 279-296, 1980.

White, R.W., R. Powell, and T.J.B. Holland, Calculation of partial melting equilibria in the system CaO-Na_2O-K_2O-FeO-MgO-Al_2O_3-SiO_2-H_2O (CNKFMASH), *J. Metam. Geol.*, 19, 139-153, 2001.

Windley, B.F., *The Evolving continents*, 526 pp., John Wiley and sons, Chester, 1995.

Winther, T.K., and R.C. Newton, Experimental melting of an hydrous low-K tholeiite: evidence on the origin of Archean cratons, *Bull. Geol. Soc. Denmark*, 39, 1991.

Wolf, M.B., and P.J. Wyllie, Dehydration-melting of solid amphibolite at 10 Kbar: textural development, liquid interconnectivity and applications to the segregation of magmas, *Contrib. Mineral. Petrol.*, 44, 151-179, 1991.

Wolf, M.B., and P.J. Wyllie, Dehydration-melting of amphibolite at 10 Kbar: the effects of temperature and time, *Contrib. Mineral. Petrol.*, 115, 369-383, 1994.

Wyllie, P.J., and M.B. Wolf, Amphibolite dehydration-melting: sorting out the solidus, in *Magmatic processes and plate tectonics*, edited by H. Prichard, T. Alabaster, N.B.W. Harris, and C. Neary, pp. 405-416, London, 1993.

Yearron, L.M., Archean granite petrogenesis and implications for the evolution of the Barberton Mountain Land, South Africa, thesis, Kingston, 2003.

Zamora, D., Fusion de la croûte océanique subductée: approche expérimentale et géochimique, Thèse d'université, Université Blaise-Pascal, Clermont-Ferrand, 2000.

J.-F. Moyen and G. Stevens. Department of Geology, University of Stellenbosch. Private bag X 01, Matieland 7602, South Africa. (moyen@sun.ac.za)

Geodynamic Modeling of Late Archean Subduction: Pressure–Temperature Constraints From Greenstone Belt Diamond Deposits

C. O'Neill and D. A. Wyman

School of Geosciences, University of Sydney, Sydney, New South Wales, Australia

The oldest known diamond deposits occur in 2.7 Ga shoshonitic lamprophyre dikes and breccias in the southern Superior Province, Canada. Given the shallow mantle sources of shoshonite magmas, the diamonds are likely to be subduction-related. During normal subduction, however, the diamond stability window lies below the source regions of typical shoshonites. Particle-in-cell finite-element modeling of normal and flat subduction indicates that shallow subduction angles optimise the conditions required for a diamond–lamprophyre association. Our results predict diamond formation in parts of the mantle wedge following flat subduction and two generations of diamonds in the subducted slab. Any of these diamond sources may be sampled by lamprophyric magmas during a short-lived period of thermal re-equilibration driven by orogeny. In contrast, normal subduction does not create a mantle wedge diamond population, and long-lived flat subduction without orogeny does not provide a means for lamprophyric magmas to access mantle wedge diamonds.

INTRODUCTION

Diamonds entrained in volatile-rich kimberlitic magmas originate beneath the deepest parts of Archean-aged cratons and therefore have been crucial in determining the early history and deep architecture of the continents [*Haggerty*, 1999]. Models of diamond formation have multiplied over the last 20 years as non-kimberlitic diamond occurrences were recognised. It is now clear that diamonds can also form at comparatively shallow depths in subducted oceanic slab and sediments or in the overlying mantle wedge. These "subduction diamonds" may be rapidly transported to the surface by magmas or converted to graphite pseudomorphs during the slow obduction of peridotite massifs [*Griffin et al.*, 2000; *Davies et al.*, 1993]. New discoveries of the oldest in situ terrestrial diamond deposits endorse the formation of diamonds

in shallow pressure–temperature (P-T) stability windows generated by subduction tectonics. The deposits are likely to provide fresh insights into the rapid growth and stabilisation of continents during the late Archean.

The ~2.7-Ga deposits are hosted by shoshonitic lamprophyre dikes and breccias in the Wawa and Abitibi greenstone belts along the southern Superior Province (Figure 1; *Ayer and Wyman* [2003]) and an apparently similar occurrence has been described from the Slave Craton [*Armstrong and Barnett*, 2003]. Tens of thousands of microdiamonds and hundreds of macrodiamonds have been recovered from the Michipicoten belt of the Wawa Subprovince, where the largest in situ diamond reported to date is 5 mm in width and 0.7 carats [*Schiller*, 2003]. In the Wawa Subprovince, the diamond-hosting lamprophyre dikes are mainly spessartites and, as is the case for most other Archean lamprophyre occurrences across the Superior Province [*Wyman and Kerrich*, 1989], they are generally metamorphosed to greenschist facies.

The dikes typically have a green, medium-grained, groundmass of actinolite, biotite, chlorite, albite, calcite,

Archean Geodynamics and Environments
Geophysical Monograph Series 164

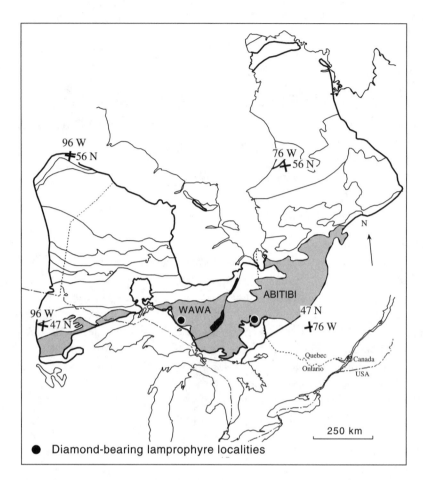

Figure 1. A location map of Archean diamondiferous lamprophyre occurrences in the southern Wawa and Abitibi Subprovinces (shaded) of the Superior Province Craton (after *Ayer and Wyman* [2003]).

epidote, and titanite [*Stone and Semenyna*, 2004]. Some samples contain amphibole or biotite macrocrysts, chromite, and clinopyroxene. The mafic to ultramafic matrix of diamond-hosting breccias is an assemblage of actinolite, chlorite, and albite with lesser amounts of epidote and titanite. Macrocrysts of magnesiohornblende and diopside occur locally [*Stone and Semenyna*, 2004]. A "pilitic" secondary pseudomorph assemblage of amphibole aggregates, with or without chlorite and carbonate, was observed in one diamond-hosting Abitibi lamprophyre [*Williams*, 2002]. These features are common in shoshonitic lamprophyres [*Rock*, 1990].

In post-Archean terranes, amphibole-rich spessartite and micaceous minettes, such as those studied in the southern Superior Province, are commonly associated with suites of larger K-rich intrusions formed by melting at comparatively shallow depths, in subduction-modified mantle [*Rock and Hunter*, 1987; *Platten*, 1991]. Field relations, along with major and trace element compositions and isotopic evidence, all indicate that the lamprophyres and the larger intrusions are co-genetic, or derived from parental melts with similar

sources [e.g., *Fowler and Henney*, 1996]. Post-Archean shoshonite suites are related to the termination of subduction and slab break off [*Atherton and Ghani*, 2002], or to other tectonic processes that involve crustal extension above subduction-modified mantle [*Carmichael et al.*, 1996].

Archean lamprophyres across the Superior Province are also spatially associated with syenites and were emplaced in orogenic settings during stabilisation of the Craton. Accordingly, they are part of a mildly alkalic igneous suite, commonly associated with conglomeratic sediments, that was emplaced in a transpressional to extensional tectonic setting near the end of greenstone belt development [*Ayer et al.*, 2002]. Evidence for their orogenic timing includes cross-cutting relationships with all volcanic assemblages, emplacement into late structures, and a titanite U–Pb age of 2674 ± 1 Ma for an Abitibi occurrence [*Wyman and Kerrich*, 1989, 1993]. The diamond-bearing occurrences specifically have also been shown to have cross-cutting relationships with the youngest volcanic sequences in the Wawa Subprovince, and one dike returned a titanite U–Pb age of 2674 ± 8 Ma [*Stott et al.*,

2002]. Diamond-bearing heterolithic breccias in the Wawa Subprovince had previously been interpreted as volcaniclastic debris flows that were contemporaneous with last stage of greenstone belt volcanism at ~2700 Ma (*Lefebvre et al.*, 2003). This interpretation would require that the breccias were petrogenetically distinct from the lamprophyre dikes. Mutually cross-cutting field relations between the dikes and breccias and possibly xenocrystic zircon ages of 2679.2 ± 1.1 Ma demonstrate, however, that the two lamprophyre facies are coeval [*Stott et al.*, 2002; *Vaillancourt et al.*, 2004]. A detailed chemical and isotopic study of the host rocks and their ultramafic xenoliths are presented elsewhere [*Wyman et al.*, 2005].

In many cases, generation, or at least final equilibration prior to ascent, of shoshonitic lamprophyre magmas occurs near pressures of 1.5 to 2.0 GPa (60–80 km depth) and temperatures of ~1150–1200 °C, as is the case for parental magmas of Pliocene Sierra Nevada minettes (MgO = 12 wt.%; *Feldstein and Lange* [1999]). Deeper origins are also documented, however, and *Righter and Carmichael* [1996] calculated equilibration conditions in excess of 3 GPa (~120 km depth) and 1200°C for a primitive minette of the Colima area, Mexico. The greatest depth yet implicated for a shoshonitic lamprophyre is associated with the Thumb minette of the Colorado Plateau. *Smith et al.* [1991] summarised studies of garnet peridotite and other xenoliths from the Thumb, some of which equilibrated at pressures between 4.5 and 5.0 GPa (~180–190 km depths) and temperatures between 1150°C and 1370°C. Despite the great depths implicated, the Thumb minette is not particularly primitive, in terms of its chemical composition (e.g., 12 wt.% MgO: *Ehrenberg*, [1982]).

The majority of diamond-hosting lamprophyres in the Wawa Subprovince are spessartitic (amphibole-rich) in character and characteristically have lower abundances of incompatible elements than do micaceous minettes. In this respect they are typical of weakly altered Archean spessartites such as those of the Yilgarn Craton [*Perring et al.*, 1989]. Many shoshonitic lamprophyre magmas evolve at the base of the crust or in crustal-level magma chambers [*Righter and Carmichael*, 1996]. This process would not allow for diamond preservation, however, and as a result the diamond-hosting examples are generally more primitive than average occurrences. Table 1 summarises representative chemical data for diamond-hosting lamprophyre dikes and breccias of the Wawa Subprovince (from *Stone and Semenyna* [2004]). Other Archean lamprophyre compositions from the southern Superior Province are given in *Wyman and Kerrich* [1989, 1993] and *Wyman et al.* [2005].

In this paper we use particle-in-cell finite-element modeling to explore the implications of diamondiferous lamprophyres for Archean tectonics, based on the geological and tectonic constraints provided by the Superior Province occurrences.

Table 1. Major- and trace-element compositions of diamond-hosting lamprophyre dikes and breccias, Wawa Subprovince [from *Stone and Semenyna*, 2004].

Sample	02DS87	02DS94	02DS98
Type	Breccia	Breccia matrix	Dike
Major element oxides and volatiles			
SiO_2	44.58	52.21	47.12
Al_2O_3	8.88	10.66	11.14
MnO	0.16	0.13	0.15
MgO	19.42	13.15	11.46
CaO	9.72	6.81	7.98
Na_2O	0.88	3.4	2.44
K_2O	0.22	1.55	3.43
TiO_2	0.69	0.66	0.89
P_2O_5	0.31	0.23	0.42
Fe_2O_3T	10.5	8.81	11.96
LOI	5.38	1.72	2.22
Total	100.75	99.34	99.22
CO_2	1.11	0.09	1.06
S	0.01	N.D.	0.01
Trace elements			
Co	76	55	61
Cr	1350	858	720
Ni	795	495	371
V	154	129	213
Sc	21	18	27
Rb	7.03	52.92	145.67
Ba	30	398	746
Sr	166.9	542	422.09
Ta	N.D.	0.34	N.D.
Hf	1.76	2.78	2.35
Ga	13	14	16
Nb	4.17	4.88	4.89
Zr	63	120	96
Y	14.53	15.99	20.16
Pb	7	10	10
Th	1.75	3.47	3.2
U	0.37	0.84	0.79
La	14.53	23.34	18.7
Ce	32.58	49.67	40.57
Pr	4.29	6.25	5.39
Nd	17.76	25.02	23.1
Sm	3.84	4.66	5.06
Eu	1.13	1.41	1.74
Gd	3.24	3.84	4.64
Tb	0.44	0.54	0.65
Dy	2.52	2.83	3.69
Ho	0.56	0.57	0.74
Er	1.5	1.56	1.93
Tm	0.22	0.22	0.29
Yb	1.33	1.47	1.77
Lu	0.197	0.212	0.271

GEOLOGICAL CONSTRAINTS
ON DIAMOND EMPLACEMENT

Based on the diamond–graphite P-T stability field, diamond formation in subduction zones can occur in cold subducted slabs or in the shallow overlying mantle, because the cold-finger effect of subduction creates a window into the low-P, low-T part of the diamond stability field (DFS) [*Barron et al.*, 1996; *Griffin et al.*, 2000]. In the case of the placer diamonds from Southeast Australia, *Barron et al.* [1996] invoked a long-lived (~100 million years) stability window at about 110 km and suggested that the diamond carriers were Jurassic to Cretaceous basanites, nephelinites, and leucitites. *Griffin et al.* [2000] suggested that the host magmas may have originated beneath a subducted slab and estimated that a low-T (~350°C) source region in static subducted slab, located at depths between 100 and 120 km, would remain in the DFS for up to a few tens of million years following subduction. Accordingly, this model implies earlier transport of diamonds (at ~200 Ma) in the New England orogen.

Figure 2A shows the thermal structure of a generic subduction zone [*Anderson et al.*, 1980] and the region at the top of the subducting slab that *Griffin et al.* [2000] suggest may correspond to the source of some placer deposit diamonds from Southeast Australia. The shallowest depth of the DFS in this figure only marginally overlaps the deepest sources implicated for shoshonitic lamprophyres. The commonly cited early Archean P-T trajectory [*Martin*, 1986; *Martin and Moyen*, 2002] would shift the typical Archean graphite–diamond transition by at least 30 km from ~170 km to depths greater than 200 km and represents a difficult scenario for the recovery of Archean subduction diamonds in shoshonitic magmas. If all other variables are kept constant and the subduction-related diamond stability window described by *Griffin et al.* [2000] is depressed only 30 km deeper, then its shallowest point would occur at depths that exceed those implicated for any shoshonitic lamprophyre. Several lines of evidence, however, suggest that the typical late-Archean geotherm was closer to the modern one than to the early Archean P-T path discussed by *Martin* [1986]. For example, the tonalite–trondhjemite–granodiorite (TTG) suite had

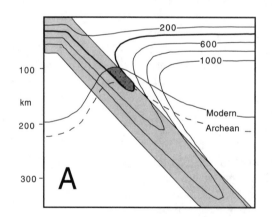

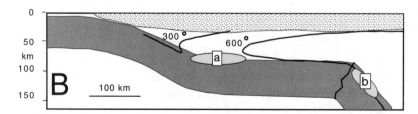

Figure 2. (A) The thermal structure of a typical subduction zone showing subducting oceanic crust (light gray) and a low-temperature region at the top of the slab (dark gray) that represents a potential source of subduction diamonds (after *Griffin et al.* [2000]). The mantle wedge above 100 km is outside of the subduction diamond stability window in this scenario. The solid "Modern" line is the present boundary between the graphite and diamond stability fields, and the dashed "Archean" line is the hypothetical Archean boundary discussed in the text. (B) The flat-subduction model of *Gutscher et al.* [2000] for central Chile with two approximate zones of diamond stability, "a" and "b". This idealised end-member model illustrates the potential for multiple zones of diamond stability in the subducting slab and the large range of temperatures that may occur in the "frozen" former mantle wedge.

evolved to nearly modern compositions by the late Archean (~2.7 Ga; *Martin and Moyen* [2002]; *Smithies* [2000]). In any case, if the geotherm was slightly steeper than the modern counterpart at the time of lamprophyre emplacement, then the top of the DFS would occur at greater depths. This effect would only make the entrainment of diamonds by shoshonitic magmas less plausible.

The regional geology of the southern Superior Province provides an indication of how the apparent obstacles to a diamond–lamprophyre association may have been avoided. If many adakite slab melts result from shallow subduction, as suggested by *Gutscher et al.* [2000], then the presence of genuinely adakitic rock types [*Smithies*, 2000] in the southern Superior Province indicates that scenarios involving normal-angle subduction may be inappropriate. The numerical modeling of *Gutscher et al.* [2000] indicated that flat subduction initially heats the slab, causing it to melt at shallow depths. Importantly, slab melting will occur for only a few million years, and subduction quickly results in the cooling of the asthenospheric mantle, eliminating much of the convecting wedge (Figure 2B). Accordingly, variations in the angle of subduction, and the location of the cold-finger effect, could be of greater significance for the generation of lamprophyre-accessible diamond stability P-T windows than the late-Archean geotherm.

Although normal subduction may never have been able to stabilise diamonds at sites above the sources of Archean (or younger) lamprophyres, shallow-angle subduction effectively creates a low-pressure diamond stability window over a wide temperature range. The low-P, low-T window is likely to extend from the subducting plate into the immediately overlying mantle. In order to better assess the role of subduction angle in the location of diamond stability in both subducting slabs and overlying mantle lithosphere, we used the Ellipsis particle-in-cell finite-element modeling code to compare different subduction scenarios.

METHODOLOGY

A particle-in-cell finite-element mantle convection code is used to simulate subduction in a mobile-lid convecting system with chemically distinct continents. The code and numerical methodology is outlined in detail in *Moresi et al.* [2003]. The approach uses a viscoplastic rheology, with a temperature-dependent viscosity defined by $\eta = Ae^{-bT}$, the Frank-Kamenetskii approximation [*Frank-Kamenetskii*, 1969], where A and b are material parameters ($1e^{24}$ and $5.233e^{-2}$, respectively), and T is temperature in °C (Table 2).

The material behaves plastically above a certain yield stress, given by $\tau_{yield} = C_0 + B_pP$ (where $C_0 = 20$ MPa is the cohesion, $B_p = 0.2$ the coefficient of friction, and P the pressure in MPa). We simulate plastic deformation with a modified viscosity (see *Moresi and Solomatov* [1998] for details). We include a damage term and strain-rate weakening [*Moresi et al.*, 2003]. Continents are modeled as chemically distinct units, with positive buoyancy and an enrichment of heat-producing elements with respect to the mantle. There is no internal stratification of heat production within the

Table 2. Model parameters.

Symbol	Description	Value
ρ_{mantle}	Density of mantle	3300 kg/m^3
ρ_{c_litho}	Density of chemically distinct lithosphere	3100 kg/m^3
ρ_{crust}	Density of crust	2900 kg/m^3
α	Thermal expansivity	3.5×10^{-5} K^{-1}
K	Thermal conductivity	4 Wm^{-1}K^{-1}
κ	Thermal diffusivity	1×10^{-6} m^{-2}s^{-1}
H$_{crust}$	Continental crustal heat production	5×10^{-10} W/kg
H$_{cont\_mantle}$	Continental lithosphere heat production	1.5×10^{-13} W/kg
H$_{mantle}$	Mantle heat production	8×10^{-12} W/kg
η	Viscosity given by Frank-Kamenetski approximation, $\eta = Ae^{-bT}$	$10^{-19} - 10^{-23}$ Pa s
A	Prefactor in Frank-Kamenetski viscosity law	10^{23} Pa s
b	Temperature coefficient in viscosity law	5.233×10^{-2} K^{-1}
C$_0$	Cohesion	20 Mpa
B$_p$	Coefficient of friction	0.2
$T_{0(solidus)}$	Peridotite solidus at atmospheric pressure	1120°C
$T_{1(solidus)}$	Slope of linearized peridotite solidus	0.114 °C/Mpa
$T_{0(liquidus)}$	Peridotite liquidus at atmospheric pressure	1720 °C
$T_{1(liquidus)}$	Slope of linearized peridotite liquidus	0.114 °C/MPa
L	Latent heat of melting	320 kJ kg^{-1}

continents. The continents are held immobile by increasing their viscosity and imposing a velocity boundary condition on the top surface.

We adopt the methodology of *de Smet et al.* [1998, 1999] to simulate melt production in our models. We define a supersolidus temperature T_{ss}, where

$$T_{ss} = \frac{T - T(P)_{solidus}}{T(P)_{liquidus} - T(P)_{solidus}} \quad (1)$$

T is the temperature of a particle tracer, P is the pressure, and $T(P)_{solidus}$ and $T(P)_{liquidus}$ are the solidus and liquidus temperatures at pressure P, for a linearized peridotite solidus and liquidus based on a number of sources [*Hirschmann*, 2000; *Takahashi*, 1990; *Takahashi and Kushiro*, 1983; *McKenzie and Bickle*, 1988]. We then calculate the melt fraction F using the third-order polynomial relationship between the supersolidus temperature T_{ss} and melt fraction F given by *McKenzie and Bickle* [1988]. We include a latent heat term in the energy equation. We calculate the regions in the DSF, using the graphite–diamond transition of *Boyd et al.* [1985].

One aspect of diamond formation we have not considered is the kinetics of the graphite–diamond transition. Available data on this transition suggests that, due to the large activation energies involved, the transition at low (<800°C) temperatures will require significant overpressure (~>3 GPa; *Sung* [2000]). There is an inherent danger in extrapolating

results from flash-heating experiments into geological contexts, in that slow nucleation and growth rates, expected in many geological contexts, will give a negative result in short-timescale experiments, and only extreme conditions that catastrophically overstep the graphite–diamond transition will result in observed diamonds. The existence of diamonds in metamorphic shear zones [*Sobolev and Shatsky*, 1990] suggests that the overpressure requirement is not insurmountable in many low-T geodynamic settings. The flat-subduction model presented here implies large shear gradients, localised shear stresses, high temperatures, and potentially catalysing fluids, and therefore the contention that diamond formation proceeds in these circumstances is not unreasonable. However, the kinetics of the transition on these timescales is, as yet, far too unconstrained to include it in any meaningful way in these models.

RESULTS

Figure 3 shows modeling results for normal subduction and illustrates the standard configuration for our simulations. The continents consist of two parts: a thin continental margin, and a cratonic region of thicker lithosphere. The thermal constraint of the continental blocks insulates the mantle beneath it and produces a thick thermal boundary (dashed line). The subducting slab is outlined in black. Regions of small-degree (0.1%) partial melting (i.e., where the mantle is supersolidus) are shown by small light-gray circles. Regions

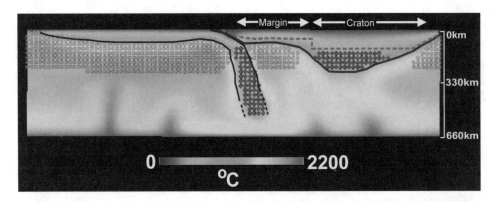

Figure 3. A conventional subduction scenario. Gray-scale denotes temperature variation, the dark gray regions corresponding to high temperatures (~2200°C) in the deep mantle and low temperatures in the shallow continental crust. Continents are modeled as two blocks of different lithospheric thickness (representing continental margin and cratonic regions). The continental lithosphere consists of the chemical boundary layer (dashed line, defining the compositional division between crust and continental lithosphere, and the mantle), and a thermal boundary layer (solid line, defining the base of the thermal lithosphere). Continents are held fixed, and a velocity condition is imposed on the oceanic lithosphere to ensure subduction at the edge of the continental margin. Upwelling plumes evolve naturally as a result of the hot thermal boundary on the lower surface. Zones of partial melting are indicated by light-gray circles. Widespread minor (<0.1%) partial melting occurs in much of the plume-fed oceanic asthenosphere. Some degree of partial melting occurs in most places where hot mantle can adiabatically rise to shallow depths. Two populations of diamonds exist in this model, one beneath the cratonic root, and one in the subducting slab (diamond stability field shown by dark-gray diamonds). As in Figures 4 and 6, each panel extends from the Earth's surface to the base of the upper mantle (~660 km).

within the DSF are shown by small dark-gray diamonds. Both the subcratonic lithosphere and subducting slab are diamondiferous in this example. However, there are no obvious mechanisms for extracting diamonds from the slab in this scenario, as (1) there is no melting beneath the diamondiferous part of the subducting slab to entrain diamonds and erupt them at the surface, and (2) the slab sinks deeper into the mantle until it eventually detaches, if subduction ceases.

Figure 4 summarises the evolution of an alternative flat-subduction scenario. Flat-subduction simulations typically assume positive buoyancy of the subducting slab—due to either the subduction of young (<20 Myr) lithosphere, which has not yet attained negative buoyancy, or the subduction of an oceanic slab with thickened crust, due to either a local hotspot or hotter (Archean) mantle temperatures [*van Hunen et al.*, 2001]. As argued above, recent adakite petrogenetic models suggest that flat subduction may have occurred in the

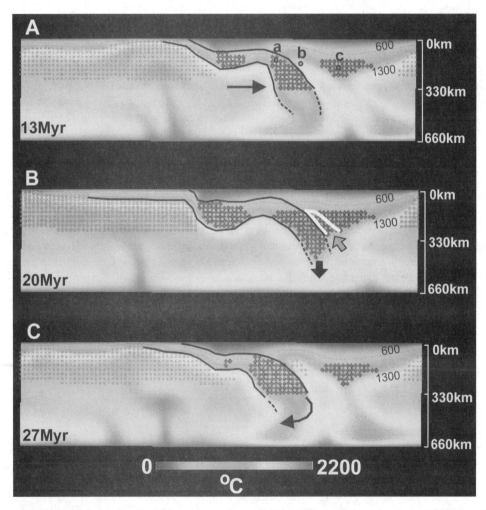

Figure 4. A flat-subduction scenario. Timescales in each panel are given in millions of years (Myr). (A) Similar to Figure 3, except that an upper mantle velocity condition was imposed to force flat subduction (see arrow). The subducting slab is outlined in black. Note there are two populations of diamonds in the subducting slab. Two isotherms are also shown, at 600°C and 1330°C (labeled). Gray-filled circles represent the position of sample points for the P-T paths shown in Figure 5. (B) As the flat-subduction system evolves, the asthenospheric wedge above the slab cools and becomes diamondiferous. Note at this time there are four distinct diamond populations: two in the subducting slab, the original cratonic diamonds, and diamonds within the mantle wedge (indicated by a solid white line). The velocity condition forcing flat subduction was turned off at this time. (C) The negative buoyancy of the subducting slab, together with its strength, results in a rolling-back of the subcontinental slab (see arrow). This results in thinning of the overlying lithosphere, intrusion of asthenospheric mantle above the slab, melting of the upwelling asthenospheric mantle, and a reheating and entrainment of the previously diamondiferous, melt-modified, mantle wedge.

southern Superior Province [*Wyman et al.*, 2002]. For our models, we simply impose a shallow mantle velocity (Figure 4A) condition to force the slab into a flat-subduction configuration (Figure 4B).

Diamond stability occurs at slightly shallower depths in the flat-subduction model but, unlike normal subduction, the subducting slab passes from the initial diamond-stability window into a zone of diamond instability and then into a second diamond-stability window as it progresses through the subduction zone. In Figure 4B, the melt-modified asthenospheric wedge (outlined in white) between the slab and the cratonic root has cooled and solidified, as predicted in the *Gutscher et al.* [2000] scenario. It has also passed into the DSF and therefore provides an analogue of the diamond-bearing peridotite of the Beni Boussara Massif prior to obduction and the resulting graphitisation of diamond [*Davies et al.*, 1993]. A notable feature of the flat-subduction scenario in Figure 4B is that the second population comprises oceanic lithosphere diamonds, cratonic diamonds, and asthenospheric diamonds. The slab becomes gravitationally unstable after it has reached the second diamond stability window and descends into the mantle (black arrow in Figure 4B).

As noted above, emplacement of diamondiferous lamprophyres occurred in an orogenic setting associated with the end of subduction. We assessed the consequences of this process on the diamond populations and subduction-related lamprophyre source regions. An effect similar to orogeny (in thermal terms) was achieved by the forced cessation of the velocity boundary condition and therefore of subduction. Detachment occurs along a suture that approximates the upper boundary of the subducted plate (shown as a black line in Figure 4B). Simultaneously, the lithosphere above the slab is thinned and the previously frozen, diamondiferous, mantle wedge is reheated to the point of generating partial melts. Warm asthenospheric mantle then flows into the space created (gray arrow in Figure 4B) and enhances melting in the old mantle wedge. For a period, diamondiferous mantle wedge material sits above ascending, melting mantle asthenosphere.

In Figure 4C, the subducting slab has rolled backwards and is fully detached from the overlying lithosphere (the frozen subduction-modified mantle wedge) and the remelting of the mantle wedge is complete. This rollback of the old slab is analogous in some ways to continental delamination models [e.g., *Zandt et al.*, 2004], and the net results, large-scale lithospheric recycling and an in-flux of warm asthenospheric material, are similar.

Figure 5 summarises the P-T paths for three sample points (a,b,c) with starting positions shown in simulation Figure 4A. The cratonic root (point c) exhibits little deformation throughout the modeling and so there is minimal variation over time in the P-T root of the cratonic lithosphere. Given that geotherms in this region are relatively stable, the lower lithosphere remains in the DSF throughout the transitions from subduction to orogeny. A sample point in the subducting slab (point a) follows a simple P-T path as it first advects laterally until it collides with the cratonic root. It then rolls back during orogeny as a result of its own negative density. There is only a gradual increase in the pressure of the sample point as the slab slowly descends into the lower mantle and the slab remains in the DSF throughout this process.

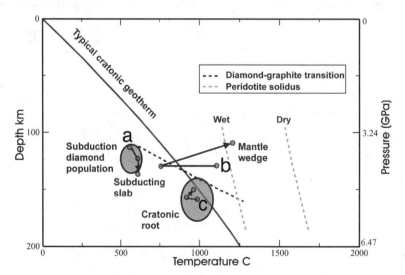

Figure 5. P-T paths for the three sample points shown in Figure 4A. Light-gray ovals denote subduction and cratonic diamond populations. The dry [*Takahashi*, 1990] and wet [*Boyd*, 1985] peridotite solidi are shown as dashed gray lines. Graphite–diamond transition, from *Boyd* [1985], is shown as a dashed black line. The average cratonic geotherm over the lifetime of a simulation is also shown.

The third sample (point b) is located in the mantle wedge above the subducting slab. This region is initially at typical upper mantle temperatures that are near, but below, the wet peridotite solidus. Subduction cools the mantle above the downgoing plate, which effectively "freezes out" the convecting asthenospheric wedge between the slab and the cratonic root. The temperature drops to within the DSF during this process. During orogeny, the lithosphere between the cratonic root and subducting slab thins and is heated as the slab rolls back. As temperatures rise, the sample point crosses the wet peridotite solidus.

Given that widespread 2.7-Ga komatiites signify large mantle plume events [*Condie*, 2004], subduction of buoyant oceanic plateau is likely to have been common at this time. As a result, flat subduction would have occurred for extended periods prior to orogeny in many Archean arcs. Accordingly, we investigated the evolution of a flat-subduction zone without imposing a subduction-terminating orogeny. This scenario is presented in Figure 6 which shows that thinning and

shearing eventually destroys the integrity of the flat portion of the subducting slab. Continued convergence also results in a thickening of oceanic lithosphere near the margin of the overriding plate. This overthickening of the thermal boundary layer near the cratonic root results in a Rayleigh–Taylor instability, which entrains much of the original slab and part of the cratonic lithosphere to produce a back-stepping of subduction. Our model does not continue to force flat subduction after the back step but it is clear that prolonged low-angle subduction would result in successive back-stepping events.

Melting in the mantle wedge above the newly active subduction zone provides a source for magmas that may migrate through the diamondiferous zone of any overlying remnants of flat slab. However, this zone of mantle melting does not correspond to a former subduction-metasomatised wedge and therefore does not represent a plausible source of lamprophyric melts. In addition, the new zone of melting (a developing mantle wedge) was not previously a site of diamond formation. According to our results, flat subduction

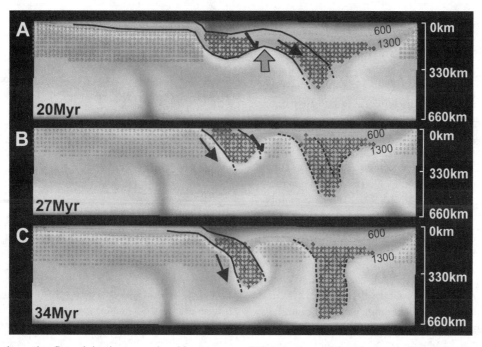

Figure 6. An alternative flat-subduction scenario without orogeny. (A) An upper mantle velocity condition is used to impose flat subduction in a fashion similar to that in Figure 4 (the nodal output of the simulation in Figure 4 was used as input for this simulation). Instead of allowing dynamic evolution of the system, however, we continue to impose a velocity on the oceanic lithosphere, resulting in ongoing convergence. This results in a necking of the subducting slab (indicated by a gray arrow), due to ongoing motion of the downwelling portion, and the development of a new shear zone (black half-arrow). (B) Continued convergence results in an overthickening of the slab near the continental margin and continued shearing of the flat slab, resulting in thermal thinning. This compromises the integrity of the slab, which becomes detached. A new subduction zone forms on the continental margin. Melting in the wedge above the new subduction zone creates a source for magmas that may sample diamonds in the adjacent lithosphere. (C) A new, convectional subduction zone forms on the continental margin. Prior thickening of the continental lithosphere (right-hand side) results in a Rayleigh–Taylor instability (cool downwelling), which entrains much of the cool thermal boundary layer.

without (or before) orogeny cannot create the conditions required to develop diamondiferous lamprophyre deposits. Nonetheless, the scenario may apply to other subduction diamond deposits, such as those of Southeast Australia, where *Griffin et al.* [2000] infer that magmas punctured the subducted slab and entrained diamonds from the top of the plate as they ascended.

DISCUSSION AND CONCLUSIONS

Results of the modeling highlight several key differences between normal- and shallow-angle subduction. Diamond stability in the slab during and after flat subduction occurs at slightly lower pressures (<4.5 GPa) than in a normal subduction. This difference might minimise difficulties of diamond entrainment in lamprophyre magmas. Flat subduction, however, moves the slab into, out of, and back into the diamond stability field so that distinct generations of slab diamonds may be present at the time of orogeny. In contrast, diamonds created at the shallowest part of the stability field during normal subduction will remain stable as the slab moves further into the mantle.

The different scenarios imply that greater diversity should be expected in slab diamond populations associated with shallow subduction. More importantly for the creation of lamprophyre-hosted deposits, shallow-angle subduction promotes diamond formation in the mantle wedge and then contributes to their preservation with a freezing effect during the subduction process (Figures 4 A,B). Formation of diamonds in the asthenospheric wedge appears negligible during normal subduction, as is illustrated in the Figure 2A and supported by our modeling results (Figure 3). This is particularly the case at the depths associated with the generation of lamprophyric magmas.

Our modeling results for orogeny after flat subduction illustrate an almost ideal scenario for lamprophyre-hosted diamond deposits. Initial reheating of the deeper parts of the frozen wedge provides a hybridised source for the lamprophyric magmas. These magmas need only ascend along structures developing in response to slab roll-back in order to rise through diamondiferous mantle. Given that zones of weakness are likely to be particularly prominent at the boundary between the retreating slab and overlying wedge, it is probable that magmas ascending up this zone will sample diamonds from both sources. The interval over which mantle wedge diamonds remain stable during orogeny may be geologically short-lived, but it corresponds closely with the period when swarms of lamprophyric magmas were emplaced into the greenstone belts in response to crustal extension [*Wyman and Kerrich*, 1989, 1993; *Ayer et al.*, 2002].

The discovery of diamonds in Archean shoshonitic lamprophyres provides novel insights and constraints into the late Archean orogen of the southern Superior Province, and into late Archean geodynamics generally. For example, recent concepts that emphasise a role for shallow-angle subduction in the generation of TTG and adakites [*Gutscher et al.*, 2000] are consistent with our models of subduction diamond genesis. Although it may be possible to construct an ad hoc model for Archean shoshonite magmas that circumvents a post-Archean association with subduction, we find it implausible that such a model could also satisfy the P-T constraints imposed by the presence of diamonds. The cold-finger effect of subduction is a distinctive feature of plate tectonics, particularly when combined with the chemical attributes of a hybridised mantle wedge.

Our results show that the previously unrecognised association of diamonds with shoshonitic magmas is a direct consequence of low-subduction angle. Although the diamondiferous lamprophyres represent a comparatively small proportion of the Archean spessartite and minette occurrences in the Superior Province, they are readily distinguished on the basis of their abundant ultramafic xenoliths and (or) their tendency toward high-Mg contents. Lamprophyre studies in Archean and younger terranes should target similar occurrences in order to characterise other ancient subduction events.

Acknowledgments. John Ayer and Ron Sage of the Ontario Geological Survey are thanked for introducing DW to the topic of Archean diamondiferous lamprophyres. The comments of two anonymous reviewers and editor Keith Benn helped to greatly clarify key points in the paper.

REFERENCES

Anderson, R.N., S.E. DeLong, and W.M. Schwarz, 1980, Dehydration, asthenospheric convection, and seismicity in subduction zones. *Journal of Geology* 88, 445-451.

Armstrong, J.P., and R.L. Barnett, 2003. The association of Zn-Chromite with diamondiferous lamprophyres and diamonds: unique compositions as a guide to the diamond potential of non-traditional diamond host rocks. 8th International Kimberlite Conference Extended Abstracts (3 pages).

Atherton, M.P., and A.A. Ghani, 2002, Slab breakoff: a model for Caledonian, Late Granite syn-collisional magmatism in the orthotectonic (metamorphic) zone of Scotland and Donegal, Ireland. *Lithos* 62, 65-78.

Ayer, J., Y. Amelin, F. Corfu, S. Kamo, J. Ketchum, K. Kwok, and N. Trowell, 2002. Evolution of the southern Abitibi greenstone belt based on U-Pb geochronology: autochthonous volcanic construction followed by plutonism, regional deformation and sedimentation. *Precambrian Research* 115, 63-95.

Ayer, J.A., and D.A. Wyman, 2003, Origin of Diamondiferous Archean Lamprophyres in the Evolution of the Michipicoten and Abitibi Greenstone Belts, 8th International Kimberlite Conference, Victoria, BC, Canada, June, 2003 Extended Abstract (5 pages).

Barron, L.M., S.R. Lishmund, G.M. Oakes, B.J. Barron, and F.L. Sutherland, 1996, Subduction model for the origin of some diamonds in the Phanerozoic of eastern New South Wales. *Australian Journal of Earth Sciences* 43, 257-267.

Boyd, F.R., J.J. Gurney, and S.H. Richardson, 1985, Evidence for a 150-200km thick Archaean lithosphere from diamond inclusion thermobarometry, *Nature* 315, 387-389

Carmichael, I.S.E., R.A. Lange, and J.F. Luhr, 1996, Quaternary minettes and associated volcanic rocks of Macota, western Mexico: a consequence of plate extension above a subduction-modified mantle wedge. *Contributions to Mineralogy and Petrology* 124, 302-338.

Condie, K.C., 2004. Supercontinents and superplume events: distinguishing signals in the geologic record. *Physics of the Earth and Planetary Interiors* 146, 319-332.

Davies, G.R., P.H. Nixon, D.G. Pearson, and M. Obata, 1993, Tectonic implications of graphitized diamonds from the Ronda peridotite massif, southern Spain. *Geology* 21, 471-474.

de Smet, J.H., A.P. van den Berg, and N.J. Vlaar, 1998, Stability and growth of continental shields in mantle convection models including recurrent melt production. *Tectonophysics* 296, 15-29.

de Smet, J.H., A.P. van der Berg, and N.J. Vlaar, 1999, The evolution of continental roots in numerical thermo-chemical mantle convection models including differentiation by partial melting. *Lithos* 48, 153-170.

Ehrenberg, S.N., 1982, Rare earth element geochemistry of garnet lherzolite and megacrystalline nodules from minette of the Colorado Plateau Province. *Earth and Planetary Science Letters* 57, 191-210.

Feldstein. S.H., and R. Lange, 1999, Pliocene potassic magmas from the Kings River region, Sierra Nevada, California: evidence for melting of a subduction-modified mantle. *Journal of Petrology* 40, 1301-1320.

Fowler, M.B., and P.J. Henney, 1996, Mixed Caledonian appinite magmas: implications for lamprophyre fractionation and high Ba-Sr granite genesis. *Contributions to Mineralogy and Petrology* 126, 199-215.

Frank-Kamenetskii, D.A., 1969, Diffusion and Heat Transfer in Chemical Kinetics, Plenum, New York, 574pp.

Griffin, W.L., S.Y. O'Reilly, and R.M. Davies, 2000, Subduction-related diamond deposits? Constraints, possibilities and new data from eastern Australia, *Reviews in Economic Geology* 11, p.291-310.

Gutscher, M.-A., R. Maury, J.P. Eissen, and E. Bourdon, 2000. Can slab melting be caused by flat subduction? *Geology* 28, 535-538.

Haggerty, S.E. 1999, A diamond trilogy: superplumes, supercontinents, and supernovae. *Science* 285, 851-860.

Hirschmann, M.M. 2000, Mantle solidus: experimental constraints and the effect of peridotite composition. Geochemistry, Geophysics, Geosystems, 1, DOI: 2000GC000070.

Martin, H. 1986, effect of steeper Archean geothermal gradient on geochemistry of subduction-zone magmas. *Geology* 14, 753-756.

Martin, H., and J.-F. Moyen, 2002, Secular changes in tonalite-trondhjemite-granodiorite composition as markers of the progressive cooling of the Earth. *Geology* 30, 319-322.

McKenzie, D.P., and M.J. Bickle, 1988, The volume and composition of melt generated by extension of the lithosphere. *Journal of Petrology* 29, 625-679.

Moresi, L., F. Dufour, and H.-B. Muhlhaus, 2003, A Lagrangian integration point finite element method for large deformation modeling of viscoelastic geomaterials. *Journal of Computational Physics* 184, 476-497.

Moresi, L., and V. Solomatov, 1998, Mantle convection with a brittle lithosphere: thoughts on the global tectonic styles of the Earth and Venus. *Geophysical Journal International* 133(3), 669-682.

Perring, C.S., N.M.S. Rock, S.D. Golding, and D.E. Roberts, 1989. Criteria for the recognition of metamorphosed lamprophyres: a case study from the Archaean of Kambalda, Western Australia. *Precambrian Research* 43, 215-237.

Platten, I.M. 1991, Zoning and layering in diorites of the Scottish Caledonian appinite suite. *Geological Journal* 265, 329-348.

Righter, K., and I.S.E. Carmichael, 1996. Phase equilibria of phlogopite lamprophyres from Western Mexico: biotite–liquid equilibria and P–T estimates for biotite-bearing igneous rocks. *Contributions to Mineralogy and Petrology* 123, 1-21.

Rock, N.M.S. 1990. Lamprophyres, Blackie and Son Ltd., Glasgow, 285pp.

Rock, N.M.S., and R.H. Hunter, 1987, Late Caledonian dyke-swarms of northern Britain: spatial and temporal intimacy between lamprophyric and granitic magmatism around the Ross of Mull pluton, Inner Hebrides. *Geologische Rundschau* 76, 805-826.

Schiller, E. 2003 Wawa diamonds. *Resource World Magazine*, June 2003, 21-25.

Smith, D., W.L. Griffin, C.G. Ryan, and S.H. Sie, 1991. Trace-element zonation in garnets from the Thumb: heating and melt infiltration below the Colorado Plateau. *Contributions to Mineralogy and Petrology* 107, 60-79.

Smithies, R.K. 2000, The Archean tonalite-trondhjemite-granodiorite (TTG) series is not an analogue of Cenozoic adakite. *Earth and Planetary Science Letters* 182, 115-125.

Sobolev, A.V., and V.S. Shatsky, 1990, Diamond inclusions in garnets from metamorphic rocks: a new environment of diamond formation. *Nature* 243, 742-746.

Stone, D., and L. Semenyna, 2004. Petrography, Chemistry and Diamond Characteristics of Heterolithic Breccia and Lamprophyre Dikes at Wawa, Ontario. Ontario Geological Survey, Open file Report 6134, 39pp.

Stott, G.M., J.A. Ayer, A.C. Wilson, and G.P.B. Grabowski, 2002. Are the Neoarchean diamond-bearing breccias in the Wawa area related to late-orogenic alkalic and "sanukitoid" intrusions? Summary of Field Work and Other Activities 2002, Ontario Geological Survey, Open File Report 6100, 9-1 to 9-10.

Takahashi, E. 1990, Speculations on the Archaean mantle: missing link between komatiite and depleted garnet peridotite. *Journal of Geophysical Research* 95, 15941-15945.

Takahashi, E., and I. Kushiro, 1983, Melting of dry peridotite at high pressure and basalt magma genesis. *American Mineralogist* 68, 859-879.

Vaillancourt, C., J.A. Ayer, S.M. Zubowski, and S.L. Kamo, 2004. Synthesis and timing of Archean geology and diamond-bearing rocks in the Michipicoten greenstone belt: Menzies and

Musquash Townships. Summary of Field Work and Other Activities 2004, Ontario Geological Survey Open File Report 6145, 6-1 to 6-9.

van Hunen J., A.P. van den Berg, and N.J. Vlaar, 2001, Latent heat effects of the major mantle phase transitions on low-angle subduction. *Earth and Planetary Science Letters* 190, 125-135.

Williams, F., 2002. Diamonds in Late Archean Calc-Alkaline Lamprophyres Ontario, Canada: Origins and Implications. Honours Thesis unpublished, University of Sydney, 82pp.

Wyman, D.A., J.A. Ayer, R.V. Conceição, and R.P. Sage, 2005, Mantle processes in an Archean orogen: evidence from 2.67 Ga diamond-bearing lamprophyres and xenoliths. Lithos (accepted).

Wyman, D.A., and R. Kerrich, 1989, Archean lamprophyre dykes of the Superior Province, Canada: distribution, petrology and geochemical characteristics. *Journal of Geophysical Research* 94, 4667-4696.

Wyman, D.A., and R. Kerrich, 1993, Archean shoshonitic lamprophyres of the Abitibi Subprovince, Canada: petrogenesis, age, and tectonic setting. *Journal of Petrology* 34, 1067-1109.

Wyman, D.A., R. Kerrich, and A. Polat, 2002. Assembly of Archean cratonic mantle lithosphere and crust: plume-arc interaction in the Abitibi-Wawa subduction-accretion complex. *Precambrian Research* 115, 37-62.

Zandt, G., H. Gilbert, T.J. Owens, M. Ducea, J. Saleeby, and C.H. Jones, 2004, Active foundering of a continental arc root beneath the southern Sierra Nevada in California. *Nature* 431, 41-46.

C. O'Neill and D. A. Wyman, School of Geosciences, Edgeworth David Building, University of Sydney, New South Wales, 2006 Australia. (dwyman@geosci.usyd.edu.au)

Reading the Geochemical Fingerprints of Archean Hot Subduction Volcanic Rocks: Evidence for Accretion and Crustal Recycling in a Mobile Tectonic Regime

A. Polat

Department of Earth Sciences, University of Windsor, Windsor, Ontario, Canada

R. Kerrich

Department of Geological Sciences, University of Saskatchewan, Saskatoon, Saskatchewan, Canada

In certain Cenozoic arcs, rare exotic volcanic rocks occur with "normal" tholeiitic to calc-alkaline lavas. These include boninites, arc picrites, adakites, high-magnesian andesites (HMA), and Nb-enriched basalts (NEB). They are associated with convergent margins characterized by high heat flow, stemming from subduction of hot young oceanic lithosphere, ridge subduction, or slab windows. We term these "hot subduction volcanic rocks". Boninites and picrites are melts of variably depleted mantle wedge induced by slab dehydration, whereas adakites reflect slab melt and mantle wedge interaction. HMA form by hybridization of adakitic melts with the mantle wedge and NEB by melting of the residue of HMA hybridization. Given the constrained geodynamic setting in which these rare volcanic rock types form, their presence in Archean greenstone belts bears on the competing models of plate tectonic subduction–accretion–recycling versus "statist" autochthonous–parautochthonous models. Boninites have recently been documented from the following areas: 3.8 Ga Isua; 3.0 Ga Mallina, Pilbara; 2.7–2.8 Ga belts in Superior Province and Dharwar Craton; and 2.8 Ga Karelian (Baltic Shield). Picrites having arc characteristics are known from the 3.8 Ga Isua, 3.0 Ivisaartoq, 2.7 Ga Wawa, and 2.55 Ga Zunhua belts, China. Structurally, many Paleoarchean to Neoarchean greenstone belts are comparable to Phanerozoic accretionary orogens, such as the Altaids and Cordilleran orogens (e.g., southern Alaska, Wrangellia). Collectively, these observations are consistent with a mobilist geodynamic regime for most Archean greenstone terranes.

1. INTRODUCTION

Archean greenstone terranes have been viewed in terms of mobilist tectonic regimes since the 1970s [*Burke et al.*, 1976; *Tarney et al.*, 1976; *Condie*, 1981]. Specifically, the

Archean Geodynamics and Environments
Geophysical Monograph Series 164
Copyright 2006 by the American Geophysical Union
10.1029/164GM13

Superior Province has been interpreted as a series of allochthonous tectonostratigraphic terranes that accreted diachronously from north to south over 2750 to 2690 Ma [*Card*, 1990; *Williams et al.*, 1991; *Stott*, 1997]. Similarly, *Şengör and Natal'in* [1996] concluded that the Yilgarn Craton grew by subduction–accretion, analogous to the Altaids of central Asia.

Recent studies suggest that some Archean greenstone belts may have developed as autochthonous sequences on older continental basement. It is explicitly recognized that some Archean volcanic successions such as the ~2.7 Ga Fortescue basalts (Pilbara Craton), Klipriviersberg basalts (Kaapvaal Craton), and Murmac Bay basalts (Rae Province) were erupted on continental crust and are autochthonous [*Goodwin*, 1991; *Arndt et al.*, 2001; *Blake et al.*, 2004]. Similarly, the Kambalda komatiite–basalt sequence in the Yilgarn Craton likely erupted on continental crust [*Lesher and Arndt*, 1995]. These continental flood basalt (CFB) sequences are plume-related and may record the initiation of continental break-up [*Blake et al.*, 2004]. In addition, three recent studies have presented autochthonous to parautochthonous models for greenstone belts in the Superior and Slave Provinces [*Ayer et al.*, 2002; *Bleeker*, 2002; *Thurston*, 2002]. However, as pointed out by *Thurston* [2002] and *Ayer et al.* [2002], the autochthonous development at the scale of greenstone assemblage does not rule out the operation of Phanerozoic-style accretionary tectonic processes in the Archean.

Geochemical data accumulated from studies of Archean greenstone belts over the last three decades are consistent with diverse types of volcanic rocks on all major Archean cratons, reflecting diverse geodynamic processes [*Dostal and Mueller*, 1997; *Polat et al.*, 1998; *Puchtel et al.*, 1998; *Polat and Hofmann*, 2003; *Manikyamba et al.*, 2004, 2005; *Sandeman et al.*, 2004; *Smithies et al.*, 2005; *Wang et al.*, 2004]. These studies have documented two major types of volcanic rock associations: (1) an oceanic plateau association composed of compositionally relatively uniform komatiites and Mg- to Fe-rich tholeiitic basalts erupted from mantle plumes, and (2) a compositionally diverse intra-oceanic island arc association. In the Abitibi belt, the former have aprons of high-Mg turbidites, whereas the latter are associated with low-Mg trench turbidites shed off bimodal arcs [*Feng and Kerrich*, 1990; *Feng et al.*, 1993]. Volcanic association in autochthonous belts is composed primarily of komatiites and tholeiitic pillow basalts [*Bleeker*, 2002; *Thurston*, 2002].

Thurston and Chivers [1990] noticed a secular variation from ~3.0 Ga platformal sequences to ~2.7 Ga Timiskaming-type sequences. Sequence is used here in the sense of tectonostratigraphic sequence. The former sequences are related to continental rifting and consist mainly of conglomerates, carbonates, and sandstones, with an overlying komatiite–tholeiite association [*Thurston and Chivers*, 1990; *Thurston*, 2002]. The latter were deposited in pull-apart basins and are composed dominantly of proximal conglomerates, sandstones, siltstones, and mudstones, intercalated with alkaline igneous rocks [*Thurston and Chivers*, 1990; *Mueller et al.*, 1996].

In addition to "normal" tholeiitic to calc-alkaline basalts, andesites, dacites, and rhyolites (BADR), there are boninites, picrites, low-Ti tholeiites (LOTI), adakites, high-magnesian andesites (HMA), and Nb-enriched basalts (NEB) in the Archean island arc association [*Kerrich et al.*, 1998; *Hollings and Kerrich*, 2000; *Polat and Kerrich*, 2001; *Boily and Dion*, 2002; *Percival et al.*, 2003; *Polat and Hofmann*, 2003; *Shchipansky et al.*, 2004; *Manikyamba et al.*, 2005; *Smithies et al.*, 2005].

Many greenstone belts are fault-bounded lithotectonic assemblages and record multiple phases of compressional, strike-slip, and extensional deformation [*Myers*, 1993; *Stott*, 1997; *de Wit*, 1998; *Krapez and Eisenlohr*, 1998; *Kusky and Polat*, 1999; *Kisters et al.*, 2003; *Daigneault et al.*, 2004; *Percival et al.*, 2004]. These features have collectively been interpreted as consistent with an Altaid-type orogen generated by successive accretion of allochthonous volcanic and sedimentary assemblages, also known as a Cordilleran type orogen [*Şengör and Natal'in*, 1996; *Polat et al.*, 1998].

In general, the geochemical characteristics of modern volcanic rocks from different tectonic settings are distinct in terms of their rare earth element (REE), large ion lithophile element (LILE), and high field strength element (HFSE) systematics [*Sun and McDonough*, 1989; *Hawkesworth et al.*, 1993; *Pearce*, 2003; and references therein]. In this study we assume that similar, uniformitarian, geochemical behaviour also prevailed in the Archean, given that certain groups of elements will behave consistently in petrogenetic processes, including source composition, residual mineralogy, partial melting, and hybridization, throughout Earth's history. However, thermal conditions in a given particular geodynamic setting were likely different in the Archean, because the Archean mantle had higher geothermal gradients than its modern counterpart [see *Fyfe*, 1978; *Bickle*, 1986]. Therefore, conditions of partial melting under the influence of a particular geothermal regime in the Archean likely differed from those in the Present [*Martin*, 1999; *Martin and Moyen*, 2002]. Accordingly, trace element systematics of ancient volcanic rocks should be used in conjunction with field characteristics to constrain their geodynamic setting.

Despite the recent significant advances in understanding Archean geology, there are still four major controversial issues regarding the origin of Archean greenstone belts: (1) geodynamic significance of boninites, picrites, adakites, HMA, and NEB [*Kerrich et al.*, 1988; *Polat et al.*, 1999, 2002; *Hollings and Kerrich*, 2000; *Polat and Kerrich*, 2001; *Smithies et al.*, 2005]; (2) petrogenetic origin of komatiites, wet or dry [*Arndt et al.*, 1998; *Grove and Parman*, 2004]; (3) origin of tonalite–trondhjemite–granodiorite suites (TTG), in terms of subducted oceanic slabs versus in situ melting of lower continental crust [*Rollinson*, 1997; *Martin and Moyen*, 2002; *Whalen et al.*, 2002; *Rapp et al.*, 2003]; and (4) autochthonous (or parautochthonous) versus allochthonous growth [*Hamilton*, 1998; *Arndt et al.*, 1998; *Ayer et al.*, 2002; *Bleeker*, 2002; *Thurston*, 2002; *Hickman*, 2004; *van Kranendonk*, 2004].

In this paper, we review the geochemical characteristics and geodynamic significance of recently recognized Archean boninites, picrites, adakites, HMA, and NEB from Paleoarchean to Neoarchean greenstone belts occurring in various Archean cratons (Tables 1–4). On the basis of the new geochemical data and structural observations, we attempt to address the first and last two issues of Archean geodynamic processes. Existing models for the origin of Archean greenstone belt are revisited.

2. CENOZOIC ISLAND ARC VOLCANIC ROCKS ASSOCIATED WITH "HOT SUBDUCTION"

Cenozoic island arc associations display great compositional diversity [*Pearce and Peate*, 1995; *Drummond et al.*, 1996; *Pearce*, 2003; *Hawkins*, 2003; and references therein]. In addition to "normal" bimodal tholeiitic to calc-alkaline BADR, and alkaline lavas, there are rare boninites, LOTI, picrites, adakites, HMA, and NEB (Table 1) [*Drummond et al.*, 1996; *Sajona et al.*, 1996; *Stern and Kilian*, 1996; *Benoit et al.*, 2002; *Pearce*, 2003]. The latter six types have been described from certain arcs, featuring hot subduction: High thermal gradients may stem from subduction of young, hot oceanic lithosphere, from ridge subduction, or slab windows (Table 1) [*Ramsay et al.*, 1984; *Yamamoto*, 1988; *Stern et al.*, 1991; *Eggins*, 1993; *Drummond et al.*, 1996; *Sajona et al.*, 1996; *Benoit et al.*, 2002; *Hawkins*, 2003; *Pearce*, 2003; *Schuth et al.*, 2004]. In addition, these rocks also occur in Paleozoic to Mesozoic suprasubduction ophiolites and accreted juvenile island arcs (Table 1).

Accordingly, in this contribution these rock types are referred to as "hot subduction" volcanic rocks. The trace element geochemical characteristics of Phanerozoic and Archean "hot subduction" volcanic rocks are compared in

Table 1. Archean Hot Subduction Volcanic Rocks and Their Phanerozoic Analogues.

Rock Type	Archean example	Phanerozoic analogue
Boninite	Abitibi [*Kerrich et al.*, 1998]; Isua [*Polat et al.*, 2002]; Frotet-Evans [*Boily and Dion*, 2002]; Karelian [*Shchipansky et al.*, 2004]; Whundo [*Smithies et al.*, 2004, 2005]; Dharwar Craton [*Manikyamba et al.*, 2005]	Troodos ophiolite [*Beccaluva and Serri*, 1988]; Papua New Guinea ophiolite [*Crawford*, 1989]; Izu-Bonin-Mariana forearc [*Pearce et al.*, 1992]; Betts Cove ophiolite [*Bédard*, 1999]; Oman ophiolite [*Ishikawa et al.*, 2002]; Ural ophiolites [*Spadea and Scarrow*, 2000]
Low-Ti tholeiite LOTI	Abitibi [*Kerrich et al.*, 1998; *Wyman*, 1999]; Karelian [*Shchipansky et al.*, 2004]	Mariana trench, Lau Basin, New Guinea [*Beccaluva and Serri*, 1988]; Tasmania [*Brown and Jenner*, 1989]
Picrite	Isua [*Polat and Hofmann*, 2003]; Wawa [*Polat and Kerrich*, 1999]; 3.0 Ga Ivisaartoq (*Polat*, unpub. data); Zunhua [*Polat et al.*, 2005]	Solomon Islands [*Ramsay et al.*, 1984; *Schuth et al.*, 2004]; Central Aleutians [*Nye and Reid*, 1986]; Japan [*Yamamoto*, 1988]; Vanuatu arc [*Eggins*, 1993]; Eastern Kamchatka [*Kamenetsky et al.*, 1995]; Lesser antilles [*Thirlwall et al.*, 1996]
Adakite	Wabigoon [*Tomlinson et al.*, 1999]; Sumozero-Kenozero [*Puchtel et al.*, 1999]; Birch-Uchi [*Hollings and Kerrich*, 2000]; Wawa [*Polat and Kerrich*, 2001]; Abitibi [*Wyman et al.*, 2002]; Yellowknife [*Cousens et al.*, 2002]; Frotet-Evans [*Boily and Dion*, 2002]; Ashuanipi complex [*Percival et al.*, 2003]; Wabigoon [*Ujike and Goodwin*, 2003]; Wutaishan [*Wang et al.*, 2004]; Vedlozero-Segozero [*Svetov et al.*, 2004]	Central America [*Defant et al.*, 1992]; Philippines [*Sajona et al.*, 1996]; Japan [*Morris*, 1995]; Southern Volcanic Zone, Andes [*Stern and Kilian*, 1996]; Baja California [*Benoit et al.*, 2002]; Northern Volcanic Zone, Andes [*Bourdon et al.*, 2002]
High-magnesian andesite (HMA)	Birch-Uchi [*Hollings and Kerrich*, 2000]; Wawa [*Polat and Kerrich*, 2001]; Abitibi [*Wyman et al.*, 2002]	Chile [*Rogers and Saunders*, 1989]; Philippines [*Sajona et al.*, 1993]; Japan [*Tatsumi and Maruyama*, 1989]; Antarctic Peninsula [*MacCarron and Smellie*, 1998]; Baja California [*Benoit et al.*, 2002]
Nb-enriched basalt (NEB)	Birch-Uchi [*Hollings and Kerrich*, 2000]; Wabigoon [*Wyman et al.*, 2000; *Ujike and Goodwin*, 2003]; Wawa [*Polat and Kerrich*, 2001]; Karelian [*Shchipansky et al.*, 2004]	Central America, Panama [*Defant et al.*, 1992]; Northern Kamchatka [*Kepezhinskas et al.*, 1996]; Philippines [*Sajona et al.*, 1996]; Baja California [*Benoit et al.*, 2002]

Table 2. Summary of the Significant Compositional and Element Ratios for the Selected Archean Boninitic Volcanic Rocks.

	3.7-3.8 Ga Isua	2.7 Ga Abitibi	2.8 Ga Frotet-Evans	2.8 Ga Karelian
SiO_2 (wt.%)	48-52	38-60	49-58	53-60
MgO	7.0-17.0	6.0-25	5.7-11.6	8.2-10.2
TiO_2	0.20-0.40	0.14-0.64	0.36-0.73	0.34-0.43
Al_2O_3	15-20	13.1-24.9	13.4-17.6	9.4-12.2
Ni (ppm)	70-610	110-760		191-452
Zr	11.0-30.0	9.7-42.0	21-71	15.8-35.2
Nb	0.15-0.80	0.27-1.90	0.90-2.3	0.67-1.0
Th	0.04-0.33	0.03-0.26	0.33-1.46	0.07-0.17
La	0.30-1.80	0.34-2.40	2.6-5.4	1.0-1.4
Y	6.0-13.7	7.4-24	9.9-22.0	9.7-14.8
Yb	0.94-1.84	1.16-3.4	1.4-2.4	1.1-1.7
$(La/Sm)_{cn}$	0.56-1.39	0.74-1.42	1.40-2.10	0.63-0.84
$(La/Yb)_{cn}$	0.16-0.66	0.21-0.58	1.0-1.9	0.57-0.75
$(Gd/Yb)_{cn}$	0.26-0.64	0.26-0.54	0.70-1.1	0.74-0.91
Zr/Y	1.4-2.4	0.9-2.2	1.6-6.1	1.8-2.4
Ti/Zr	72-116	81-118	48-110	72-151
Al_2O_3/TiO_2	45-94	26-104	18-38	26-30
$(Zr/Sm)_{pm}$	1.1-2.3	1.1-1.5	0.9-1.8	0.7-1.3
$(Nb/La)_{pm}$	0.33-0.77	0.73-1.12	0.36-0.76	0.59-0.92
$(Th/La)_{pm}$	0.57-2.10	0.59-1.02	1.0-2.2	0.37-1.2
$(Nb/Th)_{pm}$	0.26-0.63	0.83-1.20	0.22-0.75	0.77-1.75

* Data from *Polat et al.* [2002]; *Kerrich et al.* [1998]; *Boily and Dion* [2002]; and *Shchipansky et al.* [2004].

Table 3. Summary of the Significant Compositional and Element Ratios for the Selected Archean Island Arc Picritic Volcanic Rocks.

	3.7-3.8 Ga Isua	2.7 Ga Wawa	2.55-2.50 Ga Zunhua
SiO_2 (wt.%)	48-53	48-51	50.7-51.4
MgO	8.9-21.0	13.7-21.8	19.1-19.7
TiO_2	0.52-1.14	0.60-0.85	0.40-0.44
Al_2O_3	6.5-14.2	7.4-10.8	6.8-7.0
Ni (ppm)	90-830	240-1023	800-870
Zr	34-77	45-144	24.6-31.1
Nb	1.2-2.5	2.5-5.2	2.38-2.40
Th	0.25-2.2	0.80-5.7	1.61-2.18
La	1.8-6.3	6.4-53.8	11.1-11.8
Y	10.9-27.6	15.0-34.8	9.4-10.7
Yb	1.04-2.74	1.3-3.2	0.9-1.1
$(La/Sm)_{cn}$	0.66-1.56	1.40-3.05	2.6-2.7
$(La/Yb)_{cn}$	1.20-3.00	3.1-29.4	8.0-8.7
$(Gd/Yb)_{cn}$	1.29-1.87	1.57-5.10	2.0-2.2
Zr/Y	2.4-5.1	2.7-8.0	2.6-2.9
Ti/Zr	66-97	28-77	78-107
Al_2O_3/TiO_2	11.0-15.0	10.9-16.0	15-17
$(Zr/Sm)_{pm}$	0.7-1.4	0.34-0.74	0.37-0.43
$(Nb/La)_{pm}$	0.32-0.82	0.10-0.52	0.20-0.21
$(Th/La)_{pm}$	0.97-3.4	0.55-0.96	1.1-1.4
$(Nb/Th)_{pm}$	0.14-0.58	0.12-0.54	0.14-0.19

* Data from *Polat and Kerrich* [1999]; *Polat and Hofmann* [2003]; *Polat et al.* [2005].

Table 4. Summary of the Significant Compositional and Element Ratios for the 2.7 Ga Wawa Adakites, High-Magnesian Andesites (HMA), Nb-Enriched Basalts (NEB), and "Normal" Calc-Alkaline BADR*.

	Adakites	HMA	NEB	Basalts	Andesites	Dacites/Rhyolites
SiO_2 (wt.%)	64-73	55-64	50-57	46-54	57-64	68-74
MgO	1.0-3.8	3.4-7.5	3.8-5.7	3.6-6.4	1.5-2.9	0.5-1.8
TiO_2	0.28-0.80	0.47-1.83	0.63-2.24	0.6-2.1	0.52-1.72	0.16-0.41
Al_2O_3	15-18	14-17	12.3-17.1	13-17.6	14-18	13-16
Ni (ppm)	3.0-88	21-229	8.5-90.1	27-72	4.0-81	2-33
Zr	91-204	81-221	110-278	62-138	99-165	75-153
Nb	2.1-10.9	2.6-12.9	7.3-16.2	2.7-5.2	3.9-6.9	1.6-3.8
Th	0.72-9.9	1.08-6.90	1.42-3.83	0.56-3.51	1.0-4.6	0.7-3.3
La	8.2-51.5	9.6-61.6	11.9-37.2	6.7-24.6	8.3-44.6	5.0-25.0
Y	5.1-23.3	8.8-21.6	18.4-59.4	13.3-35.0	9.5-31.4	4.5-13.8
Yb	0.34-1.4	0.72-1.81	1.2-5.8	1.0-3.8	0.81-3.40	0.2-0.7
La/Sm_{cn}	2.6-4.2	2.34-3.83	1.42-2.79	1.37-2.93	1.54-3.92	2.2-5.2
La/Yb_{cn}	10.7-50.0	5.26-27.9	2.65-18.2	1.9-17.3	2.7-22.9	8.7-32.3
Gd/Yb_{cn}	1.8-5.3	1.50-4.87	1.25-4.05	1.28-3.07	1.14-3.32	1.8-3.3
Zr/Y	9.3-26.9	5.86-11.68	3.76-8.9	3.2-7.9	3.4-12.2	10.8-28.0
Ti/Zr	12.2-36.9	26-135	22.4-100	37-92	20.-90	10.7-27.9
Al_2O_3/TiO_2	21-70	9.2-30.1	6.6-24.1	6.1-25.1	8.5-30.1	34-104
$(Zr/Sm)_{pm}$	0.7-3.4	0.37-1.81	0.64-1.46	0.56-1.63	0.77-1.33	1.6-3.5
Nb/La_{pm}	0.09-0.33	0.06-0.53	0.23-0.67	0.14-0.51	0.14-0.61	0.16-0.53
Th/La_{pm}	0.81-2.96	0.51-1.77	0.54-1.43	0.59-1.1	0.63-1.0	0.89-1.88
Nb/Th_{pm}	0.08-0.40	0.09-0.69	0.39-1.03	0.13-0.86	0.15-0.67	0.11-0.51

*Modified from *Polat and Münker* [2004].

Tables 2–4 and Figures 1–5. Given the specific geodynamic regime in which the Cenozoic "hot subduction" lavas develop, the geochemical characteristics of Archean counterparts are of particular interest for understanding Archean subduction zone geodynamic and petrogenetic processes and the early evolution of the crust-mantle system [*Polat and Kerrich*, 2001; *Wyman et al.*, 2002; *Wyman*, 2003].

3. ARCHEAN "HOT SUBDUCTION" ROCKS

3.1. Boninites

The term "boninite" refers to a large variety of primary, or near-primary, magmas defined by *Crawford et al.* [1989] as $SiO_2 > 53$ wt.% and Mg# > 60. These lavas have a wide range of CaO/Al_2O_3 ratios, from <0.55 to >0.75 in conjunction with U-shaped REE patterns. The classification of boninitic rocks was initially based mainly on major elements such as CaO, K_2O, Na_2O, Al_2O_3, and SiO_2, and mineralogical and textural features [*Crawford*, 1989]. Given the fact that CaO, K_2O, Na_2O, and SiO_2 are likely to be mobile during hydrothermal alteration and metamorphism, the terms "boninitic series" or "boninite-like" have been used for metamorphosed Archean lavas that share low TiO_2, Zr, and Nb contents; high Al_2O_3/TiO_2 ratios; and U-shaped REE patterns of fresh boninites (Figures 1 and 2; Table 2)

[*Kerrich et al.*, 1998; *Polat et al.*, 2002; *Smithies et al.*, 2005], yet have lower SiO_2 contents than in the boninite definition of *Crawford et al.* [1989]. LOTI are LREE-depleted, high-Mg tholeiitic rocks associated with boninites; they have lower SiO_2 contents than boninites and have negatively fractionated REE rather than the U-shaped patterns of boninites [*Brown and Jenner*, 1989].

The distinctive U-shaped REE patterns of most boninites, with negative anomalies in Nb, Ta, P, and Ti, are considered to result from a two-stage melting process: high-degree partial melting, followed by second-stage melting of the refractory residue under hydrous conditions in convergent margins [*Sun and Nesbitt*, 1978; *Stern et al.*, 1991]. Most boninites have been reported from western Pacific arcs, but some are also known from Phanerozoic ophiolites (Table 1), such as Troodos [*Beccaluva and Serri*, 1988], Oman [*Ishikawa et al.*, 2002], and Betts Cove [*Bédard*, 1999]. LOTI may be second-stage melts, but under anhydrous conditions [*Brown and Jenner*, 1989].

In the 3800–3700 Ma Isua greenstone belt, boninitic rocks are tectonically juxtaposed to picrites having arc-like trace element patterns [*Polat and Hofmann*, 2003]. Boninitic rocks are associated with LOTI and sanukutoid intrusions in the 3010–2935 Ma Mallina Basin, Pilbara Craton, and with LOTI and tholeiitic to calc-alkaline basalts in the North Karelian and Abitibi greenstone belts (Table 1)

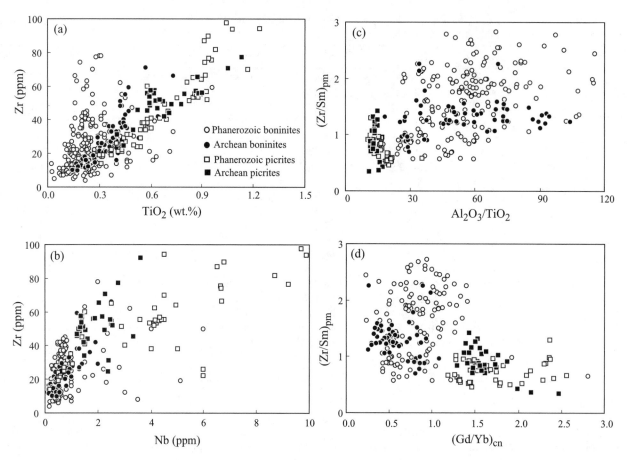

Figure 1. Variation diagrams for (a) TiO_2 versus Zr, (b) Nb versus Zr, (c) Al_2O_3/TiO_2 versus $(Zr/Sm)_{pm}$, and (d) $(Gd/Yb)_{cn}$ versus $(Zr/Sm)_{pm}$, comparing Archean boninitic and picritic volcanic rocks with those of Phanerozoic counterparts.

[*Smithies et al.*, 2002; *Shchipansky et al.*, 2004]. The 3120 Ma Whundo boninitic lavas are associated with a calc-alkaline BADR suite, whereas counterparts from the Frotet–Evans belt occur with adakites [*Boily and Dion*, 2002; *Smithies et al.*, 2005]. These are structurally overlain by tholeiitic to calc-alkaline BADR, with Th- and LREE-enriched, but Nb- and Ti-depleted primitive mantle-normalized trace element patterns. In summary, all documented Archean boninites occur with a variety of volcanic rock types that in the Phanerozoic are restricted to convergent margins.

Parman et al. [2001] argued that the geochemical characteristics of the ca. 3500 Ma Barberton komatiites are comparable to those of Phanerozoic boninites, suggesting that they formed in a subduction zone setting. The petrogenesis and geodynamic setting of these rocks, however, remain controversial [*Arndt et al.*, 1998; *Grove and Parman*, 2004]. We note that the correlated Nb/Th and Nb/U ratios of many least altered Archean komatiites, and most associated basalts, are greater than the primitive mantle values of 8 and 32, respectively, ruling out a convergent margin setting [*Kerrich and Xie*, 2002; and references therein].

3.2. Picrites

In contrast to boninitic rocks, island arc picrites display positively fractionated HREE patters $(Gd/Yb)_{cn} > 1.5$, and subchondritic to chondritic Al_2O_3/TiO_2 ratios (Figures 1 and 3; Table 3) [*Ramsay et al.*, 1984; *Yamamoto*, 1988; *Eggins*, 1993]. Picrites discussed in this paper differ from the Neoarchean plume-derived Abitibi and Enemy Lake picrites [see *Stone et al.*, 1995; *Francis et al.*, 1999] in having less fractionated HREE patterns and lower Fe_2O_3 contents. Similarly, they are distinct from plume-derived Phanerozoic picrites in terms of HFSE/REE ratios [see *Kerr et al.*, 1996].

Polat and Kerrich [1999] documented high-magnesian $(MgO = 15–22 \text{ wt.\%})$ dikes and flows associated with bimodal arc assemblages in the 2.7 Ga Wawa subprovince (Table 3). They are characterized by fractionated REE patterns and negative anomalies at Nb, Ta, P, Zr, Hf, and Ti characteristic of convergent margin magmas, but because of their primitive mantle-normalized trace element patterns and Al_2O_3/TiO_2 ratios distinct from all known Archean komatiites or boninites, *Polat and Kerrich* [1999] classified them as

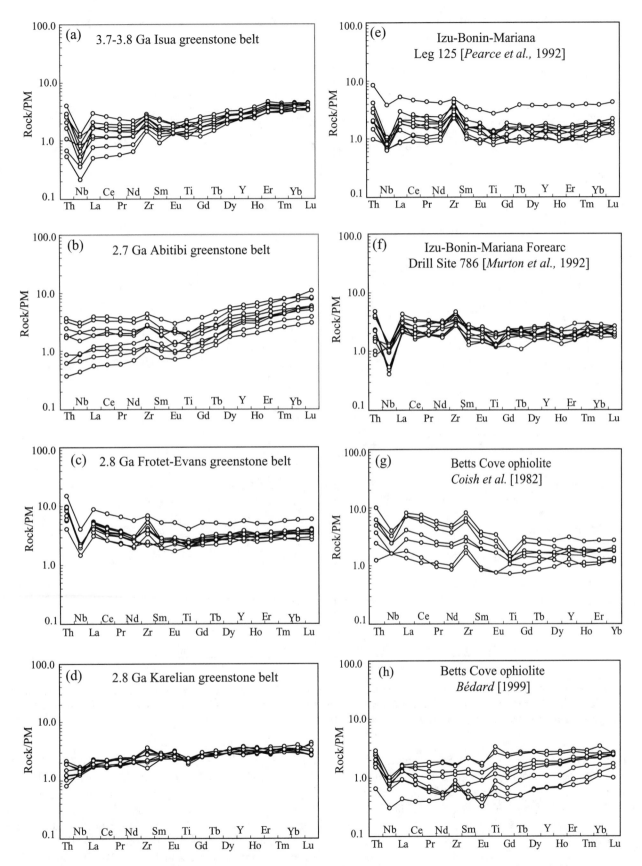

Figure 2. Primitive mantle–normalized trace element diagrams for Archean boninitic volcanic rocks and Phanerozoic counterparts. Normalization values are from *Hofmann* [1988].

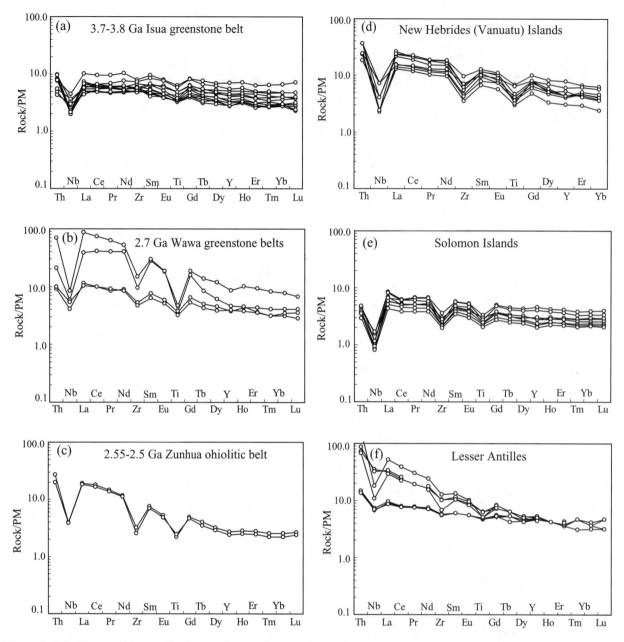

Figure 3. Primitive mantle–normalized trace element diagrams for Archean picritic volcanic rocks and Phanerozoic counterparts. Normalization values are from *Hofmann* [1988]. Data for New Hebrides Islands from *Eggins* [1993], for Solomon Islands from *Schuth et al.* [2004], and for Lesser Antilles from *Thirlwall et al.* [1996].

island arc picrites (Figure 3; Table 3). Similar types of rocks have recently been recognized in the 3800–3700 Ma Isua, 3.0 Ga Ivisaartoq, and 2.5 Ga Zunhua belts (Figure 3; Tables 1 and 3) [*Polat and Hofmann, 2003; Polat et al., 2005*].

3.3. Adakites

Adakites are Al- and Na-enriched, intermediate to felsic calc-alkaline volcanic rocks having high Sr/Y and $(La/Yb)_{cn}$

ratios (Figures 4 and 5; Table 4) [*Drummond et al., 1996; Martin, 1999; Martin et al., 2005*; and references therein]. *Rapp et al.* [1999] and *Martin et al.* [2005] distinguished two major types of adakites on the basis of SiO_2 content, such as high-SiO_2 adakites (HAS) and low-SiO_2 adakites (LSA, synonymous with HMA). In Cenozoic arcs, adakites occur as small volume flows and intrusions. Adakites are dacitic in composition but distinctive in the conjunction of extremely fractionated REE, but low Yb contents, coupled with high

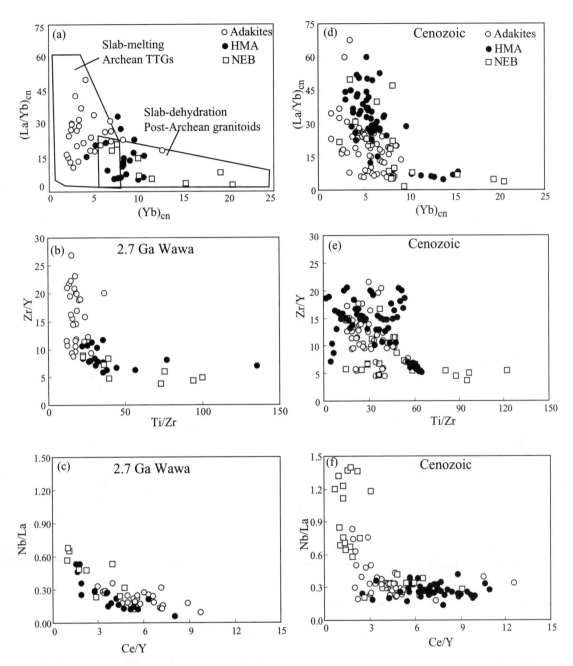

Figure 4. Selected trace element ratio variation diagrams for the 2.7 Ga Wawa adakites, HMA, and NEB and their Cenozoic counterparts. Modified from *Polat and Münker* [2004]. Chondrite normalization values are from *Sun and McDonough* [1989].

Mg#, Cr, and Ni compared to "normal" dacites. They are generally interpreted as slab melts that hybridized with peridotite when traversing the mantle wedge [*Drummond et al.*, 1996; *Martin*, 1999; *Rapp et al.*, 1999; *Smithies*, 2000; *Condie*, 2005; *Martin et al.*, 2005]. They have been described from the Western Aleutians, the Philippines, Central America, Baja California, Ecuador, and Chile (Table 1).

The presence of adakites within "normal" calc-alkaline BADR arc sequences has recently been reported from several Neoarchean greenstone belts of the Superior Province (Table 1) [*Hollings and Kerrich*, 2000; *Polat and Kerrich*, 2001; *Boily and Dion*, 2002; *Percival et al.*, 2003]. Similarly, adakites, without HMA or NEB, have been documented from bimodal arc sequences of the Yellowknife greenstone belt of

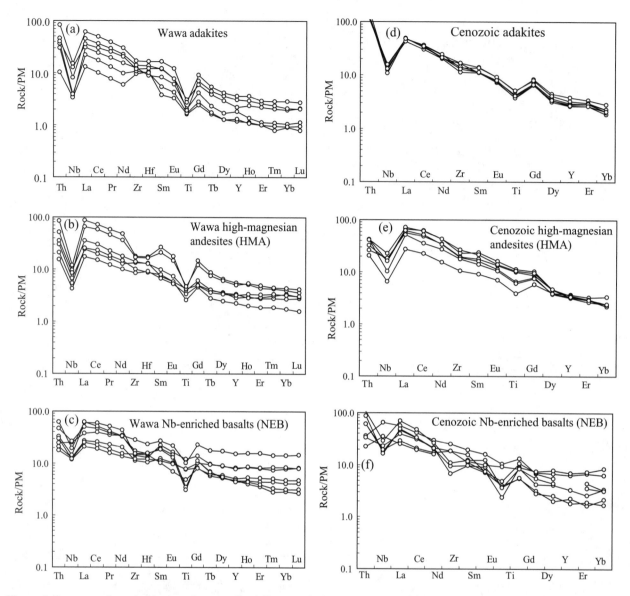

Figure 5. Representative primitive mantle–normalized diagrams for the Wawa adakites, HMA, and NEB and their Cenozoic counterparts. Data for Wawa volcanic rocks. Normalization values are from *Hofmann* [1988]. Modified from *Polat and Münker* [2004].

the Slave Province, and the Vedlozero–Segozero greenstone belt of the Baltic Shield [*Cousens et al.*, 2002; *Svetov et al.*, 2004]. Primitive basalts in these belts plot with the low Ce-Yb trend of modern intra-oceanic arcs [cf. *Hawkesworth et al.*, 1993].

3.4. HMA and NEB Association

HMA are primitive, intermediate calc-alkaline volcanic rocks, having high Mg# (>50) and high Cr and Ni contents [*Kelemen*, 1995; *Sajona et al.*, 1996]. NEB are mafic volcanic arc rocks possessing high Nb concentrations (>7 ppm

compared to <2 ppm for "normal" arc basalts) and higher $(Nb/Th)_{pm}$ and $(Nb/La)_{pm}$ ratios than "normal" arc basalts (Figures 4 and 5; Table 4) [*Defant et al.*, 1992; *Kepezhinskas et al.*, 1996].

The association of HMA and NEB with adakites in Cenozoic arcs is generally explained as follows: (1) Adakites are slab melts that are mildly contaminated or hybridized with peridotitic mantle wedge; (2) HMA are assumed to be generated by more extensive hybridization of an adakite melt, or melting of a previously adakite metasomatized mantle peridotite; and (3) NEB are melts of the residue of HMA hybridization as it is dragged to greater depth by induced convection in the mantle wedge [*Sajona et al.*, 1996; *Martin et al.*, 2005; and references therein]. HMA have been documented from the Aleutians, the Philippines, Central America, and Baja California (Table 1). The presence of the HMA and NEB association, with or without adakites, within "normal" tholeiitic to calc-alkaline BADR arc sequences has recently been reported from several Neoarchean greenstone belts of the Superior Province (Table 1).

In summary, formation of modern boninites, LOTI, picrites, adakites, HMA, and NEB typically results from dynamic interaction between subducting young plates, or spreading centres, and island arcs in the Circum-Pacific convergent margins (Tables 1 and 5). Archean equivalents of these rocks likely formed under comparable geodynamic settings. These rocks are rare in both the Archean (<5%) and Phanerozoic geological records; however, their geochemical signatures provide significant insights into petrological, thermal, and geodynamic processes [*Drummond et al.*, 1996; *Pearce*, 2003; *Schuth et al.*, 2004]. We suggest that that vigorous mantle convection and subduction of younger, hotter oceanic slabs, including ridge subduction and slab windows, may have provided optimized conditions for the production of boninites, picrites, adakites, HMA, and NEB in many Archean subduction zones [see *Hargraves*, 1986; *Martin*, 1999]. Currently it is estimated that 10% of subduction is flat [*Gutscher et al.*, 2000; *Guivel et al.*, 2003; and references therein]; this proportion is likely to have been higher in the Archean period given its higher heat flow [*Bickle*, 1986].

4. GEOCHEMICAL EVIDENCE FOR CRUSTAL RECYCLING IN ARCHEAN SUBDUCTION ZONES

Subduction zones are the major sites of modern crustal recycling, where pelagic and continentally derived sediments, and tectonically eroded forearc, are carried into the mantle by subducting oceanic slabs [*von Huene and Scholl*, 1991; *Johnson and Plank*, 1999]. Changes in the mechanisms of this process over Earth's history, however, are still poorly understood. Isotopic data from magmatic rocks and orogenic sediments are consistent with the episodic addition of juvenile material to the continental crust throughout Earth's history [*Stevenson and Patchett*, 1990; *Samson and Patchett*, 1991; *McCulloch and Bennett*, 1994]. The net mass of the continental crust, however, may not have changed significantly since the end of the Archean [*Rudnick*, 1995; *Taylor and McLennan*, 1995], suggesting that juvenile crustal accretion may be balanced by a combination of recycling of the continental crust back into the mantle and a foundering of cumulates in arcs [*DePaolo*, 1980; *Jacobsen*, 1988].

Recycling of arc and continental crust at convergent plate margins occurs by sediment subduction or subduction

erosion [*von Huene and Scholl*, 1991]. According to *von Huene and Scholl* [1991], most subducted continental material is recycled to the mantle at a rate near 1.6 km^3 yr^{-1}, equivalent to the additions of juvenile material from the mantle to the continental crust. Geochemical models of modern subduction zone magmatism include contributions to the continental crust through subducted sediments, subducted oceanic slabs, and subarc mantle wedges [*Pearce and Peate*, 1995; *Johnson and Plank*, 1999]. Crustal material from subducted sediments and oceanic crust are transferred into arcs either by hydrous metasomatic fluxes causing wedge melting or by slab melting [*Pearce and Peate*, 1995; *Johnson and Plank*, 1999].

The Paleoarchean Isua boninitic pillow basalts have an average initial ε_{Nd} value of ~+2.2, indicating a depleted mantle source [*Frei et al.*, 2004]. The LREE-enriched, U-shaped patterns of these rocks, however, imply that their mantle source must have been fertilized by LREE-enriched fluids or melts at ~3.8 Ga (Figure 2). High Os concentrations (1.9–3.4 ppb) in these flows are also consistent with a long-term depleted mantle source [*Frei et al.*, 2004]. A large positive average initial γ_{Os} value (+4.4) suggests a strongly radiogenic Os isotopic signature, consistent with the enrichment of the source region during, or prior to, melting. As a result, the Os isotopic signatures require recycling of a crustal component (probably oceanic in nature), with time-integrated superchondritic Re/Os ratios, into the mantle source. In addition, relatively high initial $^{207}Pb/^{204}Pb$ values (10–13) in the Isua volcanic rocks are consistent with the recycling of a several hundred-million-year-old oceanic crust into the mantle source of boninitic pillow basalts at about 3.8 Ga [*Kamber et al.*, 2003].

Recycling of continental material into the mantle to generate enriched domains has been proposed for several terranes in the late Archean Superior Province on the basis of Nd and Pb isotope data. Mantle residence time is estimated at 50–200 Ma [*Shirey and Hanson*, 1986; *Bédard and Ludden*, 1997; *Henry et al.*, 1998, 2000; *Ayer and Dostal*, 2000; *Tomlinson et al.*, 2002].

However, the processes and sites of Nd and Pb enrichment have not previously been well constrained. Integrated Nd-Hf isotopic data from the Wawa subprovince place important constraints on the site and origin of the enrichment [*Polat and Münker*, 2004]. The initial ε_{Hf} values of the ~2.7 Ga adakites (+3.5 to +5.2), HMA (+2.6 to +5.1), NEB (+4.4 to +6.6), and "normal" tholeiitic to calc-alkaline arc basalts (+5.3 to +6.4) are consistent with long-term–depleted mantle sources (Figure 6) [*Polat and Münker*, 2004]. The initial ε_{Nd} values in HMA (+0.4 to +2.0), NEB (+1.4 to +2.4), and "normal" arc tholeiitic to calc-alkaline basalts (+1.6 to +2.9) overlap with, but extend to lower values than, those of the slab-derived adakites (+2.3 to +2.8). The lower initial ε_{Nd}

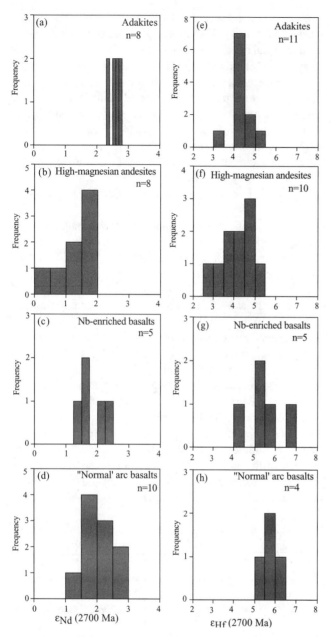

Figure 6. Histograms for the initial ϵ_{Hf} and ϵ_{Nd} values of the Wawa adakites, MA, NEB, and "normal" arc basalts]. Modified from *Polat and Münker* [2004].

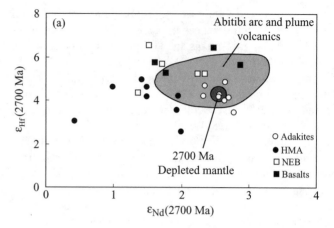

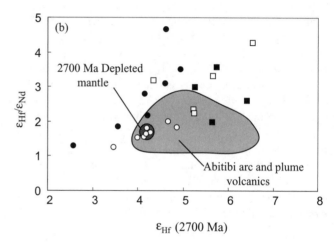

Figure 7. Variation diagrams of ϵ_{Nd} (2700 Ma) versus ϵ_{Hf} (2700 Ma) and of ϵ_{Hf} (2700 Ma) versus $\epsilon_{Hf}/\epsilon_{Nd}$ for the Wawa arc lavas. Modified from *Polat and Münker* [2004].

values in the mantle wedge–derived suites, particularly in HMA, are attributed to recycling of an Nd-enriched component with a lower ϵ_{Nd} value to the mantle wedge. As a group, the slab-derived adakites plot closest to the 2.7 Ga depleted mantle value in ϵ_{Nd} versus ϵ_{Hf} space, additionally suggesting that the Nd-enriched component in the mantle wedge did not originate from the 2.7 Ga slab-derived melts (Figure 7). Accordingly, the enriched component had been added to the

mantle wedge in variable proportions by recycling of older continental material into the upper mantle beneath the southern Superior Province, through which the Wawa arc migrated [*Polat and Kerrich*, 2002; *Polat and Münker*, 2004].

5. ARCHEAN GEODYNAMICS: AUTOCHTHONOUS VERSUS ALLOCHTHONOUS GROWTH

Models for Archean greenstone belts can be divided into two major groups. Autochthonous, or parautochthonous, models propose the development of greenstone belts on stationary, or near-stationary, continental or oceanic basement, with or without vertical displacement [*Grachev and Fedorovsky*, 1981; *Hamilton*, 1988; *Bickle et al.*, 1994; *Ayer et al.*, 2002; *Bleeker*, 2002; *Thurston*, 2002]. Alternatively, the proponents of allochthonous models argue for the formation of greenstone belts by large-scale horizontal translation in a mobile tectonic regime [*Burke et al.*, 1976; *Card*, 1990; *Desrochers et al.*, 1993; *Corfu and Stott*, 1998; *de Wit*, 1998; *Polat et al.*, 1998; *Wyman et al.*, 2002; *Percival et al.*, 2004].

According to the autochthonous models, greenstone belts were deposited unconformably as either plume-related continental rift or continental platform sequences [*Thurston*, 2002; and references therein]. Several compelling lines of evidence in the geological record are consistent with the existence of continental rift sequences in some greenstone belts of major Archean cratons (for a review, see *Hartlaub et al.* [2004]). *Ayer et al.* [2002] proposed an autochthonous origin for volcanic and sedimentary sequences in the southern Abitibi greenstone belt. They argued that plume-derived komatiite–tholeiitic basalt association, calc-alkaline BADR association, siliciclastic turbidites, and banded iron formations were deposited over 50 Ma in a near-stationary intra-oceanic island arc setting, involving plume–arc interaction and arc rifting. Both *Thurston* [2002] and *Ayer et al.* [2002] pointed out that autochthonous development at the scale of greenstone belt assemblages does not negate an accretionary origin of the Superior Province.

It is clear that some greenstone belt assemblages originated as autochthonous sequences. However, for the following reasons, we suggest that purely autochthonous models, which exclude Phanerozoic-style horizontal tectonics, have several incommensurabilities with first-order features of many Archean greenstone–granitoid terranes.

5.1. Continental Rifts

Several continental rift sequences have been recognized in Archean terranes (Table 5), including the Superior Province [*Thurston and Chivers*, 1990; *Tomlinson et al.*, 1999], Rae Province [*Hartlaub et al.*, 2004], Slave Province [*Bleeker*, 2002], Baltic Shield [*Thurston and Kozhevnikov*, 2000], Yilgarn Craton [*Swager*, 1997], Dharwar Craton [*Srivastava et al.*, 2004], and North China Craton [*Kusky and Li*, 2003]. Phanerozoic continental rifts form in various tectonic settings that result from either plate–plate or plate–plume interaction [*Şengör and Burke*, 1978; *Vine and Kearey*, 1996]. They are the products of lithospheric extension and represent continental break-up [*Storey*, 1995]. Lithospheric extension in rift systems is transformed to the neighbouring convergent plate boundaries by strike–slip (transcurrent) faults [*Şengör et al.*, 1985; *Sylvester*, 1988].

Archean rifts likely formed by similar geological processes and represent passive margins or failed rift arms (Aulacogens). *Davis and Maidens* [2003] showed that a sequence of sedimentary, volcanic, and volcaniclastic rocks was deposited in a syn-collisional orogen-parallel rift basin in the Yilgarn Craton. Neoarchean passive margin-type continental margin sedimentary sequences in the Central Orogenic Belt, North China Craton, are structurally overlain by ophiolitic mélanges and bimodal volcanic sequences, signifying an arc-continental margin collision during closure of an ocean basin [*Li et al.*, 2002; *Kusky and Li*, 2003].

5.2. Unconformities

Thurston [2002] suggested that there is a secular variation in the nature of Archean greenstone belts of the Superior Province, ranging from ~3.0 Ga unconformable platformal sedimentary rocks with overlying rift volcanics to 2.7 Ga unconformable sandstones and conglomerates deposited in pull-apart basins. On the basis of these observations, he argued that many greenstone belts in the Superior Province developed autochthonously or parautochthonously. Similarly, given the unconformable relationship between continental basement and overlying supracrustal rocks, *Bleeker* [2002] argued for autochthonous development of greenstone belts of the Slave Province. A number of unconformities have been documented in other Archean cratons [see *Goodwin*, 1991; *Blewett*, 2002].

Unconformities are an important feature of Phanerozoic orogenic belts and modern subduction–accretion complexes, indicating uplift and crustal extension, resulting from convergent tectonic processes [*Dewey and Bird*, 1971; *Dewey*, 1977; *Şengör*, 1990; *Şengör and Natal'in*, 1996; and references therein]. For example, many Mesozoic Tethyan (e.g., Oman, Kizildag, Pozanti-Karsanti ophiolites), Paleozoic Appalachian (e.g., Bay of Islands ophiolite), and Proterozoic ophiolites of Turkey—which are the most allochthonous rootless fragments in orogenic belts—were emplaced onto passive continental margin sequences [*Dewey*, 1977; *Polat et al.*, 1996; *Dilek et al.*, 1999; *Yiğitbaş et al.*, 2004; and references therein]. These continental margin sequences were deposited unconformably on older continental basement [*Dewey*, 1977]. Both Tethyan (e.g., Oman, Kizildag ophiolites) and Appalachian (e.g., Bay of Island ophiolite) ophiolites are in turn overlain unconformably by younger marine deposits [see *Dewey*, 1977; *Searle and Stevens*, 1984; *Robertson and Searle*, 1990].

Cordilleran ophiolites (e.g., Coast Range and Josephine ophiolites) in western North America are also overlain unconformably by volcano-sedimentary sequences [*Pessagno et al.*, 2000]. Similarly, there are many unconformities in the Altaid (~600–300 Ma) accretionary orogenic system, which includes numerous dismembered ophiolitic fragments [*Şengör and Natal'in*, 2004]. These unconformities and overlying strata are interpreted to reflect uplifting and/or subsidence events as a consequence of horizontal tectonic movements (such as continental break-up) and accretion of ophiolites, island arcs, oceanic plateaus, and sea mounts [*Pessagno et al.*, 2000; *Şengör and Natal'in*, 2004]. Archean counterparts probably have the same geodynamic significance.

Table 5. Interpreted Geodynamic Processes in Archean Greenstone–Granitoid Terranes and Their Phanerozoic Analogues.

Geodynamic Process	Archean examples	Phanerozoic analogues
Plume-subduction interaction	Superior Province [*Dostal and Mueller*, 1997; *Wyman*, 1999; *Hollings et al.*, 1999; *Ayer et al.*, 2002]	Tonga forearc [*Danyushevsky et al.*, 1995]; Mexican arc [*Márquez et al.*, 1999]
Plume-lithosphere interaction	Superior Province [*Tomlinson et al.*, 1999]; Pilbara Craton [*Arndt et al.*, 2001; *Blake et al.*, 2004]	Gondwana-Land [*Storey*, 1995; and references therein]; Siberian traps, Iceland-North Atlantic, Basin and Range [*Coffin and Eldholm*, 1994; *Dalziel et al.*, 2000]
Subduction accretion (Arc-arc, arc-continent, continent-continent; arc-oceanic plateau accretion)	Superior Province [*Card*, 1990; *Skulski and Percival*, 1996; *Mueller et al.*, 1996; *Polat et al.*, 1998; *Calvert and Ludden*, 1999; *Daigneault et al.*, 2004; *Percival et al.*, 2004]; Baltic Shield [*Puchtel et al.*, 1998; *Shchipansky et al.*, 2004]; Greenland [*Friend and Nutman*, 2005]; Kaapvaal Craton [*de Ronde and de Wit*, 1994]; Zimbabwe Craton [*Kusky and Kidd*, 1992; *Jelsma and Dirks*, 2002]; Yilgarn Craton [*Şengör and Natal'in*, 1996]	Himalayas, Alps, Altaids, Alaska, Appalachians, North American Cordillera [*Samson and Patchett*, 1991; *Coney and Jones*, 1985; *Kearey and Vine*, 1996; *Şengör and Natal'in*, 1996]; Japan [*Taira et al.*, 1992]; Southwestern Pacific [*Hamilton*, 1988; *Mann and Taira*, 2004]; Caledonides [*Andersen et al.*, 1990]; Andes and Central America [*Vergara*, 1995]
Thrusting and imbrication	Abitibi [*Lacroix and Sawyer*, 1995]; Wawa [*Gill*, 1992; *Corfu and Stott*, 1998]; Rainy Lake [*Poulsen et al.*, 1980]; Wabigoon [*Davis et al.*, 1988]; Hearne Domain [*MacLachlan et al.*, 2005], Bulawayo, Zimbabwe [*Jelsma and Dirks*, 2002]; North Karelian [*Shchipansky et al.*, 2004]; Warrawoona [*Kloppenburg et al.*, 2001]; Barberton [*de Ronde and de Wit*, 1994]; Yilgarn Craton [*Davis and Maidens*, 2003]	Canadian Cordillera [*Umhoefer and Schiarizza*, 1996]; American Cordillera [*Wakabayashi*, 1992]; Taurides [*Polat et al.*, 1996]; Oman [*Robertson and Searle*, 1990]; Altaids [*Şengör and Natal'in*, 1996]; Appalachians [*van Staal*, 1994]
Strike-slip faulting	Superior Province [*Mueller et al.*, 1996]; Dharwar Craton [*Chadwick et al.*, 2002]; Pilbara Craton [*Zegers et al.*, 1998]; Yilgarn Craton [*Swager*, 1997]	Alps and Himalayas [*Şengör et al.*, 1985; *Şengör*, 1990]; Altaids [*Şengör and Natal'in*, 1996]; Canadian Cordillera [*Umhoefer and Schiarizza*, 1996]; Andes [*Dewey and Lamb*, 1992]
Ridge subduction	Superior Province [*Kusky and Polat*, 1999; *Percival et al.*, 2003]; North China Craton [*Polat et al.*, 2005]	Solomon Islands [*Ramsay et al.*, 1986]; Alaska [*Kusky et al.*, 1997; *Sisson et al.*, 2003]; Chile [*Forsythe and Nelson*, 1985] South Sandwich forearc [*Barker et al.*, 1995]; Antarctic Peninsula [*Lawver et al.*, 1995]; Aleutians [*Marshak and Karig*, 1977]; Western North America [*Sisson et al.*, 2003]
Syn-subduction arc-forearc-backarc extension	Churchill Province [*Sandeman et al.*, 2004]; Karelian [*Shchipansky et al.*, 2004]; Gawler Craton [*Swain et al.*, 2005]	South Western Pacific [*Pearce et al.*, 1992; *Clift*, 1995; *Hawkins*, 2003; *Pearce*, 2003]; Alaska [*Kusky et al.*, 1997]; Antarctic Peninsula [*Hole et al.*, 1995]; Tethyan ophiolites [*Flower*, 2003]; Peru [*Petford and Atherton*, 1995]
Continental rifting	Superior Province [*Tomlinson et al.*, 1999]; Yilgarn Craton [*Swager*, 1997]; Pilbara Craton [*Blake et al.*, 2004]; Slave Province [*Bleeker*, 2002]; Rae Province [*Hartlaub et al.*, 2004]; Dharwar Craton [*Srivastava et al.*, 2004]	East Africa, South and North Atlantic, Basin and Range, Rio Grande [*Kearey and Vine*, 1996; and references therein]
Orogenic collapse	Superior Province [*Calvert et al.*, 2004]; Greenland [*Hanmer and Greene*, 2002]; Barberton [*Kisters et al.*, 2003]; Yilgarn Craton [*Davis and Maidens*, 2003]	Alps, Himalayas, Appalachians [*Dewey*, 1988]; North American Cordillera [*Vanderhaeghe and Teyssier*, 2001]; Basin and Range [*Dilek and Moores*, 1999]; Urals [*Knapp et al.*, 1998]; Caledonides [*Andersen et al.*, 1994]

5.3. Vertical and Horizontal Tectonics

Structural analyses indicate that both vertical and horizontal tectonic processes affected some Archean greenstone–granitoid terranes, resulting in a dome and basin structure [*Bouhallier et al.*, 1995; *Chardon et al.*, 1998; *Kloppenburg et al.*, 2001; *Blewett*, 2002; *Bédard et al.*, 2003; *Hickman*, 2004; *van Kranendonk*, 2004; and references therein]. Horizontal tectonics is dominantly recorded by structures in greenstone belts and subprovince boundaries, whereas vertical tectonics tends to be preserved in diapiric granitoid gneisses [*Chardon et al.*, 1998; *Bleeker*, 2002; *Bédard et al.*, 2003; *van Kranendonk*, 2004].

Dome and basin structures are typically overprinted by an anastomosing network of strike–slip faults in the Dharwar and Pilbara Cratons [*Chardon et al.*, 1998; *Hickman*, 2004]. Detailed structural analyses suggest that diapirism or vertical tectonics alone cannot account for the structural characteristics of the Pilbara Craton [*Blewett*, 2002]. Rather, these structures reflect the compressional to transpressional tectonics [*Chadwick et al.*, 2000; *Kloppenburg et al.*, 2001; *Blewett*, 2002].

Horizontal displacement, characterized by thrusting, imbrication, and overturning of folded strata, has been documented in a number of greenstone belts in several Archean cratons. Examples of thrust tectonics have been documented from greenstone belts in the Superior Province, Zimbabwe Craton, Pilbara Craton, Yilgarn Craton, Kaapvaal Craton, and Baltic Shield (Table 5). Formation of these structures was followed by regional- to continental-scale strike–slip faulting. These structures have been attributed to Phanerozoic-style accretionary and collision processes (see *de Wit* [1998]; *Kusky and Polat* [1999]).

Autochthonous models often invoke vertically acting forces to explain vertical structures characteristic of greenstone belts [*Grachev and Fedorovsky*, 1981; *Hamilton*, 1998; *Bleeker*, 2002; *Bédard et al.*, 2003]. However, asymmetrical and rootless folds, tectonic transposition, nappe and thrust structures, and vertical foliations, as shown in Figure 8, in greenstone belts are inconsistent with the dominance of vertically acting forces in the Archean. Rather they reflect horizontal compression to transpression [*Poulsen et al.*, 1980; *Gill*, 1992; *Lacroix and Sawyer*, 1995; *Swager*, 1997; *Polat and Kerrich*, 1999; *Kloppenburg et al.*, 2001; *Blewett*, 2002].

5.4. Strike–Slip Faults

In Phanerozoic orogenic belts, strike–slip faults with lengths of 100s to 1000 km accommodated large displacements and played a major role in accretion of island arcs, oceanic plateaus, and continental blocks to convergent continental margins [*Dewey*, 1977; *Sylvester*, 1988; *Şengör*, 1990; *Kearey and Vine*, 1996; *Umhoefer and Schiarriza*, 1996]. In addition, Phanerozoic orogen-parallel strike–slip faults dismembered the island arcs and fragmented continents, as manifested in escape tectonics [*Şengör et al.*, 1985; *Şengör*, 1990].

On the basis of the above observations, *Sleep* [1992], *Mueller et al.* [1996], and *Polat et al.* [1998] argued that strike–slip faulting on a similar scale in the Superior Province played an important role in lateral crustal accretion and magma emplacement. Similarly, anastomosing networks of regional-scale strike–slip faults played an important role in fragmentation of accreted crust in the Pilbara, Yilgarn, and Dharwar Cratons [*Smith et al.*, 1998; *Zegers et al.*, 1998; *Chadwick et al.*, 2000; *Kloppenburg et al.*, 2001; *Hickman*, 2004]. Pure autochthonous growth and vertical tectonic models cannot account for the presence of regional-scale strike–slip faults in Archean greenstone–granitoid terranes.

5.5. Oceanic Fragments

In Phanerozoic orogenic belts, occurrences of tectonically juxtaposed oceanic fragments (e.g., mid-ocean ridge basalts, seamounts, oceanic plateaus) with arc-derived trench turbidites are considered as evidence for ancient subduction zones (see *Kimura et al.* [1993]). A modern example is the Ontong Java ocean plateau that has been "captured", and jammed, in the New Ireland and Solomon island arcs [*Mann and Taira*, 2004]. Similarly, tectonically juxtaposed fragments of oceanic plateau komatiite–basalt with arc-derived turbidites in various Archean greenstone belts can be interpreted as Archean subduction–accretion complexes (Table 5) [*Hoffman*, 1991; *Desrochers et al.*, 1993; *Kimura et al.*, 1993; *Puchtel et al.*, 1998; *Polat and Kerrich*, 1999; *Wyman et al.*, 1999; *Shchipansky et al.*, 2004].

5.6. Evolving Systems

Autochthonous models require a static development of greenstone belts. However, poly-phase deformation, structural transposition, multiple intrusion, and greenschist to amphibolite facies metamorphism are the rule rather than the exception for most Archean greenstone belts (Figure 8) [*de Ronde and de Wit*, 1994; *Polat and Kerrich*, 1999; *Kloppenburg et al.*, 2001; *Blewett*, 2002; *Daigneault et al.*, 2004; and references therein]. These characteristics are indicative of evolving dynamic systems, in which the geometrical relationships between the tectonic units change through time as the orogens evolve [*Dewey*, 1977; *Şengör and Natal'in*, 1996, 2004; *Percival et al.*, 2003, 2004]. For example, Feng and coworkers interpreted the Pontiac subprovince of the Superior Province to have initiated as passive margin fed by a ~3 Ga terrane catchment [*Feng and Kerrich*, 1991; *Feng et al.*, 1993]. Bimodal arcs and plumes erupted

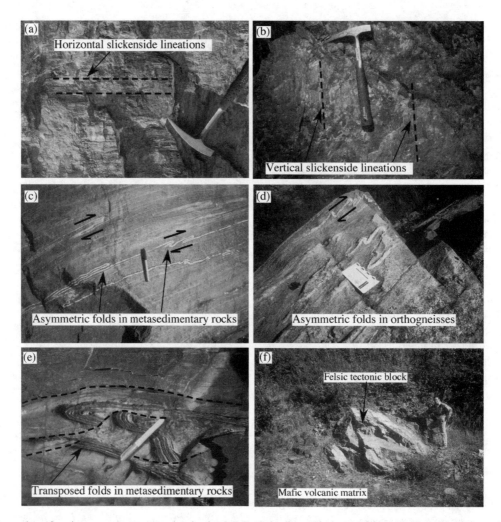

Figure 8. Examples of various structures occurring in the 2.7 Ga Schreiber–Hemlo greenstone belt: (a, b) horizontal and vertical silken side lineations, signifying a transpressional tectonic regime; (c, d) asymmetric folds consistent with a SSE tectonic transportation during accretion processes; and (e, f) transposed folds and a tectonic melange structure, signifying disturbance of stratigraphy and mixing during tectonic accretion (see *Polat and Kerrich* [1999]). All these structures are comparable to those found in Phanerozoic orogenic belts.

in the Abitibi ocean ~2.7 Ga and the Pontiac passive margin evolved into an active margin, closing the Abitibi ocean and resulting in thrusting of the Abitibi belt over the Pontiac. Both terranes were then intruded by TTG and at ~2600 Ma by intra-crustal peraluminous granites [*Feng and Kerrich*, 1991, 1992; *Feng et al.*, 1993].

5.7. Thermal Evolution

Models of heat flow predict higher heat production and mantle temperatures in the Archean than at present [*Fyfe*, 1978; *Bickle*, 1986]. This observation is difficult to reconcile with a stationary, or near-stationary, crust. A hotter Earth would have had vigorous mantle convection, leading to

longer spreading centers, numerous slowly moving smaller plates, and more ridge subduction [see *Hargraves*, 1986].

5.8. Arc Compositions

Geochemical characteristics of the extensive Archean arc tholeiitic to calc-alkaline basalts resemble those of Phanerozoic counterparts [*Hollings and Kerrich*, 2004]. Arcs do not generate magmas without plate translation [*Hamilton*, 1988]. Similarly, Archean boninitic, picritic, adakitic, HMA, and NEB compositions are consistent with hot subduction, or ridge–trench interaction, which requires migration of spreading centers to subduction zones [*Kusky and Polat*, 1999; *Percival et al.*, 2003; *Sisson et al.*, 2003; *Polat et al.*, 2005].

5.9. Enriched Domains and Crustal Recycling

Isotopic signatures of some granitoids and volcanic rocks are consistent with recycling of older continental material into the mantle source regions [*Shirey and Hanson*, 1986; *Bédard and Ludden*, 1997; *Ayer and Dostal*, 2000; *Green et al.*, 2000; *Henry et al.*, 2000; *Polat and Kerrich*, 2002; *Polat and Münker*, 2004]. Contamination of these rocks by continental crust during magma ascent, rather than contamination of their source regions by subducted crustal material, can be ruled out on the basis of several observations (see *Polat and Kerrich* [2002]).

Archean komatiites erupted from mantle plumes possess Nb/Th ratios greater than the primitive mantle value of 8 [cf. *Sun and McDonough*, 1989]. This feature is consistent with recycling of ocean crust through a convergent margin into the mantle source of the plume. Coexisting basalts have Nb/Th < 8 < Nb/Th, in keeping with recycled oceanic and continental crust [*Kerrich and Xie*, 2002].

Formation of podiform chromitites requires the generation of hydrous melts in the upper mantle [*Edwards et al.*, 2000]. The presence of podiform chromitites in the 3.0 Ga Ukrainian and 2.5 Ga Chinese harzburgites and dunites [*Li et al.*, 2002; *Gornostayev et al.*, 2004] suggests that water was recycled into Archean upper mantle. Subduction of the hydrated oceanic crust is the most effective mechanism known for transportation of water into the upper mantle [*Pearce and Peate*, 1995; *Edwards et al.*, 2000; *Poli and Schmidt*, 2002]. Recycling of continental material and water into the upper mantle can be easily achieved by subduction zone processes.

5.10. Tonalites

Tonalite–trondhjemite–granodiorite (TTG) batholiths were emplaced pre- to post-volcanic stages in greenstone belts [*Corfu and Stone*, 1998; *Chown et al.*, 2002; *Whalen et al.*, 2002; *Condie*, 2005; *Martin et al.*, 2005; and references therein]. Many of these batholiths are distinct from the volumetrically small-volume syn-arc adakites. According to *Whalen et al.* [2002], TTG of the Wabigoon subprovince are melts of overthickened amphibolite crust, as in an Andean-type arc. However, compositionally, some TTG in the Abitibi subprovince are characterized by Mg#, Cr, Co, and Ni contents too high to be melts of lower crustal basalts. Rather, they are thought to be melts of basaltic slab crust that hybridize with peridotite when traversing the mantle wedge [*Feng and Kerrich*, 1992; *Smithies*, 2000].

Martin and Moyen [2002] have shown that from 4.0 to 2.5 Ga, TTG show a trend to greater Mg# and Ni, consistent with steepening subduction as the Earth cooled, such that slab melts advect through progressively thicker mantle wedge. Concurrently, there is a secular increase in Sr content from a decreasing role of plagioclase as the depth of melting increased. In contrast, mantle plumes show a trend of decreasing Mg# and Ni through time [*Campbell and Griffiths*, 1992], such that models involving melting and delamination of overthickened crustal amphibolite (oceanic crust) are unlikely [*Zegers*, 2004; and references therein].

8. CONCLUSIONS

The geochemical characteristics of magmatism in modern subduction zones are controlled mainly by (1) the age of the subducting plate (young versus old); (2) the nature of the overriding plate (continental versus oceanic); (3) the presence or absence of sediment subduction; (4) the angle of subduction (shallow versus steep); (5) the previous history of the sub-arc mantle wedge; (6) the rate of subduction (slow versus fast); and (7) the presence or absence of a backarc basin opening [*Hawkins*, 2003; *Pearce*, 2003]. It is likely that the composition of Archean arc magmas was controlled by similar geological processes. Accordingly, we propose that Archean calc-alkaline BADR, which compositionally resemble Phanerozoic counterparts such as the Izu–Bonin–Mariana arc, also represent extensive intra-oceanic arc systems. Associated boninites, LOTI, picrites, adakites, HMA, and NEB likely reflect higher geothermal gradients of Archean subduction zones, as in Phanerozoic counterparts of "hot subduction" volcanics. Collectively, their geochemical signatures are consistent with diverse petrological processes operating in Archean subduction zones, including slab melting, slab dehydration, mantle wedge metasomatism, and crustal recycling.

If the geochemical characteristics of Archean "hot subduction" volcanic rocks have the same geodynamic significance as Phanerozoic counterparts, then they were the products of higher geothermal gradients. Higher heat flow may have stemmed from subduction of young, hot oceanic lithosphere, including ridge subduction, slab flattening, or slab windows. Higher mantle temperatures in the Archean imply much greater oceanic ridge length and slowly moving plates, with commensurately more common subduction of oceanic ridges [*Hargraves*, 1986]. Therefore, subduction of young oceanic crust in the Archean may have been an effective mechanism for the generation of boninites, picrites, adakites, HMA, and NEB.

Models for the geodynamic evolution Archean greenstone belts can be divided into two major groups: (1) autochthonous models, which assume in situ development of greenstone belts on a stationary, or near-stationary, continental or oceanic basement; and (2) allochthonous models, which argue for large-scale horizontal movements in a mobile regime. We suggest that the autochthonous models are inconsistent with combined structural, geochemical, and geochronological data produced for many greenstone belts over the last three decades, with the minor exceptions of plume lavas erupted onto rifted supercontinents and subsequently preserved as part of later supercontinent cycles.

In summary, a number of lines of evidence in the literature support allochthonous growth of many Archean greenstone belts at convergent plate margins, including the presence of (1) fold-thrust structures; (2) large-scale orogen-parallel strike–slip faults; (3) tectonically juxtaposed rocks formed at different times and in different tectonic settings; and (4) subduction zone geochemical signatures in juvenile volcanic and intrusive rocks (Table 5).

Acknowledgments. K. C. Condie, H. Martin, and J. Percival are acknowledged for their incisive and comprehensive critiques that have greatly improved the paper. This is a contribution of NSERC discovery grant 250926-03 to A.P. R.K. acknowledges an NSERC discovery grant and the George McLeod endowment to the Department of Geological Sciences, University of Saskatchewan.

REFERENCES

Andersen, T.B., P.T. Osmundsen, and L. Jolivet, Deep crustal fabrics and a model for extensional collapse of southwest Norwegian Caledonides, *J. Struct. Geol.*, 16, 1191-1203, 1994.

Andersen, T.B., K.P. Skjerlie, and H. Furnes, The Sunnfjord mélange, evidence of Silurian ophiolite accretion in the West Norwegian Caledonides, *J. Geol. Soc. London*, 147, 59-68, 1990.

Arndt, N., G. Bruzak, and T. Reischmann, The oldest continental and oceanic plateaus-geochemistry of basalts and komatiites of the Pilbara Craton, Australia, in *Mantle Plumes: Their Identification Through Time, Spec. Paper 352*, edited by K.L Buchan, and R.E Ernst, pp. 1-30, Geol. Soc. Am., Boulder, CO, 2001.

Arndt, N.T., C. Ginibre, C. Chauvel, F. Albarède, M. Cheadle, C. Herzberg, G. Jenner, and Y. Lahaye, Were komatiites wet? *Geology*, 26, 739-742, 1998.

Ayer, J., Y. Amelin, F. Corfu, S. Kamo, J. Ketchum, K. Kwok, and N. Trowell, Evolution of the southern Abitibi greenstone belt based on U-Pb geochronology: autochthonous volcanic construction followed by plutonism, regional deformation and sedimentation, *Precam. Res.*, 115, 63-95, 2002.

Ayer, J.A., and J. Dostal, Nd and Pb isotopes from the Lake of the Woods greenstone belt, northwestern Ontario: implications for mantle evolution and the formation of crust in the southern Superior Province, *Can. J. Earth Sci.*, 37, 1677-1689, 2000.

Ayres, L.D., and F. Corfu, Stacking of disparate volcanic and sedimentary units by thrusting in the Archean Favourable Lake greenstone belt, central Canada, *Precam. Res.*, 50, 221-238, 1991.

Barker, P.F., Tectonic framework of the East Scotia Sea, In: *Backarc Basins: Tectonics and Magmatism*, edited by B. Taylor, pp. 281–314, Plenum Press, New York, 1995.

Beccaluva, L., and G. Serri, Boninitic and low-Ti subduction-related lavas from intraoceanic arc-backarc systems and low-Ti ophiolites: a reappraisal of their petrogenesis and original tectonic setting, *Tectonophysics*, 146, 291-315, 1988.

Bédard, J.H., Petrogenesis of boninites from the Betts Cove Ophiolite, Newfoundland, Canada: Identification of subducted source components, *J. Petrol.*, 40, 1853-1889, 1999.

Bédard, J.H., P. Brouillette, L. Madore, and A. Berclaz, Archean cratonization and deformation in the northern Superior Province, Canada: an evaluation of plate tectonics versus vertical tectonic models, *Precam. Res.*, 127, 61-87, 2003.

Bédard, L.P., and J.N. Ludden, Nd-isotope evolution of Archean plutonic rocks in southeastern Superior Province, *Can. J. Earth Sci.*, 34, 286-298, 1997.

Benoit, M., A. Aguillon-Robles, T. Calmus, R.C. Maury, H. Bellon, J. Cotton, J. Bourgois, and F. Michaud, Geochemical diversity of late Miocene volcanism in southern Baja California, Mexico: Implication of mantle and crustal sources during the opening of an asthenospheric window, *J. Geol.*, 110, 627-648, 2002.

Bickle, M.J., Implications of melting for stabilisation of lithosphere and heat loss in the Archean, *Earth Planet. Sci. Lett.*, 80, 314-324, 1986.

Bickle, M.J., E.G. Nisbet, and A. Martin, Archean greenstone belts are not oceanic crust, *J. Geol.*, 102, 121-138, 1994.

Blake, T.S., R. Buick, S.J.A. Brown, and M.E., Barley, Geochronology of a late Archean flood basalt province in the Pilbara Craton, Australia: constraints on basin evolution, volcanic and sedimentary accumulation, and continental drift rates, *Precam. Res.*, 133, 143-173, 2004.

Bleeker, W., Archean tectonics: a review, with illustrations from the Slave craton, in *The Early Earth: Physical, Chemical and Biological Development, Spec. Publ.*, 199, edited by C.M.R. Fowler, C.J. Ebinger, and C.J. Hawkesworth, pp. 151-181, Geol. Soc. London, London, 2002.

Blewett, R.S., Archean tectonic processes: a case for horizontal shortening in the north Pilbara granite-greenstone terrane, Western Australia, *Precam. Res.*, 113, 87-120, 2002.

Boily, M., and C. Dion, Geochemistry of boninite-type volcanic rocks in the Frotet-Evans greenstone belt, Opatica subprovince, Quebec: Implications for the evolution of Archean greenstone belt, *Precam. Res.*, 115, 349-371, 2002.

Bouhallier, H., D. Chardon, and P. Choukroune, Strain patterns in Archean dome-and-basin structures: the Dharwar Craton [Karnataka, South India], *Earth Planet. Sci. Lett.*, 135, 57-75, 1995.

Bourdon, E., J.P. Eissen, M. Monzier, C. Robin, H. Martin, J. Cotton, and M.L. Hall, Adakite-like lavas from Antisana volcano [Ecuador]: Evidence for slab melt metasomatism beneath the Andean Northern Volcanic Zone, *J. Petrol.*, 43, 199-217, 2002.

Brown, A.V., and G.A. Jenner, Geological setting, petrology and chemistry of Cambrian boninite and low-Ti tholeiitic lavas in western Tasmania, in *Boninites and Related Rocks*, edited by A.J. Crawford, pp. 233-263, Unwin Hyman, London, 1989.

Burke, K., J.F. Dewey, and W.S.F. Kidd, Dominance of horizontal movements, arc and microcontinental collisions during the later permobile regime, in *The early History of the Earth*, edited by B.F. Windley, pp. 113-129, Wiley, London, 1976.

Calvert, A.J., A.R. Cruden, and A. Hynes, Seismic evidence for preservation of the Archean Uchi granite-greenstone belt by crustal scale extension. *Tectonophysics*, 388, 135-143, 2004.

Calvert, A.J., and J.N. Ludden, Archean continental assembly in the southeastern Superior province of Canada, *Tectonics*, 18, 412-429, 1999.

Campbell, I.H., and R.W. Griffiths, the changing nature of mantle hotspots through time: implications for the chemical evolution of the mantle, *J. Geol.*, 92, 497-523, 1992.

Card, K.D., A review of the Superior Province of the Canadian Shield, a product of Archean accretion, *Precam. Res.*, 48, 99-156, 1990.

Chadwick, B., V.N. Vasudev, and G.V. Hegde, The Dharwar craton, southern India, interpreted as the result of late Archean oblique convergence, *Precam. Res.*, 99, 91-111, 2000.

Chardon, D., P. Choukroune, and M. Jayananda, Sinking of the Dharwar Basin [South India]: implications for Archean tectonics, *Precam. Res.*, 91, 15-39, 1998.

Chown, E.H., R. Harrap, and A. Moukhsil, The role of granitic intrusions in the evolution of the Abitibi belt, Canada, *Precam. Res.*, 115, 291-310, 2002.

Clift, P.D., Volcanism and sedimentation in a rifting island arc terrain: an example from Tonga, SW Pacific, in *Volcanism Associated with Extension at Consuming Plate Margins, Spec. Publ.*, 81, edited by Smellie, pp. 29-51, *Geol. Soc. London*, London, 1995.

Coffin, M.F., and O. Eldholm, Large igneous provinces: crustal structure, dimensions, and external consequences, *Rev. Geophysics*, 32, 1-36, 1994.

Condie, K.C., *Archean Greenstone Belts*, 435 pp, Elsevier, Amsterdam, 1981.

Condie, K.C., *Plate Tectonics and Crustal Evolution*, 282 pp, Butterworth-Heinmann, Bath, 1997.

Condie, K.C., Episodic continental growth and supercontinents: a mantle avalanche connection? *Earth and Planet, Sci. Lett.*, 163, 97-108, 1998.

Condie, K.C., TTGs and adakites: are they both slab melts?, *Lithos*, 80, 33-44, 2005.

Coney, P.J., and Jones, D.L., Accretion tectonics and crustal structure in Alaska, *Tectonophysics*, 119, 265-283, 1985.

Corfu, F., and L.D. Ayres, Unscrambling the stratigraphy of an Archean greenstone belt: a U-Pb geochronological study of the Favourable Lake belt, northwestern Ontario, Canada, *Precam. Res.*, 50, 201-220, 1991.

Corfu, F., and D. Stone, Age structure and orogenic significance of the Berens River composite batholiths, western Superior Province, *Can. J. Earth Sci.*, 35, 1089-1109, 1998.

Corfu, F., and G.M. Stott, Shebandowan greenstone belt, western Superior Province: U-Pb ages, tectonic implications, and correlations, *Geol. Soc. Am. Bull.*, 110, 1467-1484, 1998.

Cousens, B., K. Facey, and H. Falck, Geochemistry of the late Archean Banting Group, Yellowknife greenstone belt, Slave Province, Canada: simultaneous melting of the upper mantle and juvenile mafic crust, *Can. J. Earth Sci.*, 39, 1635-1656, 2002.

Crawford, A.J., T.J. Fallon, and D.H. Green, Classification, petrogenesis and tectonic setting of boninites, in *Boninites and Related Rocks*, edited by A.J. Crawford, pp. 1-49, Unwin Hyman, London, 1989.

Daigneault, R., W.U. Mueller, and E.H. Chown, Abitibi greenstone belt plate tectonics: diachronous history of arc development, accretion and collision, in *The Precambrian Earth: Tempos and Events*, edited by P.G. Eriksson, W. Altermann, D.R. Nelson, W.U. Mueller, and O. Catuneanu, pp. 88-103, Elsevier, Amsterdam, 2004.

Dalziel, I.W.D., L.A. Lawver, and J.B. Murphy, Plumes, orogenesis, and supercontinental fragmentation, *Earth Planet. Sci. Lett.*, 178, 1-11, 2000.

Davis, B.K., and E. Maidens, Archean orogen parallel extension: evidence from the northern Eastern Goldfields Province, Yilgarn Craton, *Precam. Res.*, 127, 229-248, 2003.

Davis, D.W., R.H. Sutcliffe, and N.F. Trowell, Geochronological constraints on the tectonic evolution of a late Archean greenstone belt, Wabigoon subprovince, northwest Ontario, Canada, *Precam. Res.*, 39, 171-191, 1988.

Defant, M.J., T.E. Jackson, M.S. Drummond, J.Z. De Boher, H. Bellon, M.D. Feigenson, R.C. Maury, and R.H. Stewart, The geochemistry of young volcanism throughout western Panama and southeastern Costa Rica: an overview, *J. Geol. Soc. London*, 149, 569-579, 1992.

DePaolo, D.J., Crustal growth and mantle evolution; inferences from models of element transport and Nd and Sr isotopes, *Geochim. Cosmochim. Acta*, 44, 1185-1196, 1980.

de Ronde, E.J., and M.J. de Wit, Tectonic history of the Barberton greenstone belt, South Africa: 490 million years of Archean crustal evolution, *Tectonics*, 13, 983-1005, 1994.

Desrochers, J.P., C. Hubert, J.N. Ludden, and P. Pilote, Accretion of Archean oceanic plateau fragments in the Abitibi greenstone belt, Canada, *Geology*, 21, 451-454.

Dewey, J.F., Suture zone complexities: A review, *Tectonophysics*, 40, 53-67, 1977.

Dewey, J.F., Extensional collapse of orogens, *Tectonics*, 7, 1123-1139, 1988.

Dewey, J.F., and J. Bird, Origin and emplacement of the ophiolite suite: Appalachian ophiolites in Newfoundland, *J. Geophys. Res.*, 76, 3179-3206, 1971.

Dewey, J.F., M.R. Hempton, W.S.F. Kidd, F. Saroglu, and A.M.C. Şengör, Shortening of continental lithosphere: the Neotectonics of Eastern Anatolia – a young collision zone, *Geol. Soc. London Spec. Publication*, 19, 3-36, 1986.

Dewey, J.F., and S.H. Lamb, Active tectonics of the Andes, *Tectonophysics*, 205, 79-95, 1992.

de Wit, M.J., On Archean granites, greenstones, cratons, and tectonics: does the evidence demand a verdict? *Precam. Res.*, 91, 181-226, 1998.

Dilek, Y., and E.M. Moores, A Tibetan model for early Tertiary western United States, *J. Geol. Soc. London*, 156, 929-941, 1999.

Dilek, Y., P. Thy, B. Hacker, and S. Grundvig, Structure and petrology of Tauride ophiolites and mafic dyke intrusions (Turkey): Implications for Neothetyan ocean, *Geol. Soc. Am. Bull.*, 111, 1192-1216, 1999.

Dostal, J., and W.U. Mueller, Komatiite flooding of a rifted Archean rhyolite arc complex: geochemical signature and tectonic significance of Stoughton-Roquemaure Group, Abitibi greenstone belt, Canada, *J. Geol.*, 105, 545-563, 1997.

Drummond, M.S., M.J. Defant, and P.K. Kepezhinskas, Petrogenesis of slab-derived trondhjemite-tonalite-dacite/adakite magmas, *Trans. Royal Soc. Edinburgh Earth Sci.*, 87, 205-215, 1996.

Edwards, S.J., J.A. Pearce, and J. Freeman, New insights concerning the influence of water during formation of podiform chromitite, in *Ophiolites and Oceanic Crust: New Insights from Field Studies and Ocean Drilling Program, Spec. Paper*, 349, edited by

Y. Dilek, E. M. Moores, D. Elthon, D., and A. Nicolas, pp. 139-147, Geol. Soc. Am., Boulder, CO, 2000.

Eggins, S.M., Origin and differentiation of picritic arc magmas, Ambae [Aoba], Vanuatu, *Contrib. Mineral. Petrol.*, 114, 79-100, 1993.

Feng, R., and R. Kerrich, Geochemistry of fine-grained clastic sediments in the Archean greenstone belt, Canada: implications for provenance and tectonic setting, *Geochim. Cosmochim. Acta*, 54, 1061-1081, 1990.

Feng, R., and R. Kerrich, Single zircon age constraints on the tectonic juxtaposition of the Archean Abitibi greenstone belt and Pontiac Subprovince, *Geochim. Cosmochim. Acta*, 55, 3437-3441, 1991.

Feng, R., and R. Kerrich, Geochemical evolution of granitoids from the Archean Abitibi Southern Volcanic Zone and the Pontiac subprovince, Superior Province, Canada: Implications for tectonic history and source regions, *Chem. Geol.*, 98, 23-70, 1992.

Feng, R., R. Kerrich, and R. Mass, Geochemical, oxygen, and neodymium isotope compositions of metasediments from the Abitibi greenstone belt and Pontiac subprovince - Canada evidence for ancient crust and Archean terrane juxtaposition, *Geochim. Cosmochim. Acta*, 57, 641-658, 1993.

Flower, M.F.J., Ophiolites, historical contingency, and the Wilson cycle, in *Ophiolite Concept and the Evolution of Geological Thought, Spec. Paper,* 373, edited by Y. Dilek, and S. Newcomb, pp. 111-135, Geol. Soc. Am., Boulder, CO, 2003.

Forsythe, R.E., and E. Nelson, Geological manifestations of ridge collisions: Evidence from the Golfo de Penas-Taito Basin, southern Chile, *Tectonics*, 4, 477-495, 1985.

Francis, D., J. Ludden, R. Johnstone, and W. Davis, Picrite evidence for more Fe in Archean reservoirs, *Earth Planet. Sci. Lett.*, 167, 197-213, 1999.

Frei, R., A. Polat, and A. Meibom, The Hadean upper mantle conundrum: Evidence for source depletion and enrichment from Sm-Nd, Re-Os, and Pb isotopic compositions in 3.71 Gy boninite-like metabasalts from the Isua Supracrustal Belt, Greenland, *Geochim. Cosmochim. Acta*, 68, 1645-1660, 2004.

Friend, C.R.L., and A.P. Nutman, New pieces to the Archean jigsaw puzzle in the Nuuk region, southern West Greenland: steps in transforming a simple insight into a complex regional tectonothermal model, *J. Geol. Soc. London*, 162, 147-162, 2005.

Fyfe, W.S., The evolution of the Earth's crust: modern plate tectonics to ancient hot spot tectonics?, *Chem. Geol.*, 23, 89-114, 1978.

Gill, G.E., Structure and kinematics of a major tectonic contact, Michipicoten greenstone belt, Ontario, *Can. J. Earth Sci.*, 29, 2118-2132, 1992.

Goodwin, M.A., Precambrian Geology, The Dynamic Evolution of the Continental Crust, 666 pp, Academic press, London, 1991.

Gornostayev, S.S., R.J. Walker, E.J. Hanski, and S.E. Popovchenko, Evidence for the emplacement of ca. 3.0 Ga mantle derived mafic-ultramafic bodies in the Ukrainian Shield, *Precam. Res.*, 132, 349-362, 2004.

Grachev, A.F., and V.S. Fedorovsky, On the nature of greenstone belts in the Precambrian, *Tectonophysics*, 73, 195-212, 1981.

Green, M.G., P.J. Sylvester, and R. Buick, Growth and recycling of Early Archean continental crust: geochemical evidence from the Coonterunah and Warrawoona Groups, Pilbara Craton, Australia, *Tectonophysics*, 322, 69-88, 2000.

Grove, T.L., and S.W. Parman, Thermal evolution of the Earth as recorded by komatiites, *Earth Planet. Sci. Lett.*, 219, 173-187, 2004.

Guivel, C., Y. Lagabrielle, J. Bourgois, H. Martin, N. Arnaud, S. Fourcade, J. Cotton, and R.C. Maury, Very shallow melting of oceanic crust during spreading ridge subduction: origin of near-trench Quaternary volcanism at the Chile Triple Junction, *J. Geophys. Res. Solid Earth*, 108, Article No. 2345, 2003.

Gutscher, M.A., R. Maury, J.P. Eissen, and E. Bourdon, Can slab melting be caused by flat subduction?, *Geology*, 28, 535-538, 2000.

Hamilton, W.B., Plate tectonics and island arcs, *Geol. Soc. Am. Bull.*, 100, 1503-1527, 1988.

Hamilton, W.B., Archean tectonics and magmatism, *Int. Geol. Rev.*, 40, 1-39, 1998.

Hanmer, S., and D.C. Greene, A modern structural regime in the Paleoarchean [~3.64 Ga]: Isua greenstone belt, southern West Greenland, *Tectonophysics*, 346, 201-222, 2002.

Hargraves, R.B., Faster spreading or greater ridge length in the Archean?, *Geology*, 14, 750-752, 1986.

Hartlaub, R.P., L.M. Heaman, K.E. Aston, and T. Chacko, The Archean Murmac Bay Group: evidence for a giant Archean Rift in the Rae Province, Canada, *Precam. Res.*, 131, 345-372, 2004.

Hawkesworth, C.J., K. Gallagher, J.M. Hergt, and F. McDermott, Mantle and slab contributions in arc magmas, *Annual Rev. Earth Planet. Sci*, 21, 175-204, 1993.

Hawkins, J.W., Geology of supra-subduction zones-Implications for the origin of ophiolites, in *Ophiolite Concept and the Evolution of Geological Thought, Spec. Paper,* 373, edited by Y. Dilek, and S. Newcomb, pp. 227-268, Geol. Soc. Am. Boulder, CO, 2003.

Henry, P., R.K. Stevenson, and C. Gariépy, Late Archean mantle composition and crustal growth in the Western Superior Province of Canada: Neodymium and lead isotopic evidence from the Wawa, Quetico, and Wabigoon subprovinces, *Geochim. Cosmochim. Acta*, 62, 143-157, 1998.

Henry, P., R.S. Stevenson, Y. Larbi, and C. Gariépy, Nd isotopic evidence for early to late Archean [3.4-2.7 Ga] crustal growth in the Western Superior Province [Ontario, Canada], *Tectonophysics*, 322, 135-151, 2000.

Hickman, A.H., Two contrasting granite-greenstone terranes in the Pilbara Craton, Australia: evidence for vertical and horizontal tectonic regimes prior to 2900 Ma, *Precam. Res.*, 131, 153-172, 2004.

Hoffman, P.F., On accretion of granite-greenstone terranes, in *Nuna Conference on Greenstone Gold and Crustal Evolution*, edited F. Robert, P. A. Sheahan, and S. B. Green, pp. 32-45, *Geol. Assoc. Canada*, Val d'or, 1991.

Hofmann, A.W., Chemical differentiation of the Earth: the relationships between mantle, continental crust, and oceanic crust, *Earth Planet. Sci. Lett.*, 90, 297-314, 1988.

Hole, M.J., A.D. Saunders, G. Rogers, and M.A. Sykes, The relationship between alkaline magmatism, lithospheric extension and slab window formation along continental destructive plate margin, in *Volcanism Associated with Extension at Consuming Plate Margins, Spec. Publ.,* 81, edited by J.L. Smellie, pp. 265-285, Geol. Soc. London, London, 1995.

Hollings, P., Archean Nb-enriched basalts in the northern Superior Province, *Lithos*, 64, 1-14, 2002.

Hollings, P., and R. Kerrich, An Archean arc basalt-Nb-enriched basalt-adakite association: The 2.7 Ga Confederation assemblage of the Birch-Uchi greenstone belt, Superior Province, *Contrib. Mineral. Petrol.*, 139, 208-226, 2000.

Hollings, P., and R. Kerrich, Geochemical systematics of tholeiites from the 2.86 Ga Pickle Crow assemblage, northwestern Ontario arc basalts with positive and negative Nb-Hf anomalies, *Precam. Res.*, 134, 1-20, 2004.

Hollings, P., D. Wyman, and R. Kerrich, Komatiite-basalt-rhyolite volcanic association in Northern Superior Province greenstone belts: significance of plume-arc interaction in the generation the proto continental Superior Province, *Lithos*, 46, 137-161, 1999.

Ishikawa, T., K. Nagaishi, and S. Umino, Boninitic volcanism in the Oman ophiolite: implications for thermal condition during transition from spreading ridge to arc, *Geology*, 30, 899-902, 2002.

Jacobsen, S.B., Isotopic constraints on crustal growth and recycling, *Earth Planet. Sci. Lett.*, 90, 315-327, 1988.

Jelsma, H.A., and P.H.G.M. Dirks, Neoarchean tectonic evolution of the Zimbabwe Craton, in *The Early Earth: Physical, Chemical and Biological Development, Spec. Publ.*, 199, edited by C.M.R. Fowler, C.J. Ebinger, and C.J. Hawkesworth, pp. 183-211, *Geol. Soc. London*, London, 2002.

Johnson, M.C., and T. Plank, Dehydration and melting experiments constrain the fate of subducted sediments, *Geochem. Geophys. Geosystems* 1, Paper No. 1999GC000014, 1999.

Kamber, B.S., K.D. Collerson, S. Moorbath, and M.J. Whitehouse, Inheritance of early Archean Pb-isotope variability from long-lived Hadean protocrust, *Contrib. Mineral. Petrol.*, 145, 25-46, 2003.

Kamenetsky, V.S., A.V. Sobolev, J.-L. Joron, and M.P. Semet, Petrology and geochemistry of Cretaceous ultramafic volcanic rocks from Eastern Kamchatka, *J. Petrol.*, 36, 637-662, 1995.

Kearey, P., and F. Vine, Global Tectonics. Blackwell Scientific Publications, 302 pp, Oxford, 1996.

Kelemen, P.B., Genesis of high Mg# andesites and the continental crust, *Contrib. Mineral. Petrol.*, 120, 1-19, 1995.

Kepezhinskas, P., M.J. Defant, and M.S. Drummond, Progressive enrichment of island arc mantle by melt-peridotite interaction inferred from Kamchatka xenoliths, *Geochim. Cosmochim. Acta*, 60, 1217-1229, 1996.

Kerr, A.C., G.F. Marriner, N.T. Arndt, J. Tarney, A. Nivia, and A.D. Saunders, The petrogenesis of Gorona komatiites, picrites and basalts: new field, petrographic and geochemical constraints, *Lithos*, 37, 245-260, 1996.

Kerrich, R., and Q. Xie, Compositional recycling structure of an Archean super-plume: Nb-Th-U-LREE systematics of Archean komatiites and basalts revisited, *Contrib. Mineral. Petrol.*, 142, 476-484, 2002.

Kerrich, R., D.A. Wyman, J. Fan, and W. Bleeker, Boninite series: low Ti-tholeiite associations from the 2.7 Ga Abitibi greenstone belt, *Earth Planet. Sci. Lett.*, 164, 303-316, 1998.

Kimura, G., J.N. Ludden, J-P. Desrochers, and R. Hori, A model of ocean-crust accretion for the Superior Province, Canada, *Lithos*, 30, 337-355, 1993.

Kisters, A.F.M., G. Stevens, A. Dziggel, and R.A. Armstrong, Extensional detachment faulting and core complex formation in the southern Barberton granite-greenstone terrain, South Africa:

evidence for 3.2 Ga orogenic collapse, *Precam. Res.*, 127, 355-378, 2003.

Kloppenburg, A., S.H. White, and T.E. Zegers, Structural evolution of the Warrawoona greenstone belt and adjoining granitoid complexes, Pilbara craton, Australia, implications for Archean tectonic processes, *Precam. Res.*, 112, 107-147, 2001.

Knapp, J.H., C.C. Diaconescu, M.A. Bader, V.B. Sokolov, S.N. Kashubin, and A.V. Rybalka, Seismic reflection fabrics of continental collision and post orogenic extension in the Middle Urals, Central Russia, *Tectonophysics*, 288, 115-126, 1998.

Krapez, B., and B. Eisenlohr, Tectonic settings of Archean [3325-2775 Ma] crustal-supracrustal belts in the West Pilbara Block, *Precam. Res.*, 88, 173-205, 1998.

Kusky, T.M., D.C. Bradley, and P. Haeussler, Progressive deformation of the Chugach accretionary complex, Alaska, during a Paleogene ridge-trench encounter, *J. Struct. Geol.*, 19, 139-157, 1997.

Kusky, T.M., and W.F. Kidd, Remnants of an Archean oceanic plateau, Belingwe greenstone belt, Zimbabwe, *Geology*, 20, 43-46, 1992.

Kusky, T.M., and J. Li, Paleoproterozoic tectonic evolution of the North China Craton, *J. Asian Earth Sci.*, 22, 383-397, 2003.

Kusky, T.M., and A. Polat, Growth of granite-greenstone terranes at convergent margins, and stabilization of Archean cratons, *Tectonophysics*, 305, 43-73, 1999.

Lacroix, S., and E.W. Sawyer, An Archean fold and thrust belt in the northwestern Abitibi greenstone belt: Structural and seismic evidence, *Can. J. Earth Sci.*, 32, 97-112, 1995.

Lawver, L.A, R.A. Keller, M.R. Fisk, and J.A. Strelin, Bransfield Strait, Antarctic Peninsula active extension behind a dead arc, in *Backarc Basins: Tectonics and Magmatism*, edited by B. Taylor, pp. 315-344, Plenum Press, New York, 1995.

Lesher, M., and N.T. Arndt, REE and Nd isotope geochemistry, petrogenesis and volcanic evolution of contaminated komatiites at Kambalda, Western Australia, *Lithos*, 34, 127-157, 1995

Li, J., T.M. Kusky, and X. Huang, Archean podiform chromitites and mantle tectonites in ophiolitic mélange, North China Craton: A record of early oceanic mantle processes, *GSA Today*, 12, 4-11, 2002.

MacLachlan, K., D.W., Davis, and C. Relf, U/Pb geochronological constraints on Neoarchean tectonism: multiple compressional events in the northwestern Hearne Domain, Western Churchill Province, Canada, *Can. J. Earth Sci.*, 42, 85-109, 2005.

Manikyamba, C., R. Kerrich, S.M. Naqvi, and M.R. Mohan, Geochemical systematics of tholeiitic basalts from the 2.7 Ga Ramagiri-Hungund composite greenstone belt, Dharwar craton, *Precam. Res.*, 134, 21-39, 2004.

Manikyamba, C., S.M. Naqvi, D.V.S. Rao, M.R. Mohan, T.C. Khanna, T.G. Rao, and G.L.N. Reddy, Boninites from the Neoarchean Gadwal Greenstone belt, Eastern Dharwar Craton, India: implications for Archean subduction processes, *Earth Planet. Sci. Lett.*, 230, 65-83, 2005.

Mann, P., and A. Taira, Global tectonic significance of the Solomon Islands and Ontong Java convergent zone, *Tectonophysics*, 389, 137-190, 2004.

Márquez, A., R. Oyarzun, M. Doblas, and S.P. Verma, Alkalic (ocean-island basalt type) and calc-alkalic volcanism in the

Mexican volcanic belt: A case for plume related magmatism and propagating rift at an active margin?, *Geology*, 27, 51-54, 1999.

Marshak, R.S., and D.E. Karig, Triple junction as a cause for anomalous near-trench igneous activity between trench and volcanic arc, *Geology*, 5, 233-236, 1977.

Martin, H., Adakitic magmas: modern analogues of Archean granitoids, *Lithos*, 46, 411-429, 1999.

Martin, H., and J.-F. Moyen, Secular changes in tonalite-trondhjemite-granodiorite compositions as markers of the progressive cooling of the Earth, *Geology*, 30, 319-322, 2002.

Martin, H., R.H. Smithies, R. Rapp, J.-F. Moyen, and D. Champion, An overview of adakites, tonalite-granodiorite (TTG), and sanukitoid: relationships and some implications for crustal evolution, *Lithos*, 79, 1-24, 2005.

McCarron, J.J., and J.L. Smellie, Tectonic implications of fore-arc magmatism and generation of high-magnesian andesites: Alexander island, Antarctica, *J. Geol. Soc. London*, 155, 269-280, 1998.

McCulloch, M.T., and V.C. Bennett, Progressive growth of the Earth's continental crust and depleted mantle: geochemical constraints, *Geochim. Cosmochim. Acta*, 58, 4717-4738, 1994.

Morris, P.A., Slab melting as an explanation of Quaternary volcanism and aseismicity in southwest Japan, *Geology*, 23, 1040-1043, 1995.

Mueller, W.U., R. Daigneault, J.K. Mortensen, and E.H. Chown, Archean terrane docking: upper crustal collision tectonics, Abitibi greenstone belt, Quebec, Canada, *Tectonophysics*, 265, 127-150, 1996.

Myers, J.S., Precambrian history of the west Australian craton and adjacent orogens, *Annual Rev. Earth. Planet. Sci.*, 21, 453-485, 1993.

Nye, C.J., and M.R. Reid, Geochemistry of primary and least fractionated lavas from Okmaok volcano, central Aleutians: Implications for arc magma genesis, *J. Geophys. Res.*, 91, 10271-10287, 1986.

Parman, S.W., T.L. Grove, and J.C. Dann, The production of Barberton komatiites in an Archean subduction zone, *Geophys. Res. Lett.*, 28, 2513-2516, 2001.

Pearce, J.A., Supra-subduction zone ophiolites: The search for modern analogues, in *Ophiolite Concept and the Evolution of Geological Thought, Spec. Paper*, 373, edited by Y. Dilek, and S. Newcomb, pp. 269-293, *Geol. Soc. Am.*, Boulder, CO, 2003.

Pearce, J.A., S. van der Laan, R.J. Arculus, B.J. Murton, T. Ishii, and D.W. Peate, Boninite and Harzburgite from Leg 125 [Bonin-Mariana Fore-arc]: A case study of magma genesis during the initial stage of subduction, *Proceed. for the Ocean Drill. Prog., Sci. Res.*, 125, 623-659, 1992.

Pearce, J.A., and D.W. Peate, Tectonic implications of the composition of volcanic arc magmas, *Annual Rev. Earth Planet. Sci.*, 23, 251-285, 1995.

Percival, J.A., V. McNicoll, J.L., Brown, and J.B. Whalen, Convergent margin tectonics, central Wabigoon subprovince, Superior Province, Canada, *Precam. Res.*, 132, 213-244, 2004.

Percival, J.A., R.A. Stern, and N. Rayner, Archean adakites from the Ashuanipi complex, eastern Superior Province, Canada: geochemistry, geochronology, and tectonic significance, *Contrib. Mineral. Petrol.*, 145, 265-280, 2003.

Pessagno, E.A., D.M. Hull, and C.A. Hopson, Tectonostratigraphic significance of sedimentary strata occurring within and above the Coast Range ophiolite (California Coast Ranges) and Josephine ophiolite (Klamath Mountains), northwestern California, in *Ophiolites and Oceanic Crust: New Insights from Field Studies and Ocean Drilling Program, Spec. Paper* 349, edited by Y. Dilek, E.M. Moores, D. Elthon, and A. Nicolas, pp. 383-394, *Geol. Soc. Am.*, Boulder, CO, 2000.

Petford, N., and M.P. Atherton, Cretaceous volcanism and syn-subduction crustal extension, in *Volcanism Associated with Extension at Consuming Plate Margins, Spec. Publ.*, 81, edited by L.J. Smellie, pp. 233-248, *Geol. Soc. London*, London, 1995.

Polat, A., J.F. Casey, and R. Kerrich, Geochemical characteristics of accreted material beneath the Pozanti-Karsanti ophiolite, Turkey: Intra-oceanic detachment, assembly and obduction, *Tectonophysics*, 263, 249-276, 1996.

Polat, A., and A.W. Hofmann, Alteration and geochemical patterns in the 3.7-3.8 Ga Isua greenstone belt, West Greenland, *Precam. Res.*, 126, 197-218, 2003.

Polat, A., A.W. Hofmann, and M. Rosing, Boninite-like volcanic rocks in the 3.7-3.8 Ga Isua greenstone belt, West Greenland: geochemical evidence for intra-oceanic subduction zone processes in the Earth, *Chem. Geol.*, 184, 231-254, 2002.

Polat, A., and R. Kerrich, Formation of an Archean tectonic mélange in the Schreiber-Hemlo greenstone belt, Superior province, Canada: Implications for Archean subduction-accretion process, *Tectonics*, 18, 733-755, 1999.

Polat, A., R. Kerrich, and D.A. Wyman, The late Archean Schreiber-Hemlo and White River-Dayohessarah greenstone belts, Superior Province: collages of Oceanic plateaus, oceanic arcs, and subduction-accretion complexes, *Tectonophysics*, 294, 295-326, 1998.

Polat, A., and R. Kerrich, Magnesian andesites, Nb-enriched basalt-andesites, and adakites from late Archean 2.7 Ga Wawa greenstone belts, Superior Province, Canada: implications for late Archean subduction zone petrogenetic processes, *Contrib. Mineral. Petrol.*, 141, 36-52, 2001.

Polat, A., and R. Kerrich, Nd-isotope systematics of ~2.7 Ga adakites, magnesian andesites, and basalts-andesites, Superior Province: Evidence for shallow crustal recycling at Archean subduction zones, *Earth Planet. Sci. Lett.*, 202, 345-360, 2002.

Polat, A., and C. Münker, Hf-Nd isotope evidence for contemporaneous subduction processes in the source of late Archean arc lavas from the Superior Province, Canada, *Chem. Geol.*, 213, 403-429, 2004.

Polat, A., C. Münker, R. Rodgers, T. Kusky, J. Li, and B. Fryer, Geochemical evidence for suprasubduction zone origin of the Neoarchean (ca. 2.55-2.50 Ga) peridotites, central orogenic belt, North China craton, *Geol. Soc. Am. Bull.*, (in revision), 2005.

Poli, S., and Schmidt, M.W., Petrology of subducted slabs, *Annu. Rev. Earth Planet Sci.*, 30, 207-235, 2002.

Pouclet, A., J.S. Lee, P. Vidal, B. Cousens, and H. Bellon, Cretaceous to Cenozoic volcanism in South Korea and in the Sea of Japan: magmatic constraints on the opening of the back-ark basin, in *Volcanism Associated with Extension at Consuming Plate Margins, Spec. Publ.*, 81, edited by J.L. Smellie, pp. 169-191, *Geol. Soc. London*, London, 1995.

Poulsen, K.H., G.J. Borradaile, and M.M. Kehlenbeck, An inverted Archean succession at Rainy Lake, Ontario, *Can. J. Earth Sci.*, 17, 1358-1369, 1980.

Puchtel, I.S., A.W. Hofmann, K. Mezger, K.P. Jochum, A.A. Shchipansky, and A.V. Samsonov, Oceanic plateau model for continental crustal growth in the Archean: a case study from the Kostomuksha greenstone belt, NW Baltic Shield, *Earth. Planet. Sci. Lett.*, 155, 57-74, 1998.

Ramsay, W.R.H., A.J. Crawford, J.D. Foden, Field setting, mineralogy, chemistry, and genesis of arc picrites, New Georgia, Solomon Islands, *Contrib. Mineral. Petrol.*, 88, 386-402, 1984.

Rapp, R.P., N. Shimizu, and M.D. Norman, Growth of early continental crust by partial melting of eclogite, *Nature*, 425, 605-609, 2003.

Rapp, R.P., N. Shimizu, M.D. Norman, and G.S. Applegate, Reaction between-slab derived melts and peridotite in the mantle wedge: experimental constraints at 3.8 GPa, *Chem. Geol.*, 160, 335-356, 1999.

Robertson, A.H.F., and M.P. Searle, The northern Oman Tethyan continental margin: stratigraphy, structure, concepts and controversies, in *The Geology and Tectonics of the Oman Region, Spec. Publ.*, 49, edited by A.H.F. Robertson, M.P. Searle, and A.C. Ries, *Geol. Soc. London*, London, 3-25, 1990.

Rogers, G., and A.D. Saunders, Magnesian andesites from Mexico, Chile, and the Aleutian islands: implications for magmatism associated with ridge-trench subduction, in *Boninites and Related Rocks*, edited by A.J. Crawford, A.J. pp.416-445, Unwin-Hyman, London, 1989.

Rollinson, H., Eclogite xenoliths in west African kimberlites as residues from Archean granitoid crust formation, *Nature*, 389, 173-176, 1997.

Rudnick, R.L., Making continental crust, *Nature*, 378, 571-578, 1995.

Sajona, F.G., R. Maury, H. Bellon, J. Cotton, and M. Defant, High field strength element enrichment of Pliocene-Pleistocene island arc basalts, Zamboanga Peninsula, Western Mindanao (Philippines), *J. Petrol.*, 37, 693-726, 1996.

Samson, S.D., and P.J. Patchett, The Canadian Cordillera as a modern analogue of Proterozoic crustal growth, *Austural. J. Earth Sci.*, 38, 595-611, 1991.

Sandeman, H.A., S. Hanmer, W.J. Davis, J.J. Ryan, and T.D. Peterson, Neoarchean volcanic rocks, Central Hearne supracrustal belt, Western Churchill Province, Canada: geochemical and isotopic evidence supporting intra-oceanic, suprasubduction zone extension, *Precam. Res.*, 134, 113-141, 2004.

Schuth, S., A. Rohrbach, C. Münker, C. Ballhaus, D. Garbe-Schönberg, and C. Qopoto, Geochemical constraints on the petrogenesis of arc picrites and basalts, New Georgia Group, Solomon Islands, *Contrib. Mineral. Petrol.*, 148, 288-304, 2004.

Searle, M.P., and R.K. Stevens, Obduction processes in ancient, modern and future ophiolites, in *Ophiolites and Oceanic Lithosphere, Spec. Publ.*, 13, edited by I.G. Gass, S.J. Lippard, and A.W. Shelton, A.W., pp. 303-319, Geol. Soc. London, London, 1984.

Şengör, A.M.C., Plate Tectonics and orogenic research after 25 years: A Tethyan perspective, *Earth Sci. Rev.*, 27, 1-201, 1990.

Şengör, A.M.C., and K. Burke, Relative timing of rifting and volcanism on Earth and its tectonic implications, *Geophys. Res. Lett.*, 5, 419-421, 1978.

Şengör, A.M.C., Görür, N., and F. Saroglu, Strike-slip faulting and related basin formation in zones of tectonic escape: turkey as a case study, *Soc. Eco. Paleontolog. Mineral, Spec. Publ.* 37, 227-264, 1985.

Şengör, A.M.C., and B.A. Natal'in, Turkic-type orogeny and its role in the making of the continental crust, *Annual Rev. Earth. Planet. Sci.*, 24, 263-337, 1996.

Şengör, A.M.C., and B.A. Natal'in, Phanerozoic analogues of Archean basement fragments: Altaid ophiolites and ophirags, in *Precambrian Ophiolites and Related Rocks*, edited by T.M. Kusky, pp. 671-721, Elsevier, Amsterdam, 2004.

Shchipansky, A.A., A.V. Samsonov, E.V. Bibikova, I.I. Babarina, A.N. Konilov, K.A. Krylov, A.I. Slabunov, and M.M. Bogina, 2.8 Ga boninite-hosting partial subduction zone ophiolite sequences from the North Karelian greenstone belt, NE Baltic Shield, Russia in *Precambrian Ophiolites and Related Rocks*, edited by T. M. Kusky, pp. 425-486, Elsevier, Amsterdam, 2004.

Shirey, S.B., and G.N. Hanson, Mantle heterogeneity and crustal recycling in Archean granite-greenstone belts: Evidence from Nd-isotopes and trace elements in the Rainy Lake area, Superior Province, Ontario, Canada, *Geochim. Cosmochim. Acta*, 50, 2631-2651, 1986.

Sisson, V.B., T.L. Pavlis, S.M. Roeske, and D.J. Thorkelson, Introduction: An overview of ridge-trench interactions in modern and ancient settings, in *Geology of a Transpressional Orogen Development During Ridge-Trench Interaction Along the North Pacific Margin, Spec. Paper* 371, edited by V.B. Sisson, S.M. Roeske, and T.L. Pavlis, pp. 1-18, *Geol. Soc. Am.*, Boulder, CO, 2003.

Skulski, T., and J.P. Percival, Allochthonous 2.78 Ga oceanic plateau silvers in a 2.72 Ga continental arc sequence: Vizien greenstone belt, northeastern Superior Province, Canada, *Lithos*, 37, 163-179, 1996.

Sleep, N.H., Archean plate tectonics: what can be learned from continental geology, *Can. J. Earth Sci.*, 29, 2066-2071, 1992.

Smith, J.B., M.E. Barley, D.I. Groves, B. Krapez, N.J. McNaughton, M.J. Bickle, and H.J. Chapman, The Sholl Shear Zone, West Pilbara: evidence for a domain boundary structure from integrated tectonostratigraphic analyses, SHRIMP U-Pb dating and isotopic and geochemical data of granitoids, *Precam. Res.*, 88, 143-171, 1998.

Smithies, R.H., The Archean tonalite-trondhjemite–granodiorite [TTG] series is not an analogue of Cenozoic adakites, *Earth Planet. Sci. Lett.*, 182, 115-125, 2000.

Smithies, R.H., D.C. Champion, M.J. van Kranendonk, H.M. Howard, and A.H. Hickman, Modern-style subduction processes in the Mesoarchean: geochemical evidence from the 3.12 Ga Whundo intra-oceanic arc, *Earth Planet Sci. Lett.* 231, 221-237, 2005.

Spadea, P., and J.H. Scarrow, Early Devonian boninites from the Magnitogorsk arc, southern Urals [Russia]: implications for early development of a collisional orogen, in *Ophiolites and Oceanic Crust: New Insights from Field Studies and Ocean Drilling Program, Spec. Paper*, 349, edited by Y. Dilek, E.M. Moores, D. Elthon, D., and A. Nicolas, pp. 461-472, *Geol. Soc. Am.*, Boulder, CO, 2000.

Srivastava, R.K., R.K. Singh, and S.P. Verma, Neoarchean volcanic rocks from the southern Bastar greenstone belt, central India: Petrological and tectonic significance, *Precam. Res.*, 131, 305-322, 2004.

Stern, C.R., and R. Kilian, Role of the subducted slab, mantle wedge and continental crust in the generation of adakites from the Andean Austral Volcanic Zone, *Contrib. Mineral. Petrol.*, 123, 263-281, 1996.

Stern, R.J., J. Morris, S.H. Bloomer, and J.W. Hawkins, The source of the subduction component in convergent margin magmas: trace element and radiogenic evidence from Eocene boninites, Mariana forearc, *Geochim. Cosmochim. Acta*, 55, 1467-1481, 1991.

Stevenson, R.K., and P.J. Patchett, Implications for the evolution of continental crust from Hf isotope systematics of Archean detrital zircons, *Geochim. Cosmochim. Acta* 54, 1683-1697, 1990.

Stone, W.E., J.H. Crocket, A.P. Dickin, and M.F. Fleet, Origin of Archean ferropicrites: geochemical constraints from the Boston Creek flow, Abitibi greenstone belt, Ontario, Canada, *Chem. Geol.*, 121, 51-71, 1995.

Storey, B.C., The role of mantle plumes in continental breakup: case histories from Gondwanaland, *Nature*, 377, 301-308, 1995.

Stott, G.M., The Superior Province, Canada, in *Tectonic Evolution of Greenstone Belts, Oxford Monog. on Geol. and Geophys 35*, edited by M. de Wit, M. and L.D. Ashwal, pp. 480-507, Clarendon Press, Oxford, 1997.

Sun, S.S., and W.F. McDonough, Chemical and isotopic systematics of oceanic basalts: implications for mantle composition and processes, in *Magmatism in the Ocean Basins, Spec. Publ. 42*, edited by A.D. Saunders and M.J. Norry, pp. 313-345, *Geol. Soc. London*, London, 1989.

Sun, S.S., and R.W. Nesbitt, Geochemical regularities and genetic significance of ophiolitic basalts, *Geology*, 6, 689-693, 1978.

Svetov, S.A., H. Huhma, A.I. Svetova, and T.N. Nazarova, The oldest Aadakites of the Fennoscandian Shield, *Doklady Earth Sci.*, 397A, 878-882, 2004.

Swager, C., Tectonostratigraphy of late Archean greenstone terranes in the southern Eastern Goldfields, Western Australia, *Precam. Res.*, 83, 11-42, 1997.

Swain, G., A. Woodhouse, M. Hand, K. Barovich, M. Schwarz, and C.M. Fanning, Provenance and tectonic development of the late Archean Gawler Craton, Australia; U-Pb zircon, geochemical and Sm-Nd isotopic implications, *Precam. Res.*, 141, 106-136, 2005.

Sylvester, A.G., Strike-slip faults, *Geol. Soc. Am. Bull.*, 100, 1666-1703, 1988.

Taira, A., K.T. Pickering, B.F. Windley, and W. Soh, Accretion of Japanese island arcs and implications for the origin of Archean greenstone belts, *Tectonics*, 11, 1224-1244, 1992.

Tarney, J., I.W. Dalziel, and M.J. de Wit, Marginal Basin 'Rocas Verdes' Complex from S. Chile: A Model for Archaean Greenstone Belt Formation, in *The Early History of the Earth*, edited by B.F Windley, pp. 131-146, John Wiley, New York, 1976.

Tatsumi, Y., and S. Maruyama, Boninites and high-Mg andesites: tectonic and petrogenesis, in *Boninites and Related Rocks*, edited by A.J. Crawford, pp. 50-71 Unwin Hyman, London, 1989.

Taylor, S.R., and S.M. McLennan, The geochemical evolution of the continental crust, *Rev. Geophys.*, 33, 241-265, 1995.

Thirlwall, M.F., A.M. Graham, R.J. Arculus, R.S. Harmon, and C.G. Macpherson, Resolution of the effects of crustal assimilation, sediment subduction, and fluid transport in arc magmas:

Pb-Sr-Nd-O isotope geochemistry of Grenada, Lesser Anilles, *Geochim. Cosmochim. Acta*, 60, 4785-4810, 1996.

Thurston, P.C., Autochthonous development of Superior Province greenstone belts?, *Precam. Res.*, 115, 11-36, 2002.

Thurston, P.C., and K.M. Chivers, Secular variations in greenstone sequence development emphasizing Superior Province, Canada, *Precam. Res.*, 46, 21-58, 1990.

Thurston, P.C., and V.N. Kozhevnikov, An Archean quartz arenite-andesite association in the eastern Baltic Shield, Russia: Implications forassemblage type and shield history, *Precam. Res.*, 101, 313-340, 2000.

Tomlinson, K.Y., D.W. Davis, J.A. Percival, D.J. Hughes, and P.C. Thurston, Mafic to felsic magmatism and crustal recycling in the Obonga Lake greenstone belt, Western Superior Province: evidence from geochemistry, Nd isotopes and U-Pb, *Precam. Res.*, 114, 295-325, 2002.

Tomlinson, K.Y., D.J. Hughes, P.C. Thurston, and R.P. Hall, Plume magmatism and crustal growth at 2.9 to 3.0 Ga in the Steep Rock and Lumby Lake area, Western Superior Province, *Lithos*, 46, 103-136, 1999.

Ujike, O., and A.M. Goodwin, Origin of Archean adakites and NEBA from the Upper Keewatin assemblage, the Lake of the Woods greenstone belt, Western Wabigoon Subprovince, *Geochim. Cosmochim. Acta*, 67, A503, 2003.

Umhoefer, P.J., and P. Schiarriza, Latest Cretaceous to early Tertiary dextral strike-slip faulting on the southeastern Yalakom fault system, southeastern Coast Belt, *Geol. Soc. Am. Bull.*, 108, 768-785, 1996.

Vanderhaeghe, O., and C. Teyssier, Crustal-scale rheological transitions during late orogenic collapse, *Tectonophysics*, 335, 211-228, 2001.

van Kranendonk, M.J., Archean tectonics, A review, *Precam. Res.*, 131, 143-151, 2004.

van Staal, C.R., Brunswick subduction complex in the Canadian Appalachians: Record of the Late Ordovician to Late Silurian collision between Laurentia and Gander margin of Avalon, *Tectonics*, 13, 964-962, 1994.

Vergara, M., B. Levi, J.O. Nystrom, and A. Cancino, Jurassic and early Cretaceous island arc volcanism in the coast range of central Chile, *Geol. Soc. Am. Bull.*, 107, 1427-1440, 1995.

von Huene, R., D.V. Scholl, Observations at convergent margins concerning sediment subduction, subduction-erosion, and the growth of the continental crust, *Rev. Geophys.*, 29, 279-316, 1991.

Wakabayashi, J., Nappes, tectonics of oblique plate convergence, and metamorphic evolution related to 140 million years of continuous subduction, Franciscan Complex, California, *J. Geol.*, 100, 19-40, 1992.

Wang, Z., S.A. Wilde, K. Wang, and L. Yu, A MORB-arc basalt-adakite association in the 2.5 Ga Wutai greenstone belt: Late Archean magmatism and crust growth in the North China Craton, *Precam. Res.*, 131, 323-343, 2004.

Whalen, J.B., J.A. Percival, V.J. McNicoll, and F.J. Longstaff, A mainly crustal origin for tonalitic granitoid rocks, Superior Province, Canada: implications for late Archean Tectonomagmatic Processes, *J. Petrol.*, 43, 1551-1570, 2002.

Williams, H.R., G.M. Stott, and P.C. Thurston, Tectonic evolution of Ontario: Summary and synthesis, Part 1: Revolution in the Superior Province, in *Geology of Ontario, Spec.* Vol. 4/2, edited by P.C. Thurston, H.R. Williams, H.R. Sutcliffe, G.M. Stott, Geology of Ontario, pp. 1256-1294, *Ont. Geol. Surv.*, Sudbury, 1991.

Wyman, D.A., A 2.7 Ga depleted tholeiite suite: evidence of plume–arc interaction in the Abitibi greenstone belt, Canada, *Precam. Res.*, 97, 27-42, 1999.

Wyman, D.A., Upper mantle processes beneath the 2.7 Ga Abitibi belt, Canada: trace element perspective, *Precam. Res.*, 127, 143-165, 2003.

Wyman, D.A., J.A. Ayer, and J.R. Devaney, Niobium-enriched basalts from the Wabigoon subprovince, Canada: evidence for adakitic metasomatism above an Archean subduction zone, *Earth Planet. Sci. Lett.* 179, 21-30, 2000.

Wyman, D.A., W. Bleeker, and R. Kerrich, A 2.7 Ga komatiite, low-Ti tholeiite, arc transition, and inferred proto-arc geodynamic setting of the Kidd Creek deposit: Evidence from precise ICP MS trace element data, *Eco. Geol. Monog.*, 10, 511-528, 1999.

Wyman, D.A., R. Kerrich, and A. Polat, Assembly of Archean cratonic mantle lithosphere and crust: plume-arc interaction in the Abitibi-Wawa subduction-accretion complex, *Precam. Res.*, 115, 37-62, 2002.

Yamamoto, M., Picritic primary magma and its source mantle for Oshima-Oshima and back-arc side volcanoes, Northwest Japan, *Contrib. Mineral. Petrol.*, 99, 352-359, 1988.

Yiğitbaş, E., R. Kerrich, Y. Yücel, A. Elmas, Q.L. Xie, Characteristics and geochemistry of Precambrian ophiolites and related volcanics from the Istanbul-Zonguldak Unit, Northwestern Anatolia, Turkey: following the missing chain of the Precambrian South European suture zone to the east, *Precam. Res.*, 132, 179-206, 2004.

Zegers, T.E., Granite formation and emplacement as indicators of Archean tectonic processes, In: *The Precambrian Earth: Tempos and Events*, edited by P.G. Eriksson, W. Altermann, D.R. Nelson, W.U. Mueller, and O. Catuneanu, pp. 103-118, Elsevier, Amsterdam, 2004.

Zegers, T.E., M. Keijzer, C.W. Passchier, and S.H. White, The Mulgandinnah Shear zone: an Archean crustal scale shear zone in the eastern Pilbara, Western Australia, *Precam. Res.*, 88, 233-247, 1998.

R. Kerrich, Department of Geological Sciences, University of Saskatchewan, Saskatoon, Saskatchewan, S7N 5E2, Canada.

A. Polat, Department of Earth Sciences, University of Windsor, Windsor, Ontario, N9B 3P4, Canada. (polat@uwindsor.ca)

Late-Archean Convergent Margin Volcanism in the Superior Province: A Comparison of the Blake River Group and Confederation Assemblage

D.A. Wyman

School of Geosciences, University of Sydney, Sydney, New South Wales, Australia

P. Hollings

Department of Geology, Lakehead University, Thunder Bay, Ontario, Canada

The Blake River Group of the Abitibi Subprovince and the Confederation assemblage of the Uchi Subprovince have been characterised as subduction-related volcanic assemblages generated in oceanic and continental margins, respectively. Despite the important differences in their settings, the major element trends of tholeiitic rocks from the two areas resemble each other, and Phanerozoic arcs, more than they do tholeiites from continental rift settings that may be analogues for some Archean greenstone belts. The Blake River Group suite includes magmas generated in relict plume asthenosphere, but the chemical trends also provide evidence of slab melt metasomatism. Primitive rocks in the Confederation assemblage define trace element trends analogous to typical modern arcs, with no indication that melt-mobilized elements such as Zr and Nb have been introduced in significant amounts. Adakite-like rocks were formed as a result of local events such as arc rifting in the South Bay area, or as an indirect consequence of larger events such as global-scale mantle-plume episodes that strongly influenced the southern Abitibi Subprovince and the Blake River Group. Niobium-enriched basalts associated with crustally contaminated rhyolites in the southwest of the South Bay study area are most plausibly linked to rifting of the Uchi Subprovince proto-continent margin. Rhyolites in both areas are fractionation products of mantle-derived melts. In addition to documenting variable crustal contamination, the trace elements systematics of the rhyolites provide evidence of zircon fractionation events that occurred without significant changes in major element compositions. These results are probably attributable to extraction of rhyolitic liquids from crystal mush zones, accompanied by preferential entrainment of zircon crystals, leading to Zr fractionation.

INTRODUCTION

The broadly coeval (~2.7 Ga) Blake River Group (BRG) of the southern Abitibi Belt (Subprovince) and the Confederation Assemblage of the Birch–Uchi greenstone belt of the Uchi Subprovince in the Superior Province, Canada (Figure 1) have both been interpreted as the products of subduction tectonics [*Hollings and Kerrich*, 2000;

Archean Geodynamics and Environments
Geophysical Monograph Series 164
Copyright 2006 by the American Geophysical Union
10.1029/164GM14

Péloquin et al., 1996]. Both greenstone terranes contain a range of mafic rocks types (i.e., variably tholeiitic to calc alkaline) and host volcanogenic massive sulfide (VMS) deposits. Differences in the proportions and types of rocks in the two areas, however, suggest they represent end-members in a range of subduction settings present during the late Archean. Plume-associated komatiites are interlayered with arc-type rocks in the southcentral part of the Abitibi Subprovince that contains the BRG. This area appears devoid of crust older than ~2.75 Ga, although evidence is growing for some contributions of 2.8-Ga crust on the other margins of the Subprovince (e.g., *Thurston* [2003] and references therein). In contrast, the Confederation assemblage was not associated with plume volcanism and was situated at the margin of a proto-continent containing rocks that date back to ~3 Ga. This paper examines the geochemical characteristics of mafic rocks in both areas in order to distinguish their mantle sources. Differences in the evolutionary paths of the magmatic suites are then considered in terms of the distinctive aspects of their respective tectonic settings and a variety of crustal-level processes.

REGIONAL SETTING AND PREVIOUS WORK

The Blake River Group

The BRG has been extensively studied because it is host to numerous VMS deposits, ranging from the Horne Mine (55 Mt massive sulfide ore mined, total tonnage ~144 Mt), to the Quemont Mine along the southern margin of the Sequence (16 Mt), and relatively small ore bodies common in the Noranda Mine Sequence (e.g., 1–5 Mt), as summarised by *Gibson and Watkinson* [1990]. Other important BRG studies include those of Dimroth and coworkers [e.g., *Dimroth et al.*, 1982], *Hubert et al.* [1984], *Kerr and Gibson* [1993], and many others. Lithogeochemical studies by *Lafleche et al.* [1992a,b] and *Ludden and Péloquin* [1996] focussed on lithologies of the eastern BRG, and *Fowler and Jensen* [1989] reported data for the western BRG.

The BRG was erupted over a short period between about 2703 and 2698 Ma and therefore is among the youngest preorogenic volcanic suites in the southern Abitibi belt (*Corfu* [1993] and references therein; *Mortensen* [1993]; *Galley and van Breemen* [2002]). *Gibson and Watkinson* [1990] interpreted the stratigraphy of the eastern BRG as comprising five cycles of volcanism. *Péloquin et al.* [1996] suggested that ages of the western BRG, the Noranda area sequences, and the Rouyn-Pelletier tholeiites all overlap to some extent. Given the greater thickness of the BRG in the west, they suggest an eastward-propagating ridge scenario, where deeper sequences of the western BRG are inferred to be similar to those exposed in the Noranda area.

There is considerable variability within the BRG in terms of the geochemical signatures of volcanic rock types and stratigraphic orientation. Whereas parts of the western BRG are flat-lying with mainly calc alkaline volcanic rocks exposed [*Jensen and Langford*, 1985], the eastern BRG consists of a number of structural blocks (Figure 1E), where dips range from moderate to steep [*Péloquin et al.*, 1990], and volcanic units are variably tholeiitic, to transitional tholeiitic-calc alkaline. *Laflèche et al.* [1992a] distinguished tholeiite and calc alkaline units of the northeastern BRG on the basis of field characteristics. Tholeiitic units tend to be thicker and laterally more extensive than calc alkaline units and have other distinctive physical characteristics, such as larger pillows. Although the restricted sampling area resulted in compositionally distinct subunits on many plots, *Laflèche et al.* [1992a,b] showed that much of the variation could be accounted for by mixing of end-member tholeiitic and calc alkaline components. Rather than establish multiple compositional subgroups, *Barrett et al.* [1993] divided the compositional spectrum of Noranda area volcanic rocks into tholeiitic magmas with Zr/Y ratios less than 4.5 and transitional tholeiite–calc alkaline magmas with higher Zr/Y ratios.

Recent regional-scale models for the southern Abitibi belt have emphasized the lateral continuity of major supracrustal assemblages across the width of the belt [*Ayer et al.*, 2002]. *Wyman* [2003] showed that the High Field Strength Element (HFSE) systematics of mafic rocks in the southern Abitibi Province support such models and emphasized a role of plume-associated mantle as the source of komatiites and later as a major component in the sources of arc-style tholeiites and calc alkaline volcanic suites. Unlike typical Phanerozoic subduction zones, where the mantle wedge is metasomatised by hydrous fluids, mafic rocks of the southern Abitibi were derived from sources that were mainly modified by HFSE-bearing igneous melts. Mafic rocks of the BRG and several other southern Abitibi volcanic suites associated with crustal extension define HFSE trends that range from depleted plume-type sources to metasomatised MORB-style sources over time. On the basis of HFSE trends, *Wyman* [2003] inferred that the metasomatising igneous melts were adakitic in composition and must have been generated throughout the last 20 m.y. of the southern Abitibi belt's volcanic history, even though true adakitic units constitute only a minor component of volcanic rocks in the region.

Birch–Uchi

The Birch–Uchi greenstone belt comprises four distinct tectonic assemblages: the ~2989–2975 Ma Balmer assemblage, the ~2975–2870 Ma Narrow Lake assemblage, the ~2771 Ma Woman assemblage, and the ~2725–2745 Ma Confederation assemblage [*Rogers*, 2002a].

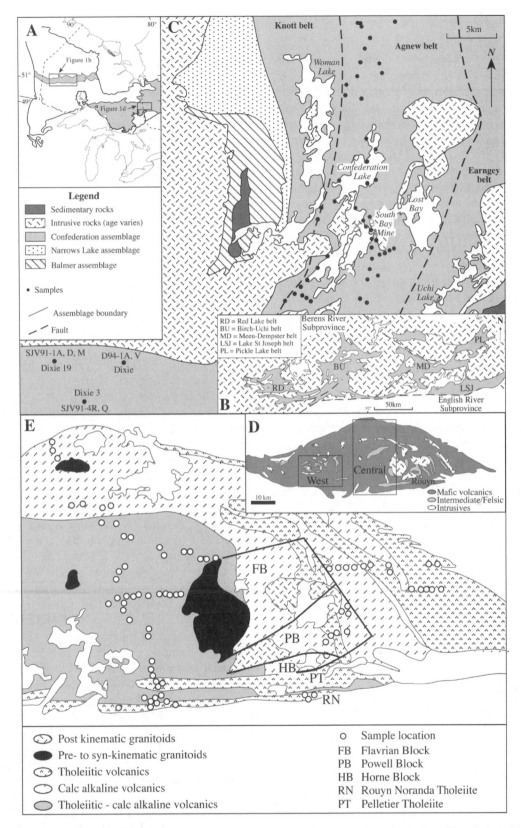

Figure 1. Location map of the two study areas. (A) Locations within the Superior Province. (B) Location of the Birch–Uchi belt [BU] in the Uchi Subprovince (modified after *Stott and Corfu* [1991]). (C) Local geology and sample locations in the South Bay study area of Confederation Assemblage, Birch–Uchi belt (modified after *Stott and Corfu* [1991] and *Rogers* [2002b]). (D) General geology of the Blake River Group, showing sample areas in the central and western parts of the Group. (E) Sample locations in the central and eastern Blake River Group (modified after *MERQ-OGS* [1983]). Additional drill core samples (not shown) were analysed from the Flavrian, Powell, and Horne Blocks.

The Confederation assemblage, which dominates the eastern portion of the Birch–Uchi greenstone belt (Figure 1C), has been mapped over several decades [*Thurston*, 1985a,b, 1986; *Good*, 1988; *Beakhouse*, 1989; *Stix and Gorton*, 1989; *Rogers*, 2002a,b]. Geochronological studies by *Nunes and Thurston* [1980] and *Noble* [1989] yielded U–Pb ages of between 2735+4/−3 and 2739±2 Ma for felsic sequences within the Confederation assemblage. More recently *Rogers* [2002b] has suggested that the Confederation assemblage can be subdivided into three north–south striking belts; from east to west these are the Knott, Agnew, and Earngey belts and all have been interpreted to have an age of ~2740 Ma. The belts are separated from each other by faults interpreted to be eastward-dipping thrust faults or high-angle reverse faults (Figure 1; *Rogers* [2002a]). *Hollings and Kerrich* [2000] reported the presence of a diverse range of rocks from what is now recognized as the Agnew belt of the Confederation assemblage, including subduction-modified tholeiites, Nb-enriched basalts (NEB), calc alkaline basalts, and adakites, which they interpreted to be consistent with the Confederation assemblage having formed in a back arc basin. The NEB are associated with felsic rocks in the southwest part of the Confederation assemblage study area and the stratigraphic position relative to the rocks of the central part of the study area is unclear.

The Agnew belt is also host to the past-producing South Bay VMS mine, which produced 1.6 million tons of ore with an average grade of 11% Zn, 2% Cu, and 2.12 ounces of Ag per ton [*Atkinson et al.*, 1990]. The felsic volcanic flows in the Confederation assemblage have been interpreted as pyroclastic deposits associated with Plinian eruptions and subsequent caldera collapse of an Archean strato volcano [*Thurston*, 1986; *Stix and Gorton*, 1989].

Tomlinson and Rogers [1999] have reported εNd$_T$ values of +2.0 to −0.4 for felsic rocks of the Agnew belt of the Confederation assemblage, whereas *Noble* [1989] determined εNd$_T$ values ranging from 1.7 to −1.2 for five felsic tuffs and sills from the Confederation assemblage in general. These data are consistent with the work of *Hollings and Kerrich* [2000], who reported elevated Th levels in some samples from the Confederation assemblage that are indicative of crustal assimilation.

Archean Felsic Rock Classifications

The origin of extensive horizons of felsic rocks in Archean greenstone belts has been attributed to a variety of causes, including direct melting of older sialic crust [e.g., *Gelinas and Ludden*, 1984], fractional crystallization of mantle-derived mafic tholeiites [e.g., *Capdevila et al.*, 1982; *Lesher et al.*, 1986], and shallow-level partial melting of a hydrated basalt source [*Hart et al.*, 2004]. The range of models reflects the variety of scenarios that have been applied to rhyolites in

general, as summarized by *Bachmann and Bergantz* [2004]. These authors proposed that crystal-poor rhyolites are generated by extraction of interstitial melts from batholithic crystal mushes.

Campbell et al. [1982] cited differing rare earth element [REE] abundances when they described two principal types of tholeiitic and calc alkaline felsic rock associated with VMS deposits. They also recognized a third suite of nonprospective rocks characterised by steep LREE fractionation and only minor Eu anomalies. *Lesher et al.* [1986] further expanded the characterization of prospective and nonprospective felsic volcanic rocks. Based principally on REE and HFSE systematics, they subdivided felsic flows into four distinct groups, based partly on the scheme proposed by *Thurston* [1981]. Nonprospective F1 dacites and rhyodacites are characterised by strongly fractionated REE (La/Yb$_n$ = 6–34), high Zr/Y ratios (9–31), and minor Eu anomalies. Variably prospective F2 rhyodacites and rhyolites display moderately fractionated REE (La/Yb$_n$ = 2–6), moderate Zr/Y ratios (6–11), and variable Eu anomalies. This calc alkaline type is found at Sturgeon Lake (Wabigoon Subprovince, Superior Province), Golden Grove (Western Australia), and the Kuroko district (Japan). The third strongly prospective tholeiitic type (F3) includes rhyolites and high silica rhyolites characterised by flat REE at high absolute abundances, pronounced negative Eu anomalies, and low Zr/Y ratios (4–7). *Lesher et al.* [1986] cited examples in the Birch–Uchi (South Bay; SB) and Noranda (BRG) study areas discussed in this paper as well as other Abitibi VMS districts (Matagami, Kam Kotia, and Kidd Creek). F3 high-silica rhyolites were further subdivided into F3a and F3b; F3a are distinguished by moderate negative Eu anomalies, intermediate HFSE, and high Sc abundances, whereas F3b display pronounced negative Eu anomalies, low abundances of Sc, but high HFSE contents. *Lesher et al.* [1986] concluded that the flat REE and negative Eu anomalies observed in F3 felsic volcanic rocks were consistent with plagioclase-dominated fractionation in a high-level magma chamber. In contrast, the fractionated REE of the F1's are attributable to low-degree partial melting of a mafic source at high pressure. *Martin* [1986] and many subsequent workers likened the F1-type rocks to adakites and inferred an origin in melted oceanic crust of Archean subduction zones. F1 magmas lack the Eu anomalies characteristic of shallow-level fractionation. The F2 rocks display compositions consistent with high-degree partial melting of a crustal source or fractional crystallisation of an intermediate parent.

ANALYTICAL METHODS AND SAMPLING METHODOLOGY

Major elements were determined by using X-ray fluorescence spectrometry (XRF) with relative standard deviations within 5%. Trace elements, including the REEs and HFSEs, were analyzed with inductively coupled mass spectrometry (ICP-MS; Perkin Elmer Elan 5000) at the Department of Geological Sciences, University of Saskatchewan, following the protocol of *Jenner et al.* [1990] and *Longerich et al.* [1990] and data reduction following the procedure of *Jenner et al.* [1990]. Thirty-one trace elements, including HFSE and REE, were determined by using $HF-HNO_3$ dissolution. Thorium, Nb, Hf, Zr, and REEs were also analyzed by using the Na_2O_2 sinter technique of *Longerich et al.* [1990] to circumvent potential problems associated with HFSE and REE in refractory minerals. Analysis of acids, distilled deionised water, and procedural blanks yielded levels of <10 ppb for Hf, Zr, Nb, and REEs. Detection limits, defined as 3σ of the procedural blank, for some critical elements are as follows (in parts per million): Th (0.01), Nb (0.006), Hf (0.008), Zr (0.004), La (0.01), and Ce (0.009). Precision (as relative standard deviations) for most elements at the concentrations present in the international reference material BIR-1 is between 2% and 4%.

Sample collection was undertaken in the east and west of the BRG and on a north–south axis to document compositional variations across the Group. In addition, more detailed sampling was undertaken in the Noranda Mine Sequences; these included representative, least altered, drill core samples of subunits provided by Inmet Mining Corporation. The immediate host sequence and hangingwall of the giant Horne Deposit was sampled along with units in the Horne structural Block, including samples provided by Noranda Exploration Ltd. Additional samples were collected along the southern margin of the BRG to characterize tholeiitic units previously distinguished as the Rouyn-Noranda and Pelletier tholeiites by *Gelinas et al.* [1984].

For ease of discussion, the BRG samples in this paper are split into several subgroups based on composition and geographic distribution. Rhyolite samples are mainly distinguished from nonrhyolite samples on the basis of Al_2O_3/TiO_2 contents ratios of 30 (SiO_2 ~70 wt.%). Pelletier and Rouyn-Noranda tholeiites are distinguished on chemical plots, and a well-defined tholeiitic fractionation suite (East Horne Block) has been identified for discussion purposes. Given that compositional variability in primitive BRG magma suites may be a function of veining by adakitic melts in their sources [*Wyman*, 2003], the ratio Nb/La rather than Zr/Y is used to subdivide the spectrum of magma compositions. Both Nb and La are present in adakites at higher concentrations than in tholeiitic basalts; these elements also have similar strongly incompatible behaviours in magmas and so are more likely to retain signature ratios than Zr/Y is. The tholeiites (Nb/La > 0.74) and mixed or transitional tholeiite-calc alkaline subgroups for unevolved samples (MgO > 5 wt.%) exhibit a good correlation between Nb/La and the Zr/Y ratio previously used by *Barrett et al.* [1993] to delineate tholeiitic and transitional suites.

Sample collection of the Confederation assemblage was undertaken from the vicinity of the South Bay VMS mine as well as from a drill core located on a broadly north–south transect across the assemblage. A total of 172 samples were analysed from the Confederation assemblage taken over an area of ~500 km² from outcrops and from the diamond drill core provided by Noranda Exploration Ltd. (Figure 1C). The petrology and geochemistry of many of the units from the Confederation assemblage are discussed in *Hollings and Kerrich* [2000]. The present paper reports new data (Tables 1 and 2) for the Confederation assemblage that complement results presented in *Hollings and Kerrich* [2000].

RESULTS

Major elements and compatible trace elements provide important constraints on the evolution of igneous suites in both study areas and establish some distinctions between the BRG and the Birch–Uchi South Bay (SB). A plot of MgO versus silica defines a broad trend for the BRG (Figure 2A). A similar plot for the SB area suggests that Mg addition has occurred in some of the felsic samples of that study area (Figure 2B). Given the proximity of massive sulfide mineralisation, addition of Mg is most likely attributable to seafloor hydrothermal processes involving extraction of Mg from seawater [*Franklin et al.*, 1981]. A variation diagram of FeO_T versus silica defines a broad continuous trend for the BRG, although compositions between 64 and 72 wt.% SiO_2 appear to be underrepresented compared to samples with higher or lower silica contents (Figure 2C). Primitive Pelletier tholeiites and transitional counterparts have similar Mg and Fe abundances at a given silica content and have overlapping Fe contents at a given molar Mg# value (not shown). These relationships rule out generation of the transitional suite via crustal contamination of tholeiites when considered in the context of trace element features such as REE contents and fractionation discussed below. An FeO_T-silica variation plot for the SB samples clearly distinguishes two distinct suites on the basis of Fe trends (Figure 2D). In the first, tholeiitic basalts define an iron enrichment trend toward evolved Febasalts and icelandites (with P_2O_5 up to 0.4 wt.%) that extends through dacites to relatively Fe-rich felsic samples. Calc alkaline basalts do not display iron enrichment but constitute the more primitive part of a second, comparatively low-Fe, trend that converges with the first at rhyolitic silica contents. The third SB suite, Nb-enriched basalts, define a subvertical trend corresponding to large variation in iron content over a small silica range.

The silica variation plots for both areas illustrate that comparatively few samples lie between about 65 and 70 wt.% SiO_2. Sampling was not undertaken proportionally based on exposure or map unit areas. Nonetheless, mapping across the Superior Province indicates that the compositional gaps

Table 1. Major and Trace Element Data for Representative Blake River Group Samples.

	Pelletier Tholeiite		Rouyn–Noranda Tholeiite	NMS Andesites			Rhyolite	NMS Rhyolites Rhyolite			Cycle 3 Tholeiites		Cycle 4		Eastern BRG		
	PEL95-2	PEL95-4	HO19	D 416	N 247	D 397	DT 3	AN 88	DT 4	DT 11	HO18-1	HO17-2	DEL9	DEL6A	ALT-3	CLR-9	ALT-2
SiO_2	44.54	45.82	50.06	55.46	56.48	55.30	58.42	73.44	71.38	77.19	76.34	74.26	76.46	75.07	48.29	53.85	69.98
TiO_2	1.03	0.84	1.72	1.26	1.00	1.74	1.54	0.54	0.58	0.23	0.25	0.25	0.27	0.28	1.34	1.27	1.17
Al_2O_3	17.51	17.74	13.97	15.78	16.28	15.17	15.34	13.16	13.62	11.13	10.92	11.58	12.22	13.38	13.78	14.05	13.17
Fe_2O_3	13.96	12.08	16.52	10.69	9.11	15.17	12.92	1.78	2.10	2.15	4.63	5.95	4.79	4.97	13.99	15.75	4.12
MnO	0.19	0.17	0.22	0.31	0.16	0.21	0.18	0.08	0.10	0.10	0.21	0.12	0.04	0.16	0.22	0.28	0.10
MgO	10.14	10.11	6.37	7.41	4.82	6.34	5.65	3.74	4.94	2.22	1.16	0.90	2.71	2.70	6.69	7.41	1.48
CaO	10.83	12.30	7.89	4.80	9.65	2.41	1.68	1.34	1.78	3.46	1.75	1.95	0.75	0.12	13.15	5.34	3.66
Na_2O	1.75	0.90	3.04	3.95	2.04	3.13	3.98	5.42	4.64	3.30	4.56	4.26	0.96	2.70	2.21	1.81	5.34
K_2O	0.00	0.00	0.06	0.16	0.30	0.09	0.07	0.40	0.73	3.30	0.13	0.68	1.76	2.70	0.18	0.12	0.73
P_2O_5	0.07	0.05	0.15	0.17	0.17	0.34	0.20	0.10	0.11	0.04	0.04	0.03	0.04	0.04	0.16	0.13	0.24
LOI	10.08	7.59	2.83	6.67	2.35	4.82	4.71	2.46	3.36	1.21	2.30	3.63	3.84	4.71	4.49	4.39	2.62
Cr	525	497	65	93	98	72	38	11	9	29	43	41	44	45	504	50	12
Co	55	55	55	33	25	31	25	2	4	1	2	1	6	1	0	0	0
Ni	175	210	68	51	34	22	25	2	3	6	1	0	4	3	3	3	0
Rb	0	1	1	4	8	2	2	7	15	51	4	13	38	55	3	3	19
Sr	136	209	107	110	173	70	53	94	78	73	64	52	39	49	174	99	139
Cs	0.25	0.21	0.24	0.23	0.20	0.14	0.14	0.12	0.28	0.47	0.11	0.59	1.22	2.40	0.05	0.31	0.33
Ba	9	8	25	103	81	36	35	144	241	534	45	162	399	422	45	37	214
Sc	53.2	42.4	47.5	32.4	28.5	38.9	35.3	15.9	18.3	9.4	12.6	11.8	2.7	6.8	34.5	47.6	21.8
V	350	324	506	286	234	279	289	17	25	0	64	52	1	1	316	343	32
Ta	0.17	0.14	0.36	0.52	0.53	0.62	0.69	1.03	1.06	1.00	1.03	1.12	1.18	1.25	0.25	0.44	0.56
Nb	2.37	2.15	6.31	7.08	7.19	10.12	10.46	15.90	15.97	15.61	19.48	18.29	19.56	24.01	3.77	7.09	9.28
Zr	38	31	103	131	111	157	170	273	281	294	260	259	279	339	48	106	135
Hf	1.10	0.88	3.14	3.85	3.24	4.49	4.99	8.24	8.97	9.09	7.21	7.43	7.86	8.97	1.10	3.15	4.03
Th	0.26	0.17	0.63	1.80	1.48	1.31	1.32	3.52	3.90	3.01	2.79	3.13	3.47	3.77	0.42	0.86	1.58
U	0.04	0.04	0.19	0.48	0.40	0.45	0.32	0.97	1.04	0.77	0.68	0.80	0.82	0.89	0.07	0.18	0.46
Y	15.6	13.9	34.6	21.9	20.5	27.0	22.2	48.1	67.3	44.4	35.7	45.0	56.7	67.6	16.3	29.0	28.2
La	2.67	1.94	6.43	13.26	13.37	12.51	14.13	18.68	31.62	22.19	22.88	25.12	24.92	26.83	4.61	7.19	12.41
Ce	7.09	5.30	17.35	32.27	33.01	32.89	37.17	49.60	79.15	55.39	54.97	60.25	59.79	66.24	11.52	18.72	28.11
Pr	1.07	0.83	2.55	4.41	4.61	4.86	5.47	7.13	10.72	7.55	6.66	7.46	7.53	8.09	1.71	2.66	3.69
Nd	5.66	4.26	10.44	17.01	17.35	20.15	22.17	27.62	40.33	29.39	29.61	33.55	33.28	36.07	8.86	12.86	17.20
Sm	1.87	1.43	4.29	3.95	3.92	5.27	5.35	6.81	9.31	6.97	7.16	8.18	8.15	8.90	2.41	3.64	4.33
Eu	0.82	0.71	1.35	1.31	1.16	1.60	1.72	1.23	1.62	1.68	1.83	2.00	1.87	2.04	0.94	0.77	1.39
Gd	2.46	1.91	4.79	4.30	4.12	6.05	5.96	8.15	10.51	8.25	7.31	8.68	8.03	8.80	2.96	4.64	5.13
Tb	0.42	0.32	0.84	0.64	0.61	0.85	0.81	1.34	1.65	1.30	1.00	1.11	1.38	1.53	0.46	0.81	0.79
Dy	2.78	2.22	5.92	4.39	4.03	5.35	4.97	9.80	12.22	8.70	6.11	7.79	9.76	10.90	3.16	5.52	5.63
Ho	0.62	0.49	1.20	0.85	0.82	1.09	0.94	1.87	2.60	1.84	1.25	1.53	2.06	2.38	0.67	1.17	1.17
Er	1.76	1.42	3.65	2.43	2.47	3.13	2.72	6.64	8.27	5.65	4.01	5.22	6.31	7.46	1.98	3.67	3.50
Tm	0.26	0.21	0.56	0.37	0.36	0.47	0.42	0.94	1.30	0.91	0.60	0.79	1.03	1.13	0.28	0.56	0.55
Yb	1.84	1.28	3.74	3.07	2.26	3.50	3.20	6.46	8.38	5.95	4.44	5.83	6.87	7.51	1.70	3.67	3.64
Lu	0.23	0.19	0.49	0.38	0.35	0.52	0.56	0.99	1.37	0.93	0.75	0.96	1.11	1.31	0.26	0.54	0.55
$[La/Yb]_N$	1.0	1.1	1.2	3.1	4.2	2.6	3.2	2.1	2.7	2.7	3.7	3.1	2.6	2.6	1.9	1.4	2.4
$[La/Sm]_N$	0.9	0.9	1.0	2.2	2.2	1.5	1.7	1.8	2.2	2.1	2.1	2.0	2.0	1.9	1.2	1.3	1.9
$[Gd/Yb]_N$	1.1	1.2	1.1	1.2	1.5	1.4	1.5	1.0	1.0	1.1	1.4	1.2	1.0	1.0	1.4	1.0	1.2
Eu/Eu^*	1.2	1.3	0.9	1.0	0.9	0.9	0.9	0.5	0.5	0.7	0.8	0.7	0.7	0.7	1.1	0.6	0.9
Al_2O_3/TiO_2	17.0	21.2	8.1	12.5	16.3	8.8	9.9	24.3	23.6	47.8	43.9	46.1	45.9	47.1	10.3	11.1	11.2
Nb/La	0.89	1.11	0.98	0.53	0.54	0.81	0.74	0.85	0.50	0.70	0.85	0.73	0.78	0.89	0.82	0.99	0.75
Zr/Y	2.4	2.2	3.0	6.0	5.4	5.8	7.7	5.7	4.2	6.6	7.3	5.8	4.9	5.0	3.0	3.6	4.8

NMS = Noranda Mine Sequence; BRG = Blake River Group; Blk = Block

Table 1 (continued).

	Eastern BRG Calc Alkaline		East Horne Blk Tholeiites		Western Horne Blk Tholeiites		Central BRG Tholeiites			Central BRG Calc Alkaline			West BRG Calc Alkaline		
	TRE-8	ALT-6	HO16	HO11-2	FLAG10	FLAG3	EV19	EV6	ARN-10	ARND	EV3	ARN-2	BNX4	BNX5	POS-3
SiO_2	49.38	58.11	53.06	65.91	48.95	78.35	48.80	60.83	52.34	49.72	52.61	66.25	50.80	57.22	74.76
TiO_2	0.83	1.01	1.40	0.63	1.98	0.14	0.64	1.51	1.47	0.69	0.85	0.66	0.87	0.88	0.31
Al_2O_3	13.08	16.70	13.64	15.02	13.49	9.67	15.68	14.41	14.13	16.09	15.56	15.00	13.62	16.48	13.09
Fe_2O_3	11.39	5.45	14.28	8.30	18.71	2.71	10.59	9.50	14.13	10.55	10.54	7.76	12.36	8.84	3.09
MnO	0.32	0.07	0.22	0.15	0.22	0.09	0.19	0.11	0.22	0.15	0.16	0.14	0.19	0.17	0.05
MgO	11.45	4.75	6.16	3.28	5.29	0.72	8.69	5.58	5.39	10.55	8.45	0.99	12.67	6.37	0.83
CaO	10.96	8.55	8.50	3.01	7.76	4.62	13.60	4.11	9.39	9.12	7.28	4.82	7.69	6.39	2.68
Na_2O	0.87	5.18	2.57	1.37	2.78	1.96	1.51	3.40	2.32	2.26	2.33	0.58	1.40	1.11	1.50
K_2O	1.33	0.05	0.03	2.20	0.71	1.71	0.26	0.29	0.51	0.81	2.09	3.63	0.25	2.44	3.61
P_2O_5	0.39	0.12	0.14	0.14	0.12	0.02	0.05	0.26	0.11	0.07	0.12	0.17	0.16	0.11	0.07
LOI	20.53	1.47	6.95	4.49	9.23	4.66	2.56	2.93	0.76	2.41	2.35	5.54	4.49	11.60	4.00
Cr	657	46	154	82	94	39	619	69	82	202	378	20	565	236	7
Co	41	20	46	14	42	3	54	24	33	41	44	9	73	33	3
Ni	120	22	46	2	67	3	127	22	26	170	205	4	434	155	2
Rb	26	1	1	70	19	64	10	10	11	30	64	116	8	79	93
Sr	131	145	144	57	136	84	157	119	162	186	200	90	168	61	80
Cs	0.81	0.04	0.10	2.36	0.68	1.23	0.03	0.04	0.32	1.03	0.72	2.50	0.57	1.42	3.37
Ba	244	66	25	341	150	206	63	101	46	111	254	371	146	359	609
Sc	33.5	33.6	45.5	17.7	66.5	11.3	114.1	78.6	48.0	21.2	43.8	15.1	33.5	12.2	7.6
V	225	299	335	52	682	1	266	298	602	164	225	31	182	190	10
Ta	0.29	0.41	0.40	0.48	0.35	0.90	0.14	0.45	0.22	0.19	0.27	0.76	0.27	0.41	1.08
Nb	4.83	4.99	6.79	8.42	5.95	14.32	2.04	7.04	2.54	3.05	4.31	11.06	4.23	5.14	13.22
Zr	93	89	101	116	85	255	29	112	40	51	72	236	61	116	241
Hf	2.87	2.68	3.03	3.10	2.47	7.61	0.85	2.90	1.28	1.44	1.94	6.68	1.53	3.09	8.60
Th	1.84	1.50	0.78	2.02	0.70	3.18	0.34	1.24	0.44	0.66	1.07	3.09	1.19	2.40	5.28
U	0.59	0.40	0.17	0.48	0.16	0.74	0.04	0.27	0.16	0.22	0.22	0.77	0.26	0.53	1.41
Y	12.3	15.1	33.3	20.2	22.5	48.2	11.4	26.2	12.6	9.9	14.7	33.3	15.7	15.3	56.2
La	13.85	14.65	7.71	10.90	6.36	17.60	1.91	9.04	3.36	5.33	7.65	21.65	11.74	11.89	29.72
Ce	32.28	33.34	20.14	25.26	16.97	44.75	5.16	23.31	8.89	12.46	18.65	49.87	30.24	25.28	72.45
Pr	4.47	4.24	2.75	3.21	2.33	5.70	0.72	2.95	1.34	1.53	2.32	6.71	3.87	2.84	9.85
Nd	20.17	15.38	13.81	13.57	11.52	26.56	3.92	13.57	6.11	5.93	9.87	25.67	16.87	11.62	37.62
Sm	4.77	3.01	4.23	3.24	3.46	6.57	1.34	3.61	1.69	1.76	2.44	6.10	3.34	2.67	9.31
Eu	1.29	0.94	1.33	0.78	1.27	1.15	0.57	0.52	0.74	0.60	0.85	1.57	0.93	1.04	1.38
Gd	3.80	3.00	5.08	3.53	3.72	6.71	1.80	4.53	2.17	1.77	2.69	6.66	2.87	2.67	10.01
Tb	0.45	0.44	0.80	0.52	0.60	1.16	0.25	0.64	0.35	0.26	0.40	0.98	0.40	0.42	1.52
Dy	2.64	2.89	5.49	3.72	4.01	7.93	2.00	4.28	2.42	1.82	2.49	6.56	2.59	2.72	10.66
Ho	0.48	0.60	1.21	0.71	0.82	1.65	0.40	0.88	0.51	0.34	0.51	1.35	0.52	0.54	2.26
Er	1.57	1.78	3.61	2.30	2.55	5.41	1.13	2.87	1.57	1.03	1.46	4.16	1.47	1.70	7.33
Tm	0.25	0.26	0.53	0.36	0.37	0.90	0.14	0.41	0.26	0.17	0.20	0.64	0.21	0.23	1.11
Yb	1.62	1.76	3.58	2.31	2.59	6.03	1.22	2.76	1.67	1.39	1.53	4.18	1.40	1.54	7.69
Lu	0.25	0.27	0.55	0.39	0.40	1.01	0.21	0.41	0.25	0.15	0.25	0.65	0.22	0.27	1.28
$[La/Yb]_N$	6.1	6.0	1.5	3.4	1.8	2.1	1.1	2.3	1.4	2.7	3.6	3.7	6.0	5.5	2.8
$[La/Sm]_N$	1.9	3.1	1.2	2.2	1.2	1.7	0.9	1.6	1.3	2.0	2.0	2.3	2.3	2.9	2.1
$[Gd/Yb]_N$	1.9	1.4	1.2	1.3	1.2	0.9	1.2	1.4	1.1	1.1	1.5	1.3	1.7	1.4	1.1
Eu/Eu*	0.9	0.9	0.9	0.7	1.1	0.5	1.1	0.4	1.2	1.0	1.0	0.8	0.9	1.2	0.4
Al_2O_3/TiO_2	15.7	16.5	9.8	23.8	6.8	70.2	24.5	9.5	9.6	23.3	18.3	22.9	15.7	18.7	41.7
Nb/La	0.35	0.34	0.88	0.77	0.94	0.81	1.07	0.78	0.76	0.57	0.56	0.51	0.36	0.43	0.44
Zr/Y	7.5	5.9	3.0	5.7	3.8	5.3	2.5	4.3	3.2	5.2	4.9	7.1	3.9	7.6	4.3

Table 2. Major and Trace Element Data for Representative South Bay Area Samples.

	South Bay mine site			Horseshoe Lake				Dixie 19		Dixie South			Dixie 18 DDH		
	F3	F3	F3	F3a	F3a	F3b	F3b	F3	F3	F3a	F3b	F3b	Dacite	Dacite	Icelandite
	SB95-4A	SB95-4E	SB95-4L	HL96-1A	HL96-1I	HL96-1Q	HL96-1T	SJV91-1L	SJV91-1N	D94-1D	D94-1B	D94-1P	SJV91-2A	SJV91-2B	SJV91-2D
SiO_2	72.01	72.34	72.81	74.29	83.75	80.26	81.94	59.19	78.04	75.73	78.13	76.95	65.48	65.53	54.27
TiO_2	0.43	0.42	0.42	0.34	0.30	0.12	0.10	0.93	0.22	0.57	0.17	0.17	0.94	0.91	1.92
Al_2O_3	10.98	10.93	11.21	10.71	8.91	11.71	10.28	15.92	12.11	11.32	11.57	11.60	11.59	11.33	12.48
Fe_2O_3	7.88	7.26	6.61	7.62	1.57	1.97	0.69	12.80	4.27	4.61	2.22	2.77	10.78	11.02	15.72
MnO	0.11	0.12	0.11	0.15	0.03	0.07	0.00	0.10	0.01	0.05	0.03	0.32	0.08	0.14	0.36
MgO	1.13	0.92	0.80	0.50	0.13	2.38	0.08	9.35	3.55	0.96	1.18	0.37	1.67	1.60	3.49
CaO	3.09	2.71	2.57	0.90	0.71	0.43	0.31	0.26	0.26	1.93	2.02	0.67	4.29	4.67	8.15
K_2O	0.88	1.49	1.11	2.57	1.04	1.60	2.62	0.87	0.89	1.72	1.30	3.46	1.45	0.76	0.60
Na_2O	3.39	3.71	4.24	2.88	3.54	1.47	4.00	0.47	0.61	3.03	3.37	3.70	3.45	3.78	2.60
P_2O_5	0.10	0.10	0.10	0.03	0.02	0.00	0.00	0.11	0.03	0.08	0.02	0.00	0.27	0.26	0.43
LOI	1.32	1.21	0.96	1.16	0.81	1.99	0.45	3.68	0.96	0.35	1.06	0.55	0.60	0.91	0.20
Cr	4	4	6	5	3.91	3.31	14	3.43	6	4	5	4.19	7	8	20
Co	9	10	9	8	1.91	3.48	1	42.29	12	16	7	6.59	29	29	53
Ni	3	2	2	1	0.92	1.52	2	0.95	1	1	1	0.30	3	3	21
Rb	16	20	13	45	12.50	53.66	50	21.49	23	38	26	58.09	30	16	6
Sr	149	98	92	64	63.33	56.02	28	30.71	54	42	169	28.69	9	123	148
Cs	0.78	0.76	0.47	1.07	0.26	0.85	0.49	3.43	1.26	1.18	0.58	0.59	4.58	1.97	0.33
Ba	172	321	332	555	251.96	204.66	414	142.92	191	211	439	564.06	222	85	193
Sc	8.15	8.25	8.13	4.86	3.81	2.14	1.86	18.10	3.00	11.83	3.23	2.67	21.99	22.67	39.80
V	4	4	4	1	1	1.78	1	12.66	4	12	7	0.59	35	35	187
Ta	1.65	1.63	1.65	2.40	1.85	2.54	1.98	2.56	1.46	2.11	2.46	3.07	1.14	1.14	0.54
Nb	29.75	31.85	29.81	42.42	34.67	34.66	26.91	54.14	21.87	34.68	35.11	52.09	19.68	17.98	9.53
Zr	628	675	631	756	609	183	144	675	245	479	233	509	390	386	183
Hf	15.46	15.49	16.14	19.38	15.49	7.54	6.00	16.98	8.29	13.24	8.71	17.89	10.68	10.46	4.58
Th	4.97	4.92	4.95	8.23	6.16	15.61	12.46	11.62	14.01	9.93	13.51	10.17	3.49	3.47	1.43
U	1.25	1.27	1.28	2.10	1.54	3.58	2.78	1.04	1.86	2.42	3.26	2.63	0.93	0.87	0.36
Y	130	133	127	173	107	94	84	60	68	112	103	168	98	75	52
La	36.22	37.12	36.42	58.53	39.92	62.08	55.52	83.18	69.84	52.16	60.36	75.74	24.05	24.13	12.84
Ce	89.57	91.41	89.39	139.15	94.84	130.85	117.55	200.65	144.04	118.81	132.93	181.59	60.98	61.23	31.81
Pr	12.26	12.62	12.36	18.50	12.73	14.85	13.47	26.13	15.97	14.86	15.70	24.15	8.72	8.56	4.46
Nd	55.61	57.07	56.72	82.13	56.31	54.74	49.28	115.12	59.84	62.39	60.91	106.60	40.97	39.88	21.28
Sm	15.70	15.83	16.25	22.12	14.76	11.95	10.71	26.38	11.33	15.21	13.78	28.12	12.05	11.39	6.46
Eu	3.71	3.63	3.80	4.00	2.86	0.76	0.55	3.70	1.02	2.49	0.95	3.64	4.35	3.39	2.31
Gd	19.13	18.86	19.15	25.20	16.63	11.35	10.70	32.41	11.15	17.74	14.08	28.67	15.72	13.43	8.12
Tb	3.27	3.34	3.38	4.32	2.79	2.14	1.91	5.40	1.76	2.90	2.68	4.98	2.74	2.13	1.32
Dy	22.55	21.92	22.38	29.07	18.77	14.81	13.20	36.33	11.43	19.52	18.13	32.45	18.00	13.04	8.90
Ho	5.02	4.80	5.01	6.29	4.11	3.45	2.93	8.00	2.52	4.16	4.06	6.96	3.83	2.90	1.93
Er	14.72	14.16	14.61	18.44	12.49	10.64	9.12	23.75	7.47	12.26	12.27	19.83	11.03	8.57	5.43
Tm	2.29	2.22	2.19	2.95	1.93	1.71	1.46	3.68	1.15	1.85	2.00	3.03	1.66	1.33	0.82
Yb	14.25	14.43	14.58	19.35	12.70	10.94	9.54	23.69	7.59	12.26	13.39	20.38	10.80	8.77	5.32
Lu	2.18	2.13	2.17	2.82	1.88	1.52	1.40	3.61	1.09	1.81	1.87	2.87	1.66	1.44	0.78
[La/Yb]$_N$	1.8	1.8	1.8	2.2	2.25	4.07	4.2	2.52	6.6	3.0	3.2	2.66	1.6	2.0	1.7
[La/Sm]$_N$	1.5	1.5	1.4	1.7	1.75	3.36	3.3	2.04	4.0	2.2	2.8	1.74	1.3	1.4	1.3
[Gd/Yb]$_N$	1.1	1.1	1.1	1.1	1.08	0.86	0.9	1.13	1.2	1.2	0.9	1.16	1.2	1.3	1.3
Eu/Eu*	0.7	0.6	0.7	0.5	0.56	0.20	0.2	0.39	0.3	0.5	0.2	0.39	1.0	0.8	1.0
Al_2O_3/TiO_2	27.6	28.2	27.5	37.3	37.7	98.2	110.8	18.0	73.6	19.7	67.9	67.8	12.1	12.4	6.6
Nb/La	0.8	0.9	0.8	0.7	1.15	1.79	0.5	1.54	0.3	0.7	0.6	1.45	0.8	0.7	0.7
Zr/Y	4.8	5.1	5.0	4.4	0.9	0.6	1.7	0.7	3.6	4.3	2.3	0.7	4.0	5.1	3.5

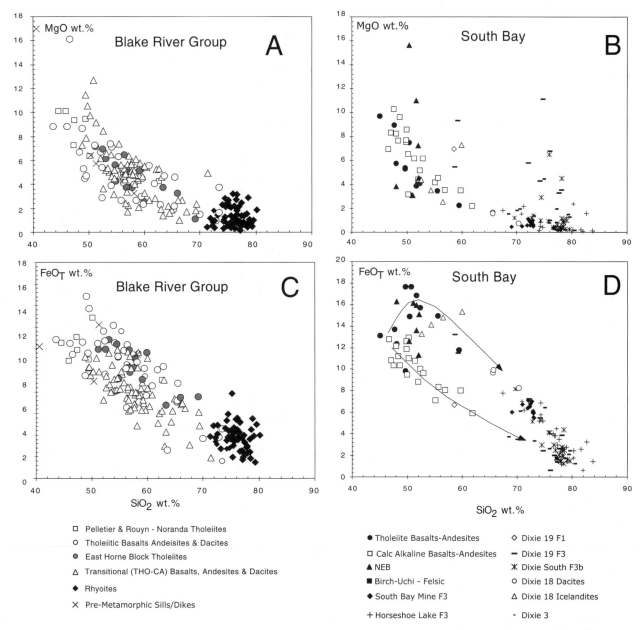

Figure 2. Silica–FeO$_T$ and silica–MgO variation diagrams for samples from the Blake River Group and South Bay study areas, including regional Birch–Uchi felsic samples from *Hollings and Kerrich* [2000].

evident on silica variation plots reflect real "bimodal" distributions in rock types and a scarcity of dacitic units in many Archean greenstone belts [e.g., *Sage et al.*, 1996].

In order to minimize the effects of alteration, variation diagrams in Figure 3 employ a ratio of two relatively immobile species, Al$_2$O$_3$/TiO$_2$. A plot of TiO$_2$ versus Al$_2$O$_3$/TiO$_2$ for the BRG sample set displays a well-defined trend, as expected from a suite of related samples (Figure 3B). A single gap in

the trend at Al$_2$O$_3$/TiO$_2 \approx 30$ corresponds to a paucity of dacitic samples. The SB samples define a slightly more complex trend because of the diverse origins of multiple SB igneous suites. Calc alkaline samples tend to have higher Ti contents at a given Al$_2$O$_3$/TiO$_2$ than do tholeiitic samples (Figure 3B). A few samples fall in the dacitic range of Al$_2$O$_3$/TiO$_2$, but a gap at ≈ 1.25 TiO$_2$ is shared by both tholeiitic and calc alkaline samples. For the BRG, moderately

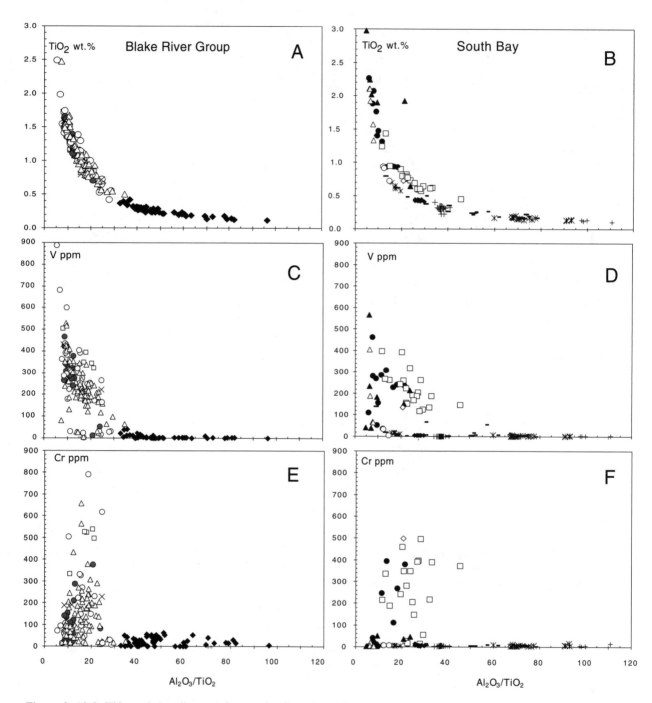

Figure 3. Al_2O_3/TiO_2 variation diagrams for samples from the Blake River Group and South Bay study areas. Symbols and data sources as in Figure 2.

compatible V and Sc (not shown) define trends that are similar to the TiO_2 plot, although rather than a smooth curve at $Al_2O_3/TiO_2 \approx 20$, V exhibits a continued steep decline in abundance (Figure 3C). The SB data for V make evident an early trend from $Al_2O_3/TiO_2 \approx 22$ (for primitive tholeiites) to lower values (8) in more evolved basalts (Figure 3D).

Following early increases in V, a sharp inflection occurs as magma evolution proceeds to markedly higher Al_2O_3/TiO_2 ratios when Ti-magnetite is removed by fractionation. Strongly compatible elements such as Cr (or Ni and MgO, not shown) decline in abundance even as Al_2O_3/TiO_2 initially decreases in the BRG tholeiite and transitional suites and in the SB tholeiite suite (Figure 3). The SB calc alkaline suite exhibits no clear trend for these elements, although abundances of both Cr and Ni fall to low levels in more evolved rock types.

Figure 4 plots variation diagrams of incompatible trace elements and trace element ratios against highly incompatible and immobile Nb. Several clear distinctions are evident between the two study areas. First, the SB data set extends to much higher incompatible element contents than the BRG. Second, whereas two trends are barely evident (at the scale plotted) in the Zr versus Nb plot for the BRG, a well-defined main trend and diffuse second trend is obvious on the SB counterpart (Figures 4 A,B). Many, but not all, of the SB samples that make up the low-Zr trend also have anomalously high Th contents and therefore are likely to have undergone substantial crustal assimilation (Figure 4D). The BRG $[La/Sm]_N$ versus Nb plot (Figure 4E) distinguishes between the transitional tholeiite–calc alkaline suite—which increases to more fractionated LREE ratios at low Nb—and the less fractionated tholeiite suite. Apart from one sample, the East Horne Block tholeiites produce a subhorizontal trend that is similar to the primitive segment of the SB tholeiite trend at $[La/Sm]_N \approx 1.5$. A plot of Nd/Zr versus Nb measures the degree and nature of HFSE anomalies versus the REE (Figures 4 G,H). Normalized Zr depletions (high Nd/Zr) may reflect hydrous mantle metasomatism or assimilation of crust produced from that environment and, in the case of felsic rocks, depletions or enrichments may reflect zircon fractionation or accumulation. Accordingly, high Nd/Zr observed in some BRG andesites may signal the late onset of hydrous mantle wedge metasomatism, as inferred by *Wyman et al.* [2002] and *Burrows and Spooner* [1989] for the southern Abitibi in general. A notable feature of the relationship between major and incompatible trace elements is summarized on plots of Zr versus Al_2O_3/TiO_2 in Figure 5. Whereas BRG rhyolites have relatively low and uniform Zr contents, SB samples encompass a wide range of Zr contents, in some instances within a single drill hole. The Horseshoe Lake samples have variable Zr at a near-constant Al_2O_3/TiO_2 of about 40, and a large range of Zr abundances occurs in Dixie South samples with $Al_2O_3/TiO_2 \approx 70$ (Figure 5B).

DISCUSSION

Geochemical Characteristics of Mafic-Intermediate Rock Types

Some differences between Archean and post-Archean mafic rocks may relate to the chemical or thermal evolution of the mantle over time. For example, Archean basalts are commonly Fe-rich compared to post-Archean examples with a similar MgO content, which in some cases may be linked to

somewhat deeper melting [*Cattel and Taylor*, 1990; *Condie*, 1985]. Nonetheless, the evolutionary trends of greenstone belt tholeiite suites help to establish (1) whether different Archean terranes formed by similar tectonic processes and (2) whether those processes resembled younger tectonic mechanisms. Proposed Archean plate tectonics analogues, such as the propagating rifted oceanic arc scenario applied to the BRG [*Péloquin et al.*, 1996] imply thin arc crust and comparatively shallow mantle melt sources with variable degrees of subduction-generated metasomatism. The scenario should be distinguishable from continental rift processes such as the mid-Proterozoic Keweenawan rifting, proposed by *Bleeker* [2002] as a possible analogue for the formation of many Archean greenstone belts. For example, much of the partial melting in this rift setting occurs at depths as great as 100 km [*Shirey et al.*, 1994].

The BRG Pelletier tholeiites correspond to a propagating rift in the model of *Péloquin et al.* [1996]. Their MgO contents of 9–10 wt.% are comparable to the maximum Lau values reported by *Pearce et al.* [1994] for the Intermediate Lau Spreading Centre (ILSC). In contrast, deeply sourced picritic magmas of the type documented in the Keewenawan rift ($\approx$18 wt.% MgO: *Klewin and Berg* [1991]) are not observed in either the BRG or SB study areas. The maximum FeO_T contents of Pelletier tholeiites and SB basalts, 12.6 wt.% and $\approx$ 15 wt.%, respectively, at $\approx$ 10 wt.% MgO [*Hollings and Kerrich*, 2000], are distinctly higher than the $\approx$ 9 wt.% FeO_T reported by *Pearce et al.* [1994] for Lau basalts at comparable Mg content. The Fe and Mg contents of the Archean basalts are similar to some Keweenawan basalts [e.g., *Walker et al.*, 2002] but the Al_2O_3/TiO_2 values of these continental rift basalts (3 to 5) are much lower than the near-chondritic ratios of the most primitive BRG and SB tholeiites (17 to 21).

The TiO_2 contents of 0.8–1 wt.% of the primitive Archean basalts are slightly lower than the 1.1–1.2 wt.% exhibited by either high-Mg lavas erupted along the Lau basin spreading centres or uncontaminated Keweenawan basalts of similar Mg content [*Shirey et al.*, 1994]. Samples of the eastern Horne Block and the Rouyn-Noranda tholeiites are less magnesian than the Pelletier examples. Given their proximity to the Pelletier Tholeiites and occurrence to the south of the main caldera sequences (Figure 1; *Kerr and Mason* [1990]), they were likely to be derived from similar sources and may represent more fractionated parts of a single suite. The three BRG suites are plotted on a TiO_2 versus MgO variation diagram (Figure 6A) along with reference magma evolutionary paths derived from a primitive sample (PEL95-4) by using the MELTS/pMELTS program of *Ghiorso et al.* [2002] and varying pressure, starting magma H_2O content, and oxidation state of the system. As is the case for the Lau basin, Ti contents reach a maximum between MgO of 4 and 5 wt.%

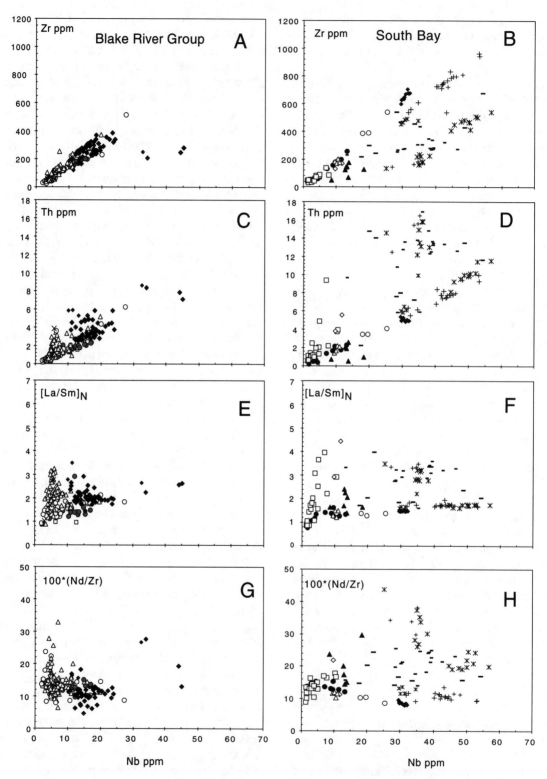

Figure 4. Trace element versus Nb variation diagrams for samples from the Blake River Group and South Bay study areas, including regional Birch–Uchi felsic samples. Symbols and data sources as in Figure 2.

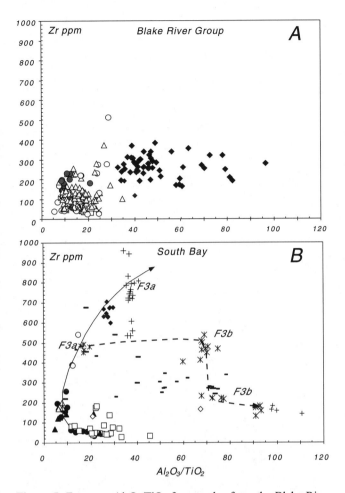

Figure 5. Zr versus Al₂O₃/TiO₂ for samples from the Blake River Group and South Bay study areas. The F3a and F3b felsic sub-types of the Horseshoe Lake and Dixie South sample sets are noted in B. Arrows show inferred magma fractionation paths from low to high Al₂O₃/TiO₂. Symbols and data sources as in Figure 2.

fractionation of SB tholeiitic magma at slightly lower pressures compared to the BRG.

The incompatible trace element systematics of mafic rocks from the two study areas distinguish their mantle sources and constrain related events in the asthenospheric mantle. For example, *Lafleche et al.* [1992a] used a plot of La/Yb versus Nb/Th to show that mixing BRG F3 rhyolite magmas and tholeiitic basaltic magmas cannot account for the spectrum of compositional variations observed in BRG volcanic suites. They inferred that early mixing of 20% end-member calc alkaline magma with tholeiite magmas, followed by fractionation, could account for the range of tholeiite compositions observed in the northeast BRG. Based on an Abitibi-wide data set, Wyman (2003) argued that adakitic magmas had variably metasomatised the mantle sources of many igneous suites in the southern Abitibi, particularly the younger sequences such as the BRG. Figure 7 demonstrates that adakitic magmas, as represented in the southern Abitibi by inner phases of the Round Lake Batholith or the "Krist Fragmental" F1-type volcanic unit of the Timmins area, provide an appropriate end-member component that could have metasomatised tholeiite sources or mixed with tholeiitic melts to produce a BRG "tholeiite–calc alkaline array". Mantle sources of the BRG and SB study areas can also be compared to possible modern subduction zone counterparts on a Nb–Zr–Y plot that largely screens out the effects of hydrous mantle metasomatism.

Pearce et al., 1994]. The maximum measured TiO₂ values, however, are lower than the maximum Lau basin levels of ≈ 2.5 wt.%. The results are consistent with fractionation at moderate depths (less than 2 kbar) and initial H₂O contents (1–2%) with oxygen fugacity near the NNO buffer.

The small numbers of SB tholeiite samples do not permit a rigorous assessment of the pressures and temperatures associated with magma evolution. The most primitive samples, however, have MgO contents (9–10 wt.%) and TiO₂ contents (0.7–0.9 wt.%) that overlap those of the primitive BRG tholeiites (Figure 6). Apart from two samples with unusually low Ti contents, the SB tholeiites define a trend that resembles that of the BRG, in that a maximum is reached at approximately 4.5 wt.% MgO, although the SB values of ≈ 2.0 wt.% are slightly higher. The result may reflect

Mantle Nb/Y Versus Zr/Y Arrays

Figure 8 summarises Nb/Y versus Zr/Y arrays for post-Archean mantle-derived rock types, the southern Abitibi belt, and select Archean greenstone belts with trends defined by samples from the BRG and SB. The slope of the mantle plume (Iceland) and modern MORB arrays (Figure 8A) is a function of the distribution coefficients of Zr, Nb, and Y in mantle phases [*Fitton et al.*, 1997]. As discussed by *Pearce et al.* [1995], mafic volcanic rocks from most Phanerozoic oceanic island arcs plot along the MORB array because hydrous metasomatism of their mantle wedge sources does not transport large quantities of Zr or Nb. Phanerozoic oceanic arc volcanic suites are illustrated in Figure 8A and include, for example, the low-K tholeiite, tholeiite, and calc alkaline series of the South Sandwich arc [*Pearce et al.*, 1995]. Data for the 2.75-Ga Manitouwadge and Winston Lake belts of Wawa Subprovince define a mantle array having a slope comparable to modern mantle sources and lying between Primitive Mantle values and modern N-MORB [*Wyman*, 2003]. This trend is consistent with mantle wedge sources that differ from modern counterparts only in the extent of time-integrated Nb depletion.

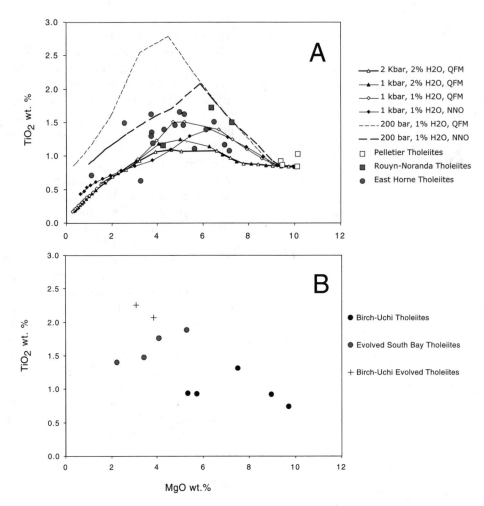

Figure 6. TiO_2 versus MgO plots for the Blake River Group and South Bay study areas. Reference lines in (A) represent modeled magma evolution paths from starting material corresponding to primitive Pelletier tholeiite sample PEL95-4 at specified pressures, initial magma H_2O contents, and oxidation states of the system. Calculated using the MELTS program of *Ghiorso et al.* [2002, and references therein]. South Bay data from *Hollings and Kerrich* [2000].

Figure 8A also summarises the results for primitive rocks of the BRG and illustrates several key features of mafic volcanic rocks in the southern Abitibi belt. In particular, the trends crosscut the plume and MORB-modern arc arrays, and the youngest rocks tend to occur at higher Zr/Y values. Trends for mafic volcanic rocks of the BRG begin in the mantle plume array and document the continued presence of plume-related asthenosphere beneath the southern Abitibi belt that was tapped during episodes of arc extension. Figure 8A shows only BRG samples with MgO > 5 wt.% (recalculated on a volatile-free basis). A comparison with data for BRG samples having MgO < 5 wt.% (Figure 8B) illustrates that magma fractionation generally has little effect on these types of plots, unless a minor phase such as zircon is fractionated [*Pearce et al.*, 1995; *Wyman*, 2003]. The plot for evolved

BRG rocks also demonstrates that magma evolution did not drive magmas toward crust-like compositions as would be expected if the southern Abitibi trends (Figure 8B) resulted from crustal contamination rather than metasomatism of their sources.

Figure 8C plots mafic rocks (>5 wt.% MgO) of the Confederation assemblage in the SB study area. Their trend passes between the composition of primitive mantle and N-MORB and is subparallel to the slope of the plume and MORB arrays. Accordingly, it resembles the trend defined by Wawa greenstone belts in Figure 8A and shows no evidence for metasomatism of mantle sources by igneous melts. The NEB exhibit variable Nb/Zr ratios that partially overlap the main SB trend. Tholeiite and calc alkaline mafic rocks are not interstratified with NEB or adakites in the southwest of the

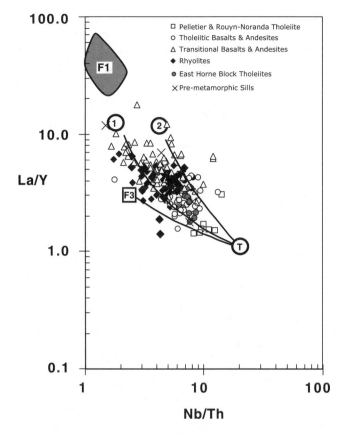

Figure 7. Mixing model for the Blake River Group magmas using La/Y versus Nb/Th. *Lafleche et al.* [1992a] noted that end-member tholeiite basalt [T] cannot be mixed with F3 rhyolite to produce the nearly continuous range of tholeiite and calc alkaline compositions that extend to points 1 and 2. Mixing [T], or its source, with adakitic magmas of the F1 field can account for the range of BRG compositions documented by *Lafleche et al.* [1992a] and the present study. The F1 field is based on the compositions of the adakitic intrusive and volcanic rocks of the southern Abitibi belt from *Feng and Kerrich* [1992] and *Wyman* [2003].

study area and therefore it is not possible to (1) gauge the effect that these magmas had on any contemporaneous mantle wedge sources or (2) unequivocally establish the age of the NEB and adakites relative to tholeiite and calc alkaline mafic rock types. Given the association of the NEB with the most contaminated rhyolites in the area, however, it is most plausible that they are linked to the early stages of the arc-rifting event. The restricted distribution of the NEB endorses the view that they are not related to a uniformly steep late-Archean geotherm.

The trend defined by less magnesian SB volcanic rocks (<5 wt.% MgO; Figure 8D) is more complex than that defined by evolved BRG samples. Basaltic to intermediate tholeiites in this category occur along the same Nb/Zr trend

as the more magnesian counterparts in Figure 8C. Calc alkaline (low-Mg) basaltic to dacitic rocks define a distinct trend that is similar to the BRG and may signify mixing of slab melts, input from tonalite–trondhjemite–granodiorite (TTG) melting at the base of the crust [e.g., *Smithies*, 2000], or direct crustal contamination. In support of the latter scenario, the Th contents of evolved SB calc alkaline rocks extends to much greater concentrations than those observed in the BRG (Figures 4 C,D). Although two rhyolites plots near the composition of average continental crust, many evolved rocks define a distinctive subhorizontal trend at Nb/Y ≈ 0.3. Given that multiple samples from individual localities occur in this trend, a mechanism is required that can progressively lower magma Zr contents while not greatly changing Nb content, most plausibly zircon fractionation.

An important implication of the differences between the ratio–ratio plots is that distinct processes generated the BRG transitional suite (Figure 8A) and the primitive SB calc alkaline suite (Figure 8C). Although primitive SB calc alkaline rocks exhibit LREE enrichment, the event that metasomatised their mantle source did not impact their Nb–Zr systematics, as is characteristic of hydrous fluids that act on the mantle wedge sources of typical modern oceanic arc basalts [*Pearce et al.*, 1995]. The trend defined by primitive BRG samples is typical for the southern Abitibi Subprovince. Similar trends are also associated with post-Archean arcs where slab melting is inferred to have occurred and with 3-Ga diorites of the Pilbara Craton, which were derived from melt-metasomatised sources (*Smithies and Champion* [2000]; Figure 8E).

Two types of metasomatic process have therefore been documented at convergent margins over a large part of Earth's history, which suggests that subduction-related metasomatism has not evolved from slab melting to slab dehydration simply as a result of cooling of the upper mantle since the (late) Archean period. Instead, melt metasomatism of mantle sources may be more directly linked to shallow subduction, as suggested by *Gutscher et al.* [2000]. Shallow subduction angles may occur as a result of several factors, including the subduction of unusually buoyant ocean plateau crust. As a result, Archean occurrences of adakite-like rocks may reflect an indirect product of regional- to global-scale mantle plume events recorded by widespread komatiites [*Wyman*, 2003; *Condie*, 2001].

Felsic Rock Types

Crustal thickness and the architecture of magma plumbing systems are likely to have more direct control on the evolution of felsic rock types. An important feature of the SB data is the variable Th and U enrichments within sample subsets derived from individual localities or drill holes. This feature

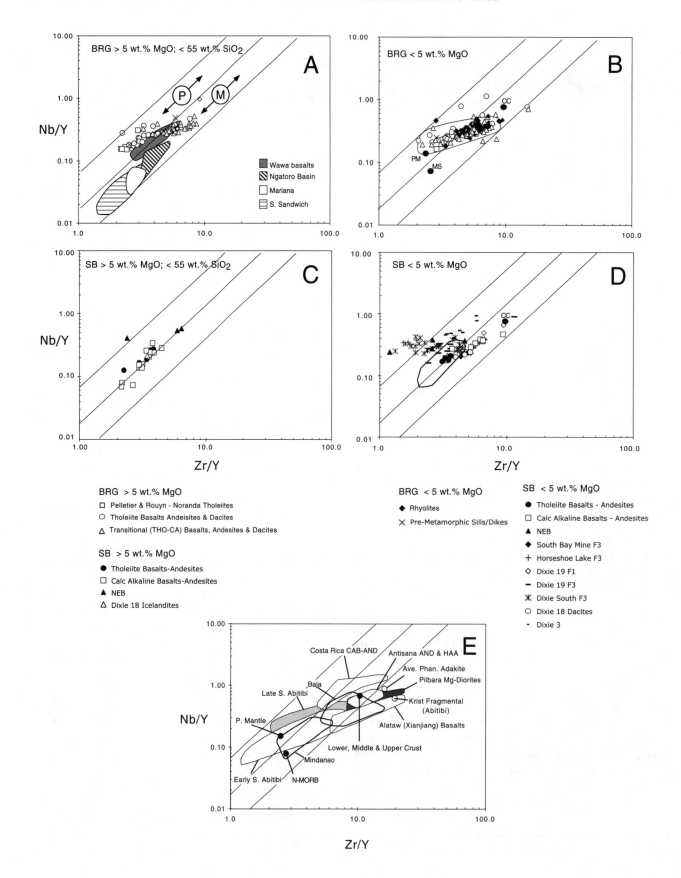

limits possible scenarios that would account for sporadic high Th contents by partial melting of anomalous rare crustal sources. Alternatively, it might be argued that high-Th samples represent small degree melts of a widespread crustal source of SB felsic rocks. If this scenario was valid, however, then high-Th samples should constitute end-members to a continuum of felsic rocks generated by variable degrees of partial melting. Instead, the high-Th samples define a "branch" that intersects the main SB trend at its mid-point (Figures 4 B,D).

A similar branch or splay is defined by many of the high-Th samples on the plot of $[La/Sm]_N$ versus Nb (Figure 4F). The main SB tholeiite trend is relatively flat compared to the calc alkaline trend, a difference which is commonly related to shallow-level closed system fractionation associated with tholeiites and deeper-level open system fractionation associated with calc alkaline magmas [Kay and Kay, 1982]. Wharton et al. [1995], for example, link the difference in tholeiite and calc alkaline fractionation styles in the Fiji arc to the degree of local crustal extension and related magma replenishment rates, etc. Given that most SB F3 rhyolites plot on an extension of the $[La/Sm]_N \approx 1.5$ trend originating with primitive basalts, their Nb-REE systematics are difficult to reconcile with melting of widespread crustal source. Instead, the data are consistent with SB felsic rocks being generated by fractionation of mantle-derived magmas with variable local input from crustal contamination, which is supported by the Confederation assemblage Sm–Nd isotopic data reported by Tomlinson and Rogers [1999] and Noble [1989].

Apart from some isolated tholeiitic centres associated with the East Horne block and the Pelletier and Rouyn-Noranda suites, trace element trends of the mafic to intermediate BRG tholeiites (defined only on the basis of Nb/La > 0.74) do not conform closely to those of the SB tholeiite suite. For example,

the slope of the $[La/Sm]_N$ versus Nb trend is more positive than that observed for corresponding SB tholeiites ($[La/Sm]_N = 2$–20 ppm Nb compared to 1.4 in the SB suite). The trace element systematics of BRG dacitic and F3 felsic rocks are also intermediate between tholeiites and typical calc alkaline trends. Incompatible elements such as Zr, Th, and U increase steadily but without the distinct second trend or branch attributed to sporadic crustal contamination in the SB suite. If many of the differences between mantle-derived tholeiitic and calc alkaline suites in modern oceanic arcs relate to the depth of melting and the crustal setting of fractionation, then the evolutionary paths of BRG magmas likely reflect intermediate settings, where magma did not generally ascend to extremely shallow levels but ponded in magma chambers established in relatively thin arc crust as a result of arc rifting similar to that proposed by Pelletier et al. [1995], Ayer et al. [2002], and others. Feng and Kerrich [1990] used the hornblende geobarometer to estimate that the syn-volcanic Flavrian pluton, preserved at the centre of the inferred Noranda cauldron, was emplaced at a pressure of about 1 kbar.

Evolution of the BRG and SB F3 rhyolites can also be considered in terms of zircon saturation temperatures (T_{Zr}) calculated on the basis of their whole-rock major element compositions and Zr contents [Watson and Harrison, 1983; Barrie, 1995]. A plot of Zr versus T_{Zr} is given in Figure 9A for comparison with plots of Nb, Th, and Al_2O_3/TiO_2 versus T_{Zr}. As expected, the Zr plot exhibits correlations between Zr and T_{Zr}. The Nb plot (Figure 9B) exhibits similar correlations with T_{Zr} that can probably be attributed to the incorporation of Nb in zircon and the generally similar behaviour of these two HFSE. In contrast, the SB felsic rocks display a complex relationship between Th contents and T_{Zr}, despite zircon's high partition coefficient for Th (Rollinson [1993] and references therein). If the melts are saturated in zircon [cf. Miller et al.,

Figure 8. Plots of Nb/Y vs Zr/Y showing the trends for modern plume (P) and MORB + Arc (M) magmas [Fitton et al., 1997]. The slope of the trends for most arc basalts are comparable to MORB because hydrous fluids transport little Zr or Nb. (A) Data for primitive BRG samples superimposed on the field for similar mafic rocks of the Manitouwadge and Winston Lake greenstone belts. Phanerozoic arc fields, which are not limited to samples with less than 5 wt.% MgO, are from Gamble et al. [1994], Pearce et al. [1995], and Elliott et al. [1997]. (B) Evolved rocks of the BRG. Reference field encloses primitive rocks of the BRG. PM = Primitive Mantle, MS = MORB; both from Sun and McDonough [1989]. CC encompasses the compositions of upper, middle, and lower crust (UC, MC, and LC, respectively) for central East China from Gao et al. [1997]. (C). Primitive rocks of the SB study area. Data from Rudnick and Fountain [1995] and Hollings and Kerrich [2000]. (D) Evolved rocks of the SB study area. Reference field encloses primitive rocks of the SB area [excluding Nb-enriched basalts (NEB)]. (E) A trend for slab melt metasomatised mantle from the Archean to the Phanerozoic. Shown are fields for Phanerozoic mafic and mafic-intermediate magmas from settings inferred to involve slab melting. Also shown are trends for the early and late (rift) assemblages of the southern Abitibi belt [Wyman, 2003]. CAB = calc alkaline basalts, AND = andesites, HAA = High-Al andesites, Phan. Adakite = average Phanerozoic adakite of Drummond et al. [1996]. Krist Abitibi = adakites. Average crust compositions from Gao et al. [1997]; Pilbara Mg-Diorite trend from data in Smithies and Champion [2000]; Alataw [Wang et al., 2003]; Antisana (Ecuador) [Bourdon et al., 2002]; Baja [Martín-Barajas et al., 1995]; Costa Rica [Defant et al., 1992]; Mindanao [Sajona et al., 2000]; Abitibi data [Wyman, 2003]. Primitive Mantle and N-NORB values from Sun and McDonough [1989]. Symbols as in Figure 2.

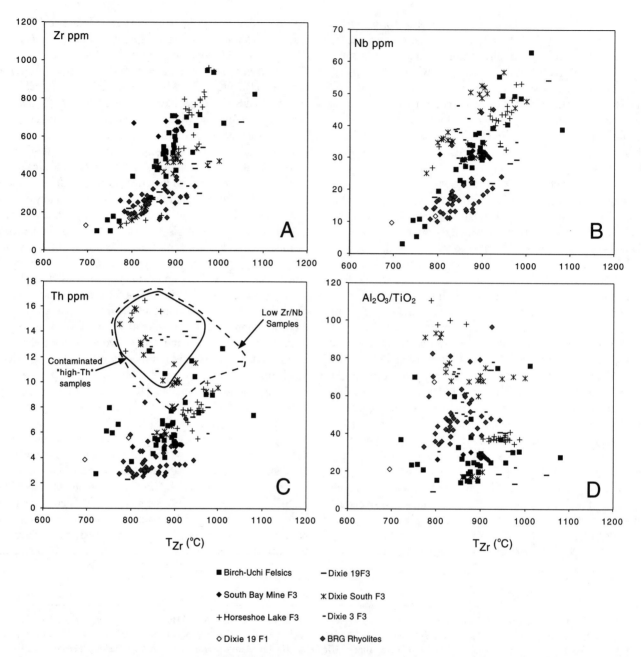

Figure 9. Variation plots for BRG and SB zircon saturation temperatures (T_{Zr}). (A, B) The expected positive correlation between Zr and T_{Zr} is matched by a correlation between Nb and T_{Zr}. (C) Th versus T_{Zr} defines two subtrends. The high-Th subtrend includes samples with anomalous Th content (shown in Figure 4D) and other samples with low Zr/Nb (shown in Figure 4B). (D) A plot of Al_2O_3/TiO_2 versus T_{Zr} displays a range of apparent temperatures at a particular Al_2O_3/TiO_2 value for groups of Horseshoe Lake ($Al_2O_3/TiO_2 \approx 35$) and Dixie South ($Al_2O_3/TiO_2 \approx 70$) samples.

2003], then two distinct compositional paths for SB rhyolites are evident with declining temperatures on this plot. An upper SB rhyolite T_{Zr} trend consists of the samples interpreted as contaminated on the basis of high-Th in Figure 4B and other samples that define the low-Nb trend of Figure 4D. The other SB T_{Zr} trend is defined by samples of the main population in Figure 4D. All BRG samples overlap this lower T_{Zr} trend.

In general, samples of the upper T_{Zr} trend correspond to those that have low Zr/Nb (Figure 4D) and include examples with clearly anomalous Th contents (Figure 4B). Therefore, the diverging Th–T_{Zr} trends are attributable to crustal contamination of some samples. We interpret the subhorizontal Al_2O_3/TiO_2 Dixie South and Horseshoe Lake trends in Figure 9D as the product of zircon removal from a magma without significant fractionation of silicate and oxide mineral phases. If this is the case, then these trends are probably indicative of some form of "two-phase flow mechanism", where interstitial melts are separated from a porous but nonconvecting crystal mush, as summarised by *Bachmann and Bergantz* [2004] for batholith-scale granodiorite intrusions. They argue that the rhyolitic liquid may entrain relatively high proportions of the available small zircon crystals compared to larger feldspar and quartz phenocrysts. It is unclear what errors this process has introduced into the T_{Zr} calculation. *Miller et al.* [2003] argue that errors associated with cumulate zircon are generally counteracted by the diluting volume of other co-existing cumulate phases. This counterbalance effect would not apply to the selective zircon entrainment process envisioned by *Bachmann and Bergantz* [2004]. At present we can only suggest that, as in Figure 5B, the Zr versus Al_2O_3/TiO_2 trend defined by Horseshoe Lake samples encompasses an approximately equal range of Zr addition and loss effects superimposed on an igneous trend ($T_{Zr} \approx 950°C$). The Dixie South trend at $Al_2O_3/TiO_2 \approx 68$ is defined by samples that occur along the main Nb–Th trend (Figure 4D) but are among the low Zr/Nb samples (Figures 4B and 9C).

The Dixie South data set includes samples with Zr $\approx$ 475 ppm with low Al_2O_3/TiO_2 ratios that could be classified as F3a types in terms of their TiO_2 (0.6 wt.%) and Sc (12 ppm) contents and Eu/Eu* (0.5). Dixie South samples with higher Al_2O_3/TiO_2 values more closely conform to the F3b category with lower TiO_2 (0.15 wt.%) and Sc (2.5 ppm), combined with more pronounced Eu anomalies (Eu/Eu* $\approx$ 0.3). The transition from F3a to F3b carries the hallmarks of extensive plagioclase fractionation but also corresponds to a contamination event that generated high-Th rhyolites. Four F3b Horseshoe Lake samples lie on an extension of the Dixie South trend and may be correlative with this suite.

The Th systematics of BRG rhyolites suggests they did not undergo contamination to the extent observed in some SB suites, which is consistent with the setting of the southern Abitibi belt prior to orogeny. It is not possible with our BRG data set to identify fluctuations in HFSE comparable to those found in SB drill hole data sets. A key physical requirement of the batholith–rhyolite model of *Bachmann and Bergantz* [2004] is that crystal abundance in the magma reaches a "rheological locking point" at $\approx$ 50 vol.% that hinders convection. This requirement is likely to have been met by the source of BRG rhyolites during development of the Flavrian pluton in the Noranda area, although the intrusive phase corresponding to the pre-eruptive source of BRG rhyolites is poorly constrained. The pluton is composite in nature and

comprises diorite sills, several trondhjemite phases, and tonalite. *Galley* [2003] suggests a relatively deep rhyolite source beneath the recognised caldera intrusive phases. In any case, the crystal-poor felsic volcanic rocks of the BRG represent plausible candidates for extracted liquids, and the distinct gap on a BRG plot of TiO_2 versus Al_2O_3/TiO_2 is consistent with this process.

CONCLUSIONS

Our results highlight many differences between the BRG and SB but do not contradict the previously proposed subduction-related scenarios. The evolution of tholeiitic suites in both areas are similar to those documented in (rifted) arc environments but are clearly distinct from the more deeply sourced volcanic suites found in intracontinental rift settings. In the case of the BRG, comparatively few tholeiitic units such as the Pelletier and Rouyn-Noranda examples are free of a significant slab melt input. Some of the differences between the areas relate to the lingering effect of mantle plume ascent on the sub-BRG mantle, although the volcanic suite is not itself associated with komatiitic volcanism, which peaked 15 to 20 m.y. earlier [*Wyman et al.*, 2002]. A similar slab melt contribution may have occurred in low-Mg calc alkaline rocks of the SB but more plausibly results from mixing with TTG magmas generated at the base of the crust or direct crustal contamination. In contrast, the HFSE systematics of relatively primitive tholeiitic and calc alkaline rocks of the SB area (>5 wt.% MgO) demonstrate that their mantle sources closely resemble those of normal Phanerozoic arcs, despite the presence of NEB in the south of the study area and minor occurrences of F1 type felsic rocks closer to the mine site. The lack of evidence for HFSE metasomatism in the sources of tholeiitic and calk alkaline mafic rocks indicates that metasomatism of the sub-arc mantle was dominated by hydrous fluids. Therefore, slab melting occurred not in response to a pervasive steep geotherm but to specific geodynamic events, which in this case were probably linked to the early phases of arc rifting along the continental margin.

Although the tholeiite magmas cannot have been generated at the depths associated with intracontinental rifts, the presence of significant existing crust is readily apparent in the Th–HFSE systematics of the SB suite. A minor amount of intra-assemblage contamination via crustal assimilation may have impacted the BRG rhyolites, given that some rhyolites plot at higher Th contents than a main BRG trend. Most of these occurrences, however, could be accounted for by shallow mixing of tholeiite and transitional magmas, as suggested by *Lafleche et al.* [1992a]. Major element indices of fractionation, such as silica contents or maximum Al_2O_3/TiO_2, are comparable for the two suites. In contrast, the maximum enrichment of incompatible elements in the SB suite greatly exceeds that of the BRG. This difference may be partly ascribed to the extent to which magma supply created open or closed plumbing systems. The effects of zircon entrainment during fluid extraction episodes also extended the range of HFSE contents in SB felsic rocks.

Our results provide no evidence in support of a separate origin for either the SB or BRG felsic magmas via crustal melting. Rhyolites of the SB suite lie along compositional trends (e.g., $[La/Sm]_N$) defined by the more primitive SB tholeiitic rock types and cannot be derived from sources with strongly fractionated REE, such as many granitoid crustal suites. Thorium systematics monitor crustal input in the SB suite but suggest sporadic contamination of a mantle-derived magma rather than wholesale melting of older crust. In some models, felsic samples with high-Th contents and pronounced Eu anomalies might correspond to small degree melts of mafic crust that left residual plagioclase in their sources. If so, then they should also be most enriched in incompatible HFSE such as Zr and Nb; however, the opposite occurs for Dixie South and Horseshoe Lake samples sets of the SB area. The very minor compositional gap that is evident on a BRG plot of TiO_2 versus Al_2O_3/TiO_2 would be a fortuitous one if it signifies that evolution of mantle-derived magmas stopped just short of compositions generated by a crustal melting process. In contrast, the liquid extraction process outlined by *Bachmann and Bergantz* [2004] provides a plausible mechanism for creating minor compositional gaps in a suite of rocks related by fractionation, with or without minor assimilation. This finding does not negate the concept of rift-related thermal corridors, which may be an important requirement for the generation of major massive sulfide mining districts [*Galley*, 2003] and the crustal melting mechanism may also be applicable in some other terranes.

Acknowledgments.   The authors acknowledge financial support for this study from the Canadian Mining Industry Research Organization (CAMIRO) and a matching Natural Sciences and Engineering Research Council of Canada Industry Oriented Research grant. B. Morgan is thanked for assistance with ICP-MS analyses. Editor Kent Condie and anonymous reviewers are thanked for their detailed critiques of this manuscript.

REFERENCES

Atkinson, B.T., Parker, J.R. and Storey, C.C., 1990, Red Lake Resident Geologist's District-1990. In *Report of Activities 1990*, Resident Geologists, Ontario Geological Survey, Miscellaneous Paper 152, 31-66.

Ayer, J., Amelin, Y., Corfu, F., Kamo, S., Ketchum, J., Kwok, K., Trowell, N., 2002. Evolution of the Southern Abitibi Greenstone Belt Based on U-Pb Geochronology: Autochthonous Volcanic Construction Followed by Plutonism, Regional Deformation and Sedimentation. Precambrian Research 115, 63-95.

Bachmann, O. and Bergantz, G.w., 2004. On the origin of crystal-poor rhyolites: Extracted from batholithic crystal mushes. *Journal of Petrology*, 45, 1565-1582.

Barrett, T.J., MacLean, W.H., Cattalani, S. and Hoy, L., 1993, Massive sulfide deposits of the Noranda area, Quebec. V. The Corbet mine, *Canadian Journal of Earth Sciences*, 30, 1934-1954.

Barrie, C.T., 1995, Zircon thermometry of high-temperature rhyolites near volcanic-associated massive sulfide deposits, Abitibi subprovince, Canada. *Geology*, 23, 169-172.

Beakhouse, G.P., 1989. Geology of the Western Birch Lake area, Kenora District, Patricia Portion, Ontario Geological Survey, *Open File Report 5700*, 106p.

Bleeker, W., 2002. Archaean tectonics; a review, with illustrations from the Slave Craton In; Fowler, C., Ebinger, C., Hawkesworth, C. (Eds) The early Earth; physical, chemical and biological development. Geological Society Special Publications, vol.199, pp.151-181, 2002.

Bourdon, E., Eissen, J.-P., Monzier, M., Robin, C., Martin, H., Cotton, J., & Hall, M.L., 2002, Adakite-like lavas from Antisana Volcano (Ecuador): Evidence for slab melt metasomatism beneath the Andean Northern Volcanic Zone. *Journal of Petrology*, 43, 199-217.

Burrows, D.R., Spooner, E.T.C., 1989. Relationships between Archean gold quatz vein-shear zone mineralisation and igneous intrusions in the Val d'Or and Timmins areas, Abitibi Subprovince, Canada. *Economic Geology Monograph*, 6, 424-444.

Campbell, I.H., Lesher, C.M., Coad, P., Franklin, J.M., Gorton, M.P. and Thurston, P.C., 1982, Rare earth elements in volcanic rocks associated with Cu-Zn massive sulphide mineralization: a preliminary report, *Canadian Journal of Earth Sciences*, 19, 619-623.

Capdevila, R., Goodwin, A.M., Ujike, O., and Gorton, M.P., 1982. Trace-element geochemistry of Archean volcanic rocks and crustal growth in southwestern Abitibi Belt, Canada. *Geology*, 10, 418-422.

Cattell, A.C. and Taylor, R.N., 1990. Archaean basic magmas. In: Hall, R.P. and Hughes, D.J., (ed.s) Early Precambrian Basic Magmatism, Blackie, Glasgow, pp.11-39.

Condie, K.C., 1985, Secular variation in the composition of basalts: an index to mantle evolution. *Journal of Petrology*, 26, 545-563.

Condie, K.C., 2001, Mantle Plumes and Their Record in Earth History, Cambridge University Press, Cambridge, U.K., 306 p.

Corfu, F., 1993, The evolution of the Southern Abitibi greenstone belt in light of precise U-Pb geochronology, *Economic Geology*, 88, 1323-1340.

Defant, M.J., Jackson, T.E., Drummond, M.S., De Boher, J.Z., Bellon, H., Feigenson, M.D., Maury, R.C., & Stewart, R.H. 1992. The geochemistry of young volcanism throughout western Panama and southeastern Costa Rica: an overview. *Journal of the Geological Society of London*, 149, 569-579

Dimroth, E., Imreh, L., Rocheleau, M. and Goulet, N., 1982, Evolution of the south-central part of Archean Abitibi belt. Part I: Stratigraphy and paleogeographical model, *Canadian Journal of Earth Sciences*, 19, 1729-1758.

Drummond, M.S., Defant, M.J., and Kepezhinsksas, P.K., 1996, Petrogenesis of slab-derived trondhjemite-tonalite-dacite/adakite magmas. *Transactions of the Royal Society of Edinburgh*, 87, 205-215.

Elliott, T., Plank, T., Zindler, A., White, W., and Bourdon, B., 1997, Element transport from slab to volcanic front at the Mariana arc. *Journal of Geophysical Research*, 102B, 14991-15019.

Feng, R. and Kerrich, R., 1990, Geobarometry, differential block movements, and crustal structure of the southwestern Abitibi greenstone belt. *Geology*, 18, 870-873.

Fitton, J.G., Saunders, A.D., Norry, M.J., Hardarson, B.S., Taylor, R.N., 1997. Thermal and Chemical Structure of the Iceland Plume. *Earth and Planetary Science Letters*, 153, 197-208.

Fowler, A.D. and Jensen, L.S., 1989. Quantitative Trace-Element Modelling of the Crystallization History of the Kinojevis and Blake River Groups, Abitibi Greenstone Belt, Ontario. *Canadian Journal of Earth Science*, 26, 1356-1367.

Franklin, J.M., Sangster, D.F. and Lydon, J.W., 1981, Volcanic-associated massive sulphide deposits: *Economic Geology*, Anniversary Volume, 485-627.

Galley, A.G., 2003, Composite synvolcanic intrusions associated with Precambrian VMS-related hydrothermal systems. *Mineralium Deposita*, 38, 443-473.

Galley, A.G., and van Breemen, O., 2002. Timing of synvolcanic magmatism in relation to base-metal mineralization, Rouyn-Noranda, Abitibi volcanic belt, Quebec. Current Research - Geological Survey of Canada, Report: 2002-F8, 9 pp., 2002. http://www.nrcan.gc.ca/gsc/bookstore/free/cr_2002/F08.pdf

Gamble, J.A., Wright, I.C., Woodhead, J.D., and McxCulloch, M.T., 1994, Arc and back-arc geochemistry in Smellie, J.L., (Ed.) the southern Kermadec arc-Ngatoro Basin and offshore Taupo Volcanic Zone, SW Pacific. In Geological Society of London Special Publication 81, 193-212.

Gao, S., Luo, T.-C., Zhang, B.-R., Zhang, H.-F., Han, Y.-W., Zhao, Z.-D., and Hu, Y.-K., 1997, Chemical composition of the continental crust as revealed by studies in East China. *Geochimica and Cosmochimica Acta*, 62, 1959-1975.

Gelinas, L. and Ludden, J.N., 1984, Rhyolitic volcanism and the geochemical evolution of an Archean central ring complex: the Blake River Group volcanics of the southern Abitibi belt, Superior Province, *Physics of Earth and Planetary Interiors*, 35, 77-88.

Gelinas, L., Trudel, P., Hubert, C., 1984. Chemostratigraphic division of the Blake River Group, Rouyn-Noranda area, Abitibi, Quebec. Canadian Journal of Earth Sciences 21, 220-231.

Ghiorso, M.S., Hirschmann, M., Reiners, P.W., and Kress, V., 2002. The pMELTS; a revision of MELTS for improved calculation of phase relations and major element partitioning related to partial melting of the mantle to 3 GPa. *Geochemistry, Geophysics, Geosystems*, vol. 3, doi:10.1029/2001GC000217.

Gibson, H.L. and Watkinson, D.H., 1990, Volcanogenic massive sulphide deposits of the Noranda shield volcano and cauldron, Quebec. In, Rive, M., Verpaelst, P., Gagnon, Y., Lulin, J.M., Riverin, G. and Simard, A.(eds.), The northwestern Quebec polymetallic belt, The Canadian Institute of Mining and Metallurgy, Special Volume 43, 119-132.

Good, D.J., 1988. Geology of the east half of the Birch Lake area, District of Kenora, Patricia Portion. Ontario Geological Survey, Open File Report 5685, 1331p.

Gutscher, M., Maury, R., Eissen, J.-P., and Bourdon, E., 2000. Can slab melting cause flat subduction? *Geology*, 28, 535-538.

Hart, T.R., Gibson, H.L., Lesher, C.M., 2004. Trace element geochemistry and petrogenesis of felsic volcanic rocks associated with volcanogenic massive Cu-Zn-Pb sulfide deposits. *Economic Geology and the Bulletin of the Society of Economic Geologists*, 99, 1003-1013.

Hollings, P. and Kerrich, R., 2000. An Archean arc basalt - Nb-enriched basalt - adakite association: The 2.7 Ga Confederation assemblage of the Birch-Uchi greenstone belt, Superior Province. *Contributions to Mineralogy and Petrology*, 139, 208-226.

Hubert, C., Trudel, P. and Gelinas, L., 1984, Archean wrench fault tectonics and structural evolution of the Blake River Group, Abitibi belt, Quebec: *Canadian Journal of Earth Sciences*, 21, 1024-1032.

Jenner, G.A., Longerich, H.P., Jackson, S.E. and Fryer, B.J., 1990. ICP-MS - A powerful tool for high precision trace-element analysis in earth sciences: Evidence from analysis of selected USGS reference samples. *Chemical Geology*, 83, 133-148.

Jensen, L.S. and Langford, F.F., 1985. Geology and petrogenesis of the Archean Abitibi Belt in the Kirkland Lake area, Ontario. *Ontario Geological Survey Miscellaneous Paper*, vol.123, 130 pp.

Kay, S.M., Kay, R.W. and Citron, G.P., 1982, Tectonic controls on tholeiitic and calc-alkaline magmatism in the Aleutian arc, *Journal of Geophysical Research*, 87, 4051-4072.

Kerr, D.J. and Gibson, H.L., 1993, A comparison of the Horne volcanogenic massive sulfide deposit and intracauldron deposits of the Mine Sequence, Noranda, Quebec, *Economic Geology*, 88, 1419-1442.

Kerr, D.J. and Mason, R., 1990, A re-appraisal of the geology and ore deposits of the Horne Mine Complex at Rouyn-Noranda, Quebec, In; Rive, M., Verpaelst, P., Gagnon, Y., Lulin, J.M., Riverin, G. and Simard, A.(eds.), The northwestern Quebec polymetallic belt, The Canadian Institute of Mining and Metallurgy, Special Volume 43, 153-166.

Klewin, K.W., and Berg, J.H., 1991. Petrology of the Keweenawan Mamainse Point lavas, Ontario; petrogenesis and continental rift evolution. *Journal of Geophysical Research*, B96, 457-474.

Lafleche, M.R., Dupuy, C. and Dostal, J., 1992a, Tholeiitic volcanic rocks of the Late Archean Blake River Group, southern Abitibi greenstone belt: Origin and geodynamic implications, *Canadian Journal of Earth Sciences*, 29, 1448-1458.

Lafleche, M.R., Dupuy, C., Bougault, H., 1992b. Geochemistry and Petrogenesis of Archean Mafic Volcanic Rocks of the Southern Abitibi Belt, Quebec. *Precambrian Research*, 57, 207-241.

Lesher, C.M., Goodwin, A.M., Campbell, I.H. and Gorton, M.P., 1986. Trace-element geochemistry of ore-associated and barren, felsic metavolcanic rocks in the Superior Province, Canada. *Canadian Journal of Earth Sciences*, 23, 222-237.

Longerich, H.P., Jenner, G.A., Fryer, B.J. and Jackson, S.E., 1990. Inductively coupled plasma-mass spectrometric analysis of geological samples: a critical evaluation based on case studies. *Chemical Geology*, 83, 105-118.

Ludden, J.N. and Péloquin, S., 1996. A geodynamic model for the evolution of the Abitibi Belt; implications for the origins of volcanic massive sulphide (VMS) deposits. In; Wyman, D., (Ed.) Trace element geochemistry of volcanic rocks: applications for massive sulphide exploration. Short Course Notes - Geological Association of Canada 12, pp.205-237.

Martin, H., 1986. Effect of steeper Archean geothermal gradient on geochemistry of subduction zone magmas. *Geology*, 14: 753-756

Martín-Barajas, A., Stock, J.M., Layer, P., Hausback, B., Renne, P., & López-Martínez, Margarita, 1995, Arc-rift transition volcanism in the Peurtecitos Volcanic Province, northeastern Baja California, Mexico. *Geological Society of America Bulletin*, 107, 407-424.

MERQ-OGS, 1983, Lithostratigraphic map of the Abitib Subprovince; Ontario Geological Survey/Ministere de l'…nergie et des Ressources, Québec, 1:500 000, Map 2484.

Miller, C.F., McDowell, S.M., and Mapes, R.W., 2003, Hot and cold granites? Implications for zircon satuation temperatures and preservation of inheritance. *Geology*, 31, 529-532.

Mortensen, J.K., 1993, U-Pb geochronology of the eastern Abitibi Subprovince. Part 2: Noranda-Kirkland Lake area, *Canadian Journal of Earth Sciences*, 30, 29-41

Noble, S.R., 1989. Geology, geochemistry and isotope geology of the Trout Lake batholith and the Uchi-Confederation lakes greenstone belt, northwestern Ontario, Canada. Unpublished PhD thesis, University of Toronto, Ontario, 288p.

Noble, S.R., Krogh, T.E. and Evensen, N.M., 1989. U-Pb age constraints on the evolution of the Trout Lake-Uchi-Confederation lakes granite-greenstone terrane, Superior Province, Canada. Geological Association of Canada-Mineralogical Association of Canada, Joint Annual Meeting, Program with Abstracts 14, p.A56.

Nunes, P.D. and Thurston, P.C., 1980. Two hundred and twenty million years of Archean evolution: a zircon U-Pb age stratigraphic study of the Uchi-Confederation lakes greenstone belt, Northwestern Ontario. *Canadian Journal of Earth Sciences*, 17, 710-721.

Pearce, J.A., Baker, P.E., Harvey, P.K., and Luff, I.W., 1995, Geochemical evidence for subduction fluxes, mantle melting and fractional crystallization beneath the South Sandwich Island arc. *Journal of Petrology*, 36, 1073-1109.

Pearce, J.A., Ernewein, M., Bloomer, S.H., Parson, L.M., Murton, B.J. and Johnson, L.E., 1994, Geochemistry of Lau Basin volcanic rocks: Geological Society, London, Special Publications, 81, 53-75.

Péloquin, S., Potvin, R., Paradis, S., Lafleche, M., Verpaelst, P., and Gibson, H., 1990. The Blake River Group, Rouyn-Noranda area, Québec; a stratigraphic synthesis. In: Rive, M., Verpaelst, P., Gagnon, Y., Lulin, J-M., Riverin, G., Simard, A. (Eds.). The northwestern Quebec polymetallic belt; a summary of 60 years of mining exploration. Special Volume - Canadian Institute of Mining and Metallurgy, vol.43, pp.107-118.

Péloquin, A.S., Verpaelst, P. and Ludden, J.N., 1996, Spherulitic rhyolites of the Archean Blake River Group, Canada: Implications for stratigraphic correlation and VMS exploration: *Economic Geology*, 91, 343-354.

Rogers, N., 2002a. Whole-rock chemical analyses from the Birch-Uchi greenstone belt, Superior Province. Geological Survey of Canada, Open File Report 4271.

Rogers, N., 2002b. Geology, Confederation Lake, Ontario. Geological Survey of Canada, Open File 4265, scale 1:50,000.

Rollinson, H., 1993, Using geochemical data: Evaluation, presentation, interpretation, Longman, Harlow, England, 352p.

Rudnick, R.L. and Fountain, D.M., 1995, Nature and composition of the continental crust: a lower crustal perspective: *Review of Geophysics*, 33, 267-309.

Sage, R.P., Lightfoot, P.C., Doherty, W., 1996. Bimodal cyclical Archean basalts and rhyolites from the Michipicoten (Wawa) greenstone belt, Ontario: geochemical evidence for magma contributions from the asthenospheric mantle and ancient continental lithosphere near the southern margin of the Superior Province. *Precambrian Research*, 76, 119-153.

Sajona, F.G., Maury, R.C., Pubellier, M. Leterrier, J., Bellon, H., & Cotton, J., 2000, Magmatic sourse enrichment by slab-derrived meltsin a young post-collision sertting, central Mindanao (Phillipines). *Lithos*, 54, 173-206.

Shirey, S., Lewin, K., Berg, J., and Carlson, R., 1994. Temporal changes in the sources of flood basalts: Isotopic and trace element evidence from the 1100 Ma old Keweenawan Mamainse Point Formation, Ontario, Canada. *Geochimica et Cosmochimica Acta*, 58, 4475-4490.

Smithies, R.H., and Champion, D.C., 2000, The Archaean High-Mg Diorite Suite: links to tonalite-trondhjemite-granodiorite magmatism and implications for early Archaean crustal growth. *Journal of Petrology*, 41, 1653-1671.

Stix, J. and Gorton. A.P., 1989. Physical and chemical processes of Archean subaqueous pyroclastic rocks. in Geoscience Research Grant Program, Summary of Research 1988-1989, Ontario Geological Survey, Miscellaneous Paper 143, 231-238.

Stott, G.M. and Corfu, F., 1991, Uchi Subprovince, *In*; Geology of Ontario, Ontario Geological Survey, Special Volume 4, Part 1, 145-238.

Sun, S.-s. and McDonough, W.F., 1989. Chemical and isotopic systematics of oceanic basalts: implications for mantle composition and processes. In: A.D. Saunders and M.J. Norry (Editors), Magmatism in the ocean basins. Geological Society Special Publication, 313-345.

Thurston, P.C., 1981. Economic evaluation of Archaean felsic volcanic rocks using REE geochemistry. In: Glover, J. E. and Groves, D. I., Archaean geology; Second international symposium. Special Publication - Geological Society of Australia, no.7, pp.439-450.

Thurston, P.C., 1985a, Physical volcanology and stratigraphy of the Confederation Lake area, District of Kenora (Patricia Portion); Ontario Geological Survey, Report 236, 117p.

Thurston, P.C., 1985b. Geology of the Earngey-Costello area, District of Kenora, Patricia Portion; Ontario Geological Survey, Report 234, 125p.

Thurston, P.C., 1986. Volcanic cyclicity in mineral exploration; the caldera cycle and zoned magma chambers; in Volcanology and Mineral deposits, Ontario Geological Survey, Miscellaneous Paper 129, 104-123.

Thurston, P.C., 2003, Autochthonous development of Superior Province greenstone belts? *Precambrian Research*, 115, 11-36.

Tomlinson, K.Y., and Rogers, N., 1999. Neodymium-isotopic characteristics of the Uchi-Confederation Lakes region, northwestern Ontario. Current Research - Geological Survey of Canada, Report: 1999-E, pp.91-99.

Walker, J.A., Gmitro, T.T., Berg, J.H., 2002, Chemostratigraphy of the Neoproterozoic Alona Bay Lavas, Ontario. *Canadian Journal of Earth Sciences*, 39, 1127-1142.

Wang, Q, Zhao Z., Bai Z., Bao Z., Xu J., Xiong X., Mei H., & Wang Y., 2003 Carboniferous adakites and Nb-enriched arc basaltic rocks association in the Alataw Mountains, north Xinjiang: interactions between slab melt and mantle peridotite and implications for crustal growth. *Chinese Science Bulletin*, 48, 2108-2115.

Watson, E.B. and Harrison, T.M., 1983, Zircon saturation revisited: temperature and compositional effects in a variety of crustal magma type, *Earth and Planetary Science Letters*, v. 64, p. 295-304.

Wharton, M.R., Hathway, B., and Colley, H., 1994, Volcanism associated with extension in an Oligocene-Miocene arc, southwestern Viti Levu, Fiji, In: Smellie, J.L. (ed.), Volcanism associated with extension at consuming plate margins, Special Publication of the Geological Society of London 81, 95-114.

Wyman, D.A., 2003, Upper mantle processes beneath the 2.7 Ga Abitibi belt, Canada: a trace element perspective. *Precambrian Research*, 127, 143-165.

Wyman, D.A., Kerrich, R., and Polat, A., 2002, Assembly of Archean cratonic mantle lithosphere and crust: plume-arc interaction in the Abitibi-Wawa subduction-accretion complex, *Precambrian Research*, 115, 37-62.

P. Hollings, Department of Geology, Lakehead University, Thunder Bay, Ontario, P7B 5E1, Canada.

D. A. Wyman, School of Geosciences, Edgeworth David Building, University of Sydney, Sydney, New South Wales, 2007 Australia. (dwyman@geosci.usyd.edu.au)

High-Pressure Intermediate-Temperature Metamorphism in the Southern Barberton Granitoid–Greenstone Terrain, South Africa: A Consequence of Subduction-Driven Overthickening and Collapse of Mid-Archean Continental Crust

Johann Diener[1], Gary Stevens, and Alex Kisters

Department of Geology, University of Stellenbosch, Stellenbosch, South Africa

The Early- to Mid-Archean Barberton granitoid–greenstone terrain of South Africa consists of the low-grade metamorphic Barberton greenstone belt that is surrounded by a composite granitoid–gneiss domain. High-grade greenstone lithologies along the southern margin of the belt and within the adjacent tonalite–trondhjemite–granodiorite (TTG)–greenstone domain record evidence of high-P intermediate-T metamorphism, with estimates of 9–12 kbar at 650–700 °C, 6 kbar at 490 °C, and 7.4 kbar at 580 °C documented from different areas of the domain. Pseudosection modeling of metamorphic assemblages suggests that the domain experienced a high-P clockwise P–T path, involving burial to depths of 35–45 km and heating to peak temperatures while at depth, followed by exhumation along a near-isothermal decompression path. Metamorphism coincides with the main accretionary event in the Barberton greenstone belt at ca. 3.23 Ga. The trajectory of the P–T path and the depths of burial are consistent with the subduction of this continental fragment during the early stages of collision. Exhumation of the domain occurred by extensional detachment faulting during the subsequent collapse of the overthickened orogen. The high-grade TTG–greenstone domain represents a section through the deeper segments of a Mid-Archean collisional belt. The P–T path preserved in these rocks may indicate the presence of relatively cold and rigid continental crust that was able to sustain crustal overthickening before the onset of gravitational collapse.

INTRODUCTION

High-pressure (high-P) metamorphic rocks such as blueschists and eclogites are characteristic of zones of convergence and collision in modern crustal environments [*Ernst*, 1973, 1988; *Le Fort*, 1975; *England and Thompson*, 1984] and the preservation of these rocks in older terrains can be used to infer the presence of ancient subduction zones and continental sutures [e.g., *Eide and Lardeaux*, 2002]. The oldest crustal remnants preserved on Earth occur in granitoid–greenstone terrains that consist of isolated greenstone belts surrounded by extensive granitoid plutons and gneiss domains. Examples of classic granitoid–greenstone terrains include Isua (West Greenland) [*McGregor*, 1973], the Pilbara Craton (Western Australia) [*Hickman*, 1983], Barberton (South Africa) [*Viljoen and Viljoen*, 1969; *Anhaeusser*, 1973], the Slave Province (Canada) [*Henderson*, 1981], and the Superior Province (Canada) [*Stott*, 1997]. The supracrustal greenstone belts typically

[1] Now at School of Earth Sciences, University of Melbourne, Victoria, Australia.

Archean Geodynamics and Environments
Geophysical Monograph Series 164
10.1029/164GM15

consist of low-grade regional metamorphic rocks, and it is only in close proximity to the granitoid plutons and gneisses that these rocks exhibit higher metamorphic grades. The higher-grade rocks occasionally preserve evidence of high-P metamorphism, as suggested by isolated occurrences of kyanite in rocks from Isua [*Boak and Dymek*, 1982], the Pilbara [*Collins and Van Kradendonk*, 1999], the Superior Province [*Benn et al.*, 1994], and Barberton [*De Wit et al.*, 1983; *Diener et al.*, in press]. However, some of these higher-grade domains have equally been interpreted to be the result of a contact metamorphic overprint that occurred during the emplacement of the granitoid plutons [*Anhaeusser*, 1984; *Wilkins*, 1997; *Collins and Van Kradendonk*, 1999]. Consequently, whether much of the Archean high-grade metamorphic record can be regarded as resulting from regional plate tectonic processes remains in question.

In addition, the general absence of regionally metamorphosed high-P rocks in Archean terrains from across the world has prompted suggestions that plate convergence and collision did not take place during the early evolution of the Earth [e.g., *Hamilton*, 1998]. Non-uniformitarian models also propose that the Archean crust was not rigid enough to support the stresses arising from lateral plate-boundary processes and that the dominant means of crustal evolution

was by magmatic and/or solid-state diapirism, i.e., partial convective overturn [*Collins et al.*, 1998]. Alternatively, even if horizontal plate tectonics did occur, the higher crustal heat flow and elevated geothermal gradients in the Archean crust [*Sandiford*, 1989; *Choukroune et al.*, 1995; *Sandiford and McLaren*, 2002] would have precluded the formation and preservation of relatively low-temperature (low-T) blueschist- and eclogite-facies rocks.

It is also suggested that, because of the higher heat flow in the Archean, the crust could have suffered a significant rheological weakening [e.g., *Marshak*, 1999; *Chardon et al.*, 2002]. The weaker crustal column would not have been able to sustain large vertical loads and would therefore have collapsed prior to the significant crustal stacking that is necessary for the formation of the high-P rocks typical of more modern orogenic belts [*Grotzinger and Royden*, 1990; *Marshak et al.*, 1992; *Marshak*, 1999].

The 3.5–3.1 Ga Barberton granitoid–greenstone terrain of southern Africa (Figure 1) is one of the oldest and best-preserved examples of Archean geology in the world and has, in a number of previous studies, served as a natural laboratory to investigate the processes that shaped the early continents [e.g., *Viljoen and Viljoen*, 1969; *De Wit et al.*, 1983, 1987, 1992; *Lowe*, 1994, 1999; *De Ronde and De Wit*, 1994;

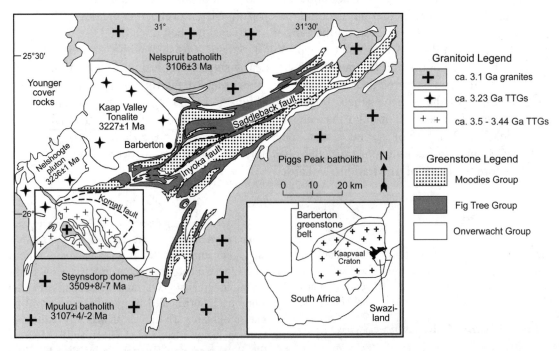

Figure 1. Simplified geological map of the Barberton greenstone belt and surrounding granitoid domain, modified after *Anhaeusser et al.* [1981] and *De Ronde and De Wit* [1994]. Cited ages are crystallization ages of the granitoid plutons, taken from *Kamo and Davis* [1994] and *De Ronde and Kamo* [2000]. The Saddleback–Inyoka fault is a major domain boundary that separates the northern and southern domains of the BGB. The southern TTG–greenstone domain is separated from the BGB by the Komati fault, and the extent of the investigated area in the southern TTG–greenstone domain is indicated by the box and presented in more detail in Figure 2.

Kamo and Davis, 1994]. Recent metamorphic studies in the Barberton terrain have begun to reveal the characteristics of ca. 3.23 Ga geodynamic processes associated with the emergence of the Kaapvaal Craton as a stable lithospheric block [*Dziggel et al.*, 2002; *Kisters et al.*, 2003; *Diener et al.*, in press]. Like most granitoid–greenstone terrains, the Barberton terrain consists of a predominantly greenschist-facies metamorphic greenstone sequence of the Barberton greenstone belt (BGB) that is surrounded and intruded by tonalite–trondhjemite–granodiorite (TTG) plutons and gneisses, some of which contain high-grade metamorphic greenstone remnants (Figures 1 and 2).

It is within these higher-grade remnants, as well as along the higher-grade marginal zones of the greenstone belt, that evidence has emerged to suggest the existence of high-P, intermediate-T amphibolite-facies metamorphic conditions

associated with 3.23 Ga tectonism [*Dziggel et al.*, 2002; *Kisters et al.*, 2003; *Diener et al.*, in press]. The peak metamorphic P–T pairs documented by these studies are exceptionally rare in the Archean rock record, but the low average apparent geothermal gradients recorded in all cases (ca. 20 °C/km) are consistent with the products of a collisional tectonic process. Consequently these rocks may record direct P–T evidence of Archean subduction and collisional processes, including evolution through the blueschist- and/or eclogite-facies.

This paper reviews the evidence for high-P intermediate-T metamorphism in these rocks and, whereas the previous studies primarily focused on constraining peak metamorphic conditions via conventional thermobarometry, we provide additional insights into the P–T path evolution of this domain via pseudosection modeling.

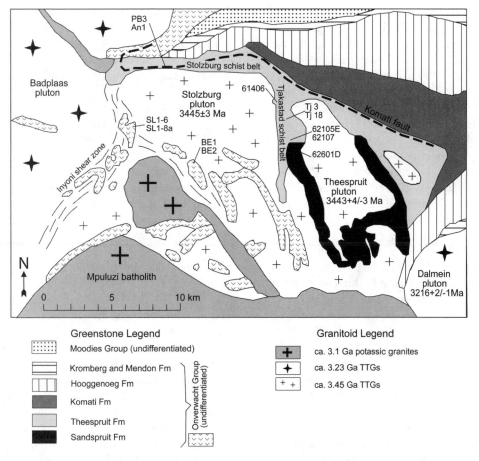

Figure 2. Simplified geological map and stratigraphic column of the southern BGB and the adjacent TTG–greenstone domain, modified after *Anhaeusser et al.* [1981]. Cited ages are from *Kamo and Davis* [1994], except for the age of the Stolzburg pluton, which is from *Kröner et al.* [1991]. The southern TTG–greenstone domain is separated from the BGB by an extensional detachment that encompasses the Komati fault and bounds the domain to the north. The Inyoni shear zone separates the southern TTG–greenstone domain from the younger granitoid domain to the west. The location of samples referred to in the text is also indicated.

REGIONAL GEOLOGY

The stratigraphy of the BGB consists of a lower succession of mafic–ultramafic lavas of the 3550–3300 Ma Onverwacht Group that are overlain by clastic marine sediments and felsic volcanics of the 3260–3225 Ma Fig Tree Group. The stratigraphy is completed by the continentally derived coarse-clastic sediments of the 3225–3215 Ma Moodies Group [*Viljoen and Viljoen*, 1969; *SACS*, 1980; *Armstrong et al.*, 1990; *Kröner et al.*, 1991, 1992, 1996; *Kamo and Davis*, 1994; *Byerly et al.*, 1996]. The southern part of the BGB comprises the lower formations of the Onverwacht Group, namely, the Sandspruit, Theespruit, and Komati Formations (Figure 2). The Sandspruit and Theespruit Formations occur along the margins of the BGB and as remnants within the adjacent TTG–greenstone domain, and they are separated from the main body of the BGB by the Komati fault (Figures 1 and 2) [*De Wit et al.*, 1983; *Armstrong et al.*, 1990; *Diener et al.*, in press]. The Sandspruit Formation consists of metabasites and intercalated chert horizons, whereas the Theespruit Formation consists of similar lithologies in addition to voluminous felsic volcaniclastic units, all of which have been metamorphosed to amphibolite-facies grades [*Viljoen and Viljoen*, 1969]. The low-grade metamorphic Komati Formation consists predominantly of peridotitic komatiite and forms the lowermost part of the relatively intact greenstone sequence above the Komati fault [*Viljoen and Viljoen*, 1969; *Dann*, 2000].

Early workers proposed that the BGB consists of a relatively intact stratigraphic sequence and attributed the bulk of deformation and metamorphism exhibited by the belt to have occurred during the emplacement of the granitoid plutons into this terrain [e.g., *Viljoen and Viljoen*, 1969; *Anhaeusser*, 1973, 1984]. The plutons were envisaged to have been emplaced as forced diapirs that sheared the pluton margins and wallrocks during "dynamic" contact metamorphism [*Anhaeusser*, 1984], thereby giving the high-grade marginal zones the appearance of regional metamorphic rocks.

Subsequent structural and geochronological investigations have revealed that the BGB is made up of distinct structural and stratigraphic domains assembled during two main accretionary episodes that were accompanied by periods of voluminous TTG magmatism [*Williams and Furnell*, 1979; *De Wit*, 1982; *De Wit et al.*, 1983, 1987, 1992; *Lowe et al.*, 1985, 1999; *Kröner et al.*, 1992, 1996; *Lowe*, 1994, 1999; *De Ronde and De Wit*, 1994; *Kamo and Davis*, 1994]. The first accretionary epsiode, D_1, occurred at ca. 3445 Ma and is recognized in the southern parts of the BGB, where it affected the lower formations of the Onverwacht Group [*De Wit et al.*, 1983; *De Ronde and De Wit*, 1994]. It is proposed that during D_1 the Komati Formation was thrust onto the high-grade greenstones along the Komati fault, coincidental with synkinematic emplacement of TTGs such as the Stolzburg and Theespruit plutons into the rocks below the Komati fault (Figure 2). In this scenario, the Sandspruit and Theespruit Formations are interpreted to have been principally deformed and metamorphosed during D_1 at ca. 3445 Ma [*De Ronde and De Wit*, 1994].

The main assembly of the BGB occurred during a second accretionary episode (D_2), when the northern and southern domains of the BGB were amalgamated along the central Saddleback–Inyoka fault system (Figure 1). D_2, a short-lived tectonic event that occurred between 3229 and 3226 Ma, was accompanied by a second episode of felsic magmatism typified by the Nelshoogte and Kaap Valley plutons that occur to the north of the BGB (Figure 1) [*De Ronde and De Wit*, 1994; *Kamo and Davis*, 1994; *De Ronde and Kamo*, 2000]. During D_2, the TTG–greenstone domain to the south of the BGB was metamorphosed to peak P–T conditions in excess of 11 kbar and 680 °C before it was exhumed and juxtaposed against the BGB by extensional detachment faulting along the Komati Fault during the collapse of the D_2 orogen (see below) [*Dziggel et al.*, 2002; *Kisters et al.*, 2003; *Diener et al.*, in press]. The evolutionary history of the BGB was concluded with the emplacement of the ca. 3110 Ma Nelspruit, Mpuluzi, and Piggs Peak batholiths to the north and south of the BGB (Figure 1).

ORIGIN OF THE SOUTHERN
TTG–GREENSTONE DOMAIN

Intrusive contacts of the Theespruit and Stolzburg plutons with the Sandspruit and Theespruit Formations, at the base of the BGB, illustrate the magmatic assembly of the southern TTG–greenstone domain at 3.45 Ga. Volcanic activity in the BGB was contemporaneous with the TTG plutonism as is manifest by 3453 Ma-old pyroclastic and volcanoclastic felsic units that are common in the Theespruit Formation [*Armstrong et al.*, 1990]. Widely developed magmatic breccias along the TTG–greenstone contacts and sharply cross-cutting relationships with little evidence for intrusion-related wall-rock strains are common [*Viljoen and Viljoen* 1969; *Anhaeusser et al.*, 1981; *De Wit et al.*, 1983]. These and other intrusive features point to the relatively shallow emplacement levels of the granitoids [*Kisters and Anhaeusser*, 1995]. The sum of these features, as well as the mainly felsic nature of TTG plutonism, leads to the proposal that the southern TTG–greenstone domain exposes a shallow-crustal section through a 3.45 Ga magmatic arc [*De Wit et al.*, 1987; *Armstrong et al.*, 1990; *Kisters and Anhaeusser*, 1995].

The 3.45 Ga TTG plutonism also attests to the formation of juvenile continental crust at that time [*De Wit et al.*, 1992]. Older (> 3.5 Ga) TTG gneisses are exposed to the east in the Steynsdorp dome and in the Ancient Gneiss Complex in

Swaziland [*Kröner et al.*, 1991]. This suggests that the location of this 3.45 Ga magmatic arc is along the margins of a pre-existing continental block. The 3.45 Ga TTG plutons and associated volcanosedimentary sequences may, therefore, track the lateral growth of the earliest continental nucleus through magmatic accretion along plutonic arc systems [e.g., *De Wit et al.*, 1992].

It should be emphasized that most studies concur that the southern TTG–greenstone domain is entirely allochthonous with respect to the rest of the structurally overlying BGB [e.g., *De Wit et al.*, 1983, 1987; *Armstrong et al.*, 1990; *De Ronde and De Wit*, 1994; *Lowe et al.*, 1999; *Kisters et al.*, 2003; *Diener*, 2004; *Diener et al.*, in press]. However, the timing and kinematics of the Komati fault that separates these two domains are controversial, as recent metamorphic, geochronological, and structural studies have suggested that the actual juxtaposition of the two domains only occurred at ca. 3.23 Ga, during the exhumation of the high-grade southern TTG–greenstone domain (see below) [*Kisters et al.*, 2003; *Diener et al.*, in press].

The currently exposed extent of the 3.45 Ga magmatic arc is limited by bounding shear zones and gneiss belts that are mainly related to the 3.23 Ga exhumation of the domain. In the north, the TTG–greenstone domain is juxtaposed against the bulk of the low-grade BGB along an extensional detachment that encompasses the Komati fault (Figure 2) [*Kisters et al.*, 2003; *Diener et al.*, in press]. In the west, the high-grade rocks of the Stolzburg pluton are bounded by the Inyoni shear zone, a northerly-striking subvertical gneiss belt up to 2 km wide (Figure 2) [*Belcher et al.*, 2004]. A NW–SE trending zone of vertical, banded gneisses and interleaved amphibolite-facies greenstones in the south may represent the southern boundary, but structural relationships are largely obscured by the sheet-like, 3.1 Ga Mpuluzi batholith. Our regional mapping did not identify high-strain zones to the east. The kyanite-bearing assemblages in the Theespruit Formation around the Steynsdorp dome may indicate that the 3.23 Ga high-grade metamorphism possibly extends beyond the southern margin of the Barberton granitoid–greenstone terrain, encompassing the older TTG–gneiss complexes of the Steynsdorp dome and the Ancient Gneiss Complex in Swaziland.

The bulk of the southern TTG–greenstone domain is characterized by relatively low strain intensities. Primary igneous textures and intrusive relationships are, for the most part, preserved. Apart from the bounding shear zones, deformation of the rocks has resulted in only weak to moderately developed, mainly prolate fabrics, which can be related to the late-stage exhumation of the domain [*Kisters et al.*, 2003; *Diener et al.*, in press]. The low strain intensities in these rocks suggest that the southern TTG–greenstone domain has a preserved structural integrity and represents a relatively intact crustal block with a minimum aerial extent of 400–500 km^2.

THE P–T AND TECTONIC EVOLUTION OF THE SOUTHERN TTG–GREENSTONE DOMAIN

3.23 Ga High-P Amphibolite-Facies Metamorphism of the Greenstone Xenoliths

Previous interpretations of the high-grade metamorphism in the southern TTG–greenstone domain as the consequence of either dynamic contact metamorphism [*Anhaeusser*, 1984] or D_1 regional metamorphism [*De Ronde and De Wit*, 1994] were based largely on field observations and qualitative estimates of peak metamorphic conditions. Quantitative peak metamorphic estimates and the P–T–t path followed by these rocks were not constrained by these studies.

The first detailed metamorphic petrological investigation examined the greenstone remnants preserved in the southern part of the Stolzburg pluton (Figure 2) and revealed that the metabasites in this area contain well-equilibrated peak metamorphic assemblages of Grt-Cpx-Pl-Qtz ± Hbl (Table 1; Figure 3a) [*Dziggel et al.*, 2002; mineral abbreviations follow *Kretz*, 1983]. The peak metamorphic assemblages documented in this study are overprinted by the typical greenschist-facies assemblage Act-Ab-Ep, but equilibration during retrogression appears to have been incomplete (Figure 3b). Conventional thermobarometry of the peak metamorphic mineral chemistry from a number of different samples yielded P–T estimates of 8–11 kbar and 650–700 °C (Table 1) [*Dziggel et al.*, 2002]. These conditions are well within the kyanite stability field and are typical of the pressures and temperatures attained at the deeper levels of modern collisional belts [e.g., *Spear*, 1993]. Titanite and zircon that crystallized during metamorphism were dated at 3229 ± 9 and 3227 ± 7 Ma, respectively [*Dziggel*, 2002], indicating that high-grade metamorphism post-dates the magmatic assembly of the southern TTG–greenstone domain by more than 200 Ma. Similar metamorphic ages have been documented for titanite from TTGs throughout the high-grade domain [*Kamo and Davis*, 1994].

Both the style and timing of metamorphism unequivocally indicate that peak metamorphism occurred during a distinct tectonic event and is in no way related to the assembly of the southern TTG–greenstone domain at ca. 3.45 Ga. The timing of metamorphism coincides with that of the D_2 accretionary episode and, consequently, it was proposed that metamorphism occurred in response to D_2 crustal thickening and domain accretion in the BGB [*Dziggel et al.*, 2002]. The southern TTG–greenstone domain therefore records the deep-crustal manifestation of this collisional tectonic episode.

Metamorphism, Retrogression, and Uplift of the Theespruit Formation

In contrast to the high-grade conditions experienced by the southern TTG–greenstone domain, the BGB to the north of

Table 1. Summary of the Mineral Assemblages, Published Peak and Retrograde P–T Estimates, and Bulk Composition of Selected Samples from the Southern TTG–Greenstone Domain.

Location/reference	Sample number	Peak assemblage	Retrograde assemblage	Peak P–T estimates	Retrograde P–T estimates	Pseudosection composition (mol%)
Greenstone xenoliths (a)	SL1-8a	grt-cpx-hbl-pl-qtz	ep	8.8–11.8 kbar		$SiO_2 = 75.07$, $Al_2O_3 = 6.59$, $FeO = 3.10$, $MgO = 3.14$, $CaO = 9.87$, $Na_2O = 2.23$
Greenstone xenoliths (a)	SL1-6	grt-cpx-pl-qtz	ep-act	8.7–11 kbar, 640–750 °C		$SiO_2 = 68.14$, $Al_2O_3 = 5.89$, $FeO = 4.63$, $MgO = 5.90$, $CaO = 11.82$, $Na_2O = 3.57$
Greenstone xenoliths (a)	BE1	cpx-hbl-pl-qtz	ep	8.2–12.0 kbar, 630–753 °C		
Greenstone xenoliths (a)	BE2	hbl-pl-qtz	ep-act	600-705 °C		
Theespruit Formation (SSB) (b)	An1	grt-hbl-bt-pl-qtz	act-ep-chl	5.5 ± 0.9 kbar, 491 ± 40 °C		
Theespruit Formation (SSB) (b)	PB3	grt-hbl-bt-ep-pl-qtz	act-ep-qtz	6.3 ± 1.5 kbar, 492 ± 40 °C		
Theespruit Formation (TSB) (c)	61205E; Tj 18	grt-st-bt-chl-pl-qtz	ms-chl ± grt ± pl	7.9 ± 1.1 kbar, 556 ± 19 °C	3.8 ± 1.3 kbar, 543 ± 20 °C	$SiO_2 = 74.67$, $Al_2O_3 = 8.78$, $FeO = 5.71$, $MgO = 7.29$, $CaO = 1.73$, $Na_2O = 1.269$, $K_2O = 0.55$
Theespruit Formation (TSB) (c)	62601D	ky-st-bt-chl-pl-qtz	sil-ms-chl			$SiO_2 = 77.30$, $Al_2O_3 = 11.15$, $FeO = 3.46$, $MgO = 5.06$, $CaO = 0.76$, $Na_2O = 1.27$, $K_2O = 0.99$
Theespruit Formation (TSB) (c)	Tj 3; 62107; 61406	grt-hbl-ep-pl-qtz	chl ± grt ± hbl ± pl	7.0 ± 1.2 kbar, 537 ± 45 °C	6.1 ± 2.7 kbar, 569 ± 42 °C	

Sample localities are shown in Figure 2; the data are from (a) *Dziggel et al.* [2002], (b) *Kisters et al.* [2003], and (c) *Diener et al.* [in press].

the Komati fault experienced only lower greenschist-facies metamorphism during D_2, with maximum P–T estimates of 2.6 ± 0.6 kbar and 360 ± 50 °C [*Cloete*, 1999]. Coupled to estimates from the high-grade domain, these estimates indicate that a metamorphic break equivalent to a crustal column of more than 25 km exists between the southern TTG–greenstone domain and the BGB. Therefore, this domain must have been effectively decoupled from the BGB during D_2 and could only have been juxtaposed against the greenstone belt subsequent to 3.23 Ga peak metamorphism [*Kisters et al.*, 2003]. The transition between the high-grade domain and the low-grade greenstone sequence coincides with the Theespruit Formation that occurs along the margins of the Stolzburg and Theespruit plutons [*Viljoen and Viljoen*, 1969; *Kisters et al.*, 2003]. The Theespruit Formation is intensely deformed and exhibits an extremely high apparent metamorphic gradient, with amphibolite-facies lithologies close to the pluton margins that are replaced by lower-grade greenschist-facies rocks within a distance of ca. 1500 m.

In the Stolzburg schist belt (Figure 2), amphibolite-facies portions of the Theespruit Formation preserve peak metamorphic mineral assemblages of Grt-Hbl-Bt-Pl-Qtz ± Ep in certain metamafic horizons (Table 1). The mineral chemistry of these assemblages yields peak metamorphic conditions of ca. 6 kbar and 490 ± 40 °C that were attained at 3219 ± 9 Ma, contemporaneous with peak metamorphism in other areas of the southern TTG–greenstone domain (Table 1) [*Kisters et al.*, 2003].

Similar conditions have been documented further to the east, where the Theespruit Formation in the Tjakastad schist belt (Figure 2) contains metamafic horizons exhibiting peak assemblages of Grt-Hbl-Ep-Pl-Qtz [*Diener et al.*, in press]. The Tjakastad schist belt also contains rare clastic metasediments consisting of Grt-St-Bt-Chl-Pl-Qtz and Ky-St-Bt-Chl-Pl-Qtz assemblages that proved to be particularly useful in constraining P–T conditions (Table 1; Figures 3c–f). The compositions of the near-peak metamorphic mineral generations in these rocks yield P–T estimates of 7.4 ± 1.0 kbar

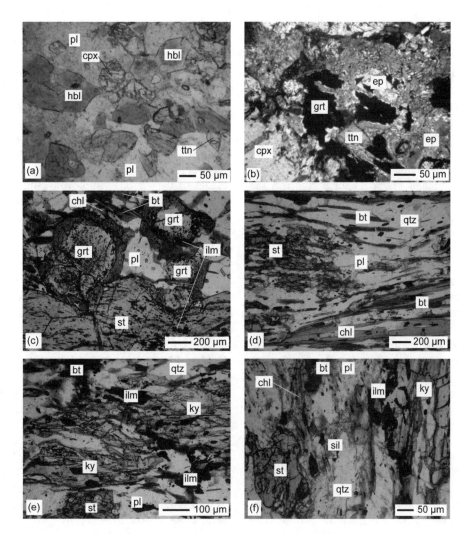

Figure 3. Photomicrographs of samples from the southern TTG–greenstone domain. (a) Texturally equilibrated Hbl-Cpx-Pl-Qtz assemblage in sample SL1-8a (from *Dziggel et al.* [2002]; plane-polarized light). (b) Peak assemblage of Grt-Cpx in sample SL1-6 being partially replaced by retrograde epidote (from *Dziggel et al.* [2002]; crossed nicols). (c) Grt-St-Bt-Chl-Pl-Qtz assemblage in metasediments from the Theespruit Formation in the Tjakastad schist belt (sample 62105F, plane-polarized light). (d) Deformed and recrystallized St-Bt-Chl-Pl-Qtz assemblage, illustrating the post-peak metamorphic deformation that affected the rocks in the Theespruit Formation (sample Tj 18 from the Tjakastad schist belt, plane-polarized light). (e) Coarse-grained and deformed peak-metamorphic Ky-St-Bt-Pl-Qtz assemblage in sample 62601D from the Tjakastad schist belt (plane-polarized light). (f) Fibrous sillimanite overgrowing the kyanite-bearing peak assemblage in sample 62601D (plane-polarized light).

and 560 ± 20 °C for this part of the Theespruit Formation (Table 1) [*Diener et al.*, in press]. An age of 3229 ± 25 Ma was obtained from metamorphic titanite and confirms that peak metamorphic conditions were attained during D$_2$.

Whereas the greenstone xenoliths commonly preserve peak metamorphic assemblages with limited retrogression, the high-grade lithologies of the Theespruit Formation were subjected to pervasive retrogression and major shearing subsequent to the P–T peak [*Kisters et al.*, 2003; *Diener et al.*, in press]. This caused the recrystallization and re-equilibration of peak metamorphic minerals and also led to the generation

of retrograde assemblages (Figures 3d and f). The retrograde minerals define the same fabric as peak metamorphic minerals, suggesting that deformation was initiated under peak or near-peak metamorphic conditions and continued during retrogression [*Kisters et al.*, 2003; *Diener et al.*, in press]. The retrograde minerals are characteristic of greenschist-facies metamorphic conditions and are typically Chl-Ab assemblages in metamafic rocks and Chl-Ms assemblages in felsic volcaniclastic horizons. Plagioclase coronas on garnet in metamafic rocks as well as sillimanite overgrowths on peak metamorphic kyanite in metasediments from the

Tjakastad schist belt (Figure 3f) are interpreted as retrograde occurrences related to decompression (see below).

Further evidence of decompression during retrogression has been documented from sites of garnet breakdown, where peak metamorphic garnet was consumed in a reaction that produced a new generation of compositionally distinct retrograde minerals. In the Tjakastad schist belt, this process produced the retrograde assemblages Grt-Hbl-Chl-Pl-Qtz and Grt-Chl-Ms-Pl-Qtz in certain metamafic and metasedimentary samples, respectively (Table 1). These retrograde assemblages yielded P–T estimates (Table 1) suggesting that the retrograde path of the Tjakastad schist belt involved an initial period of near-isothermal decompression of ca. 4 kbar prior to cooling into the greenschist-facies [*Diener et al,.* in press]. The retrograde path is consistent with rapid uplift, and *Diener et al.* [in press] estimated an average uplift rate of 2–5 mm/a for the rocks of the Tjakastad schist belt. The structures and fabrics associated with the retrograde deformational event, coupled to the rapid uplift rate, suggest that exhumation was accomplished by extensional detachment faulting [*Kisters et al.*, 2003]. This led *Kisters et al.* [2003] to propose that the high-grade domain is a metamorphic core complex [e.g., *Davis*, 1980; *Lister and Davis*, 1989] that was exhumed during the collapse of the D$_2$ orogen [e.g., *Dewey*, 1988].

PSEUDOSECTION MODELING AND P–T PATH DETERMINATIONS

The P–T estimates from previous studies of the southern TTG–greenstone domain were obtained by conventional geothermobarometry and, in general, were able to constrain only the near-peak metamorphic conditions experienced by these rocks. In an attempt to gain possible insights into the pro- and retrograde evolutionary paths, and to confirm the peak metamorphic estimates, we constructed pseudosections [e.g., *Harris et al.*, 2004; *Zeh et al.*, 2004] for selected samples from different areas of the high-grade domain.

Pseudosections were constructed with THERMOCALC (Version 3.25) [*Holland and Powell*, 1998], using the internally consistent thermodynamic dataset of *Holland and Powell* [1998] in the form of the April 2005 upgrade. The mineral solution models used are those presented in *White et al.* [2000, 2001], except for muscovite [*Coggon and Holland*, 2002], the feldspars [*Holland and Powell*, 2002], the amphiboles [*Dale et al.*, in review], and the clinopyroxenes [*Holland and Powell*, 1996]. The bulk compositions used in the calculations were obtained by bulk-rock X-ray fluorescence (XRF) analysis [*Dziggel et al.*, 2002; *Diener*, 2004] and calculations were performed in NCKFMASH for clastic metasediments and NCFMASH for metabasites. Almost all Ti in both rock types is contained in ilmenite, and

the XRF analyses were adjusted for Ti by subtracting the appropriate amount of Fe to effectively "project" from ilmenite. Similarly, the bulk of Mn is locked away in the cores of garnet porphyroblasts that do not participate in the high-grade metamorphic evolution of these samples. Therefore, the XRF analyses were also adjusted to remove the composition of the garnet cores from the effective bulk composition that was used in the pseudosections. All calculations assumed the presence of a pure water fluid phase. The location, effective bulk compositions, and mineral assemblages of the various samples investigated are presented in Figure 2 and Table 1, and the pseudosections constructed for these samples are presented in Figure 4.

Greenstone Xenoliths

Pseudosections of samples from the greenstone xenoliths show that the observed peak assemblage of Grt-Cpx-Hbl-Pl-Qtz in sample SL1-8a is stable at 8.2–9 kbar for temperatures of 620–700 °C and at slightly lower pressures for temperatures above 700 °C (Figure 4a). The observed peak assemblage of Grt-Cpx-Pl-Qtz in sample SL1-6 occurs only at pressures above 10.5–12 kbar and temperatures above 630 °C (Figure 4b). Even though there is a separation between the pressure estimates from the two samples, the results from the pseudosections are within the range of published peak metamorphic estimates and confirm that this part of the high-grade domain experienced pressures in excess of 10–12 kbar at temperatures of ca. 650–700 °C [*Dziggel et al.*, 2002]. The ubiquitous occurrence of epidote inclusions in peak metamorphic garnet as well as the retrograde replacement of garnet by amphibole—plagioclase assemblages suggests that the P–T evolution of these samples occurred via a clockwise P–T loop (see below).

Theespruit Formation

Pseudosections were constructed for garnet- and kyanite-bearing metasedimentary samples from the Theespruit Formation in the Tjakastad schist belt. The garnet-bearing peak assemblage is stable at 5–8.5 kbar and 600–650 °C (sample 62105F; Figure 4c), whereas the kyanite-bearing assemblage occurs at 5.4–8.4 kbar and 600–650 °C (sample 62601D; Figure 4d). Temperature estimates from both pseudosections are 20–40 °C higher than the peak estimate of ca. 560 °C obtained by conventional thermobarometry. This discrepancy is likely to be an artifact of the reset, near-peak metamorphic mineral compositions that were used in the thermobarometry calculations. However, there is generally a good correlation between pseudosection estimates and published P–T estimates, confirming that the Theespruit Formation in the Tjakastad schist belt experienced peak metamorphic conditions of ca. 7.4 kbar at temperatures of 560–600 °C [*Diener et al.*, in press].

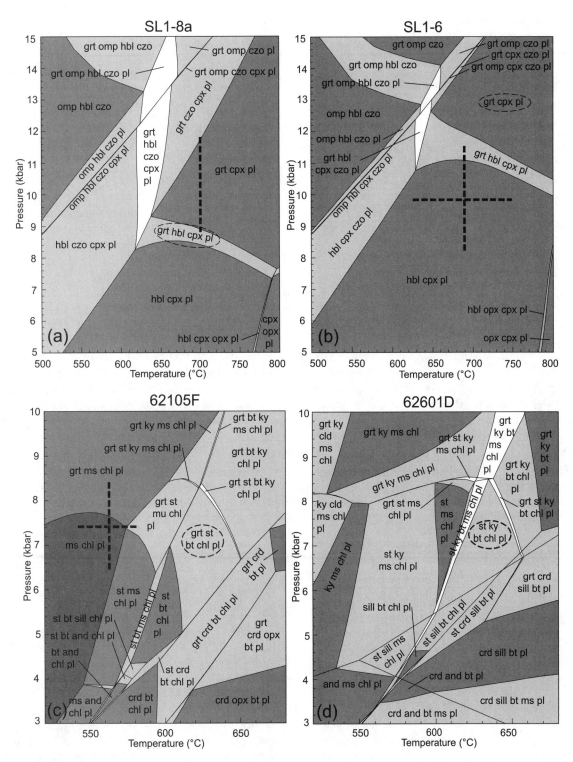

Figure 4. Pseudosections constructed for (a) sample SL1-8a, (b) sample SL1-6, (c) sample 62105F, and (d) sample 62601D from the southern TTG–greenstone domain. Refer to Table 1 for a description of individual samples and the bulk compositions used in the calculations. Divariant fields are shown in white and higher-variance fields are indicated by progressively darker shading. All fields contain quartz in addition to the minerals shown. Note that not all fields are labeled and that adjacent fields of the same variance are, in fact, separated by small lower-variance fields. The observed mineral assemblage in each sample is highlighted by a dashed circle; published P–T estimates obtained by conventional thermobarometry are shown as thick dashed lines.

Rare occurrences of sillimanite overgrowths on kyanite have been reported from sample 62601D (Figure 3f) [*Diener et al.*, in press]. None of the other peak metamorphic mineral phases in this sample show any evidence of reaction textures or destabilization, suggesting that they could form stable assemblages in conjunction with both kyanite and sillimanite. This is confirmed by the pseudosection of this sample (Figure 4d), which exhibits a field of Sil-St-Bt-Chl-Pl-Qtz stability adjacent to the low-P boundary of the peak metamorphic Ky-St-Bt-Chl-Pl-Qtz stability field. The transition between these two fields can only be accomplished by near-isothermal decompression, supporting the proposal of *Diener et al.* [in press] that the retrograde path in these rocks involved an initial period of isothermal decompression prior to cooling into the greenschist-facies.

The data presented here confirm and substantiate the published estimates of metamorphic conditions in different areas of the high-grade TTG–greenstone domain to the south of the BGB. The similar style and timing of metamorphism from different areas of the domain highlight that all these rocks manifest the same high-P intermediate-T regional metamorphic episode, albeit at slightly different grades (Figure 5). P–T pairs consistently point to almost identical low average apparent geothermal gradients of ca. 18–22 °C/km from all areas of the high-grade domain. The occurrence of the same style of metamorphism across the whole of the southern TTG–greenstone domain underlines the structural integrity of this crustal block.

In contrast to the low apparent geothermal gradients recorded in the high-grade domain during D_2, considerably higher apparent geothermal gradients of ca. 40 °C/km are recorded in the lower greenschist-facies greenstones of the BGB during D_2 (Figure 5) [*Cloete*, 1999; *Diener et al.*, in press]. The contrasting styles of metamorphism highlight the fact that metamorphism of the southern TTG–greenstone domain is not a higher-grade continuation of metamorphism in the BGB, and indicates that the two domains experienced very different tectono-metamorphic histories.

CONJECTURES ON THE SETTING OF THE 3.23 GA HIGH-P METAMORPHISM

The main D_2 collisional event of the BGB saw the amalgamation of a northern domain with a southern domain at ca. 3229–3226 Ma [*De Ronde and De Wit*, 1994; *De Ronde and Kamo*, 2000]. Peak metamorphic conditions in the northern and southern supracrustal domains to either side of the Saddleback–Inyoka fault did not exceed lower-greenschist facies grades [*Xie et al.*, 1997; *Cloete*, 1999], and both the metamorphism and deformation record a relatively shallow, thin-skinned type tectonism [*De Wit et al.*, 1983; 1992]. On the basis of structural evolution and lithological assemblages, *De Ronde and De Wit* [1994] concluded that the D_2 tectonism recorded in the BGB was the result of the collision of two island arcs in an arc-trench setting.

At the same time, the southern TTG domain underwent high-P intermediate-T metamorphism, recording the burial of this continental sliver to depths of 35–45 km [*Dziggel et al.*, 2002; *Diener et al.*, in press]. The high-P nature of the metamorphism in the southern TTG domain, as well as the common occurrence of prograde epidote inclusions within garnet in the metamafic rocks, argues for a clockwise P–T path, and a significant component of heating while at maximum pressure. Figure 6 illustrates portions of the likely P–T path followed by the central and marginal portions of the domain, as constrained by the pseudosection modeling and previously discussed geothermobarometry and mineral textural evidence. It appears probable that the transition from an epidote-hornblende–bearing assemblage to a hornblende-free garnet-plagioclase–bearing assemblage in sample SL1-6 would have required heating through the fairly narrow pressure interval of 11–13 kbar at 620–650 °C (Figures 4b and 6). Such high-P, clockwise P–T trajectories are typically interpreted to be the product of collisional tectonics [e.g., *Rotzier et al.*, 1998; *Borghi et al.*, 2003], and it is common in such scenarios that peak temperature is attained during the exhumation of the rocks, as thermal re-equilibration is generally slower than the rate of the tectonic processes. However, heating of the southern TTG–greenstone domain at mid- to lower-crustal levels (Figure 6) suggests that thermal re-equilibration of the deeply buried continental rocks occurred prior to the isostatic rebound. This may indicate a finite residence time of the rocks at mid-crustal levels for the re-establishment of the disrupted isotherm to occur. This supposition is supported by the inferred retrograde path that is characterized by near-isothermal decompression, with no evidence for significant heating during exhumation. The kyanite-staurolite–bearing assemblage in sample 62601D from the Tjakastad schist belt would have been particularly sensitive to any heating during decompression, as it would have resulted in the crystallization of cordierite (Figures 4f and 6). The absence of cordierite in this sample therefore argues that these rocks did not evolve via a decompression-heating trajectory.

The exhumation of high-P continental rocks is determined by a variety of factors such as external forces, i.e., the structurally driven extrusion of rocks, buoyancy of the thickened crustal column, erosion rates, the onset and degree of extension in response to either changing boundary conditions (e.g., slab roll-back), and/or the gravitational collapse of the thickened orogen. The prolate fabrics recorded in large parts of the southern TTG–greenstone domain record the orogen-parallel extrusion of the rocks in response to lateral, subhorizontal shortening during plate convergence

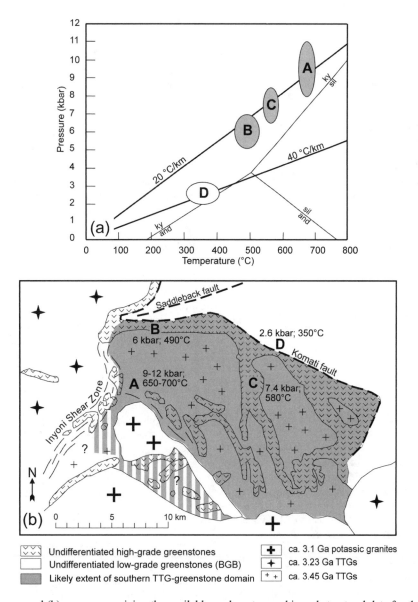

Figure 5. (a) P–T diagram and (b) map summarizing the available peak metamorphic and structural data for the D_2 collisional event in the BGB. P–T estimates from different areas of the southern TTG–greenstone domain (points A–C in (a) and (b)) plot along a similar apparent geothermal gradient of ca. 20 °C/km, testifying to the structural integrity of this domain (shaded area). Estimates of D_2 metamorphic conditions in the BGB (point D) highlight that a metamorphic break of some 4 kbar existed between the BGB and the southern TTG–greenstone domain at that time.

and concomitant vertical thickening of the crust, signifying the gravitational collapse of the thickened orogen [*Kisters et al.,* 2003]. The overprinting of these earlier, high-grade fabrics by non-coaxial, retrograde mylonites in the northern extensional detachment indicates the exhumation of the southern TTG–greenstone domain from mid- and deep-crustal levels during the syn- to late-orogenic collapse and extension of the crust. The onset of the Moodies Group sedimentation in small, fault-bounded basins shortly after the

main D_2 collisional phase may represent the shallow-crustal expression of this late-collisional extension [*Heubeck and Lowe,* 1994]. Thus, exhumation of the deep-crustal continental sliver was probably achieved by buoyancy during crustal extension.

As yet, the structures responsible for the prograde burial of these continental rocks have not been identified. Considering the relatively low-strain intensities and overall structural integrity of the high-grade TTG–greenstone domain, the

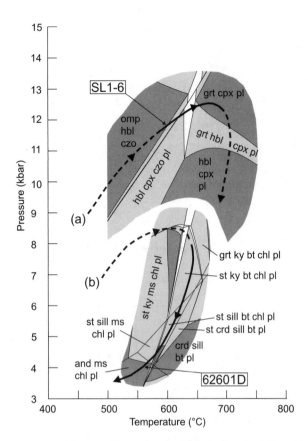

Figure 6. P–T diagram illustrating the likely P–T paths followed by (a) the central and (b) the marginal zones of the southern TTG–greenstone domain as constrained by pseudosection modeling of samples SL1-6 and 62601D. The prograde transition from an hornblende–epidote–bearing to hornblende-free garnet-plagioclase–bearing assemblage in sample SL1-6 occurred by heating at pressures of ca. 12 kbar, whereas the retrograde evolution of sample 62601D involved a transition from kyanite to sillimanite stability that was achieved by near-isothermal decompression. A combination of these segments strongly advocates that the evolution of the southern TTG–greenstone domain occurred by a high-P clockwise P–T trajectory typical of collisional tectonics. Refer to Figure 3 for the assemblages present in fields that are not labeled.

burial of the rocks was most likely accomplished along bounding structures. The preservation of high-P assemblages and the melange-like character of the western boundary of the southern TTG–greenstone domain, represented by the 2–3 km-wide gneiss belt of the Inyoni shear zone (Figure 2) [*Dziggel et al.*, 2002], may point to the location of this composite gneiss belt close to the paleosuture. The structures along the northern margin of the domain may have initially been active during the burial of the domain, but they have been completely overprinted and obliterated by shear zones that accommodated the subsequent exhumation of the rocks [*Kisters et al.*, 2003; *Diener et al.*, in press].

Irrespective of the original geometry and spatial distribution of structures that led to the burial of the rocks, the trajectory of the P–T path, the depth of burial, and the rate of exhumation are consistent with a subduction of the 3.45 Ga continental sliver. The similar geotherms and the structural integrity of the currently exposed southern TTG–greenstone domain suggest that it underwent burial and exhumation as a coherent block. Burial depths of 35–45 km imply that this part of the shallow-crustal 3.45 Ga magmatic arc was accreted and underplated to the base of a colliding continental block during the 3.23 Ga collision (Figure 7). The 3.45 Ga arc domain may have represented the leading edge of a microcontinent situated largely to the southeast and brought into the 3.23 Ga D_2 subduction zone with a northern continent (Figure 7). The short-lived nature of the D_2 collisional event between 3229 and 3226 Ma [*De Ronde and Kamo*, 2000] suggests an abortion of crustal convergence at the time when the continental block of the southern TTG–greenstone domain was subducted.

The corresponding deeper levels of the northern continent have not yet been identified. The recognition of such rocks is certainly more difficult in this part of the Barberton terrain, since the northern parts of the BGB are intruded by extensive syn- to late-D_2 TTGs of the Nelshoogte and Kaap Valley plutons, which may represent the last vestiges of TTG magmatism produced by melting of the subducting oceanic slab (Figure 7). Large parts of the northern margin of the BGB are, furthermore, blanketed by late-stage 3.1 Ga tabular plutons such as the Nelspruit batholith.

IMPLICATIONS FOR ARCHEAN GEODYNAMIC MODELS

The pressures in excess of 10–12 kbar that are recorded in the southern TTG–greenstone domain are comparable to pressures estimated from modern retrograde blueschist terrains in continental collision zones such as the Western Alps and New Caledonia [*Ernst*, 1988, and references therein]. However, temperature estimates from the TTG–greenstone domain are ca. 150–200 °C higher than typical temperature estimates from these younger terrains. The peak P–T pairs from the TTG–greenstone domain are, however, nearly identical to P–T conditions reported from ophiolites in the continental suture between the Congo and Kalahari Cratons within the ca. 520 Ma Zambezi Belt [*Johnson and Oliver*, 2004]. These similarities may suggest that the P–T history of the southern TTG–greenstone domain could conceivably have involved evolution through the blueschist-facies, and one might speculate that these rocks could represent the Mid-Archean equivalent of modern-day blueschists or eclogites.

The significantly depressed apparent geothermal gradients that are recorded in the southern TTG–greenstone domain are

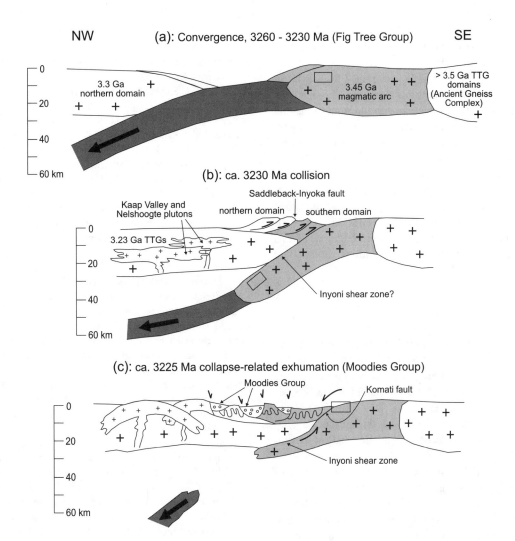

Figure 7. Schematic cross-section illustrating the evolution of the southern TTG–greenstone domain as envisaged by this work. (a) Convergence of the northern and southern domains of the BGB during the deposition of the Fig Tree Group from ca. 3260 to 3230 Ma. The box indicates the position of the southern TTG–greenstone domain within the ca. 3.45 Ga magmatic arc at the leading edge of the southern continental block. (b) Collision of the northern and southern domains at ca. 3230 Ma. The supracrustal sequences of the ca. 3.45 Ga arc are tectonically scraped off and accreted to the northern domain, corresponding to the thin-skinned tectonics and D₂ deformation observed in the BGB [*De Ronde and De Wit*, 1994]. The bulk of the 3.45 Ga arc (including the southern TTG–greenstone domain) is subducted and underplated to the northern domain, resulting in doubly thickened crust and burial to depths of 35–45 km. Syn- to late-collisional TTGs such as the 3236 Ma Nelshoogte and 3227 Ma Kaap Valley plutons [*Kamo and Davis, 1994*; *De Ronde and Kamo*, 2000] are emplaced into the northern domain. (c) Subduction is aborted and the doubly thickened crust collapses, leading to the exhumation of the southern TTG–greenstone domain. Extension of the upper crust leads to the formation of isolated, fault-bounded molasse basins into which the Moodies Group is deposited [*Heubeck and Lowe*, 1994]. Corresponding extension at lower crustal levels causes flow of the ductile crust, resulting in uplift and doming of the basement domains in the northern and southern continental blocks. The doming results in the tightening and steepening of structures and fabrics in the BGB and the formation of dome-and-keel structures.

likely to be a function of two, somewhat compensatory, factors; first, that the earlier assembled continental fragment represented by the southern TTG–greenstone domain was characterized by a relatively low geothermal gradient prior to the orogeny, and second, that burial and exhumation of the domain was relatively rapid. The exhumation rates of

2–5 mm/a constrained from the Theespruit Formation by *Diener et al.* [in press] suggest the relatively rapid uplift of this domain. However, given the supposition that the southern TTG–greenstone domain appears to have resided at mid- to lower-crustal levels long enough for at least some re-establishment of the disrupted geotherm to occur, it would suggest that the crust was relatively cool and sufficiently strong to support the overthickened crustal column before it underwent gravitational collapse.

The recognition of the regionally developed high-P intermediate-T metamorphic assemblages in this Mid-Archean continental fragment are at variance with interpretations that the Archean crust was characterized by markedly higher geothermal gradients that, in turn, would have led to the rheological weakening of the crust and thereby prevention of significant crustal thickening [Marshak et al., 1992; Marshak, 1999]. The results presented here provide insight into the deeper sections of an Archean granitoid–greenstone belt and suggest substantial thickening of Mid-Archean continental crust and subsequent gravitational collapse of this crustal column similar to orogenic evolutions recorded in younger Proterozoic and Phanerozoic belts.

CONCLUSIONS

High-P intermediate-T metamorphic assemblages that are regionally developed in supracrustal remnants in the TTG–greenstone domain south of the Barberton greenstone belt document the burial of this 3.45-Ga-old, initially shallow-crustal magmatic arc to depths of 35–45 km. The burial of this domain occurred at ca. 3.23 Ga, during the main D_2 collisional event that is recorded in the shallow-crustal greenstone belt.

Pseudosection modeling of metamafic and metasedimentary metamorphic assemblages allows some constraints to be placed on the form and trajectory of the P–T path followed by this domain. The clockwise burial path, burial depths, and rates of burial are consistent with the subduction of the magmatic arc and possibly a doubling of thickness of continental crust in the D_2 collisional belt. The subsequent near-isothermal decompression path followed during the rapid exhumation of the high-grade domain, together with the onset of extensional tectonics, indicates the gravitational collapse of the overthickened orogen.

The evolution of this continental fragment preserved in the Barberton granitoid–greenstone terrain indicates that the Mid-Archean continental crust was sufficiently cool and rigid to undergo and sustain significant crustal stacking.

Acknowledgments. This material is based on work supported by the National Research Foundation (South Africa) under grant numbers NRF 2050238 and NRF 2060045. We thank Annika Dziggel for providing some of the photomicrographs presented in this paper and Roger Powell for providing a copy of the amphibole model. This paper benefited from constructive reviews by Keith Benn and two anonymous reviewers, which are greatly appreciated.

REFERENCES

Anhaeusser, C.R. (1973) The evolution of the Precambrian crust of southern Africa, *Phil. Trans. R. Soc. Lond.*, A 2731, 359-388.

Anhaeusser, C.R. (1984) Structural elements of Archaean granite-greenstone terranes as exemplified by the Barberton Mountain Land, southern Africa, in *Precambrian Tectonics Illustrated*, edited by A. Kröner and R. Greiling, Stuttgart, 418pp.

Anhaeusser, C.R., L.J. Robb and M.J. Viljoen (1981) Provisional geological map of the Barberton Greenstone Belt and surrounding granitic terrane, Eastern Transvaal and Swaziland, Geol. Soc. S. Afr., scale 1:250 000.

Armstrong, R.A., W. Compston, M.J. De Wit and I.S. Williams (1990) The stratigraphy of the Barberton Greenstone Belt revisited: a single zircon ion microprobe study, *Earth Planet Sci. Lett.*, 101, 90-106.

Belcher, R.W., A.F.M. Kisters, J.D. Westraat, J.F.A. Diener and G. Stevens (2004) The architecture of the southern Barberton granite-gneiss terrain. Geoscience Africa 2004, Abstract Volume, Univ. Witwatersrand, Johannesburg, South Africa, 55-56.

Benn, K.,W. Miles, M.R. Ghassemi and J. Gillett (1994) Crustal structure and kinematic framework of the northwestern Pontiac Subprovince, Quebec: an integrated structural and geophysical study, *Can. J. Earth Sci.*, 31, 271-281.

Boak, J.L. and R.F. Dymek (1982) Metamorphism of the ca. 3800 Ma supracrustal rocks at Isua, West Greenland: implications for early Archean crustal evolution, *Earth Planet. Sci. Lett.*, 59, 155-176.

Borghi, A., D. Castelli, B. Lombardo and D. Visona (2003) Thermal and baric evolution of garnet granulites from the Kharta region of S. Tibet, E. Himalaya, *Eur. J. Mineralogy*, 15, 401-418.

Byerly, G.R., A. Kröner, D.R. Lowe, W. Todt and M.M. Walsh (1996) Prolonged magmatism and time constraints for sediment deposition in the early Archean Barberton greenstone belt: evidence from the Upper Onverwacht and Fig Tree Groups, *Precambrian Res.*, 78, 125-138.

Chardon, D., J.J. Peucat, M. Jayananda, P. Choukroune and C.M. Fanning (2002) Archaean granite-greenstone tectonics at Kolar, (South India): Interplay of diapirism and bulk inhomogeneous contraction during juvenile magmatic accretion. *Tectonics*, 21, 700-717.

Choukroune, P.H., A. Bouhallier and N.T. Arndt (1995) Soft lithosphere during periods of Archaean crustal growth or crustal reworking, in *Early Precambrian Processes*, edited by M.P. Coward and A.C. Ries, Geol. Soc. Lond. Spec. Publ. 95, 67-86.

Cloete, M. (1999) Aspects of volcanism and metamorphism of the Onverwacht Group lavas in the southwestern portion of the Barberton Greenstone Belt, *Mem. Geol. Surv. S. Afr.*, 84, 232pp.

Coggon, R. and T.J.B. Holland (2002) Mixing properties of phengitic micas and revised garnet-phengite thermobarometers, *J. Metamorphic Geol.*, 20, 683-696.

Collins, W.J. and M.J. Van Krandendonk (1999) Model for the development of kyanite during partial convective overturn of Archaean granite-greenstone terranes: the Pilbara Craton, Australia, *J. Metamorphic. Geol.*, 17, 145-156.

Collins, W.J., M.J Van Kradendonk and C. Teyssier (1998) Partial convective overturn of the Archaean crust in the eastern part of the Pilbara Craton, Western Australia: 2. Driving mechanisms and tectonic implications, *J. Struct. Geol.*, 20, 1405-1424.

Dale, J., R. Powell, R.W. White, F.L. Elmer and T.J.B. Holland. A thermodynamic model for Ca-Na clinoamphiboles in NCFMASHO for petrological calculations, submitted to *J. Metamorphic Geol.*

Dann, J.C. (2000) The 3.5 Ga Komati Formation, Barberton Greenstone Belt, South Africa: part I: new maps and magmatic architecture, *S. Afr. J. Geol.*, 103, 47-68.

Davis, G.A. (1980) Structural characteristics of metamorphic core complexes, southern Arizona, *Geol. Soc. Amer. Mem.*, 153, 35-77.

De Ronde, C.E.J. and M.J. De Wit (1994) The tectonic history of the Barberton greenstone belt, South Africa: 490 million years of Archaean crustal evolution, *Tectonics*, 13, 983-1005.

De Ronde, C.E.J. and S.L. Kamo (2000) An Archaean arc-arc collisional event: a short-lived (ca 3 Myr) episode, Weltevreden area, Barberton Greenstone Belt, South Africa, *J. Afr. Earth. Sci.*, 30, 219-248.

Dewey, J.F. (1988) Extensional collapse of orogens, *Tectonics*, 7, 1123-1139.

De Wit, M.J. (1982) Gliding and overthrust nappe tectonics in the Barberton greenstone belt, *J. Struct. Geol.*, 4, 117-136.

De Wit, M.J., R.E.P. Fripp and I.G. Stanistreet (1983) Tectonic and stratigraphic implications of new field observations along the southern part of the Barberton greenstone belt, *Spec. Pub. Geol. Soc. S. Afr.*, 9, 21-29.

De Wit, M.J., R.A. Armstrong, R.J. Hart and A.H. Wilson (1987) Felsic igneous rocks within the 3.3 to 3.5 Ga Barberton greenstone belt: high-level equivalents of the surrounding tonalite trondhjemite terrain, emplaced during thrusting, *Tectonics*, 6, 529-549.

De Wit, M.J., C. Roering, R.J. Hart, R.A. Armstrong, C.E.J. De Ronde, R.W.E. Green, M. Tredoux, E. Peberdy and R.A. Hart (1992) Formation of an Archaean continent, *Nature*, 357, 553-562.

Diener, J.F.A. (2004) The tectono-metamorphic evolution of the Theespruit Formation in the Tjakastad schist belt and surrounding areas of the Barberton greenstone belt, South Africa, M.Sc thesis, 210pp, Univ. Stellenbosch, South Africa.

Diener, J.F.A., G. Stevens, A.F.M. Kisters and M. Poujol (in press) Metamorphism and exhumation of the basal parts of the Barberton greenstone belt, South Africa: Constraining the rates of Mesoarchaean tectonism, *Precambrian Res.*

Dziggel, A. (2002) The petrogenesis of 'lower' Onverwacht Group clastic metasediments and related metavolcanic rocks in the southern part of the Barberton Mountain Land, South Africa, Ph.D. thesis, 230pp, Univ. Witwatersrand, Johannesburg, South Africa.

Dziggel, A., G. Stevens, M. Poujol, C.R. Anhaeusser and R.A. Armstrong (2002) Metamorphism of the granite-greenstone terrane south of the Barberton greenstone belt, South Africa: an insight into the tectono-thermal evolution of the 'lower' portions of the Onverwacht Group, *Precambrian Res.*, 114, 221-247.

Eide, E.A. and J.-M. Lardeaux (2002) A relict blueschist in meta-ophiolite from the central Norwegian Caledonides – discovery and consequences, *Lithos*, 60, 1-19.

England, P.C. and A.B. Thompson (1984) Pressure-temperature-time paths of regional metamorphism I. Heat transfer during the evolution of regions of thickened continental crust, *J. Petrol.*, 25, 894-928.

Ernst, W.G. (1973) Blueschist metamorphism and P-T regimes in active subduction zones, *Tectonophysics*, 17, 255-272.

Ernst, W.G. (1988) Tectonic history of subduction zones inferred from retrograde blueschist P-T paths, *Geology*, 16, 1081-1084.

Grotzinger, J.P. and L. Royden (1990) Elastic strength of the Slave Craton at 1.9 Gyr and implications for the thermal evolution of the continents, *Nature*, 347, 64-66.

Hamilton, W.B. (1998) Archaean magmatism and deformation were not products of plate tectonics, *Precambrian Res.*, 91, 143-179.

Harris, N.B.W., M. Caddick, J. Kosler, S. Goswami, D. Vance and A.G. Tindle (2004) The pressure-temperature-time path of migmatites from the Sikkim Himalaya, *J. Metamorphic Geol.*, 22, 249-264.

Henderson, J.B. (1981) Archean basin evolution in the Slave province, Canada, in *Plate tectonics in the Precambrian*, edited by A. Kröner, Elsevier, Amsterdam, pp. 213-235.

Heubeck, C. and D.R. Lowe (1994) Late syndepositional deformation and detachment tectonics in the Barberton Greenstone Belt, South Africa, *Tectonics*, 13, 1514-1536.

Hickman, A.H. (1983) Geology of the Pilbara Block and its environs, *Geol. Survey Western Aust. Bull.*, 127, 268pp.

Holland, T.J.B. and R. Powell (1996) Thermodynamics of order-disorder in minerals. 2. Symmetric formalism applied to solid solutions, *Amer. Miner.*, 81, 1425-1437.

Holland, T.J.B. and R. Powell (1998) An internally consistent thermodynamic data set for phases of petrological interest, *J. Metamorphic Geol.*, 16, 309-343.

Holland, T.J.B. and R. Powell (2002) Activity-composition relations for phases in petrological calculations: an asymmetric multicomponent formulation, *Contrib. Mineral. Petrol.*, 145, 492-501.

Johnson, S.P. and G.J.H. Oliver (2004) Tectonothermal history of the Kaourera Arc, northern Zimbabwe: implications for the tectonic evolution of the Irumide and Zambezi Belts of south central Africa, *Precambrian Res.*, 130, 71-97.

Kamo, S.L. and D.W. Davis (1994) Reassessment of Archaean crustal development in the Barberton Mountain Land, South Africa, based on U-Pb dating, *Tectonics*, 13, 167-192.

Kisters, A.F.M. and C.R. Anhaeusser (1995) Emplacement features of Archaean TTG plutons along the southern margin of the Barberton Greenstone Belt, South Africa, *Precambrian Res.*, 75, 1-15.

Kisters, A.F.M., G. Stevens, A. Dziggel and R.A. Armstrong (2003) Extensional detachment faulting and core-complex formation in the southern Barberton granite-greenstone terrain, South Africa: evidence for a 3.2 Ga orogenic collapse, *Precambrian Res.*, 127, 355-378.

Kretz, R. (1983) Symbols for rock-forming minerals, *Amer. Miner.*, 68, 277-279.

Kröner, A., G.R. Byerly and D.R. Lowe (1991) Chronology of early Archaean granite-greenstone evolution in the Barberton Mountain land, South Africa, based on precise dating by single zircon evaporation, *Earth Planet. Sci. Lett.*, 103, 41-54.

Kröner, A., E. Hegner, G.R. Byerly and D.R. Lowe (1992) Possible terrane identification in the early Archaean Barberton Greenstone Belt, South Africa, using single zircon geochronology, *Eos Trans. AGU*, 73, 6-16.

Kröner, A., E. Hegner, J.I. Wendt and G.R. Byerly (1996) The oldest part of the Barberton granitoid-greenstone terrain, South Africa, evidence for crust formation between 3.5 and 3.7 Ga, *Precambrian Res.*, 78, 105-124.

Le Fort, P. (1975) Himalaya: the collided range. Present knowledge of the continental arc, *Amer. J. Sci.*, 275A, 1-44.

Lister, G.S. and G.A. Davis (1989) The origin of metamorphic core complexes and detachment faults formed during Tertiary continental extension in the northern Colorado River region, U.S.A., *J. Struct. Geol.*, 11, 65-94.

Lowe, D.R. (1994) Accretionary history of the Archaean Barberton Greenstone Belt (3.55 – 3.22 Ga), southern Africa, *Geology*, 22, 1099-1102.

Lowe, D.R. (1999) Geologic evolution of the Barberton Greenstone Belt and vicinity, in *Geologic Evolution of the Barberton Greenstone Belt, South Africa*, edited by D.R. Lowe and G.R. Byerly, Spec. Paper Geol. Soc. Amer., 329, 287-312.

Lowe, D.R., G.R. Byerly, B.L. Ransom and B.R. Nocita (1985) Stratigraphic and sedimentological evidence bearing on structural repetition in Early Archaean rocks of the Barberton Greenstone Belt, South Africa, *Precambrian Res.*, 27, 165-186.

Lowe, D.R., G.R. Byerly and C. Heubeck (1999) Structural divisions and development of the west-central part of the Barberton Greenstone Belt, in *Geologic Evolution of the Barberton Greenstone Belt, South Africa*, edited by D.R. Lowe and G.R. Byerly, Spec. Paper Geol. *Soc. Amer.*, 329, 37-82.

Marshak, S. (1999) Deformation style way back when: thoughts on the contrasts between Archaean/Paleoproterozoic and contemporary orogens, *J. Struct. Geol.*, 21, 1175-1182.

Marshak, S., F.F. Alkmin and H. Jordt-Evangelista (1992) Proterozoic crustal extension and the generation of dome-and-keel structure in an Archaean granite-greenstone terrane, *Nature*, 357, 491-493.

McGregor, V.R. (1973) The early Precambrian gneisses of the Godthab district, West Greenland, *Phil. Trans. R. Soc. Lond.*, A273, 343-358.

Rotzier K., R. Schumacher, W.V. Maresch and A.P. Willner (1998) Characterization and geodynamic implications of contrasting metamorphic evolution in juxtaposed high-pressure units of the western Erzgebirge (Saxony, Germany), *Eur. J. Mineralogy*, 10, 261-280.

SACS (South African Committee for Stratigraphy) (1980) *Stratigraphy of South Africa, Part 1, Lithostratigraphy of the Republic of South Africa, South West Africa/Namibia and the Republics of Bophutatswana, Transkei and Venda*, compiled by L.E. Kent, Handbook Geol. Surv. S. Afr., 8, 690pp.

Sandiford, M. (1989) Secular trends in the thermal evolution of metamorphic terrains, *Earth Planet. Sci. Lett.*, 95, 85-96.

Sandiford, M. and S. McLaren (2002) Tectonic feedback and the ordering of heat producing elements within the continental lithosphere, *Earth Planet. Sci. Lett.*, 204, 133-150.

Spear, F.S. (1993) *Metamorphic phase equilibria and pressure-temperature-time paths*, Mineral. Soc. Amer. Monograph, Washington D.C, 799pp.

Stott, G.M. (1997) The Superior Province, Canada, in *Greenstone Belts*, edited by M.J. de Wit and L.D. Ashwal, Oxford University Press, 480-507.

Viljoen, M.J. and R.P. Viljoen (1969) An introduction to the geology of the Barberton granite-greenstone terrain, *Spec. Pub. Geol. Soc. S. Afr.*, 2, 9-28.

White, R.W., R. Powell, T.J.B. Holland and B. Worley (2000) The effect of TiO_2 and Fe_2O_3 on metapelite assemblages at greenschist and amphibolite-facies conditions: mineral equilibria calculations in the system K_2O-FeO-MgO-Al_2O_3-SiO_2-H_2O-TiO_2-Fe_2O_3, *J. Metamorphic Geol.*, 18, 497–511.

White, R.W., R. Powell and T.J.B. Holland (2001) Calculation of partial melting equilibria in the system CaO-Na_2O-K_2O-FeO-MgO-Al_2O_3-SiO_2-H_2O (CNKFMASH), *J. Metamorphic Geol.*, 19, 139-153.

Wilkins, C. (1997) Regional and contact metamorphism, in *Greenstone Belts*, edited by M.J. De Wit and L.D. Ashwal, Oxford University Press, 126-163

Williams, D.A.C. and R.G. Furnell (1979) Reassessment of part of the Barberton type area, South Africa, *Precambrian Res.*, 9, 325-347.

Xie, X., G.R. Byerly and R.E Ferrell (1997) IIb trioctahedral chlorite from the Barberton Greenstone Belt: crystal structure and rock composition constraints with implications to geothermometry, *Contrib. Mineral. Petrol.*, 126, 275-291.

Zeh, A., R. Klemd, S. Buhlmann and J.M. Barton (2004) Pro- and retrograde P-T evolution of granulites of the Beit Bridge Complex (Limpopo Belt, South Africa): constraints from quantitative phase diagrams and geotectonic implications, *J. Metamorphic Geol.*, 22, 79-96.

J. Diener, School of Earth Sciences, University of Melbourne, Victoria 3010, Australia. (j.diener@pgrad.unimelb.edu.au)

A. Kisters and G. Stevens, Department of Geology, University of Stellenbosch, Private Bag X1, Matieland 7602, South Africa.

Reflections of the Neoarchean: A Global Perspective

Arie J. van der Velden and Frederick A. Cook

Department of Geology and Geophysics, University of Calgary, Calgary, Canada

Barry J. Drummond and Bruce R. Goleby

Geoscience Australia, Canberra, Australia

Deep seismic reflection data from the Superior and Slave cratons in Canada and the Yilgarn craton in Australia are characterized by subhorizontal to shallowly dipping reflections in the crust and upper mantle that are interpreted as ~2.8–2.6 Ga structural fabrics. When considered in the context of regional geology, these fabrics indicate that peak orogenesis leading to cratonization was dominated by horizontally directed forces and plate interactions. The orientation of reflections beneath gneiss- to greenstone-domain transitions are inconsistent with models of vertical tectonics involving a convective overturn of the crust. In at least three different areas, reflections that project from the lower crust across the Moho into the mantle are interpreted as relict subduction zones or terrane boundaries. Mid-crustal reflections resembling "shingles" are interpreted as imbricates and/or tectonically underplated material. These reflection patterns occur in Neoarchean cratons worldwide and are remarkably similar to reflection patterns beneath Proterozoic and Phanerozoic orogens. Thus, Neoarchean cratons appear to be products of rigid plate behaviour in which horizontal forces were dominant.

1. INTRODUCTION

Regional seismic reflection profiles from Archean cratons around the world exhibit fundamentally similar geometry of crustal, and in some cases upper mantle, structure. These patterns serve as a test for models of Archean tectonism, which are often envisaged to be dominated by either vertical or horizontal (plate-like) tectonics.

Vertical tectonism involves a convective overturn of the crust. Early studies on the formation of Archean granite–greenstone terranes emphasized the role of igneous activity, particularly the extrusion of high-temperature komatiitic volcanic rocks and the intrusion of granitoid domes. Tectonic models that were derived for these terranes

therefore emphasized the addition of igneous rocks to the crust [e.g., *Campbell and Hill*, 1988; *Zegers and van Keken*, 2001]. The formation of high density mafic and ultramafic volcanic supracrustal sequences overlying less dense, usually felsic, crust implied the likelihood of convective overturn caused by gravitational instability [e.g., *Anhaeusser*, 1973; *Mareschal and West*, 1980; *Collins et al.*, 1998; *Bleeker*, 2002; *Hamilton*, 1998, 2003]. In this model, diapiric upwelling of the granitic rocks beneath sinking mafic material is accommodated by magmatic and/or ductile flow. Granite–gneiss domains are envisioned as mushroom-shaped, with steep to overhanging flanks, and are surrounded by keel-shaped greenstone domains that project deep within the crust. When faults are present, they are likely to be deeply penetrating near-vertical structures.

Horizontal tectonism, on the other hand, involves thickening of crust due to horizontally directed forces, presumably caused by plate interactions [e.g., *Barley et al.*, 1989; *Kusky*, 1989; *Percival et al.*, 1994; *Calvert et al.*, 1995; *de Wit*, 1998;

Archean Geodynamics and Environments
Geophysical Monograph Series 164
10.1029/164GM16

Blewett, 2002]. In this model, crustal thickening is caused by processes such as (1) structural repetition and folding above shallowly dipping detachments; (2) ductile flow within weak layers, and formation of nappes; and (3) accretion of terranes through obduction and tectonic underplating. Low-angle extensional faults that exhume core complexes may also exist [e.g., *James and Mortensen*, 1992; *Pehrsson et al.*, 2000]. The structural style is similar to that of modern orogens although some differences may be expected due to higher heat production in the Archean compared to today [*Bickle*, 1978; *Hoffman and Ranalli,*1988; *Pollack*, 1997].

The fundamental result that derives from deep seismic reflection profiling is the geometry of crustal boundaries, including those that are compositional and structural. Crustal structure should be drastically different depending on whether the present configuration of the crust is a result of vertical or horizontal tectonism. Thus, the seismic reflection technique is well suited to address these distinctly different configurations of the crust. In a crust deformed by vertical tectonism, boundaries would typically be steep and project deep into the crust, crustal-scale synforms and antiforms would be prevalent, and there would be little or no continuity of mid-crustal reflections beneath dome-to-keel transitions. The seismic profiles would likely exhibit, at a crustal scale, reflection patterns similar to those from sedimentary basins that are disrupted by salt or shale diapirs. On the other hand, a crust thickened and stabilized by horizontal tectonism would likely be dominated by subhorizontal structures. The seismic profiles would resemble those acquired across modern orogens and might exhibit gently dipping reflections along shear zones, with structural truncations and changes in reflection fabrics above and below the detachments.

It has been suggested that the seismic reflection method is capable of imaging shallowly dipping reflectors only, whereas more steeply dipping features are not detected [e.g., *Hamilton*, 1998; *Bleeker*, 2002]. If this were the case, our suggestion that Archean cratons formed by horizontally dominated processes may seem like a foregone conclusion. However, reflections with dips of 50° or more have been imaged in Archean cratons by using high stacking velocities, prestack time migration [*van der Velden and Cook*, 2002], and dip-moveout (DMO) [*Goleby and Drummond*, unpublished data] processing techniques. Even steeper features such as subvertical faults are clearly visible as steep zones of reflection truncations (see section 3.6). Hence, if steep synformal and antiformal structures dominated the crust, they would be readily detected. Furthermore, the observed subhorizontal reflection fabrics are a real and pervasive feature of all of the Archean cratons examined in this paper, and they are not artifacts of the acquisition and processing methods. Thus the presence of subhorizontal reflectivity precludes the interpretation of steep boundaries younger than the reflectors

in those regions; for example, subhorizontal reflections may be used to interpret a depth limit to a steeply dipping structure that occurs at the surface above the reflections.

In this paper we describe regional seismic reflection profiles across four regions of exposed Archean rocks (Figure 1). Two of these cross parts of the largest Archean craton in the world, the Superior Province in Canada, one is on the Yilgarn craton in Australia, and the fourth is on the southwestern part of the Slave craton in Canada. In all of these cases, the observed reflection patterns are consistent with the interpretation that these Archean cratons were assembled above shallowly dipping detachments and that horizontal forces were dominant during the final cratonic assembly. The similarity of crustal and upper mantle reflectivity to that observed within Proterozoic and Phanerozoic orogens provides evidence that, by the end of the Archean, tectonic processes were manifested as subhorizontal structures throughout much of the crust and upper mantle and thus that some early form of lithospheric plate tectonics was operating at ~2.8–2.6 Ga. Other regional profiles that exhibit similar reflectivity characteristics have been described from the western Kaapvaal craton in South Africa [*de Wit and Tinker*, 2004] and the Baltic shield in Finland [*Kukkonen et al.*, 2004] but are not specifically included here because we do not presently have access to the digital data to facilitate comparative analyses. In this paper the seismic results are described in the context of the regional geology, and we summarize detailed analyses and interpretations that have been presented previously in the referenced publications. For each area, we initiate the descriptions with a brief summary of major geological features that are relevant to the story.

2. SEISMIC REFLECTION DATA

2.1. Eastern Superior Province: Opatica–Abitibi Domains

The Lithoprobe Opatica–Abitibi profile in northwestern Quebec crosses the transition from the amphibolite-facies Opatica orthogneiss domain on the north to the greenschist-facies Abitibi granite–greenstone domain on the south [*Calvert and Ludden*, 1999*; Benn,* this volume]. In the Opatica, tonalites formed at ~2.74–2.70 Ga developed a gneissic foliation and zones of mylonitization at ~2.70–2.69 Ga (D1) and were foreshortened along D2 shear zones at ~2.69 Ga [*Mortensen*, 1993; *Sawyer and Benn*, 1993; *Davis et al.*, 1995]. On the north, the older (~2.82 Ga) Lac Rodayer pluton is separated from the Opatica orthogneisses by a 5-km-wide north-dipping D2 thrust [*Sawyer and Benn*, 1993]. The Abitibi consists of ~2.75–2.70 Ga mafic volcanic rocks, scattered felsic volcanic edifices, and synvolcanic plutons and is characterized by positive (juvenile) εNd isotopic signatures [*Chown et al.*, 1992; *Mortensen*, 1993; *Ayer et al.*, 2002].

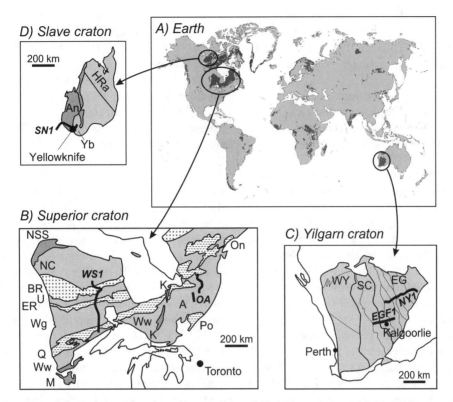

Figure 1. (A) World map. Outcrop of Archean rocks indicated in dark shading. (B) Superior craton, after *Card* [1990]. Dashed areas, metasedimentary domains; crossed areas, Berens River arc; light shading, granite–greenstone domains; dark shading, high-grade gneiss domains. A, Abitibi; BR, Berens River arc; ER, English River; K, Kapuskasing structural uplift; M, Minnesota River superterrane; NC, North Caribou; NSS, Northern Superior superterrane; OA, Opatica–Abitibi seismic profile; On, Opinaca; Po, Pontiac; Q, Quetico; U, Uchi; Wg, Wabigoon–Winnipeg River; WS1, Western Superior seismic profile; Ww, Wawa. (C) Yilgarn craton, after *Myers* [1995] and *Goleby et al.* [2004]. EG, Eastern Goldfields superterrane; EGF1, Eastern Goldfields seismic profile; NY1, Northern Yilgarn seismic profile; SC, Southern Cross superterrane; WY, Western Yilgarn superterrane. (D) Slave craton, after *Kusky* [1989]. An, Anton complex; HRa, Hackett River arc; SN1, Snorcle 1 seismic profile; Yb, Yellowknife basin.

It is uncertain whether the Abitibi represents a collage of smaller oceanic terranes [*Desrochers et al.*, 1993; *Kimura et al.*, 1993; *Jackson and Cruden*, 1995; *Mueller et al.*, 1996; *Polat et al.*, 1998] or formed as a coherent tectonic unit [*Ayer et al.*, 2002; *Benn*, this volume]. The northern Abitibi was foreshortened at ~2.71–2.69 Ga by north–south thrusting and folding. Peak metamorphism in the Opatica and northern Abitibi took place at ~2.69–2.68 Ga, resulting in migmatization of orthogneiss in the core of the Opatica domain [*Powell et al.*, 1995; *Sawyer and Benn*, 1993]. Both domains were dissected by a series of post-orogenic dextral strike-slip faults after ~2.69 Ga [*Davis et al.*, 1995].

Approximately 200–300 km west of the Opatica–Abitibi profile, the Proterozoic Kapuskasing uplift exposes an oblique cross section of Abitibi crust. It consists of a greenstone succession ~10 km thick, followed by a section of tonalitic gneiss (Wawa gneiss) ~10 km thick, which grades into a layered paragneiss and anorthosite sequence that is at amphibolite to granulite facies [*Moser*, 1994; *Percival and West*, 1994]. Structural and geochronological studies of the Wawa gneiss indicate that compressional deformation peaked at ~2.67–2.66 Ga and was followed by extension [*Krogh*, 1993; *Moser*, 1994]. Thus, faulting continued in the mid-crust while the upper crust had stabilized at ~2.67 Ga, a pattern consistent with underthrusting of material from the south [*Krogh*, 1993; *Moser*, 1994].

Seismic reflection data were collected in the Abitibi in 1988 and 1990–1991, and the Opatica was crossed in 1993 [*Percival and West*, 1994; *Lacroix and Sawyer*, 1995; *Calvert et al.*, 1995; *Bellefleur et al.*, 1997; *Calvert and Ludden*, 1999; *Benn*, this volume]. The Abitibi middle crust is characterized by prominently layered, gently north-dipping reflections resembling "shingles" (Foldouts 1a, 2a). This reflection pattern is interpreted to represent a series of imbricate slices thrust from the south beneath the Abitibi [*Calvert and Ludden*, 1999], which is consistent with studies of the correlative Wawa gneiss domain that indicate deformation becomes progressively younger with increasing depth.

Beneath the Opatica, dipping reflections extend from the Moho at ~11 s (~33 km depth) into the upper mantle, where they disappear at ~19 s (~65 km depth) (Foldouts 1a, 2a). These reflections are interpreted to represent a relict subduction zone [*Calvert et al.*, 1995]. *Calvert et al.* [1995] and *Calvert and Ludden* [1999] suggested that the mantle reflections are related to convergence between the Abitibi and Opatica domains, and that the suture may be the intervening Casa–Beradi tectonic zone (Foldouts 1a, 2a). North–south foreshortening occurred at the same time in the Opatica and Abitibi suggesting that the two domains were joined by ~2.71 Ga [*Calvert and Ludden*, 1999]. On the other hand, synvolcanic plutons in the Abitibi and orthogneisses of the Opatica have similar ages, suggesting that the Opatica may represent a deeper level of exposure of the Abitibi and may not be separate terranes [*Benn*, this volume]. Other possible candidates for correlative sutures are the Porcupine–Destor fault zone, which separates the northern and southern Abitibi domains [*Mueller et al.*, 1996], and the Kirkland-Cadillac fault zone, which separates the Abitibi from the Pontiac metasedimentary domain to the south (not shown here, see *Calvert and Ludden*, 1999; *Benn*, this volume). It is also possible that the mantle reflectors are an intraplate thrust fault [*Benn*, this volume].

The mantle reflections are aligned with the northern limit of the shingle reflections. Above the mantle reflections, crustal reflectivity in the Opatica is broadly synformal, and is interpreted to represent a fan-shaped deformation pattern as expected above a subduction zone [*Calvert and Ludden*, 1999]. Dextral post-orogenic strike-slip faults that transect the Opatica–Abitibi are visible as near-vertical reflections and zones of reflection truncations. For example, in the Opatica, the Nottaway River shear zone projects to ~25 km, where it flattens (Foldouts 1a, 2a) [*Calvert et al.*, 1995; *Calvert and Ludden*, 1999].

2.2. Western Superior Province

The western Superior Province records the ~2.73–2.68 Ga accretion of several composite terranes, Winnipeg River–Wabigoon, Wawa–Abitibi, and Minnesota River, to the southern margin of the North Caribou protocraton (Figure 1). These composite terranes contain Mesoarchean (~3.5–3.0 Ga) rocks as well as Neoarchean greenstone and granite domains, reflecting a complex internal history before being accreted by progressive southward underthrusting. The high-grade syn-orogenic English River and Quetico metasedimentary belts separate the composite terranes and may obscure sutures [e.g., *Langford and Morin*, 1976; *Card*, 1990; *Williams*, 1990; *White et al.*, 2003].

The ~500-km-wide North Caribou terrane consists of ~3.00 Ga juvenile crust [e.g., *Corfu and Stone*, 1998]. Its southern margin is represented by the Uchi greenstone belt, which records 2.99–2.98 Ga plume (rift-related) magmatism on North Caribou basement followed by ~2.75–2.73 Ga arc magmatism [*Sanborn-Barrie et al.*, 2001]. The Berens River arc is an Andean-type arc that developed on the North Caribou–Uchi margin at ~2.75–2.71 Ga [*Corfu and Stone*, 1998], prior to ~2.72–2.70 Ga collision of the Uchi margin with the Winnipeg River–Wabigoon composite terrane (Uchian orogeny) [*Stott and Corfu*, 1991; *Sanborn-Barrie et al.*, 2001]. Sediments of the English River terrane were deposited during the Uchian orogeny and were metamorphosed to amphibolite, migmatite, and granulite facies at ~2.69 Ga [*Breaks*, 1991]. The Wawa–Abitibi terrane collided at ~2.69 Ga (Shebandowanian orogeny). The Quetico terrane contains synorogenic (~2.70–2.69 Ga) sediments that were metamorphosed to amphibolite-granulite facies at ~2.67–2.65 Ga [*Percival and Williams*, 1989; *Williams*, 1990]. The poorly exposed Minnesota River terrane was accreted at ~2.68 Ga (Minnesotan orogeny).

Seismic reflection data were collected along corridor 1 of the Lithoprobe Western Superior transect in 1996–1997. The ~600-km-long profile crosses the Superior Province from the North Caribou terrane in the north to the Wawa terrane in the south (Foldouts 1b, 2b). Prominent reflectivity within the North Caribou terrane is subparallel, exhibits folding with a wavelength of ~40 km, and appears truncated by the reflection Moho at ~12.5 s (~38 km depth). The Berens River arc is poorly reflective, a characteristic of inactive continental magmatic arcs in Proterozoic and Phanerozoic orogens [*Cook and van der Velden*, in press]. At the surface, the boundary between the English River–Wabigoon and North Caribou–Uchi is the Sydney Lake fault, a late strike-slip fault which is delineated by a steep zone of reflection truncations that extends to ~7 s (~21 km depth) [*White et al.*, 2003]. At depth, the boundary between these terranes likely projects into the upper mantle at S1 (Foldouts 1b, 2b), outlining an interpreted ~2.72–2.70 Ga paleosubduction zone that accommodated convergence during the Uchian orogeny [*White et al.*, 2003]. Above these mantle reflections, crustal reflections outline a bivergent pattern, which is consistent with deformation above a subduction zone.

South of the Sydney Lake fault, the lowermost crust exhibits layered reflectivity. Coincident seismic refraction data indicate that this lower crust has very high velocities (7.4–7.5 km/s) and is highly anisotropic (8%); accordingly, it has been interpreted as a mafic oceanic slab [*White et al.*, 2003; *Musacchio et al.*, 2004]. This lower crustal slab projects into the upper mantle at S2, outlining a second inferred paleosubduction zone that was likely to have been active at ~2.69-2.68 Ga. It is uncertain whether this slab connects to the Wawa or Minnesota River superterranes [*White et al.*, 2003]. Above the inferred oceanic slab, the middle crust of the English River–Wabigoon superterrane is characterized by north-dipping "shingle" reflections (Sh, Foldouts 1b, 2b), which are interpreted to represent imbricates and/or a subduction complex [*White et al.*, 2003]. These structures likely folded an overlying layer –Winnipeg River(Wabigoon), which is characterized by gently undulating subparallel reflectivity that is discordant with the underlying imbricate structures.

Further geophysical evidence for subduction is provided by (1) teleseismic data, which indicate that the mantle beneath the accreted terranes has east–west-oriented anisotropy while the mantle beneath the North Caribou terrane is isotropic [*Kendall et al.*, 2002]; and (2) magnetotelluric surveying, which has imaged a steeply north-dipping zone of high resistivity in the mantle that is broadly aligned with the S2 mantle reflections [*Craven et al.*, 2004].

2.3. Yilgarn Craton

The Yilgarn craton is divided into the Western Yilgarn, the Southern Cross, and the Eastern Goldfields superterranes [*Myers*, 1995; 1997] (Figure 1). The older Southern Cross superterrane contains exposures of ~3.00–2.95 Ga greenstone assemblages, which are metamorphosed to upper amphibolite facies. The Southern Cross superterrane is thought to represent a deeper structural level than the Eastern Goldfields superterrane, where greenstone assemblages range in age from ~2.71 to 2.62 Ga [*Swager*, 1997]. Terranes within the Eastern Goldfields are fault-bounded, and their lithostratigraphies do not correlate in detail across the faults [*Swager*, 1997]. Deformation included an early extensional phase, north–south shortening (D1); a regional D2 event, which involved east–west shortening that imposed the prominent north–south regional pattern to the granitoids and greenstone belts; and the development of a series of regional strike-slip and transpressional (D3) shear zones that transect the province [*Swager*, 1997; *Blewett et al.*, 2004]. Significant gold deposits occur in second- and higher-order splays off several of these shear zones. Many of the shear zones had a complex history. For example, the Ida fault is interpreted to have formed during D2 thrusting, subsequently reactivated as a D3 strike slip or transpressional fault, and then again remobilized as a normal fault, downdropping the Eastern Goldfields superterrane beside the Southern Cross superterrane [*Drummond et al.*, *2000*]. Voluminous syn- to posttectonic granitoid plutonism continued until ~2.62 Ga.

Seismic reflection data were collected across the southwestern Yilgarn Province in 1991 [*Drummond et al.*, 2000]

and the northeastern Yilgarn Province in 1999 [*Goleby et al.*, 2004]. The profiles, separated ~170 km along-strike, have been merged into a regional profile by correlating the Emu fault on the 1991 traverse with the Ockerbury Shear zone on the 2001 traverse (Foldouts 1c, 2c). The profile forms a 700-km-long transect that crosses most of the Yilgarn Province at a high angle to regional D2 strike; due to this projection, however, some features such as the Moho appear offset in traveltime from one line to the other. The small difference in Moho depth between the two traverses indicates a slight southerly dip of the Moho beneath the Yilgarn craton. Additionally, the use of widely spaced explosive sources on the 1991 profile versus tightly spaced Vibroseis sources on the 2001 profile has resulted in some differences in reflection character; for example, reflectivity in the upper crust is better delineated on the Vibroseis profile, whereas lower crustal reflectivity is better defined on the dynamite data.

Many of the D2 shear zones are highly reflective, dip to the east, and can be followed from surface exposures to the lower crust. For example, the Laverton and Yamarna shear zones can be followed from the near-surface along dipping reflections to the lower crust, where they appear to become listric and flatten into a subhorizontal reflection fabric [*Goleby et al.*, 2004]. Reflections from the Bardoc, Keith Kilkenny, and Celia shear zones can be followed to ~4 s (12 km depth), where they merge with the regional east-dipping reflection fabric. Beneath the Southern Cross Province, shingle reflections have been identified in the middle crust and are characterized by fault-plane reflections and hanging-wall cutoffs (Sh; Foldouts 1c, 2c) [*Drummond et al.*, 2000]. The orientation and spacing of these structures is similar to those observed within the Eastern Goldfields Province to the northeast, i.e., to the east of the Ockerbury fault, indicating that they are likely a continuation of the same deformation event. Thus, the seismic reflection data support the interpretation that regional crustal deformation of the Yilgarn craton was accommodated along fault systems that detached along regional decollements within the lowermost crust.

2.4. Slave Province

The Slave Province is characterized by scattered outcrops of ~4.1–3.0 Ga granites and gneisses, a thin ~2.92–2.81 Ga metasedimentary–volcanic cover sequence, ~2.72–2.70 Ga mafic volcanic rocks, and ~2.68–2.66 Ga metasedimentary rocks [e.g., *Isachsen and Bowring*, 1994; *Bleeker et al.*, 1999]. Rocks older than ~2.8 Ga are confined to the western Slave Province, and the rocks of the eastern Slave Province show juvenile isotopic signatures [*Davis and Hegner*, 1992]. The eastern Slave Province contains arc assemblages, and convergence between the eastern arcs and western basement complex is thought to be responsible for ~2.66–2.58 Ga deformation, metamorphism, and plutonism observed in the Slave Province [*Kusky*, 1989; *Davis and Hegner*, 1992; *Relf et al.*, 1999; *van der Velden and Cook*, 2002].

The Lithoprobe SNORCLE 1 profile traverses the southwestern tip of the Slave Province in the vicinity of Yellowknife (Figure 1) and crosses the transition of the Anton plutonic complex on the west to the Yellowknife basin on the east [*Cook et al.*, 1999; *van der Velden and Cook*, 2002]. Along the profile, the Anton complex consists entirely of post-deformational granites, but studies of amphibolite- and granulite-facies rocks to the north indicate that the Anton complex is an imbricated stack of supracrustal rocks bounded by an outward-dipping extensional fault on the east [*Pehrsson and Villeneuve*, 1999; *Pehrsson et al.*, 2000]. Steeply inclined mafic volcanic rocks are exposed at Yellowknife, which are overlain on the east a thick sequence of turbidites. Although the base of the greenstones is not exposed at Yellowknife, basement rocks were likely present beneath the greenstones when they formed, and the Yellowknife greenstone belt is commonly interpreted as a back-arc or marginal basin [*Helmstaedt and Padgam*, 1986; *Fyson and Helmstaedt*, 1988]. An early phase of folding (D1) occurred at ~2.66–2.63 Ga, followed by a major folding event (D2) at 2.61–2.59 Ga [*Davis and Bleeker*, 1999]. Gold occurs in D2 shear zones at Yellowknife. All these rocks are intruded by voluminous syn- to post-deformational granitoids.

The seismic data reveal gently west-dipping reflections on both sides of Yellowknife –Foldouts 1d, 2d). These reflections have arcuate reflections above them that are truncated by the west-dipping planar reflections, suggesting the planar reflections are faults with hanging wall cutoffs above them [*Cook et al.*, 1999; *van der Velden and Cook*, 2002]. The

interpretation of thrust-and-fold structures within the Anton plutonic domain is consistent with structural and geochronologic studies along-strike to the north, which also indicate crustal thickening by structural imbrication [*Pehrsson and Villeneuve*, 1999]. A poorly reflective zone beneath Yellowknife is caused by seismic noise from the Yellowknife goldmine and townsite and by truncations along the subvertical Yellowknife River fault [*van der Velden and Cook*, 2002]. The character and depth extent of the planar reflections on both sides of Yellowknife is the same, suggesting that these reflections correlate [*Cook et al.*, 1999; *van der Velden and Cook*, 2002]. Therefore, the transition between the Anton plutonic domain and the Yellowknife greenstone domain is likely underlain by gently west-dipping dipping faults. This interpretation differs from that of *Bleeker* [2002], who suggested that the Yellowknife area represents a dome-to-keel transition with ~30 km of structural relief.

3. DISCUSSION

3.1. Comparison and Interpretation

The most striking similarity amongst these widely separated profiles from different Archean regions is that the crust, particularly the lower crust, is dominated by subhorizontal layered reflectivity in all cases. These layers persist beneath many major structures, including granite–greenstone boundaries. Furthermore, structures commonly flatten into deeper layering. Taken together, these observations support the concept of horizontal tectonism.

The interpretation of the crustal reflectivity patterns in Foldouts 1 and 2 in terms of mapped geological units implies that most of the reflectivity is caused by deformed

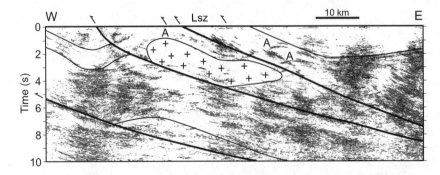

Figure 2. Detail of the Yilgarn data set (Foldouts 1c, 2c). The Laverton shear zone (Lsz) is highly reflective and can be followed from mapped surface exposures along gently east-dipping reflections to the lower crust. Significant gold deposits occur in splays of this shear zone. Antiformal reflections (A) above the planar east-dipping reflections are interpreted as hanging wall cutoffs within greenstones above the faults. In other regions where direct correlations to surface structures are not possible (e.g., Slave Province), such reflection patterns (planar reflections, with arched and truncated reflections above) are used to infer faults and hanging wall cutoffs. A tabular granite (+ pattern) is interpreted above a shear zone based on a lack of reflections and density considerations from gravity modeling [*Goleby et al.*, 2004].

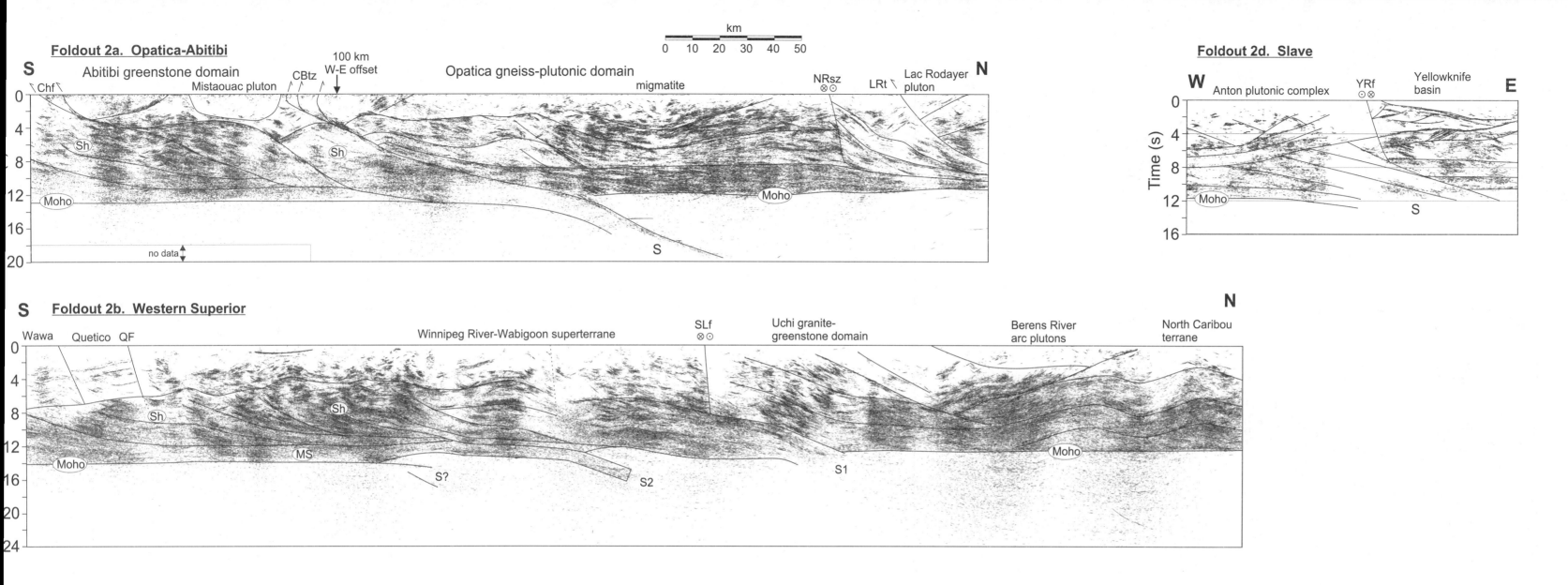

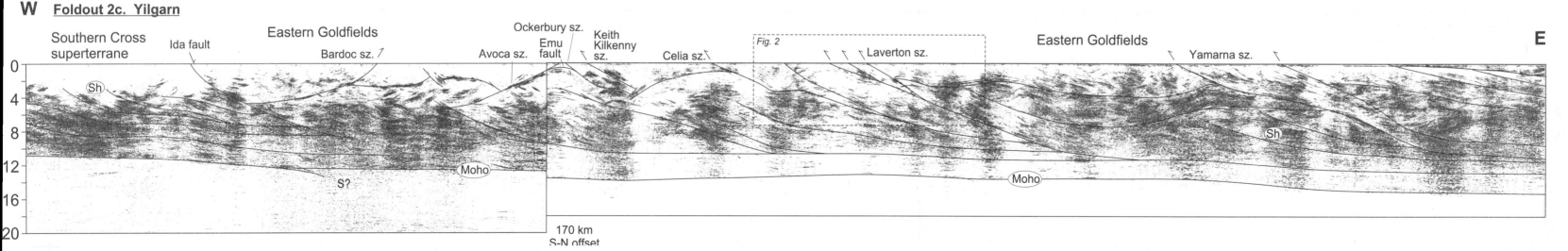

Interpretation of the seismic reflection profiles. (a) Interpretation of the Opatica–Abitibi profile, modified from *Calvert et al.* [1995], *Lacroix and Sawyer* [1995], and *Calvert and Ludden* [1999]. CBtz, Casa–Beradi tectonic zone; Chf, Chicobi fault zone; LRt, Lac Rodayer thrust; NRsz, Nottaway River shear zone; S, Subduction zone; Sh, shingle reflections. [int]erpretation of the Western Superior profile, after *White et al.* [2003]. MS, mafic lower crustal slab; QF, Quetico fault; Sh, shingle reflections; SLf, Sydney Lake fault; S1 and S2, subduction zones. (c) Interpretation of the Yilgarn profile, simplified from *Drummond et al.* [2000] and *Goleby et al.* [2004]. Sh, shingle reflections. Boxed area is enlarged in Figure 2. [Inter]pretation of the Slave Profile, after *Cook et al.* [1999] and *van der Velden and Cook* [2002]. YRf, Yellowknife River fault.

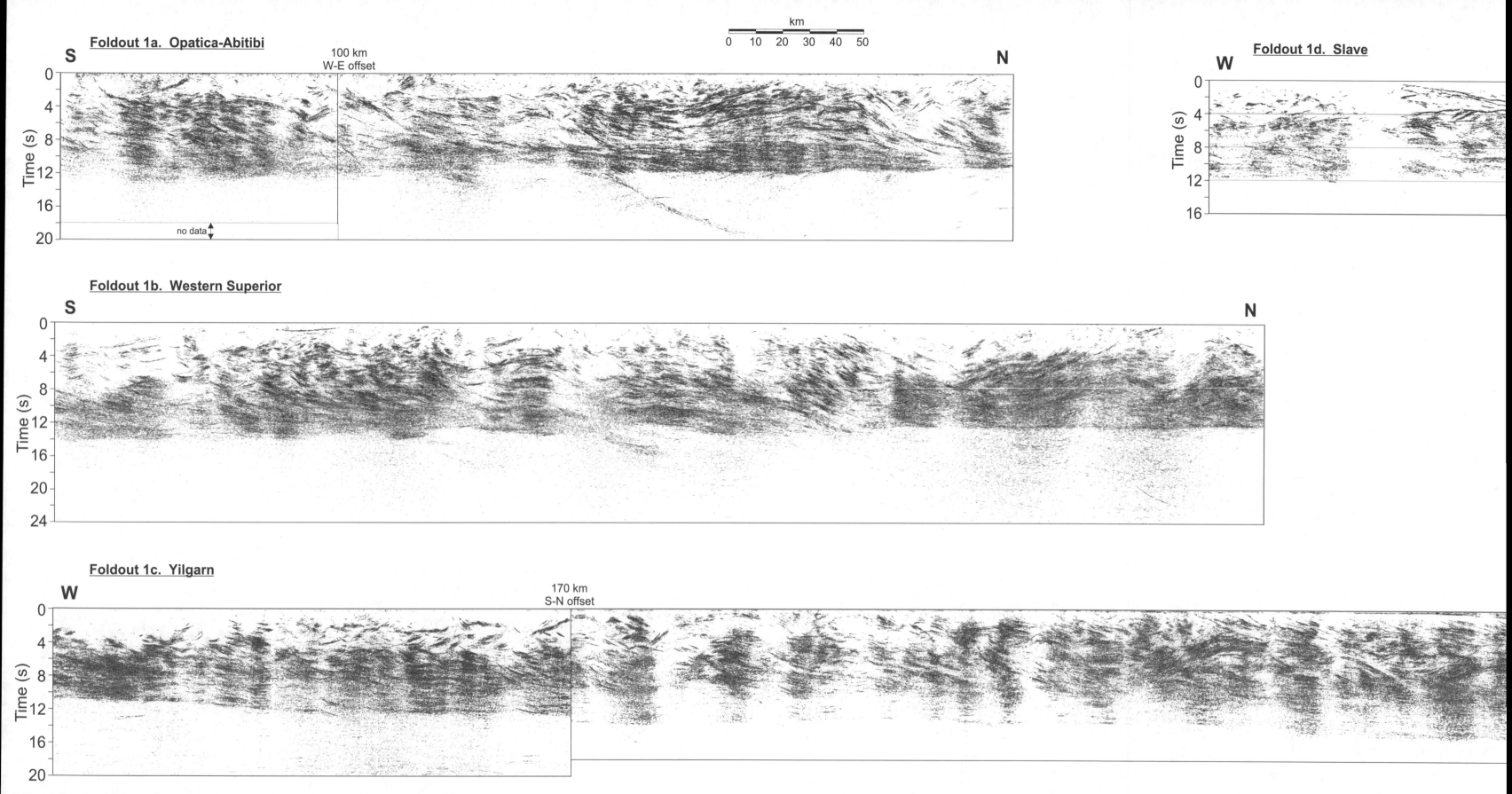

Foldout 1a. Opatica-Abitibi

S   N

100 km
W-E offset

km
0   10   20   30   40   50

Time (s)
0
4
8
12
16
20

no data

Foldout 1d. Slave

W

Time (s)
0
4
8
12
16

Foldout 1b. Western Superior

S   N

Time (s)
0
4
8
12
16
20
24

Foldout 1c. Yilgarn

W

170 km
S-N offset

Time (s)
0
4
8
12
16
20

Foldout 1. Seismic reflection profiles across Archean cratons. The data have been migrated and coherency filtered and are plotted at 1:1, assuming a velocity of 6000 m/s. Locations of large (>20 km) along-strike projections are indicated along the top of the profiles.

compositional layering within those units. In some areas these reflections can be correlated directly to faults (e.g., Eastern Goldfields, Figure 2); elsewhere faulting is inferred from hanging wall and footwall cutoffs along apparent fault plane reflections (e.g., Slave Province, Western Superior) or correlation of three-dimensional orientation of reflections to surface structures (e.g., Opatica–Abitibi [*Bellefleur et al.*, 1997]). In all cases, most of the reflectivity appears to have been formed and/or realigned during the most significant deformation event that led to cratonization. Some reflections may represent pre-deformational unconformities or stratigraphy, or post-deformational mafic sills [*Cook et al.*, 1999; *Bleeker*, 2002], but the vast majority of reflectivity is related to ~2.8–2.6 Ga deformation in each of the cratons examined. Thus, the reflection patterns provide a snapshot of the nature of Neoarchean tectonism.

3.2. Structure of Granitoids

Structures related to peak orogenic activity are preserved, even though many areas have experienced substantial post-orogenic plutonism. The observation that reflections are continuous beneath post-orogenic plutons as well as beneath areas that did not experience plutonism provides supporting evidence for the following inferences: (1) Plutons are typically tabular or uprooted and do not project deeply into the crust as diapirs [*Cruden*, 2005], and (2) granitic magmas did not significantly disrupt the reflectors on ascent, perhaps by migrating along relatively narrow fissures [*Petford et al.*, 1993]. Similar relationships between reflectivity patterns and post-orogenic plutons are observed in Proterozoic and Phanerozoic orogens [*Cook et al.*, 1999; *Cook and van der Velden*, in press; *van der Velden et al.*, 2004].

Vertical mass flux, particularly in the form of mafic and ultramafic volcanics and granitoid intrusions, are important elements in Archean crustal building. It is possible that uprooted plutons represent decapitated diapirs and that granitoids transposed into structural sheets had diapiric origins. However, recent seismic data show them in the structural context of tectonically thickened crust. For example, granite is interpreted above a fault in the Eastern Goldfields (Figure 2), and the Mistaouac pluton of the Abitibi is likely underlain by detachments (Foldout 2a).

3.3. Shingle Reflections

Sigmoidal reflection patterns in the middle crust resembling "shingles" [*Cook*, 1986] are commonly interpreted to represent structural imbrication. They are observed within the Winnipeg River–Wabigoon and Abitibi domains of the Superior Province and throughout the Yilgarn craton –Sh in Foldout 2). They typically occur within arc terranes above an inferred subducted plate. Similar reflection patterns are observed in the Archean Kaapvaal craton in South Africa [*de Wit and Tinker*, 2004], in the Proterozoic Hottah arc terrane in the Northwest Territories [*Cook et al.*, 1999], in the Neoproterozoic–Lower Paleozoic Ganderia arc of the Newfoundland Appalachians [*van der Velden et al.*, 2004], and above the subducting Juan de Fuca plate beneath Vancouver Island [*Clowes et al.*, 1987]. It is often difficult to tell whether such a reflection pattern indicates local structural imbrication of parautochthonous material, or whether the region is a subduction complex consisting of tectonically underplated slices of exotic material. Nevertheless, the implications of shingling reflection patterns are threefold: (1) The crust was likely thickened by horizontally directed compression (horizontal tectonics) in these areas; (2) the direction of dip of the shingles may be an indicator of subduction polarity; and (3) mid-crustal shingles may be an indicator of downward-younging deformation; i.e., thrusting took place beneath a superstructure that had previously stabilized.

3.4. Gneiss to Greenstone Transitions

The reflection profiles cross a number of transitions from lower-grade greenstone domains to higher-grade gneiss-plutonic domains. Beneath the Opatica–Abitibi and the Anton–Yellowknife basin transitions, reflections that underlie the transition verge up toward the greenstones. This is opposite to what would be expected in a diapiric model; i.e., one would expect reflectors to be deflected up toward and be disrupted beneath the gneiss-plutonic domain. Instead, the structural style resembles that of Phanerozoic orogens, where gneiss domes are typically underlain by thrust ramps (e.g., Monashee complex, Canadian Cordillera [*Varsek and Cook*, 1994]; Meelpaeg allochthon, Newfoundland [*van der Velden et al.*, 2004]; Kangmar dome, Tibet [*Alsdorf et al.* 1998]). This geometry suggests that the gneiss-plutonic domains likely formed by contraction above thrust ramps. In the western Superior Province, the transition between the North Caribou–Berens River gneiss-plutonic domain to Uchi greenstone domain is underlain by gently dipping, undulating reflections (Foldout 2b) and therefore the transition likely does not penetrate deeply into the crust.

Gneiss domains may be amplified by extension along outward-dipping normal faults. One example is the Ida fault of the Yilgarn Province, which dips away from the higher-grade rocks of the Southern Cross Province. Extensional faulting in the Anton domain of the Slave Province by extension has been documented by *Pehrsson et al.* (2000) and elsewhere in the Slave Province by *James and Mortensen* [1992]. Extension may have occurred at the Berens River–Uchi transition of the Western Superior Province [*Calvert et al.*, 2004].

This structural style resembles core complexes in modern orogens [e.g., *Coney*, 1980].

3.5. Archean Subduction

The Archean cratons from which we have regional reflection data exhibit reflections that project from the lower crust and/or Moho into the upper mantle. A seminal example occurs beneath the Opatica (Foldout 2a) [*Calvert et al.*, 1995], and several others occur beneath the Abitibi [*Calvert and Ludden*, 1999], the Western Superior Province (S1 and S2; Foldout 2b) [*White et al.*, 2003], and the Slave Province (S; Foldout 2d) [*Cook et al.*, 1999; *van der Velden and Cook*, 2002]. Based on their geometry (dipping into the upper mantle) and spatial relationships to crustal reflection fabrics and regional geological features, these reflections are interpreted to represent remnants of lithospheric boundaries that have accommodated convergence between plates [*van der Velden and Cook*, 2005]. Thus, the observation that one plate appears to be thrust beneath another leads to the conclusion that these reflections represent relict subduction zones. Similar mantle reflections observed beneath Proterozoic and Phanerozoic orogens are also interpreted as relict subduction zones [*van der Velden and Cook*, 2005; *Cook and van der Velden*, in press]. Two reflections with this geometry that have not previously been interpreted this way are illustrated here (S?, Yilgarn Province Foldout 2c; S?, Western Superior Province, Foldout 2b). As an alternative hypothesis, *Hamilton* [1998] suggested the mantle reflections beneath the Opatica might be caused by a delaminating slab of mafic restite that was arrested while sinking into the mantle.

3.6. Steep Faults

The seismic reflection profiles cross a number of steeply dipping faults. These faults are post-orogenic, orogen-parallel strike-slip faults that appear to have played little role in the thickening and accumulation of crust. They typically have only small vertical displacements and are likely unrelated to the formation of granitoid domes. Steep faults are characterized by subvertical zones of reflection truncations and abrupt lateral changes in reflectivity fabric [*Lemiszki and Brown*, 1988; *Hajnal et al.*, 1996]. For example, the Quetico and Sydney Lake faults in the Superior Province (Foldout 2b) [*White et al.*, 2003], the Nottaway River shear zone in the Opatica (Foldout 2a) [*Calvert et al.*, 1995], and the Yellowknife River fault (Foldout 2d) [*Cook et al.*, 1999; *van der Velden and Cook*, 2002] are all well imaged. They typically project to the mid- to lower crust where they merge with low-angle detachment zones. Similar reflection patterns are found at strike-slip faults in Proterozoic orogens [e.g., Trans-Hudson; *Hajnal et al.*, 1996], and Phanerozoic orogens [e.g., Baie Verte line, Appalachians; *van der Velden et al.*, 2004].

4. CONCLUSIONS

Neoarchean cratons are characterized by gently dipping, often listric, reflections and regionally extensive subhorizontal reflections that are interpreted to represent structural fabrics established mainly by horizontal compression and structural repetition. Dipping "shingle" reflections in the middle crust are interpreted as imbricate structures and/or slices of accreted material. Dipping reflections that project from the lower crust into the upper mantle are inferred to represent relict subduction zones. Steep post-orogenic strike-slip faults are characterized by zones of reflection truncations. These reflection patterns, observed in Neoarchean cratons throughout the world, are remarkably similar to those observed within Proterozoic and Phanerozoic orogens, which are interpreted to be products of plate tectonics. These reflection patterns cannot be easily explained as a geometric anomaly that is specific to a single craton or locality. Overall, these reflection patterns suggest that ~2.8–2.6 Ga cratons appear to be products of some form of early plate tectonic processes in which horizontal forces were prevalent. The next logical step for future seismic reflection profiling is to traverse even older cratons.

Acknowledgments. Funding for the Lithoprobe work in Canada was provided by the Natural Sciences and Engineering Research Council and the Geological Survey of Canada. The Lithoprobe data were processed at the Lithoprobe Seismic Processing Facility at the University of Calgary with the assistance of K. Hall, E. Lynn, R. Maier, and K. Vasudevan. The Yilgarn data were processed at the ANSIR facility with the assistance of T. Barton, T. Fomin, D. Johnstone, and L. Jones. We thank Lesley Chorlton for the world map. Discussions with R. Blewett and M. de Wit helped shape the ideas presented here. BJD and BRG publish with the permission of the Chief Executive Officer of Geoscience Australia.

REFERENCES

Alsdorf, D., L. Brown, K.D. Nelson, Y. Makovsky, S. Klemperer, and W. Zhao (1998), Crustal deformation of the Lhasa terrane, Tibet plateau from project Indepth deep seismic reflection profiles, *Tectonics, 17*, 501-519.

Anhaeusser, C.R. (1973), The evolution of the early Precambrian crust of southern Africa, *Phil. Trans. R. Soc. Lond.*, A273, 359-388.

Ayer, J., Y. Amerlin, F. Corfu, S. Kamo, J. Ketchum, K. Kwok, and N. Trowell (2002), Evolution of the southern Abitibi greenstone belt based on U-Pb geochronology: autochthonous volcanic construction followed by plutonism, regional deformation and sedimentation, *Precambrian Res., 115*, 63-95.

Barley, M.E., B.N. Eisenlohr, D.I. Groves, C.S. Perring, and J.R. Vearncombe (1989), Late Archaean convergent tectonics and gold mineralization: A new look at the Norseman Wiluna Belt, Western Australia, *Geology*, 17, 448452.

Bellefleur, G., A.J. Calvert, and M.C. Chouteau (1997), A link between deformation history and the orientation of reflective structures in the 2.68-2.83 Ga Opatica belt of the Canadian Superior province, *J. Geophys. Res.*, 102, B7, 15243-15257.

Bickle, M.J. (1978), Heat loss from the Earth: Constraints on Archean tectonics from the relation between geothermal gradients and the rate of plate production, *Earth Plan. Sci. Lett.*, 40, 301-315.

Bleeker, W. (2002), Archaean tectonics: A review, with illustrations from the Slave craton, in *The Early Earth: Physical, Chemical and Biological Development*, edited by C.M.R. Fowler, C.J. Ebinger and C.J. Hawkesworth, Geological Society of London Special Publications, vol. 199, 151-181.

Bleeker, W., J.W.F. Ketchum, V.A. Jackson, and M.E. Villeneuve (1999), The Central Slave Basement Complex, Part 2: Its structural topology and autochthonous cover, *Can. J. Earth Sci.*, 36, 1111-1130.

Blewett, R.S. (2002), Archaean tectonic processes: A case for horizontal shortening in the north Pilbara granite-greenstone terrane, western Australia, *Precambrian Res.*, 113, 87-120.

Blewett, R.S., K.F. Cassidy, D.C. Champion, P.A. Henson, B.S. Goleby, L. Jones, and P.B. Groenewald (2004), The Wangkathaa orogeny: An example of episodic regional D2 in the late Archaean Eastern Goldfields province, western Australia, *Precambrian Res., 130*, 139-159.

Breaks, F.W. (1991), English River subprovince, in *Geology of Ontario*, edited by P.C. Thurston, H.R. Williams, R.H. Sutcliffe, and G.M. Stott, Ontario Geological Survey Special Volume 4, part 1, 239-277.

Calvert, A.J. and J.N. Ludden (1999), Archean continental assembly in the southeastern Superior Province in Canada, *Tectonics*, 18, 412-429.

Calvert, A.J., A.R. Cruden, and A. Hynes (2004), Seismic evidence for preservation of the Archean Uchi granite-greenstone belt by crustal-scale extension, *Tectonophysics*, 388, 135-143.

Calvert, A.J., E.W. Sawyer, W.J. Davis, and J.N. Ludden (1995), Archean subduction inferred from seismic images of a mantle suture in the Superior Province, *Nature*, 375, 670-674.

Campbell, I.H. and R.I. Hill (1988), A two-stage model for the formation of the granite–greenstone terrains of the Kalgoorlie–Norseman area, Western Australia, *Earth Plan. Sci. Lett.*, 90, 1125.

Card, K.D. (1990), A review of the Superior Province of the Canadian Shield, a product of Archean accretion, *Precambrian Res.*, 48, 99-156.

Chown, E.H., R. Daigneault, W. Mueller, and J.K. Mortensen (1992), Tectonic evolution of the northern volcanic zone, Abitibi belt, Quebec, *Can. J. Earth Sci.*, 29, 2211-2255.

Clowes, R.M., M.T. Brandon, A.G. Green, C.J. Yorath, A. Sutherland Brown, E.R. Kanasewich, and C. Spencer (1987), Lithoprobe - southern Vancouver Island: Cenozoic subduction complex imaged by deep seismic reflections, *Can. J. Earth Sci.*, 24, 31-51.

Collins, W.J., M.J. van Kranendonk, and C. Teysier (1998), Partial convective overturn of Archean crust in the east Pilbara craton, western Australia: driving mechanisms and tectonic implications, *J. Struct. Geol.*, 20, 1405-1424.

Coney, P.J. (1980), Cordilleran metamorphic core complexes: A review, in *Cordilleran Metamorphic Core Complexes*, edited by M.D. Crittenden Jr., P.J. Coney, and G.H. Davis, Geol. Soc. America Memoir 153, 7-31.

Cook, F.A. (1986), Continental evolution by lithospheric shingling, in *Reflection Seismology: the Continental Crust*, edited by M. Barazangi and L. Brown, AGU Geodynamics Series 14, 13-19.

Cook, F.A., and A.J. van der Velden (2005), Crustal seismic reflection profiles of collisional orogens, in *Regional Geology of the World*, edited by A.W. Bally and D.G. Roberts, in press.

Cook, F.A., A.J. van der Velden, K.W. Hall, and B.J. Roberts (1999), Frozen subduction in Canada's Northwest Territories: Lithoprobe deep seismic reflection profiling of the western Canadian shield, *Tectonics*, 18, 1-24.

Corfu, F., and D. Stone (1998), Age, structure, and orogenic significance of the Berens River composite batholiths, western Superior Province, *Can. J. Earth Sci.*, 35, 1089-1109.

Craven, J.A., T. Skulski, and D. White (2004), Lateral and vertical growth of cratons: seismic and magnetotelluric evidence from the western Superior transect, in *Lithoprobe Celebratory Conference*, Lithoprobe report 86, University of British Columbia.

Cruden, A.R. (2005), Emplacement and growth of plutons: Implications for rates of melting and mass transfer in continental crust, in *Evolution and Differentiation of the Continental Crust*, edited by M. Brown and T. Rushmer, Cambridge University Press, in press.

Davis, D.W., N. Machado, C. Gariepy, E.W. Sawyer, and K. Benn (1995), U-Pb geochronology of the Opatica tonalite-gneiss belt and its relationship to the Abitibi greenstone belt, Superior Province, Quebec, *Can. J. Earth Sci.*, 32, 113-127.

Davis, W.J., and W. Bleeker (1999), Timing of plutonism, deformation, and metamorphism in the Yellowknife domain, Slave Province, Canada, *Can. J. Earth Sci.*, 36, 1169-1187.

Davis, W.J., and E. Hegner (1992), Neodynium isotopic evidence for the tectonic assembly of Late Archean crust in the Slave Province, *Contrib. Mineral. Petrol.*, 111, 493-504.

Desrochers, J.P., C. Hubert, J.N. Ludden, and P. Pilote (1993), Accretion of Archean oceanic plateau fragments in the Abitibi greenstone belt, Canada, *Geology*, 21, 451-454.

de Wit, M.J. (1998), On Archaean granites: does the evidence demand a verdict? *Precambrian Res.*, 91, 143-179.

de Wit, M., and J. Tinker (2004), Crustal structures across the central Kaapvaal craton from deep-seismic reflection data, *S. African J. Geol.*, 107, 185-206.

Drummond, B.J., B.R. Goleby, and C.P. Swager (2000), Crustal signature of late Archean tectonic episodes in the Yilgarn craton, western Australia: Evidence from deep seismic sounding, *Tectonophysics*, 329, 193-221.

Fyson, W.K., and H. Helmstaedt (1988), Structural patterns and tectonic evolution of supracrustal domains in the Archean Slave Province, Canada, *Can. J. Earth Sci.*, 25, 301-315.

Goleby, B.R., R.S. Blewett, R.J. Korsch, D.C. Champion, K.F. Cassidy, L.E.A. Jones, P.B. Groenewald, and P. Henson (2004), Deep seismic profiling in the Archaean northeast Yilgarn craton, western Australia: Implications for crustal architecture and mineral potential, *Tectonophysics*, 388, 119-133.

Hajnal, Z., S. Lucas, D. White, J. Lewry, S. Bezdan, M.R. Stauffer, and M.D. Thomas (1996), Seismic reflection images of high-angle faults and linked detachments in the Trans-Hudson orogen, *Tectonics*, 15, 427-439.

Hamilton, W.B. (1998), Archean magmatism and deformation were not products of plate tectonics, *Precambrian Res.*, 91, 143-179.

Hamilton, W.B. (2003), An alternative Earth, *GSA Today*, 13, no. 11, 4-12.

Helmstaedt, H., and W.A. Padgham (1986), A new look at the stratigraphy of the Yellowknife Supergroup at Yellowknife, NWT. Implications for the age of gold-bearing shear zones and Archean basin evolution, *Can. J. Earth Sci.*, 23, 454-475.

Hoffman, P.F., and G. Ranalli (1988), Archean oceanic flake tectonics, *Geoph. Res. Lett.*, 15, 1077-1080

Isachsen, C.E., and S.A. Bowring (1994), Evolution of the Slave craton, *Geology*, 22, 917-920.

Jackson, S.L., and A.R. Cruden (1995), Formation of the Abitibi greenstone belt by arc-trench migration, *Geology*, 23, 471-474.

James, D.T., and J.K. Mortensen (1992), An Archean metamorphic core complex in the southern Slave Province: Basement - cover relations between the Sleepy Dragon complex and the Yellowknife greenstone belt, *Can. J. Earth Sci.*, 29, 2133-2145.

Kendall, J.M., S. Sol, C.J. Thomson, D.J. White, I. Asudeh, C.S. Snell, and F.H. Sutherland (2002), Seismic heterogeneity and anisotropy in the western Superior Province, Canada: insights into the evolution of a craton, in *The Early Earth: Physical, Chemical and Biological Development*, edited by C.M.R. Fowler, C.J. Ebinger and C.J. Hawkesworth, Geological Society of London Special Publications, vol. 199, 27-44.

Kimura, G., J.N. Ludden, J.P. Desrochers, and R. Hori (1993), A model for ocean-crust accretion for the Superior province, Canada, *Lithos*, 30, 337-355.

Krogh, T.E. (1993), High precision U-Pb ages for granulite metamorphism and deformation in the Archean Kapuskasing structural zone, Ontario: implications for structure and development of the lower crust. *Earth Plan. Sci. Lett.*, 119, 1-18.

Kukkonen, I.T., P. Heikkinen, E. Ekdahl, S.E. Hjelt, A. Korja, R. Lahtinen, J. Ylniemi, R. Berzin, and FIRE working group (2004), FIRE transects: New images of the crust in the Fennoscandian shield, in *The 11th international symposium on deep seismic reflection profiling of the continents and their margins, programme and abstracts*, edited by D.B. Snyder and R.M. Clowes, Mont Tremblant, Quebec, Canada, Sept 25 Oct 1, 2004, Lithoprobe report 84, p. 65.

Kusky, T.M. (1989), Accretion of the Slave Province, *Geology*, 17, 63-67.

Lacroix, S., and E.W. Sawyer (1995), An Archean fold-thrust belt in the northwestern Abitibi greenstone belt: structural and seismic evidence, *Can. J. Earth Sci.*, 32, 97-112.

Langford, F.F., and J.A. Morin (1976), The development of the Superior Province by merging island arcs, *American Journal of Science*, 276, 1023-1034.

Lemiszki, P.J., and L.D. Brown (1988), Variable crustal structure of strike slip fault zones observed on deep seismic reflection profiles, *Geol. Soc. Am. Bull.*, 100, 665-676.

Mareschal, J.C., and G.F. West (1980), A model for Archean tectonism; part 2, numerical models of vertical tectonism in greenstone belts, *Can. J. Earth Sci.*, 17, 60-71.

Mortensen, J.K. (1993), U-Pb geochronology of the eastern Abitibi subprovince, part 1: Chibougamau-Matagami-Joutel region, *Can. J. Earth Sci.*, 30, 11-28.

Moser, D.E. (1994), The geology and structure of the mid-crustal Wawa gneiss domain: a key to understanding tectonic variation with depth and time in the late Archean Wawa orogen, *Can. J. Earth Sci.*, 31, 1064-1080.

Mueller, W.U., R. Daigneault, J.K. Mortensen, and E.H. Chown (1996), Archean terrane docking: upper crust collision tectonics, Abitibi greenstone belt, Quebec, Canada, *Tectonophysics*, 265, 127-150.

Musacchio, G., D.J. White, I. Asudeh, and C.J. Thomson (2004), Lithospheric structure and composition of the Archean western Superior Province, *J. Geophys. Res.*, 109, B03304.

Myers, J.S. (1995), The generation and assembly of an Archaean supercontinent: evidence from the Yilgarn craton, western Australia, in *Early Precambrian Processes*, edited by M.P. Coward and A.C. Ries, Geol. Soc. Spec. Publ. 95, 143-154.

Myers, J.S. (1997), Preface: Archaean geology of the Eastern Goldfields of western Australia: regional overview, *Precambrian Res.*, 83, 1-10.

Pehrsson, S.J., and M.E. Villeneuve (1999), Deposition and imbrication of a 2670-2629 Ma supracrustal sequence in the Indin Lake area, southwestern Slave Province, Canada. *Can. J. Earth Sci.*, 36, 1149-1168.

Pehrsson, S.J., T. Chaco, M. Pilkington, M.E. Villeneuve, and K. Bethune (2000), The Anton terrane revisited: Late Archean exhumation of a moderate-pressure granulite terrane in the western Slave Province, *Geology*, 28, 1075-1078.

Percival, J.A., and G.F. West (1994), The Kapuskasing uplift: A geological and geophysical synthesis, *Can. J. Earth Sci.*, 31, 1256-1286.

Percival, J.A., and H.R. Williams (1989), Late Archean Quetico accretionary complex, Superior province, Canada, *Geology*, 17, 23-25.

Percival, J.A., R.A. Stern, T. Skulski, K.D. Card, J.K. Mortensen, and N.J. Begin (1994), Minto block, Superior Province: Missing link in deciphering assembly of the craton at 2.7 Ga, *Geology*, 22, 839-842.

Petford, N., R.C. Kerr, and J.R. Lister (1993), Dike transport of granitoid magmas, *Geology*, 21, 845-848.

Polat, A., R. Kerrich, and D.A. Wyman (1998), The late Archean Schreiber-Hemlo and White River -Dayohessarah greenstone belts, Superior province: collages of oceanic plateaus, oceanic arcs, and subduction-accretion complexes, *Tectonophysics*, 289, 295-326.

Pollack, N.H. (1997), Thermal characteristics of the Archean, in *Greenstone Belts*, edited by M.J. de Wit and L.D. Ashwal, Oxford University Press, 223-232.

Powell, W.G., D.M. Carmichael, and C.J. Hodgson (1995), Conditions and timing of metamorphism in the southern Abitibi greenstone belt, Quebec, *Can. J. Earth Sci.*, 32, 787-805.

Relf, C., H.A. Sandeman, and M.E. Villeneuve (1999), Tectonic and thermal history of the Anialik River area, northwestern Slave Province, Canada, *Can. J. Earth Sci.*, 36, 1207-1226.

Sanborn-Barrie, M., T. Skulski, and J. Parker (2001), Three hundred million years of tectonic history recorded by the Red Lake greenstone belt, Ontario, *Geological Survey of Canada, Current Research*, 2001-C19, http://cgc.rncan.gc.ca/librairie/cr/2001/win_e.php

Sawyer, E.W., and K. Benn (1993), Structure of the high-grade Opatica Belt and adjacent low-grade Abitibi Subprovince,

Canada; an Archaean mountain front, *Journal of Structural Geology*, 15, 1443-1458.

Stott, G.M., and F. Corfu (1991), Uchi subprovince, in *Geology of Ontario*, edited by P.C. Thurston, H.R. Williams, R.H. Sutcliffe, and G.M. Stott, Ontario Geological Survey Special Volume 4, part 1, 145-236.

Swager, C.P. (1997), Tectono-stratigraphy of late Archaean greenstone terranes in the southern Eastern Goldfields, western Australia, *Precambrian Res.*, 83, 11-42.

van der Velden, A.J., and F.A. Cook (2005), Relict subduction zones in Canada, *J. Geophys. Res.*, B8302.

van der Velden, A.J., and F.A. Cook (2002), Products of 2.65-2.58 Ga orogenesis in the Slave Province correlated with Slave-Northern Cordillera Lithospheric Evolution (SNORCLE) seismic reflection patterns, *Can. J. Earth Sci.*, 38, 1189-1200.

van der Velden, A.J., C.R. van Staal, and F.A. Cook (2004), Crustal structure, fossil subduction, and the tectonic evolution of the Newfoundland Appalachians: evidence from a reprocessed seismic reflection survey, *Geol. Soc. Am. Bull.*, 116, 1485-1498.

Varsek, J.L., and F.A. Cook (1994), Three-dimensional crustal structure of the Eastern Cordillera, southwestern Canada and northwestern United States, *Geol. Soc. Am. Bull.*, 106, 803-823.

White, D.J., G. Musacchio, H.H. Helmstaedt, R.M. Harrap, P.C. Thurston, A. van der Velden, and K. Hall (2003), Images of a lower-crustal oceanic slab: Direct evidence for tectonic accretion in the Archean western Superior Province, *Geology*, 31, 997-1000.

Williams, H.R. (1990), Subprovince accretion tectonics in the south-central Superior Province, *Can. J. Earth Sci.*, 27, 570-581.

Zegers, T.E., and P.E. van Keken (2001), Middle Archean continent formation by crustal delamination, *Geology*, 29, 1083-1086.

F. A. Cook and A. J. van der Velden, Department of Geology and Geophysics, University of Calgary, Calgary, Canada. (dvajvan@ucalgary.ca)

B.J. Drummond and B.R. Goleby, Geoscience Australia, Canberra, Australia.

Tectonic Delamination of the Lower Crust During Late Archean Collision of the Abitibi–Opatica and Pontiac Terranes, Superior Province, Canada

Keith Benn

Ottawa-Carleton Geosciences Centre and Department of Earth Sciences, University of Ottawa, Ottawa, Canada

The southeastern Superior Province of the Canadian Shield preserves a very complete record of the accretion and deformation of juvenile crust during a 100 Ma period, 2750 Ma through 2650 Ma. The region includes, from north to south, the Opatica granite–gneiss belt, the Abitibi granite–greenstone Subprovince, and the Pontiac metasedimentary Subprovince. The nature of the terrane suture and the crustal structure in the southeastern Superior Province are evaluated based on a synthesis of lithological, structural, geochronological, and geophysical data, including a reinterpretation of Lithoprobe deep seismic reflection profiles. All geological and geophysical data are consistent with the Opatica belt being contiguous with middle crust that underlies greenstones of the Abitibi Subprovince. The Opatica belt can now be considered part of the Abitibi Subprovince, which represents one large tectonic terrane, the Abitibi–Opatica terrane. The boundary between the greenstones of the Abitibi Subprovince and the metasedimentary rocks of the Pontiac Subprovince is a Late Archean terrane suture. The reinterpretation of the seismic reflection profiles suggests the Pontiac–Abitibi terrane suture involved wedging of older crust, underlying the Pontiac Subprovince, into the middle crust of the younger Abitibi Subprovince, resulting in delamination of the Abitibi lower crust. Deformation related to collision resulted in folding of the upper and middle crust of the Abitibi–Opatica plate for 250 km inboard of the terrane suture. The crustal deformation style, delamination, and thrusting of the stronger lower crust and the large-scale folding of the softer upper-middle crust are compatible with calculated strength profiles that include a lower crust composed of mafic granulite. The rheological profile of the Abitibi–Opatica plate, and the folding of the upper and middle crust during plate collision, may have resulted from a combination of radiogenic heating due to the abundance of granitoids in the middle crust and syncollisional plutonism.

1. INTRODUCTION

The Superior Province of the Canadian Shield is the largest known Archean craton and forms the core of the North American continent. It has classically been divided into a number of subprovinces, based on differences in structural trends, metamorphic grades, lithological makeup, and geochronology [*Card and Poulsen*, 1998]. The southern part of the Superior Province is represented by E–W elongate, Late Archean geological subprovinces, each characterized by a predominance of metavolcanic, metasedimentary, or metaplutonic rocks (Figure 1) and by predominantly E–W trending structural lineaments.

Archean Geodynamics and Environments
Geophysical Monograph Series 164
Copyright 2006 by the American Geophysical Union
10.1029/164GM17

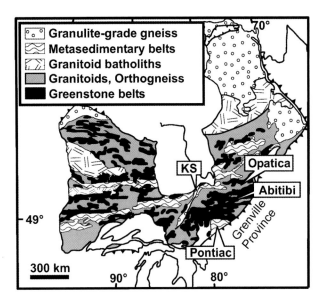

Figure 1. Geological map of the Archean Superior Province in the Canadian Shield. KS, Kapuskasing Structure.

The region considered in this paper is located in the southeastern Superior Province, in Ontario and Quebec. It represents a surface area of roughly 2.4×10^5 km^2 made up of Late Archean crust that was accreted and deformed between 2750 and 2650 Ma. The study area includes the Opatica metaplutonic belt; the greenstones, mainly metavolcanics, of the Abitibi Subprovince; and the predominantly metasedimentary Pontiac Subprovince (Figure 1). A wealth of recent geological, geophysical, and geochronological data are available for the region, much of it generated by the Abitibi–Grenville transect of Canada's Lithoprobe program, which was completed in the previous decade [*Ludden and Hynes*, 2000].

The SE Superior Province is a key region for developing and testing ideas and conceptual models for Late Archean geodynamics and tectonics [e.g., *Dewit*, 1998]. The present study area has also been compared to results of quantitative geodynamic models of terrane collisions [*Ellis and Beaumont*, 1999], based upon a previous interpretation of the crustal structure, which is revisited and revised in the present paper.

Most tectonic models for the SE Superior Province call for actualistic plate tectonics and include the presence of one or more allochthonous terranes, accreted during collisions between lithospheric plates at subduction zones. Identifying and documenting the terranes that are present, and studying the nature of the terrane sutures, are of critical importance for our understanding of the geodynamic processes that were operative during the Late Archean.

In this paper, the principal existing datasets for the SE Superior Province are synthesized, including those for the geology, structure, geophysics, and geochronology. A new

interpretation of the Lithoprobe deep-seismic reflection profiles through the Pontiac and Abitibi subprovinces and the Opatica belt is also presented and compared to the previous interpretation. It is concluded that the Abitibi Subprovince and the Opatica belt represent a single tectonic terrane, here referred to as the Abitibi–Opatica terrane, which, on its northern boundary, is separated from older, pre-2800 Ma crust by a ductile shear zone. To the south of the Opatica belt, one terrane suture is recognized and corresponds to the boundary between the low-grade southern Abitibi Subprovince and the Pontiac Subprovince, which includes amphibolite-grade (kyanite-bearing) metasedimentary rocks and anatectic granites.

The synthesis of the geological and geophysical data and the reinterpretation of the seismic reflection profiles lead to the interpretation that, during plate collision, the lower crust of the southern Abitibi Subprovince was delaminated by a wedge of older (pre-3000 Ma) crust underlying the Pontiac Subprovince. The rheological implications of collision-related delamination of the lower crust of the Abitibi are discussed. Collision also resulted in broadly distributed shortening of the overriding Abitibi–Opatica plate, which was accommodated mainly by folding of the upper and middle crust and by delamination (in the suture) and thrusting of the comparatively stiff lower crust, presumably composed largely of mafic granulites.

2. GEOLOGY AND GEOCHRONOLOGY

The study area includes, from north to south, the Opatica belt, the Abitibi Subprovince, and the Pontiac Subprovince (Figure 1, Plate 1). The three regions are differentiated according to the most abundant lithologies as well as by the peak metamorphic grades. Plate 1 is a lithological map, where the metavolcanic rocks of the Abitibi Subprovince are assigned to four units by ages. That was done for simplicity of presentation and also because no recent correlation exists for the metavolcanic units of the Abitibi in Ontario and Quebec. The metasedimentary units in the Abitibi and Pontiac Subprovinces, the Porcupine Assemblage, the Pontiac Group and the Timiskaming Assemblage, are well established. For the Opatica belt, the rocks are grouped into two units, older orthogneiss and younger plutons that were mapped and dated along three traverses in Quebec [*Benn et al.*, 1992; *Sawyer and Benn*, 1993; *Davis et al.*, 1995].

Figure 2 displays high-resolution geochronological data for plutonic units in the study area; references to the sources of the data are provided in the figure caption. Reference is made mainly to the tonalite and granodiorite intrusions that were emplaced prior to and during orogeny.

2.1. Opatica Belt

The Opatica belt is made up of tonalite orthogneiss and several suites of tonalite, granodiorite, and granite plutons.

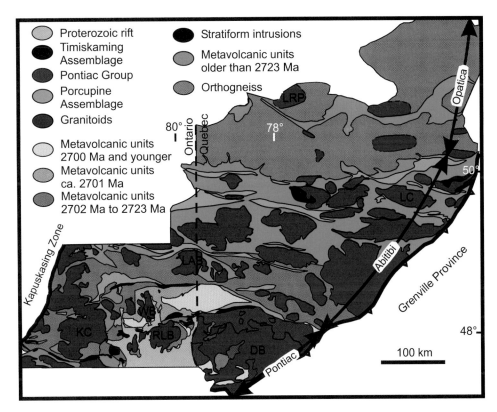

Plate 1. Lithological map of the study area. The metavolcanic rocks in the Abitibi and Pontiac Subprovince are grouped according to ages. Granitoid complexes, batholiths, and plutons discussed in the text are labeled as follows, with the boundary between the provinces of Ontario and Quebec shown for reference: KC, Kenogamissi complex; LC, Lapparent complex; RLB, Round Lake batholith; WB, Watabeag batholith; LAB, Lake Abitibi batholith; DB, Décelles batholith; LRP, Lake Rodayer pluton.

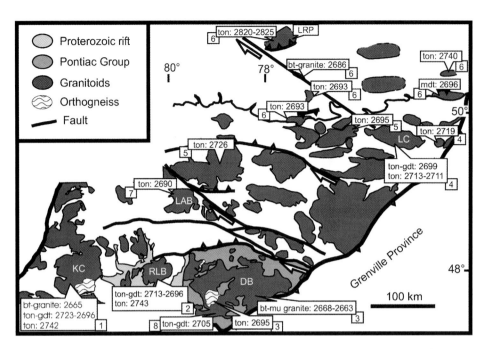

Figure 2. U-Pb zircon dates for granitic rocks. The granitoid complexes, batholiths, and plutons are labeled as in Plate 1. Dates are in millions of years. The literature sources for the geochronological data are keyed by the numbers, as follows: 1, *Heather and Shore*, 1999; *Ayer et al.*, 2002b; 2, *Mortensen*, 1993a; 3, *Mortensen and Card*, 1993; 4, *Mortensen*, 1993b; 5, *Davis et al.*, 2000; 6, *Davis et al.*, 1995; 7, *Mortensen*, 1987; 8, *Machado et al.*, 1991, 1993. Codes for dated rock types: ton, tonalite; gdt, granodiorite; mdt, monzodiorite; bt-granite, biotite granite; bt-mu granite, biotite-muscovite granite.

Some of the tonalite plutons have diorite components. The tonalite and granodiorite plutons have been deformed to varying degrees. Deformation in the Opatica belt occurred under greenschist-grade conditions along the border with the Abitibi Subprovince and under amphibolite-grade conditions further to the north [*Benn et al.*, 1992]. Migmatized tonalite orthogneiss is documented in the central part of the belt, where the peak metamorphic temperature reached 760°C [*Sawyer*, 1998].

The oldest dated plutonic rocks of intermediate to felsic composition in the study area are found in the north of the Opatica belt, where diorite and tonalite samples from the Lac Rodayer pluton yielded dates from 2825 Ma to 2820 Ma. That pluton is separated from the Opatica rocks to the south by a ductile thrust (Figure 2) [*Benn et al.*, 1992; *Sawyer and Benn*, 1993]. To the south of the thrust, the oldest unit in the Opatica belt is tonalite orthogneiss, dated 2740 Ma (an older date of 2761 Ma has a very large associated error [*Davis et al.*, 1995]). A younger suite of plutons in the Opatica belt is represented by monzodiorite and tonalite plutons that provided dates ranging from 2696 to 2693 Ma.

The youngest suite of plutonic rocks in the Opatica belt are biotite granites that crosscut all other units and are dated 2690 to 2686 Ma. The biotite granites may have been generated during anatexis of the older tonalite orthogneiss [*Sawyer*, 1998].

2.2. Abitibi Subprovince

The regional metamorphism of the Abitibi Subprovince is subgreenschist- to greenschist-grade [*Powell et al.*, 1995; *Easton*, 2000]. The stratigraphy is made up predominantly of metavolcanic rocks, consisting of komatiites, komatiitic basalts, and abundant tholeiitic volcanic rocks of mafic to felsic composition, overlain by a suite that includes calcalkaline rocks in the southern part of the subprovince (the youngest metavolcanic unit in Plate 1). The metavolcanic units are (possibly unconformably) overlain by graywackes of the Porcupine Assemblage, which are themselves unconformably overlain by conglomerates and sandstones of the Timiskaming Assemblage (Plate 1).

In Ontario, the stratigraphy of the Abitibi Subprovince is interpreted to represent a period of nearly continuous volcanism and minor sedimentation that lasted some 50 Ma, between ca. 2750 Ma and ca. 2697 Ma, followed by roughly 20 Ma of siliciclastic sedimentation, represented by the Porcupine and Timiskaming assemblages [*Ayer et al.*, 2002b]. That interpretation is herein extended to the Abitibi in Quebec as it also conforms to the geochronological data for that region.

The ca. 50 Ma of predominantly volcanic stratigraphy in the Abitibi contains elements of plume-related magmatism and also of tholeiitic to calcalkaline supra-subduction zone magmatism that included some minor melting of a northward dipping slab [*Polat and Kerrich*, 2001]. Field relationships and petrology of lavas indicates intimate temporal relationships between those two petrogenetic elements [*Lafleche et al.*, 1992; *Scott et al.*, 2002]. Interlayering of boninite lavas with komatiite–tholeiite lavas suggests plume interaction with an active subduction zone [*Kerrich, et al.*, 1998]. One interpretation is that the plume–subduction interaction may have occurred between 2720 Ma plume impingement and the beginning of arc development at 2716 Ma [*Wyman*, 1999]. However, the eruption of plume-related komatiitic lavas is documented to have continued from 2750 Ma until 2703 Ma [*Sproule et al.*, 2002], suggesting a much longer duration for the apparent plume–subduction interaction.

The Porcupine Assemblage contains zircons that yielded dates from 2825 Ma to 2685 Ma [*Bleeker and Parrish*, 1996; *Ayer et al.*, 2002b]. It crops out within synclines throughout the Abitibi Subprovince, suggesting that much of the region may have remained submerged during deposition of that unit. The Timiskaming Assemblage consists of conglomerate, sandstone, and turbidite that record deposition in nonmarine, submarine fan, and eolian environments [*Hyde*, 1980], suggesting deposition in a locally emergent shallow-water environment. Outcrops of that youngest assemblage are mostly (but not entirely) confined to narrow lineaments within tight synclines, in some cases situated near regional fault zones [*Benn and Peschler*, 2005].

Several suites of tonalite and granodiorite plutons intruded the supracrustal units and form composite batholiths such as the Round Lake, Lake Abitibi, and Watabeag batholiths and the Kenogamissi and Lapparent complexes (Plate 1). The batholiths include orthogneiss components that represent highly deformed tonalite plutons of different ages. Gravity models suggest that the batholiths are no more than 6 km thick and have generally tabular shapes in profile [*Peschler et al.*, 2004].

The intermediate to felsic plutonic rocks in the Abitibi Subprovince represent a range of compositions and apparent crystallization ages similar to those of the Opatica belt. The oldest tonalite plutons are dated 2742 Ma (within the Kenogamissi complex) and 2743 Ma (within the Round Lake batholith). Younger tonalite and granodiorite plutons have yielded a range of dates from 2726 Ma to 2695 Ma (Figure 2), which represents a range of apparent crystallization ages that is broader than, but overlaps with, the ages of plutons in the Opatica belt. It may be that the broader range of dates for tonalite and granodiorites in the Abitibi Subprovince is due to more dating having been completed there than in the Opatica belt.

Also, like in the Opatica belt, the youngest major plutonic suite is made up of biotite granites that were emplaced very late in the regional tectonic history. A date of 2665 Ma for that suite, in the Kenogamissi complex, suggests that the

granites may be some 20 Ma younger in the Abitibi than in the Opatica.

In the Abitibi Subprovince, tonalite and granodiorite plutons and orthogneiss that are older than ca. 2710 Ma are typically interpreted to represent synvolcanic intrusions, genetically associated with volcanic units of similar ages [*Sutcliffe et al.*, 1993; *Chown et al.*, 2002]. Here the interpretation is that the ca. 2740 tonalite plutons dated in the Opatica belt are also synvolcanic intrusions, i.e., magma chambers that were genetically associated with a metavolcanic upper crust similar to the Abitibi Subprovince, but which has now been eroded away.

The above interpretation implies that the Opatica belt would be representative of the middle crust of intermediate to felsic composition that presently underlies the Abitibi belt, as previously suggested by *Benn et al.* [1992]. That, however, is at odds with the interpretation that the Opatica belt and the Abitibi Subprovince would represent distinct tectonic terranes, an interpretation that was based on a previous interpretation of the Lithoprobe seismic reflection profiles [*Calvert et al.*, 1995] and that will be revisited further on in this paper.

2.3. Pontiac Subprovince

The principal supracrustal unit in the Pontiac Subprovince is the Pontiac Group. It is composed of metapsammites and metapelites that have been metamorphosed to greenschist grade in the north, near to the Abitibi Subprovince, and to kyanite grade in the central part of the subprovince [*Benn et al.*, 1994], where the Pontiac Group is intruded by anatectic biotite muscovite (± sillimanite) granites of the Decelles batholith (Plate 1). Greenschist-grade metavolcanic rocks crop out around the margins of the Pontiac Subprovince, mainly near its southern margin (Plate 1) and structurally overlie the Pontiac Group. Orthogneiss that structurally underlies the Pontiac Group crops out in the northwest and southwest of the Pontiac Subprovince (Plate 1).

Detrital zircon dates from the Pontiac Group range from 3028 Ma to 2683 Ma [*Mortensen and Card*, 1993; *Davis*, 2002]. The youngest of the dates provides a bracket for the age of turbidite sedimentation, which is very similar to the ages of youngest detrital zircons in the Porcupine Assemblage, in the Abibiti Subprovince. The oldest detrital zircon dates from the Pontiac Group suggest a source region for clastic sediment that has not been identified anywhere in the Pontiac Subprovince, or anywhere to the north or west of it, in the SE Superior Province.

It is interpreted that the source for the pre-3000 Ma zircons dated from the Pontiac Group was within crust that is older than the Pontiac, the Abitibi, and the Opatica and is now situated within the Proterozoic Grenville orogenic province

(Plate 1). The results of a study of isotopic compositions of granitic rocks in the Pontiac and Abitibi subprovinces indicated the Pontiac Subprovince would be underlain by crust older than 3000 Ma [*Carignan et al.*, 1993], so some of the source region for the oldest zircons may also now underlie the Pontiac Subprovince.

Tonalite and granodiorite plutons in the southern part of the Pontiac Subprovince have yielded a date of 2705 Ma, an apparent crystallization age that is slightly older than for the nearby metavolcanic rocks (Plate 1), which have yielded dates of 2700 to 2680 Ma [*Machado et al.*, 1991, 1993]. Outcrops of an orthogneissic tonalite pluton in the SW Pontiac Subprovince are dated 2695 Ma [*Mortensen and Card*, 1993] (Figure 2). Hence the tonalite and granodiorite plutons of the Pontiac Subprovince are very similar in age to many of the plutons of the same composition in both the Opatica belt and the Abitibi Subprovince.

The late tectonic Decelles batholith (Plate 1) is made up mainly of anatectic biotite muscovite granites that locally contain sillimanite. The granite has been dated 2663 Ma, and an associated pegmatite yielded a date of 2668 Ma.

3. STRUCTURAL GEOLOGY

The precollisional structural history of the region is not well understood, perhaps due to overprinting of early structures by collision-related deformation. In the greenstones of the NE Abitibi Subprovince, an early generation of upright, open, gently plunging N–S trending folds has been documented and interpreted to represent flexural folds associated with local subsidence of the Abitibi volcanic basin [*Daigneault et al.*, 1990]. Likewise, in the southern Abitibi Subprovince, the existence of early, upright, open, E–W trending folds was attributed to flexure of the volcanic strata during basin subsidence [*Dimroth et al.*, 1983]. Evidence for tectonic events that may have predated the main collisional event discussed below is, so far at least, lacking.

Here, the structures related to plate collision are synthesized to aid with interpretation of the geophysical data. The principal regional structures are folds and fault zones that are interpreted here to be the result of a single N–S shortening associated with a collisional tectonic event. Folding was the principal response of the upper and middle crust to collisional tectonics.

3.1. Folds

The outcrop pattern of the SE Superior Province is controlled mainly by regional, E–W trending folds (Figure 3). Cross-sections [*Benn*, 2005; *Benn and Peschler*, 2005] and gravity modeling [*Peschler et al.*, 2004] in the southern Abitibi Subprovince indicate that the regional folds are upright to slightly overturned, have amplitudes ranging from

3 to 10 km, and have wavelengths on the order of 10 to 30 km. The Blake River syncline (Figure 3) is a first-order fold in the southern Abitibi Subprovince.

In the NE Abitibi Subprovince, structural profiles in the greenstones also show fold amplitudes that vary from a few km to about 8 km [*Daigneault, et al.*, 1990]. Regional-scale upright folds (Figure 3) deformed every unit of the supracrustal stratigraphy in the study area; therefore, deformation continued until after deposition of the Timiskaming Assemblage, ca. 2677 Ma [*Corfu et al.*, 1991]. Here, it is speculated that the early stages of folding may have exercised at least a partial control on the locations and depths of basins where siliciclastic sedimentation occurred.

Tonalite and granodiorite units in the study area have also been deformed during the collisional event. Figure 3 shows that many of the axial surface traces of regional folds are mapped as continuous into granitoid batholiths. Mapping in the Kenogamissi complex in the SW Abitibi Subprovince reveals that older (> 2720 Ma) plutons were folded under amphibolite-grade metamorphic conditions whereas younger (< 2715 Ma) plutons were folded while still partially molten [*Benn*, 2005]. Therefore, in the southern Abitibi Subprovince, the regional shortening event that gave rise to folding had begun by 2715 Ma. The youngest plutonic units, biotite granites, are in general undeformed or weakly deformed, and therefore the regional shortening event, though probably diachronous across strike, was waning by ca. 2670 Ma.

In the Pontiac Subprovince, the upright E–W trending regional folds deform an earlier generation of isoclinal recumbent folds (not shown in Figure 3 for reasons of scale) in the metasedimentary Pontiac Group [*Benn et al.*, 1994]. The early recumbent folds are interpreted to record thickening of the sedimentary units during thrusting of the southern Abitibi over the Pontiac basin, which led to kyanite-grade metamorphism in the central part of the Pontiac Subprovince [*Benn et al.*, 1994].

3.2. Fault Zones

Fault zones, some continuous for hundreds of kilometers (Figure 3), represent the other principal structural element that deformed the SE Superior Province at the regional scale. The fault zones have received a great deal of attention in the literature, in part because they tend to be loci for large gold deposits. Some regional-scale fault zones, notably, the Kirkland–Cadillac fault zone and the Porcupine–Destor fault zone (Figure 3), have previously been interpreted to be potential tectonic terrane boundaries [*Kerrich and Feng*, 1992; *Mueller et al.*, 1996; *Daigneault et al.*, 2002].

The kinematics and the internal structures of the regional fault zones in the Opatica belt and in the Abitibi Subprovince are discussed in detail elsewhere [*Robert*, 1989; *Benn et al.*, 1992; *Wilkinson et al.*, 1999; *Benn and Peschler*, 2005].

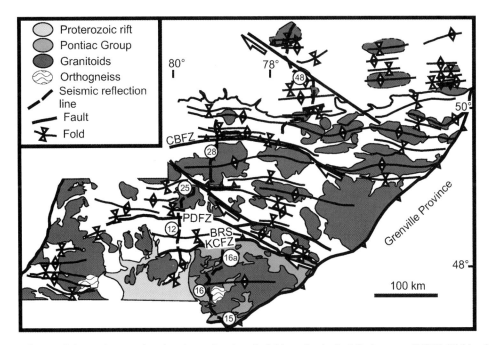

Figure 3. Structural map of the study area showing the regional-scale folds and principal fault zones. KCFZ, Kirkland–Cadillac fault zone; PDFZ, Porcupine–Destor fault zone; CBFZ, Casa Berardi fault zone; BRS, Blake River syncline. The Lithoprobe deep seismic reflection profile traces are also shown and labeled as in Plate 2B. Structures are compiled from the following sources—*Benn et al.*, 1992, 1994; *Chown et al.*, 1992; *Heather and Shore*, 1999; *Benn*, 2005—and from 1:100 000-scale digital geological maps published by the Ontario Geological Survey.

Briefly, E–W striking fault zones and segments of fault zones have accommodated severe flattening and locally preserve down-dip extension lineations. SE-striking fault zones and fault zone segments preserve dextral kinematic indicators, though the amount of strike-slip displacement does not appear to be large, judging by the apparent offsets of geological units. Vertical components of displacement on the fault zones within the Abitibi Subprovince are limited to 2 km, as revealed by metamorphic studies in the greenstones [*Powell et al.*, 1995].

The Kirkland–Cadillac fault zone that marks the boundary between the Pontiac and Abitibi subprovinces in Québec, but which transects the southern Abitibi greenstones in Ontario (Figure 3), is discussed at greater length in the reinterpretation of the Lithoprobe seismic reflection profiles.

4. GEOPHYSICS

As a result of the Abitibi–Grenville Lithoprobe transect [*Ludden and Hynes*, 2000] and later studies [*Peschler et al.*, 2004], a large geophysical data set and several interpretations of the geophysical data are available for the SE Superior Province. The geophysical data set includes deep seismic reflection lines that together provide a profile across the Pontiac and Abitibi subprovinces that extends into the

Opatica belt (Figure 3). Other geophysical data of regional importance include seismic refraction, gravity, and heat flow. Here, the geophysical data are first considered at the scale of the study area, followed by a summary of previous interpretations regarding the composition and structure of the underlying lithosphere. Then, the reinterpretation of the seismic reflection profiles offered attempts to tie the seismic reflectivity patterns to the structural geology that is well documented by surface mapping.

4.1. Geophysical Data and Models: Previous Interpretations

Crustal velocity models calculated from seismic refraction profiles across the Abititi and the Pontiac indicate a three-layer crust, interpreted to be composed of a 12-km-thick heterogeneous upper crust with P-wave velocities of 5.6 to 6.4 km/s, a middle crust from 12 to 30 km depth with very homogeneous velocities of 6.4 to 6.6 km/s, and a 10-km-thick (30 to 40 km depth) lower crust with velocities varying from 6.9 to 7.3 km/s [*Grandjean et al.*, 1995]. The homogeneous nature of the middle crust is also indicated by magnetotelluric data from the Pontiac Subprovince [*Kellett et al.*, 1992]. The upper crustal layer is obviously composed of metavolcanic and metasedimentary rocks, granitic plutons, and batholiths. The homogeneous seismic velocities of the middle crust are interpreted to represent orthogneiss and paragneiss [*Grandjean et al.*, 1995]; the lower crustal layer has velocities typical of mafic granulite-grade metamorphic rocks [*Rudnick and Fountain*, 1995], which is consistent with the 30 to 40 km depth of the lower crustal layer.

The results of the crustal velocity models were interpreted to indicate that an allochthonous veneer of upper crust in the Abitibi Subprovince would overlie an imbricated, autochthonous middle and lower crust [*Grandjean et al.*, 1995]. That interpretation leaves open the question of the tectonic mechanisms that would be responsible for approximately 400 km of horizontal transport (across-strike width of the Abitibi and Opatica) of metavolcanic and metasedimentary rocks that would have been emplaced upon the autochthonous basement. The presently available paleomagnetic data for the region do not allow that intriguing interpretation to be tested.

Previous interpretations of Lithoprobe deep seismic reflection data are consistent with the three-layer crustal model. Based on variations in reflectivity below the northern Abitibi Subprovince, it was suggested that the upper crust of the Abitibi overlies metaplutonic rock of the Opatica belt, whereas the middle crust of the Abitibi would be composed of imbricated slices of metavolcanic, metasedimentary, and deformed plutons that were thrust under the Opatica [*Sénéchal et al.*, 1996; *Bellefleur et al.*, 1998]. That interpretation is represented by the large wedge of "arc plutonic" material indicated below the northern Abitibi in Plate 2A.

The seismic reflection data gathered along line 48 show the presence of a prominent reflector below the Moho, within the lithospheric mantle (Plate 2B). The mantle reflector was interpreted to represent a preserved subducted slab, suggesting the presence of a paleosubduction zone (Plate 2A) [*Calvert et al.*, 1995]. The implication of this interpretation is that the Opatica belt, to the north, and the Abitibi Subprovince, to the south, would represent distinct tectonic terranes and that collision would have resulted in wedging of the Opatica crust into the Abitibi crust and in the obduction of the upper crust of the northern Abitibi onto the Opatica belt.

Seismic reflector patterns in the southern Abitibi were interpreted to indicate a synformal shape at depth, below the Blake River syncline [*Jackson et al.*, 1995; *Bellefleur et al.*, 1998]. On the other hand, the reflectivity pattern below the northern Abitibi is suggestive of horizontal to shallowly northward dipping structures [*Bellefleur et al.*, 1998] (Plates 2A,B). The different reflectivity patterns in the northern and southern Abitibi Subprovince were interpreted to indicate different precollisional tectonic histories, with the synformal pattern beneath the southern Abitibi having resulted from the formation of a rift basin prior to the ultimate collisional event that caused upright folding of the Abitibi greenstones [*Bellefleur et al.*, 1998].

As shown in Plate 2B, seismic reflection profiles suggest a thickening of the crustal section below the southern Abitibi Subprovince and the northern Pontiac Subprovince. Apparent crustal thickening is also documented below the Opatica–Abitibi boundary (5 km thicker than below the northern Abitibi) by gravity modeling [*Telmat et al.*, 2000]. Another interpretation of first-order importance, based on the same gravity modeling results, is that no density contrast is required within the lithospheric mantle below the Opatica–Abitibi boundary in that region where the north-dipping reflector cross-cuts the Moho. Therefore, if a subducted high-density slab were preserved there [*Calvert et al.*, 1995; *Ludden and Hynes*, 2000], it would need to be very thin [*Telmat et al.*, 2000].

4.2. Reinterpretation of Lithoprobe Seismic Reflection Profiles

A new interpretation of the seismic reflection profiles that traverse the study area is shown as a line drawing overlaid on the migrated seismic profiles in Plate 2C; a full interpretation is shown in Plate 2D. The bulk three-layer crustal model is retained but the interpretation differs from previous ones in three fundamentally important ways. First, a different interpretation is adopted for the deformation of the middle crust in the Opatica belt and Abitibi Subprovince, one that is more easily reconciled with known geological structures. Second, no terrane suture is recognized between the Opatica belt and the Abitibi Subprovince. On the other hand, the boundary between the Pontiac and Abitibi Subprovinces that crops out as the Kirkland–Cadillac fault zone (Plate 2D) is identified as a terrane boundary. Third, the proposed interpretation does not include a crustal slice of "unknown affinity", as in Plate 2A.

Regarding the structure of the middle crust, the interpretation in Plate 2D suggests that the pattern of reflectors can be interpreted primarily as upright to slightly overturned folds rather than as complex arrays of shear zones as shown in Plate 2A. The new interpretation has two advantages. First, the deformation of the middle crust is explained as the result of the same folding event that is well documented by extensive field mapping and that involves plutons and orthogneiss as well as greenstones (Figure 3). Mechanically, that interpretation implies that the middle crust, as well as the upper crust, were sufficiently ductile that the main response to crustal shortening was folding. Second, the greenstones of the upper crust are not interpreted as allochthonous slices displaced hundreds of kilometers upon thrusts but rather as the autochthonous upper layer of a differentiated crust.

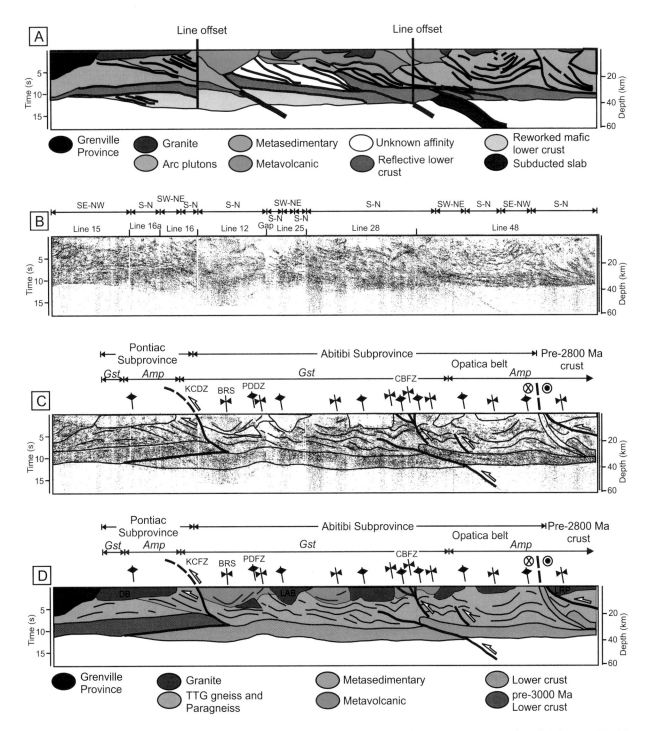

Plate 2. Compilations of the Lithoprobe deep seismic reflection profiles located in Figure 3. (A) Interpretation of *Calvert and Ludden* [1999]. (B) The migrated profiles, with the line orientations indicated above the diagram. (C) Line drawing of the interpretations proposed in this paper, overlaid on the migrated profiles. (D) The full interpretation proposed in this paper. KCFZ, Kirkland–Cadillac fault zone; PDFZ, Porcupine–Destor fault zone; CBFZ, Casa Berardi fault zone; BRS, Blake River syncline; Gst, greenschist-grade; Amp, amphibolite-grade; DB, Décelles batholith; LAB, Lake Abitibi batholith; LRP, Lac Rodayer pluton.

The geological and structural syntheses provided above indicate there is no compelling geological evidence to suggest that the Opatica belt and the Abitibi Subprovince represent two distinct tectonic terranes. Rather, the similarity of the plutonic histories of the Opatica and the Abitibi, especially the fact that the earliest known stages of tonalitic plutonism in both regions have essentially the same age, ca. 2740 Ma, permits the two to be included in one contiguous tectonic terrane. The Opatica belt can be included in the Abitibi Subprovince and considered to be representative of the amphibolite-grade middle crust underlying the northern Abitibi greenstones.

Plate 2D shows the presence of a terrane suture between the Pontiac and Abitibi subprovinces. In outcrop, the suture corresponds to the Kirkland–Cadillac fault zone, in the plane of the profile. At depth, the suture corresponds to a wedge of highly reflective lower crust of the Pontiac Subprovince that delaminated the lower crust of the Abitibi Subprovince from its middle crust. The lower crust of the Pontiac Subprovince would correspond to the pre-3000 Ma crust under the Pontiac Subprovince, the existence of which is indicated by isotopic studies of granitic plutons [*Carignan et al.*, 1993]. Also, the overthrusting of the Pontiac Subprovince metasediments by the southern Abitibi belt, along with the thickening of the metasedimentary rocks by recumbent folding [*Benn, et al.*, 1994], would explain the kyanite-grade metamorphism in the Pontiac Group.

The folding of the upper and middle crust in the Abitibi Subprovince and the Opatica belt does not appear to have strongly affected the highly reflective lower crust below either region, except possibly where the lower crust may be deformed into very open folds below the Blake River syncline (Plate 2D). Instead, the lower crust was sufficiently stiff that tectonic shortening resulted in localization of strain and the formation of at least one shear zone that displaced the Moho. In Plate 2D, that shear zone corresponds to the reflector in the mantle lithosphere previously interpreted as a subducted slab. As indicated by gravity modeling across the Opatica–Abitibi boundary, there is no evidence in the gravity field for a dense subducted slab underlying that region [*Telmat et al.*, 2000]. The mantle reflector is here interpreted to be a mylonitic shear zone preserved in the mantle lithosphere. Uplift of the crust above the mantle thrust may have also caused the nucleation of a steep thrust above it that crops out as several closely spaced splays in the northern Abitibi Subprovince [*Lacroix and Sawyer*, 1995]. Those thrusts are collectively labelled the Casa Berardi Fault Zone in Figure 3 and Plate 2D.

The reinterpretation of the Lithoprobe seismic profiles presented here does not, of course, illustrate possible variations in crustal thickness or composition across the region, perpendicular to the profiles. Heat flow in the western part of the Abitibi Subprovince is much higher than in its eastern part, where heat flow is the lowest in North America [*Mareschal et al.*, 2000; *Cheng et al.*, 2002], thus suggesting that the composition of the crust may vary greatly from west to east. Whether the variation in the compositions of crustal sections in the western and eastern parts of the Abitibi is the result of differing magmatic histories, or the result of the regional structural evolution, or both, remains an open question.

A word is required regarding the interpretation of the Kirkland–Cadillac fault zone as a terrane suture in Québec. On regional maps, the fault zone is shown as continuous westward, into Ontario, where it clearly is not a terrane suture because it transects the greenstones of the southern Abitibi. A recent reinterpretation of the Porcupine–Destor and Kirkland–Cadillac fault zones in Ontario suggests they may represent linear belts of detachment folding, rather than crustal-scale faults [*Benn and Peschler*, 2005]. The fact that the Pontiac Subprovince is bounded on both its northern and western sides by the Abitibi Subprovince indicates that the Pontiac likely represents a tectonic indenter. The nature of the Kirkland–Cadillac fault zone must change fundamentally where its Ontario segment links to the northwestern edge of the Pontiac indenter.

5. IMPLICATIONS: CRUSTAL RHEOLOGY AND LATE ARCHEAN OROGENY

The interpretations of the crustal structure have two important implications for the rheology of the Late Archean crust in the study area and perhaps for interpretations of other Late Archean orogenic belts as well. The implication for crustal rheology arises from the nature of the Pontiac–Abitibi suture, where the lower crust of the southern Abitibi Subprovince is delaminated. The implication for the structural styles of Archean orogeny arise from the interpretation of folding to the base of the middle crust.

5.1. Crustal Rheology

It is well known that lithospheric rheological profiles have important implications for the evolutions of orogens and the crustal structures that develop in orogenic belts, where lithospheric plates collide [*Ranalli and Murphy*, 1987]. There remains considerable debate on the nature of rheological profiles in different geodynamic environments [*Hoffman and Ranalli*, 1988; *Ranalli*, 1997; *Axen et al.*, 1998]. Some conclusions regarding the rheological profiles of the Abitibi–Opatica plate and the Pontiac plate can be based on the interpretations of crustal structures presented above.

The interpretation in Plate 2D suggests that the lower crust of the 2750 Ma to 2680 Ma Abitibi–Opatica plate was delaminated as the result of wedging by the crust underlying the Pontiac Subprovince. Also, the middle and upper crust of the Abitibi–Opatica plate responded to collision by folding, whereas the lower crust responded by delamination and thrusting. It follows that, during collision of the Pontiac plate with the southern Abitibi, the rheological profile of the Abitibi–Opatica lithosphere included a relatively stiff lower crust underlying a softer, more ductile middle crust. It also follows that the Pontiac crust was strong enough to wedge into the middle crust below the Abitibi.

To quantitatively test that interpretation, strength profiles were calculated for the postulated lithospheres that would have corresponded to the Abitibi and the Pontiac in Late Archean time (Figure 4). The goal of the calculations was to show whether the thermo-mechanical structure of the two plates would be consistent with the delamination model, given reasonable interpretations and assumptions regarding the compositions and physical parameters of the lithospheric sections underlying the Abitibi and Pontiac subprovinces.

The profiles were prepared based on the interpretation of a collision between a magmatically active (or recently active) and therefore relatively hot Abitibi–Opatica plate with an older (> 3000 Ma), colder Pontiac plate. The physical parameters chosen to calculate the geothermal gradients and the strength profiles for the two lithospheres are given in Tables 1 and 2. The heat production value for the upper-middle crust was chosen to give a surface heat flow comparable to Phanerozoic orogenic belts when combined with a reasonable mantle heat flow for the Late Archean Abitibi (Table 1). This value is roughly twice the crustal heat production values determined for the present-day western Abitibi [*Pinet et al.*, 1991; *Mareschal and Jaupart*, this volume]. The other values for heat production and thermal conductivity are from *Rudnick* [1992] and *Afonso and Ranalli* [2004].

The two modeled lithospheres are of the same three-layer composition: a granitic upper-middle crust, a lower crust composed of mafic granulite, and a peridotitic mantle. Allowing for several kilometers of erosion off the Abitibi greenstones, a 35 km thickness was chosen for the upper layer, whereas the lower crustal layer was modeled with a thickness of 10 km. The geothermal gradients for the three-layer models were calculated using equations 3a,b,c of *Afonso and Ranalli* [2004] (see also *Sandiford et al.* [2004]), which account for crustal heat production as well as mantle heat flow. The two calculated geothermal gradients in Figure 4A indicate that temperatures at the base of the lower crust would be 690°C for the older Pontiac lithosphere and 960°C for the younger Abitibi one.

For simplicity, the geothermal gradients were varied by choosing different values for the mantle heat flow below the two crusts (q_m in Table 1). Crustal heat production values were the same in the two calculations (Table 1). The higher value for q_m in the Abitibi geotherm calculation falls within the range of values from young orogenic belts [*Lenardic and*

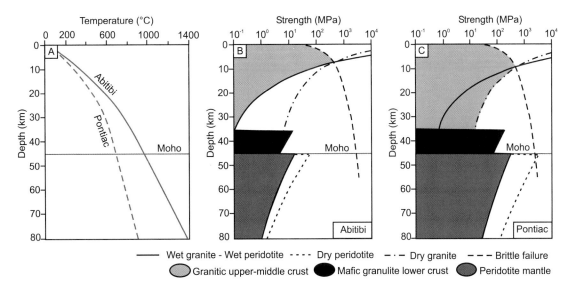

Figure 4. (A) Calculated geothermal gradients for "hot" and "cold" lithospheres, used here to approximate the Abitibi and the Pontiac regions, ca. 2700 Ma. (B, C) Lithospheric strength profiles, calculated using the geothermal gradients in (A). In (B, C), the shaded regions represent results that best explain the delamination of the lower crust of the Abitibi–Opatica terrane, as interpreted in Plate 2 and discussed in the text. Thermal and material parameters used for calculating the geothermal gradients and the strength profiles are provided in Tables 1 and 2.

Table 1. Parameters Used for Calculations of Geothermal Gradients. H: heat production; k: thermal conductivity; h: thickness of crustal layer; q_m: heat flow from the mantle. Choices of values and literature sources are discussed in the text.

Layer	H (μW m^{-3})	k (W m^{-1}K^{-1})	h (km)
Upper-middle *Tonalite*	1.43	2.50	35
Lower *Mafic granulite*	0.28	2.10	10
Mantle *Peridotite*	0.006	3.00	
Terrane	q_m (mW m^{-2})		
Abitibi	30		
Pontiac	15		

Table 2. Creep Parameters Used for Calculations of Strength Profiles. All values are from *Ranalli* [1997].

Rock type/Layer	A MPa^{-n} s^{-1}	n	E kJ mol^{-1}
Wet granite/ Upper-middle crust	2.0×10^{-4}	1.9	137
Dry granite/ Upper-middle crust	1.8×10^{-9}	3.2	123
Mafic granulite/ Lower crust	1.4×10^{4}	4.2	445
Wet peridotite/Mantle	2.0×10^{3}	4.0	471
Dry peridotite/Mantle	2.5×10^{4}	3.5	532

Kaula, 1995] and the lower value used for the Pontiac calculation is comparable to those of some stable cratonic areas [*Jaupart and Mareschal*, 1999]. Keep in mind that the values of the thermal parameters used here are chosen to represent an older, stable lithosphere and a younger, hotter lithosphere. The calculations are meant to approximate geothermal gradients in the study area ca. 2700 Ma; they not are proposed as highly accurate assessments of the geothermal gradients at that time.

The strength of the uppermost crust is modeled using the frictional failure criterion [*Ranalli and Murphy*, 1987],

$$\sigma_1 - \sigma_3 = \beta \rho g z (1 - \lambda),$$

where the differential stress ($\sigma_1 - \sigma_3$) represents rock strength, the parameter $\beta = 3.0$ (corresponding to thrust faulting), λ is the ratio of pore fluid pressure to lithostatic load (fixed at 0.36), g is the acceleration of gravity, ρ is rock density (2800 kg m^{-3}), and z is depth. Steady-state creep strengths were calculated according to

$$\sigma = (\dot{\varepsilon}/A)^{1/n} \exp(E/nRT)$$

using the values given in Table 2 for the material parameters A, n, and E. The strain rate, $\dot{\varepsilon}$, was fixed at 10^{-12} s^{-1}. That strain rate may be high by present-day plate tectonic standards; however, it may be appropriate for the hotter Archean Earth, when plate displacements were likely more vigorous [*Nisbet and Fowler*, 1983].

For the granitic upper-middle crust, calculations were performed using material parameters for both wet and dry granite. Similarly, the strengths of mantle composed of wet and dry peridotite were calculated (Table 2). The lower crust was interpeted to be mafic granulite, in accordance with crustal seismic velocity models [*Grandjean et al.*, 1995].

The calculated geotherms for the Abitibi and Pontiac regions, and the corresponding strength profiles, are presented in Figure 4. The results are somewhat akin to lithospheric strength profiles calculated for "cold" and "hot" geotherms by *Ranalli and Murphy* [1987]. As shown in Figures 4B and C, however, the strength profiles include a strong lower crustal layer, composed of mafic granulite. The strength profiles are necessarily somewhat speculative but they appear to satisfy conditions imposed by the structural interpretations.

The strength profile calculated for the (hotter) Abitibi geotherm (Figure 4B) results in a very weak middle crust, especially if the results for wet granite are considered. In that case, a large contrast in strength occurs at the base of the granitic middle crust, where it overlies a much stronger granulitic lower crust. The strength profiles for the colder geotherm (Figure 4C) also show a middle crust that is relatively soft compared to the underlying lower crustal layer. Overall, however, the lithosphere depicted in Figure 4C is much stronger than the one in Figure 4B, particularly if the strength of the dry granite example is considered (shaded in Figure 4C). These results show that the interpretation of wedging of the older, colder, and stronger Pontiac crust into the middle crust underlying the Abitibi, and the resulting delamination of the Abitibi lower crust, is reasonable, given appropriate and realistic compositions and physical parameters for the two modeled lithospheres.

Lithospheric wedging has been documented in seismic reflection profiles across Precambrian sutures between Proterozoic and Archean terranes. Examples are the Slave–Hottah terrane collision in northern Canada, documented by Lithoprobe [*Cook et al.*, 1998; *Snyder*, 2002] and the Karelian–Svecofennian terrane suture, as revealed by the BABEL seismic reflection survey [*Snyder*, 2002]. In each of those examples, wedging of the older (Slave, Karelian) craton into the younger (Hottah, Svecofennian) lithosphere resulted in delamination of the younger mantle lithosphere from the overlying, softer lower crust.

In the Abitibi–Opatica plate, wedging and delamination did not involve mantle lithosphere. Instead, a relatively stiff lower crust, which is interpreted to be composed of mafic granulite-grade metamorphic rocks, was delaminated from the middle crust. Crustal delamination as described here has not been described from other Archean terranes. However, the rheological consequences of the model can potentially explain the deformation style for the Abitibi–Opatica plate, which is not unique, and therefore the crustal delamination model may be applicable to some other Archean terranes.

5.2. Rheological Profiles and Structural Style of the Orogen

The present interpretation of the crustal structure of the Abitibi Subprovince (including the Opatica belt) suggests that deformation, in response to collision with the Pontiac Subprovince, was characterized by folding of the upper and middle crust and by delamination and thrusting of the lower crust. This in turn implies an important strength contrast between the lower and middle crust that has not been previously documented in the study area or in other Archean terranes. The proposed rheological profile for the Abitibi–Opatica plate (Figure 4B) therefore requires some discussion of its possible origin.

The upright folding of the Abitibi greenstones and the Opatica belt, which is continuous for hundreds of kilometers across strike, is not unique amongst Archean terranes. A similar structural style is documented in the Sino–Korean craton and also in the Dharwar craton of India [*Choukroune et al.*, 1995] and upright folding is also typical of much of the Yilgarn craton, Australia. One possible explanation for such ductile behaviour of the upper and middle crust, resulting in distributed deformation over hundreds of kilometers across strike, is that the lithosphere in those regions was softened prior to plate collision—for instance, by mantle heat input (e.g., from mantle plumes) [*Choukroune et al.*, 1995].

If a strength profile resembling the one in Figure 4B were the result of an anomalously high mantle heat addition to the crust, then explanation of the apparent rigidity of the lower crust through which the added heat would be conducted, compared to the softer middle crust underlying the Abitibi, would be required. Whether heat input from a mantle plume, perhaps much greater than the mantle heat flow used in the geotherm calculations in this paper, could result in significant softening of the middle granitic crust while a lower, quartz-poor granulitic crust remained relatively stiff would need further modeling to obtain an answer. An alternative, or possibly complementary, explanation offered here calls for a soft middle and upper crust due to a combination of radiogenic heat concentration in the middle crust and syntectonic magmatism. Analytical simulations of the mechanical response of the lithosphere to variations in the distribution of heat-producing

elements (HPE) suggest that lithospheric strength profiles are strongly influenced by the distribution of HPE in the crust and that the distribution of HPE evolves with both magmatic and tectonic activity [*Sandiford and McLaren*, 2002]. Burial of radiogenic, granitoid-rich middle crust below several kilometers of overlying volcanic rocks can lead to important increases in middle to deep crustal temperatures (hundreds of degrees), resulting in decreases in the effective viscosities by an order of magnitude or more [*Sandiford et al.*, 2004]. In the presence of tectonic stresses, such crustal softening could also lead to high rates of distributed ductile strain in the softer part of the crust compared to the more rigid parts [*Sandiford and McLaren*, 2002; *Bodorkos and Sandiford*, this volume].

The precollisional geological history of the Abitibi Subprovince is summarized as approximately 50 Ma of more or less continuous volcanism, followed by 20 Ma of clastic sedimentation [*Ayer et al.*, 2002a]. Seismic velocity models of the middle crust of the Abitibi Subprovince suggest it is composed largely of granitic material, which is consistent with the interpretation presented in this paper that the Opatica belt is representative of the Abitibi middle crust. If the middle crust of the Abitibi Subprovince was also built up during 50 Ma of plutonism and is composed in large part of tonalite and granodiorite plutons, then "conductive incubation" [*Sandiford et al.*, 2004] of the middle crust could lead to its softening and perhaps to softening of the base of the upper crust as well.

Such internal (to the crust) radiogenic heat production and temperature increases may have led to softening of the middle to upper crust of the Abitibi–Opatica plate that allowed it to deform mainly in a ductile fashion, by folding, for 250 km inboard of the Abitibi–Pontiac terrane suture. Ongoing tonalite–granodiorite plutonism during orogeny and folding [*Benn*, 2005] could certainly have contributed to the softening of the middle and upper crust as well [*Jackson and Cruden*, 1995]. If this interpretation is correct, it might also be considered as a potential explanation of large-amplitude, large-wavelength folding of other Archean granite–greenstone belts.

6. CONCLUSIONS

A new interpretation of the tectonic terranes and crustal structures of the SE Superior Province has been presented that is based on a synthesis of geological, geochronological, and geophysical data. The study includes the Opatica belt, the Abitibi Subprovince, and the Pontiac Subprovince, a region that represents approximately 2.4×10^5 km^2 of Late Archean terranes composed of greenstones and high (amphibolite)-grade metasedimentary and metaplutonic rocks.

As a result of this synthesis and a reinterpretation of Lithoprobe deep seismic reflection profiles, it is suggested that the Opatica belt and the Abitibi Subprovince, which were previously interpreted to represent distinct tectonic terranes, are in fact one contiguous terrane. The Opatica would represent amphibolite-grade middle crust similar to that which underlies the Abitibi greenstones. Collision with an older lithospheric plate to the south, represented by older crust underlying the Pontiac Subprovince, resulted in folding that affected the upper and middle crust of the Abitibi–Opatica plate for 250 km inboard of the Abitibi–Pontiac terrane suture. Older crust also lies to the north of the Opatica and is separated from it by a ductile shear zone.

During collision between the older, pre-3000 Ma crust that underlies the Pontiac Subprovince and the younger, 2750 to 2700 Ma crust of the southern Abitibi Subprovince, the lower crust of the Pontiac Subprovince was wedged into the Abitibi crust, resulting in delamination of the Abitibi lower crust. Whereas the upper and middle crust of the Abitibi–Opatica plate deformed in a highly distributed ductile fashion, the lower crust responded by delamination and localized thrusting.

The interpreted crustal structures imply a rheological profile for the Abitibi–Opatica plate that had a soft middle crust overlying a stronger lower crust. Strength profiles were calculated and presented that support the interpretation, given reasonable interpretations of lithospheric compositions and mechanical properties. The proposed strength profiles potentially also explain the structural style of in the Abitibi Subprovince, which involves extensive folding of the upper and middle crust.

The structural style of the Abitibi Subprovince is not unique in Archean granite–greenstone belts. The rheological profiles, and perhaps the crustal delamination model, proposed in this paper may be compatible with the tectonic evolutions of other Archean terranes, particularly Late Archean terranes.

Acknowledgments. The research was funded by a Natural Sciences and Engineering Research Council of Canada Discovery Grant awarded to the author. Comments on an earlier version of the manuscript by A.R. Cruden and J.-C. Mareschal were helpful in improving the work and are greatly appreciated.

REFERENCES

Afonso, J.C., and G. Ranalli (2004), Crustal and mantle strengths in continental lithosphere: is the jelly sandwich model obsolete? *Tectonophysics*, 394, 221-232.

Axen, G.J., *et al.* (1998), If the strong crust leads, will the weak crust follow? *GSA Today*, 8, 1-8.

Ayer, J., *et al.* (2002a), Evolution of the southern Abitibi greenstone belt based on U-Pb geochronology: autochthonous volcanic construction followed by plutonism, regional deformation and sedimentation, *Precam. Res.*, 115, 63-95.

Ayer, J.A., *et al.* (2002b), New geochronological and neodymium isotopic results from the Abitibi greenstone belt, with emphasis on the timing and the tectonic implications of Neoarchean sedimentation and volcanism, in *Summary of field work and other activities, Ontario Geological Survey, Open file report 6100*, edited, pp. 5.1-5.16.

Bellefleur, G., *et al.* (1998), Crustal geometry of the Abitibi Subprovince, in light of three-dimensional seismic reflector orientations, *Can. J. Earth Sci.*, 35, 569-582.

Benn, K. (2005), Late Archean Kenogamissi complex, Abitibi Subprovince, Ontario: doming, folding and deformation-assisted melt remobilization during syntectonic batholith emplacement, *Trans. Roy. Soc. Edinburgh: Earth Sci.*, 95, 297-307.

Benn, K., *et al.* (1992), Orogen parallel and transverse shearing in the Opatica belt, Quebec: implications for the structure of the Abitibi Subprovince, *Can. J. Earth Sci.*, 29, 2429-2444.

Benn, K., *et al.* (1994), Crustal structure and kinematic framework of the northwestern Pontiac Subprovince, Quebec: an integrated structural and geophysical study, *Can. J. Earth Sci.*, 31, 271-281.

Benn, K., and A.P. Peschler (2005), A detachment fold model for fault zones in the Late Archean Abitibi greenstone belt, Ontario, *Tectonophysics*, 400, 85-104.

Bleeker, W., and R.R. Parrish (1996), Stratigraphy and U-Pb zircon geochronology of Kidd Creek: implications for the formation of giant volcanogenic massive sulphide deposits and the tectonics history of the Abitibi greenstone belt, *Can. J. Earth Sci.*, 33, 1213-1231.

Calvert, A.J., *et al.* (1995), Archean subduction inferred from seismic images of a mantle suture in the Superior Province, *Nature*, 375, 670-674.

Calvert, A.J., and J.N. Ludden (1999), Archean continental assembly in the southeastern Superior Province of Canada, *Tectonics*, 18, 412-429.

Card, K.D., and K.H. Poulsen (1998), Geology and mineral deposits of the Superior Province of the Canadian Shield, in *Geology of the Precambrian Superior and Grenville Provinces and Precambrian fossils in North America*, edited by S. B. Lucas and M. R. St.-Onge, pp. 13-204, Geological Survey of Canada, Ottawa.

Carignan, J., *et al.* (1993), Pb isotopic geochemistry of granitoids and gneisses from the late Archean Pontiac and Abitibi subprovinces of Canada, *Chem. Geol.*, 106, 299-316.

Cheng, L.Z., *et al.* (2002), Simultaneous inversion of gravity and heat flow data: constraints on thermal regime, rheology and evolution of the Canadian Shield crust, *J. Geodynamics*, 34, 11-30.

Choukroune, P., *et al.* (1995), Soft lithosphere during periods of Archean crustal growth or crustal reworking, in *Early Precambrian Processes*, edited by M. P. Coward and A. C. Ries, pp. 67-86, Geological Society of London, London.

Chown, E.H., *et al.* (1992), Tectonic evolution of the Northern Volcanic Zone, Abitibi belt, Quebec, *Can. J. Earth Sci.*, 29, 2211-2225.

Chown, E.H., *et al.* (2002), The role of granitic intrusions in the evolution of the Abitibi belt, Canada, *Precam. Res.*, 115, 291-310.

Cook, F.A., *et al.* (1998), Tectonic delamination and subcrustal imbrication of the Precambrian lithosphere in northwestern Canada mapped by Lithoprobe, *Geology*, 26, 839-842.

Corfu, F., *et al.* (1991), U-Pb ages and tectonic significance of Late Archean alkalic magmatism and nonmarine sedimentation—Timiskaming Group, Southern Abitibi Belt, Ontario, *Can. J. Earth Sci.*, 28, 489-503.

Daigneault, R., *et al.* (1990), Tectonic evolution of the northeast portion of the Archean Abitibi Greenstone-Belt, Chibougamau Area, Quebec, *Can. J. Earth Sci.*, 27, 1714-1736.

Daigneault, R., *et al.* (2002), Oblique Archean subduction: accretion and exhumation of an oceanic arc during dextral transpression, Southern Volcanic Zone, Abitibi Subprovince Canada, *Precam. Res.*, 115, 261-290.

Davis, D.W. (2002), U-Pb geochronology of Archean metasedimentary rocks in the Pontiac and Abitibi subprovinces, Quebec, constraints on timing, provenance and regional tectonics, *Precam. Res.*, 115, 97-117.

Davis, W.J., *et al.* (1995), U-Pb geochronology of the Opatica tonalite-gneiss belt and its relationship to the Abitibi greenstone belt, Superior Province, Quebec, *Can. J. Earth Sci.*, 32, 113-127.

Davis, W.J., *et al.* (2000), Geochronology and radiogenic isotope geochemistry of plutonic rocks from the central Abitibi subprovince: significance to the internal subdivision and plutono-tectonic evolution of the Abitibi belt, *Can. J. Earth Sci.*, 37, 117-133.

Dewit, M.J. (1998), On Archean granites, greenstones, cratons and tectonics—does the evidence demand a verdict? *Precam. Res.*, 91, 181-226.

Dimroth, E., *et al.* (1983), Evolution of the south-central part of the Archean Abitibi Belt, Quebec. Part II: Tectonic evolution and geomechanical model, *Can. J. Earth Sci.*, 20, 1355-1373.

Easton, R.M. (2000), Metamorphism of the Canadian Shield, Ontario, Canada. I. The Superior Province, *Canadian Mineralogist*, 38, 287-317.

Ellis, S., and C. Beaumont (1999), Models of convergent boundary tectonics: implications for the interpretation of Lithoprobe data, *Can. J. Earth Sci.*, 36, 1711-1741.

Grandjean, G., *et al.* (1995), Crustal velocity models for the Archean Abitibi greenstone belt from seismic refraction data, *Can. J. Earth Sci.*, 32, 149-166.

Heather, K.B., and G.T. Shore (1999), Geology of the Swayze greenstone belt, Ontario. Open file 3384a, Geological Survey of Canada, Ottawa.

Hoffman, P.F., and G. Ranalli (1988), Archean flake tectonics, *Geophys. Res. Lett.*, 15, 1077-1080.

Hyde, R.S. (1980), Sedimentary facies in the Archean Timiskaming Group and their tectonic implications, Abitibi greenstone belt, northeastern Ontario, *Precam. Res.*, 12, 161-195.

Jackson, J. (2002), Strength of the continental lithosphere: time to abandon the jelly sandwich? *GSA Today*, 12, 4-10.

Jackson, S.L., and A.R. Cruden (1995), Formation of the Abitibi greenstone belt by arc-trench migration, *Geology*, 23, 471-474.

Jackson, S.L., *et al.* (1995), A seismic-reflection-based regional cross section of the southern Abitibi greenstone belt, *Can. J. Earth Sci.*, 32, 135-148.

Jaupart, C., and J.C. Mareschal (1999), The thermal structure and thickness of continental roots, *Lithos*, 48, 93-114.

Kellett, R.L., *et al.* (1992), A model for lower crustal electrical anisotropy for the Pontiac Subprovince of the Canadian Shield, *Geophys. J. Int.*, 111, 141-150.

Kerrich, R., and R. Feng (1992), Archean geodynamics and the Abitibi-Pontiac collision: implications for advection of fluids at transpressive collisional boundaries and the origin of giant quartz vein systems, *Earth-Sci. Rev.*, 32, 33-60.

Kerrich, R., *et al.* (1998), Boninite series: low Ti-tholeiite associations from the 2.7 Ga Abitibi greenstone belt, *Earth Planet. Sci. Lett.*, 164, 303-316.

Lacroix, S., and E.W. Sawyer (1995), An Archean fold-thrust belt in the northwestern Abitibi Greenstone Belt: structural and seismic evidence, *Can. J. Earth Sci.*, 32, 97-112.

Lafleche, M.R., *et al.* (1992), Geochemistry and petrogenesis of Archean mafic volcanic-rocks of the southern Abitibi belt, Quebec, *Precam. Res.*, 57, 207-241.

Lenardic, A., and W.M. Kaula (1995), Mantle dynamics and the heat-flow into the Earth's continents, *Nature*, 378, 709-711.

Ludden, J., and A. Hynes (2000), The Lithoprobe Abitibi-Grenville transect: two billion years of crust formation and recycling in the Precambrian Shield of Canada, *Can. J. Earth Sci.*, 37, 459-476.

Machado, N., *et al.* (1991), Géochronologie U-Pb du territoire québecois: fosses du Labrador et de l'Ungava et sous-Province du Pontiac. *Ministére de l'Énergie et des Ressources du Québec, Rapport interimaire MB 91-07.*

Machado, N., *et al.* (1993), Géochronologie U-Pb du territoire québecois: fosses de l'Ungava et du Labrador, Province de Grenville et sous-pronvices de Pontiac et de l'Abitibi, *Ministère de l'Énergie et des ressources du Québec, Rapport intermimaire GM 59899.*

Mareschal, J.C., *et al.* (2000), Heat flow and deep thermal structure near the southeastern edge of the Canadian Shield, *Can. J. Earth Sci.*, 37, 399-414.

Mortensen, J.K. (1987), U-Pb zircon ages for volcanic and plutonic rocks of the Noranda-Lake ABitibi area, Abitibi Subprovince, Quebec., in *Current Research, part A, Geological Survey of Canada paper 87-1A*, pp. 581-590.

Mortensen, J.K., and K.D. Card (1993), U-Pb age constraints for the magmatic and tectonic evolution of the Pontiac Subprovince, Quebec, *Can. J. Earth Sci.*, 30, 1970-1980.

Mortensen, J.K. (1993a), U-Pb geochronology of the eastern Abitibi Subprovince. Part 2: Noranda-Kirkland Lake area, *Can. J. Earth Sci.*, 30, 29-41.

Mortensen, J.K. (1993b), U-Pb geochronology of the Lapparent Massif, northeastern Abitibi Belt—basement or synvolcanic pluton, *Can. J. Earth Sci.*, 30, 42-47.

Mueller, W.U., *et al.* (1996), Archean terrane docking: upper crust collision tectonics, Abitibi greenstone belt, Quebec, Canada, *Tectonophysics*, 265, 127-150.

Nisbet, E.G., and C.M.R. Fowler (1983), Model for Archean plate tectonics, *Geology*, 11, 376-379.

Peschler, A.P., *et al.* (2004), Insights on Archean continental geodynamics from gravity modelling of granite-greenstone terranes, *Journal of Geodynamics*, 38, 185-207.

Pinet, C., *et al.* (1991), Heat-flow and structure of the lithosphere in the eastern Canadian Shield, *J. Geophys. Res.-Solid Earth*, 96, 19941-19963.

Polat, A., and R. Kerrich (2001), Geodynamic processes, continental growth, and mantle evolution recorded in late Archean greenstone belts of the southern Superior Province, Canada, *Precam. Res.*, 112, 5-25.

Powell, W.G., *et al.* (1995), Conditions and timing of metamorphism in the southern Abitibi greenstone belt, Quebec, *Can. J. Earth Sci.*, 32, 787-805.

Ranalli, G. (1997), Rheology of the lithosphere in space and time, in *Orogeny through Time*, edited by J.-P. Burg, and M. Ford, pp. 19-37, Geological Society of London Special Publication No. 121.

Ranalli, G., and D.C. Murphy (1987), Rheological stratification of the lithosphere, *Tectonophysics*, 132, 281-295.

Robert, F. (1989), Internal structure of the Cadillac tectonic zone southeast of Val d'Or, Abitibi greenstone belt, Quebec, *Can. J. Earth Sci.*, 26, 2661-2675.

Rudnick, R.L. (1992), Xenoliths—samples of the lower continental crust, in *Continental Lower Crust*, edited by D. M. Fountain, *et al.*, pp. 269-316, Elsevier, Amsterdam.

Rudnick, R.L., and D.M. Fountain (1995), Nature and composition of the continental-crust—a lower crustal perspective, *Rev. Geophys.*, 33, 267-309.

Sandiford, M., and S. McLaren (2002), Tectonic feedback and the ordering of heat producing elements within the continental lithosphere, *Earth Planet. Sci. Lett.*, 204, 133-150.

Sandiford, M., *et al.* (2004), Conductive incubation and the origin of dome-and-keel structure in Archean granite-greenstone terrains: a model based on the eastern Pilbara Craton, Western Australia, *Tectonics*, 23, doi:10.1029/2002TC001452.

Sawyer, E.W. (1998), Formation and evolution of granite magmas during crustal reworking—the significance of diatexites, *J. Petrol.*, 39, 1147-1167.

Sawyer, E.W., and K. Benn (1993), Structure of the high-grade Opatica Belt and adjacent low-grade Abitibi Subprovince, Canada: an Archean mountain front, *J. Struct. Geol.*, 15, 1443-1458.

Scott, C.R., *et al.* (2002), Physical volcanology, stratigraphy, and lithogeochemistry of an Archean volcanic arc: evolution from plume-related volcanism to arc rifting of SE Abitibi Greenstone Belt, Val d'Or, Canada, *Precam. Res.*, 115, 223-260.

Sénéchal, G., *et al.* (1996), Integrated geophysical interpretation of crustal structures in the northern Abitibi belt: constraints from seismic amplitude analysis, *Can. J. Earth Sci.*, 33, 1343-1362.

Snyder, D.B. (2002), Lithospheric growth at margins of cratons, *Tectonophysics*, 355, 7-22.

Sproule, R.A., *et al.* (2002), Spatial and temporal variations in the geochemistry of komatiitic basalts in the Abitibi greenstone belt, *Precam. Res.*, 115, 153-186.

Sutcliffe, R.H., *et al.* (1993), Plutonism in the southern Abitibi Subprovince: a tectonic and petrogenetic framework, *Economic Geology*, 88, 1359-1375.

Telmat, H., *et al.* (2000), Crustal models of the eastern Superior Province, Quebec, derived from new gravity data, *Can. J. Earth Sci.*, 37, 385-397.

Wilkinson, L., *et al.* (1999), Timing and kinematics of post-Timiskaming deformation within the Larder Lake-Cadillac deformation zone, southwest Abitibi greenstone belt, Ontario, Canada, *Can. J. Earth Sci.*, 36, 627-647.

Wyman, D.A. (1999), A 2.7 Ga depleted tholeiite suite: evidence of plume-arc interaction in the Abitibi Greenstone Belt, Canada, *Precam. Res.*, 97, 27-42.

K. Benn, Ottawa-Carleton Geosciences Centre and Department of Earth Sciences, University of Ottawa, Ottawa, Ontario K1N 6N5, Canada. (kbenn@uottawa.ca)

The Early Record of Life

Frances Westall

Centre de Biophysique Moléculaire, CNRS, Orléans, France

Gordon Southam

Department of Earth Sciences, University of Western Ontario, London, Ontario, Canada

Over the last 3.5 Gyr, the biosphere has had a profound influence on the atmosphere, hydrosphere, and lithosphere, at or near the Earth's surface. However, the conditions when life originated were very different from those of the present-day planet. At that period the physical and chemical conditions of the early Earth "controlled" the biosphere. Global temperatures may have been higher; the oxygen and carbon dioxide partial pressures were lower and higher, respectively; the amount of UV radiation striking the surface would have been higher; and the oceans were most likely more acidic. However, as soon as life began, it started to exert its influence, reducing the entropy of the environments it inhabited. Structural and chemical fossils found in Early Archean habitats demonstrate that the biosphere was already in an advanced evolutionary state since much of the strata preserved from this period appears to have been colonized by morphologically and biochemically diverse bacteria.

INTRODUCTION

The search for traces of early life in the oldest rocks on Earth is unfortunately limited by the fact that well-preserved ancient terrains are few and far between. This is because plate tectonics has eliminated all but a few vestiges of the ancient crust. Of those remaining, most are heavily metamorphosed and therefore the structure and chemistry of potential biosignatures is altered. The terrains containing the oldest well-preserved sediments occur in the Early-Mid Archean (3.5–3.3 Ga) Greenstone Belts of Barberton in East South Africa and the Pilbara in Northwest Australia. Investigations for traces of life in these rocks are historically based on previous studies that commenced in the 1950s with the discovery of coccoid and filamentous microfossils in thin sections of stromatolites from the 1.9-Gyr-old Gunflint Formation,

Ontario [*Tyler and Barghoorn*, 1954]. These chert-embedded carbonaceous microfossils physically resembled modern cyanobacteria, suggesting that they were the remnants of these oxygenic photosynthetic organisms. The early studies were, however, strongly influenced by the methods of the previous investigations (light-microscope investigations of thin sections), as well as the presumed cyanobacterial nature of the first microfossils found. This methodological approach and the microbiological assumptions are the cause of much of the recent debate on the interpretation of biosignatures in the most ancient sediments [*Brasier et al.*, 2001; *Westall*, 2005], since the vast majority of bacteria[1] are very small, on the order of 1 μm, and are thus individually below the resolution of light microscopes. Other, more powerful, techniques are necessary to observe their fossilized forms.

Below we give a brief historical outline of the main controversies to highlight some of the problems involved in this

---

[1] The term "bacteria" is used here in a generic sense; i.e., it is synonymous with the term "prokaryotes" and indicates microorganisms from both the Bacteria and the Archaea Domains.

Archean Geodynamics and Environments
Geophysical Monograph Series 164
Copyright 2006 by the American Geophysical Union
10.1029/164GM18

type of research, followed by a discussion of biosignatures and biogenicity. We then present the record of biosignatures in the ancient rocks and discuss their significance and the implications for the diversity, distribution, and early evolution of life.

BRIEF HISTORICAL REVIEW AND CONTROVERSIES

Southwest Greenland (Isua and Akilia)

The oldest known sediments occur in small enclaves of metamorphosed rocks (amphibolite to granulite facies) in the Isua and Akilia Greenstone Belts in southwest Greenland [3.8 Ga; *McGregor and Mason*, 1977]. Microfossil investigations based on light-microscope observation of polished thin sections and scanning/transmission electron microscopy of the carbonaceous residues of acid-dissolved samples began in the late 1970s [*Pflug*, 1979, 2001; *Pflug and Jaeschke-Boyer*, 1979; *Robbins*, 1987, *Robbins et al.*, 1987; *Robbins and Iberall*, 1991]. These studies produced descriptions of a variety of microfossils from banded iron formation (BIF) rocks, including "yeasts" [*Isuasphera isua pflug*; *Pflug and Jaeschke-Boyer*, 1979] and microbial colonies [*Apellela ferrifera*; *Robbins*, 1987]. Later investigations of the carbon isotope ratios of graphite inclusions from interpreted Isua and Alkilia metasediments range from −11‰ to −49‰, suggesting that methanogenic and oxygenic photosynthetic microorganisms abounded in the early oceans [*Schidlowski*, 1988, 2001; *Mojzsis et al.*, 1996; *Rosing*; 1999, *Rosing and Frei*, 2004; *Ueno et al.*, 2002]. In addition, $\delta^{15}N$ values ranging from −1.7‰ to +5.9‰ were also obtained from the Isua rocks by *Pinti et al.* [2001] and *Papineau et al.* [2005], who tentatively linked the isotope ratios to biogenic activity.

Problems with the interpretations of these biosignatures started to arise when the origins of the "metasediments" and the graphite and its associated $\delta^{13}C$ signature were questioned [*Myers*, 2001; *Fedo and Whitehouse*, 2002; *Lepland et al.*, 2002; *Zuilen et al.*, 2002, 2003; *Schoenberg et al.*, 2002]. However, Fe isotopes have been used to support the sedimentary origin of the strata in question [*Dauphas et al.*, 2004]. As for the graphite particles, their physical occurrence in the sediments of Akilia has been recently contested [*Lepland et al.*, 2005], whereas those in the Isua sediments could have an abiogenic origin. *Van Zuilen et al.* [2002, 2003] proposed an hydrothermal origin for both the formation of the graphite and their isotopic signature; *Schoenberg et al.* [2002] suggested an extraterrestrial origin for the carbon (the carbon isotope ratios of a carbonaceous chondrite, such as Murchison, range from −5‰ to −20‰ $\delta^{13}C$ [*Sephton et al.*, 2003]). A further complication in the search for biosignatures is that these rocks have been recently contaminated by endolithic fungi and cyanobacteria [*Hayes et al.*, 1983;

van Zuilen et al., 2002; *Westall and Folk*, 2003]. *Bridgewater et al.* [1981] also noted that some of the microfossil-like objects in the rocks were purely abiogenic precipitations. Finally, reviewing the nitrogen isotope data for the Isua rocks, *van Zuilen et al.* [2005] conclude that the data were not a sufficiently robust biosignature in rocks of this age.

The Barberton and Pilbara Greenstone Belts

In contrast to the Isua and Akilia greenstone belts in Southwest Greenland, large areas of the Early-Mid Archean terrains in Australia and South Africa have undergone only low-grade metamorphism (uppermost prehnite–pumpellyite to lowermost greenschist-grade) and thus, many volcanic and sedimentary sequences are superbly preserved. Following the work of *Tyler and Barghoorn* [1954], investigations were made of macroscopic, stromatolite-like structures [*Walter*, 1976, 1983; *Byerly et al.*, 1986; *Hofmann et al.*, 1999] and, using light-microscope examination of thin sections, of presumptive microfossils and microbial mats [*Schopf and Walter*, 1983; *Walsh and Lowe*, 1985; *Schopf and Packer*, 1987; *Schopf*, 1993, 1998; *Walsh*, 1992, 2004; *Ueno et al.*, 2001a,b; *Schopf et al.*, 2002; *Tice and Lowe*, 2004]. The biogenic origin of the stromatolite structures, and of the spheroidal and filamentous "microfossils", was questioned by *Lowe* [1994] and *Buick* [1990], respectively. More recently a controversy has arisen concerning the interpretation of certain carbonaceous filaments from one location in the Pilbara—the Apex Chert from the Towers Formation [*Brasier et al.*, 2002]. Whereas *Schopf* [1993] had identified the filaments as fossil cyanobacteria, i.e., oxygenic photosynthetic bacteria, new field studies and optical microscope observations have shown that the original sample was obtained from a hydrothermal vein and that many of the filamentous structures formed part of larger, more diffuse carbonaceous structures, which have no morphological resemblance to fossil cyanobacteria. *Brasier et al.* [2002] therefore concluded that the microstructures were not fossilized microorganisms and that the carbonaceous material, with its −16‰ to −26‰ $\delta^{13}C$ signature, was probably abiogenic in origin, having been formed by Fischer–Tropsch processes in a hydrothermal setting. An alternative hypothesis was put forward by *Westall* [2003, 2004] on the basis of her investigations of similarly aged Early Archean sediments containing probable microfossils from the Kitty's Gap Chert in the nearby Coppin Gap greenstone belt. She suggested that hydrothermal fluids traversing sediment horizons that contain fossilized microbial remains could have entrained some of the carbonaceous matter, subsequently depositing it higher up the vein. Penecontemporaneous chert dyke formation within the Early-Mid Archean strata had previously been suggested by *Lowe and Byerly* [1986b]. This alternative explanation presupposes that the microbial remains are present in the sediments adjacent to the vein [see discussion below and *Westall et al.*, 2005].

This brief introduction to some of the controversies surrounding the search for, and the interpretation of, biosignatures in the most ancient rocks on Earth underlines the necessity of a multiscale and multidisciplinary approach [*Westall*, 2005]: Large-scale geologic context information is necessary for understanding the nature and distribution of potential habitats, whereas microscale understanding of the nature of microorganisms and their interactions with their microenvironment and their fossilization are of equal importance. Finally, knowledge of the subsequent diagenetic and metamorphic changes to the sedimentary deposits is essential for reaching a probability of biogenicity of potentially biologic structures.

Bacterial Signatures in the Geologic Record

When searching for bacterial signatures in the geologic record, the spatial and temporal perspective of the bacteria vs. the geology of their environment must be considered. These scales differ by several orders of magnitude. First, the small size of microorganisms (μm-scale) necessitates observation using high-resolution microscopes in order to study any fine structure. Beyond an individual bacterium, cell–cell interactions, i.e., the number of planes of cell division, becomes important. A bacterium possessing a single plane of cell division will produce chains of cells or a filament; two planes of cell division will produce "sheets" of cells; and more than two planes of cell division will produce clusters of cells. The resulting bacterial colonies, tens of micrometers in diameter and more, can occur individually or in association with other microcolonies within biofilms on sediment surfaces. However, where conditions are stable for a duration of time long enough for their development or where nutrients and favourable redox conditions exist for rapid growth, these "biofilms" can reach centimetres or even meters in plan view and millimetres or more in thickness, forming significant microbial mats [*Krumbein et al.*, 1994; *Stahl*, 1994]. Under Early-Mid Archean conditions, where grazing by higher organisms (a selective pressure against mat formation today) did not occur, the size limit for mat formation would have matched the geological conditions suitable for growth.

Second, the time it takes for a cell to divide can be very short, on the order of hours to days (depending on environment and nutritional conditions). For example, biofilms can form rapidly on sediment surfaces in mud flat environments between tides [*Stahl*, 1994], and although the division times of the individual microorganisms may be rapid, large-scale microbial biofilms or mats can persist for many years, e.g., the meter-high stromatolites in Shark Bay, Northwest Australia, which are about 3000 years old [*Burne*, 1991].

With times scales potentially as short as hours or days, how can such ephemeral phenomena be preserved in the rock record? Given the right conditions, i.e., sufficient mineral ions in the aqueous medium, microorganisms and their biofilms or mats can be rapidly fossilized, e.g., in hot spring environments [*Cady and Farmer*, 1996; *Phoenix et al.*, 2000; *Konhauser et al.*, 2003]. A number of experimental simulations of the fossilization process have demonstrated that fossilization can start within hours [*Ferris et al.*,1986; *Westall et al.*, 1995; *Toporski et al.*, 2002; *Benning et al.*, 2002]. Fossilization takes place either (1) by the complexation of mineral ions to functional groups in the organic material, gradually replacing the organic structure and at the same time trapping the degrading organic molecules in the polymerising mineral matrix, or (2) when amorphous minerals formed in the aqueous phase bind to the cell envelope [*Konhauser et al.*, 2003; *Rancourt et al.*, 2005]. The organic matter in the fossilized microorganisms is degraded, but parts of it may remain trapped in the mineral matrix. Alternatively, in modern, oxidising hydrothermal environments, the organic matter can be completely oxidised, although the microbial cast still remains [e.g., *Cady and Farmer*, 1996].

In natural systems, microorganisms typically produce copious quantities of extracellular polymeric substances (EPS) that serve many purposes, including protection of the colonies from the external environment and providing a unique microenvironment or habitable niche in which the organisms can control the redox and pH conditions [*Charaklis and Wilderer*, 1989; *Westall et al.*, 2000]. EPS can also be fossilized; in fact, they are more easily preserved than the microorganisms. Thus, a fossilized microbial colony, biofilm. or mat will include not only the organisms themselves but also EPS that may be only loosely associated with the cells (i.e., bacterial slime). As well as linking individual microbial cells and colonies, EPS-rich biofilms can also play an important role in the stabilization of sediment surfaces by cementing mineral particles together, [*Krumbein et al.*, 1994], thus prolonging the stability of a particular habitat.

Microbial interactions occur on the molecular to global scale [*Newman and Banfield*, 2002]. Microorganisms such as chemoautolithotrophs obtain their carbon and energy from mineral sources. Enzymes released during their metabolic processes are thus implicated in the corrosion and early diagenetic alteration of minerals. Likewise, microorganisms can either directly or indirectly precipitate minerals, such as minute magnetite crystals in the cells of magnetotactic bacteria [*Bazylinski*, 1996], or carbonate as a by-product of their metabolic processes (e.g., by oxygenic photosynthetic cyanobacteria [*Chekroun et al.*, 2004]). They can also stock elements, such as sulphur, in their cells [*Madigan et al.*, 2000]. The meter-scale stromatolites, so characteristic of the Proterozoic era [*Awramik and Sprinkle*, 1999], are a product of particle trapping and binding together with carbonate precipitation by oxygenic photosynthesising bacterial mats. Microbial activity probably also had an important influence of the formation of banded iron deposits [*Konhauser et al.*, 2002]. Finally, on a global scale, the presence of oxygen in the atmosphere has been attributed, at least partly, to the activity of these oxygenic photosynthetic organisms [*Walker*, 1977; *Holland*, 1984, 1994].

Establishing Biogenicity

The "thorniest" issue in early life studies is the establishment of syndepositional biogenicity. Although compatibility of the environment of deposition with the existence of living organisms is an important prerequisite, a decade of work on extremophilic microorganisms has shown that very few environments, such as hot lavas and hot hydrothermal systems, cannot be colonised by microbes [*Rothschild and Mancinelli*, 2001]. The temperature limit for life appears to be around 121 °C [*Kashefi and Lovley*, 2003]. With some shielding from solar UV radiation (for example, either by clay or an outer layer of dead cells), microorganisms can survive up to 6 years in space conditions (i.e., vacuum, no water, UV radiation, and sharp temperature excursions [*Horneck et al.*, 2002]).

Yet another point of discussion in the early life debate centers on the methodology used in observing the microfossil remains. As noted above, the traditional method was based on optical microscope study of thin sections. However, a major problem with optical microscope observation is the question of scale. Most microorganisms are of the order of 1 μm in size and, under the low nutrient conditions encountered in many natural environments (vs. nutrient-rich laboratory culture conditions), they can be much smaller, even down to ~150 nm [*Madigan et al.*, 2000]. The early Archean cherts consist of microcrystalline quartz in the size range of less than 1 μm to a few microns. Therefore, any eventual microfossils of the same size would be indistinguishable from the microcrystalline quartz in thin section. Such a situation has been revealed in both a modern hot spring environment, which showed that the silicified microorganisms in a siliceous matrix were in fact invisible by light microscopy [*Handley*, 2004], and in early Archean cherts from the Pilbara, where microfossils <1 μm were indistinguishable from microcrystalline quartz in the same size range [*Westall et al.*, 2005]. On the other hand, cyanobacteria can be an order of magnitude larger than other prokaryotes and their fossilized remains are more readily visible in optical microscopy. Observational methods that possess a higher spatial resolution, e.g., scanning electron microscopy (SEM), have been used to observe microbial structures liberated from their quartz matrix in the ancient cherts by etching with hydrofluoric (HF) acid [*Westall and Gerneke*, 1998; *Westall*

et al., 2001, 2005; *Westall*, 2003, 2004]. However, this technique has been criticised on two grounds: the possibility of the production of artefacts (i.e., bacteriomorph structures or biomimics) [*Schopf and Walter*, 1983; *Cady et al.*, 2003], and the difficulties in ascertaining that the microstructures were really formed in situ in the rock.

To date, there has been no single observation or measurement that, on its own, is proof of biogenicity. *Cady et al.* [2004], *Westall et al.* [2004, 2005], and *Westall* [2005] advocate an integrative approach to the identification of past traces of life based on many lines of evidence. These criteria vary depending on the species of microorganism; for example, criteria used for investigating typical prokaryote microorganisms are different in many respects from those used for studying the well-known group of prokaryotes, the cyanobacteria, which have unique morphological characteristics, such as size, shape, cell structure and differentiation, and biogeochemical (organic and inorganic) signatures, such as different organic macromolecules associated with sheaths and membranes.

The most direct evidence of past life is the existence of the structural remains of the microorganism, i.e., the individual cell, colonies of microorganisms, microbial biofilms, or mats. Structural features include:

(1) Morphological phenomena such as, size, shape, evidence for life processes as in cell division, evidence for death as in cell collapse, and evidence of flexibility for the filamentous structures.
(2) Colony and biofilm/mat characteristics, such as association with a number of other organisms of the same species, association with different species (a consortium), association with EPS, and formation of biofilms and mats.

One of the main problems with structural characteristics is the potential for confusion with abiogenic organic and inorganic bacteriomorphs [e.g., *Yushkin et al.*, 2000; *Garcia-Ruiz et al.*, 2003]. However, no bacteriomorphs have yet been identified that exhibit the *totality* of the structural characteristics of microorganisms in colonies, biofilms or mats.

Evidence of biogeochemical aspects of the life processes of microorganisms is also a critical aspect in the evaluation of the biogenicity of bacterial fossils. They include:

(1) The macromolecular composition of the cell envelopes of microorganisms and their EPS (or their degradation products). Organic macromolecules are potentially very strong biomarkers because they can be specific to certain components of certain species. However, this type of analysis is fraught with contamination issues [*Summons et al.*, 1999; *Brocks et al.*, 1999].
(2) The association of important bioelements, such as N, S, and P with potential microorganisms; however, labile organic material is rapidly lost or recycled during aging,

and these elements may not survive several billions of years of alteration through diagenesis and metamorphism.

(3) The association of heavy metals and rare earth elements (REEs) with microbes and their biofilms. Organic matter readily complexes toxic heavy metals [*Brantley et al.,* 2001], which can be preserved in fossilized microbial structures.

(4) The fractionation by microbial processes of the isotopes of certain elements, such as C, S, N, and Fe [*Lyons et al.,* 2003]. As noted above, distinguishing between a biogenic isotope ratio and that produced by abiogenic processes is not always self-evident [*van Zuilen et al.,* 2002, 2003].

Evaluation of the biogenicity of a potentially biogenic feature needs to be based on a suite of structural and chemical signatures, and the evidence for past life should be best expressed in terms of probability. The data for the following analysis of the implications of the evidence relating to early life (Table 1) was obtained by various methods, including macroscopic observation of stromatolite-like structures, thin section, and SEM observation of bacteriomorph structures; C, N, S, and Fe isotopes analyses; and continuous scanning electron ionization mass spectroscopy and Raman spectroscopic analysis of carbonaceous compounds. Many of these studies have concentrated on only one or a few chemical or morphological characteristics. Individually, the relevance of the data may be debated, as noted above, because other, non-biogenic processes can produce similar signatures. Despite this, when all the data are compared (Table 1), the individual analyses appear to be consistent with each other, thus allowing evaluation of the level of evolution of life and its distribution in the Early-Mid Archean period. However, before describing the evidence for early life, a brief resumé of the global environmental conditions is necessary in order to place the microenvironments and their life forms into context.

TRACES OF EARLY-MID ARCHEAN LIFE

The Early-Mid Archean Environment

Thurston and Ayres [2004] note a variety of geological settings recorded in Archean greenstone belts, including shallow water volcanic platforms with sediments, shallow and deep water basaltic provinces, shallow water to emergent arc sequences with bimodal volcanics, and subaerial pull-apart basins. These are all geological settings that occur on Earth today. The sediments were typically of volcanic or chemical origin. Quartzitic sediments, indicating preservation of eroded granitic continental crust, did not appear or were not preserved until after 3.2 Ga [*Erikssen et al.,* 1994;

Heubeck and Lowe, 1999]. As a result of the higher heat flow and greater volcanic activity on the early Earth, flushing of the upper crust by circulating hydrothermal fluids was an important process that altered the volcanic and sedimentary deposits, transporting certain elements and precipitating others [*Paris et al.,* 1985; *Duchač and Hanor,* 1987; *de Wit and Hart,* 1993; *Orberger et al.,* 2005]. Contemporaneous, post-depositional, and hydrothermal silicification episodes affected both the volcanic rocks and the sediments in the Barberton and the Pilbara terrains. Furthermore, many of the chemical sediments are, at least partly, of hydrothermal origin, e.g., some cherts and bedded barite deposits [e.g., *Paris et al.,* 1985; *Nijman et al.,* 1998; *van Kranendonk et al.,* 2004].

In other respects, however, the Earth presented a very different habitat compared with the modern planet (see reviews in *Nisbet and Sleep* [2001]; *Nisbet and Fowler* [1999, 2004]; *Westall* [2003, 2004]). The hotter mantle meant that there was a greater amount of volcanic and hydrothermal activity, as well as faster recycling of the crust [*Arndt,* 1994]. The atmosphere was composed of mostly CO_2, together with CO, NH_3, N_2, and probably a certain amount of CH_4 (the latter being possibly at least partly biological [*Pavlov et al.,* 2001]). Oxygen levels were low (<0.2% of present atmospheric levels [*Kasting,* 1993]), the oxygen being produced by photolysis of water vapour in the upper atmosphere. Given the high CO_2 concentrations in the atmosphere, the oceans were probably slightly acidic [*Grotzinger and Kasting,* 1993]. These acidic conditions, combined with the rapid cycling of hydrothermal fluids, would have resulted in extensive mineral weathering. Higher salinity [*Ronde and Ebbessen,* 1996] and possibly higher temperatures (50–80 °C according to *Knauth and Lowe* [2003]) were also characteristic of the Early-Mid Archean oceans. The lack of significant oxygen in the atmosphere meant that UV radiation levels were high [*Cockell,* 2000]. However, the deleterious effects of UV radiation on the biosphere may have been mitigated by large amounts of dust and aerosols in the atmosphere from volcanic eruptions, meteoritic impacts [cf. *Lowe and Byerly,* 1986a; *Lowe et al.,* 2003; *Kyte et al.,* 2003], and possibly water vapour [*Lammer et al.,* 2005] or CH_4 smog [*Lovelock,* 1988].

However, from a microbiological point of view, it is the microenvironments in these different settings that are important, along with the existence of substances that can provide a source of carbon and other essential elements, such as H, N, O, P, and S, and a source of energy, which controlled the success (or not) of the early biosphere ("abundant", i.e., non-limiting, trace elements were presumably provided by active carbonic acid weathering). Microbes can obtain their energy from inorganic (lithotrophs) or organic (organotrophs) chemical compounds, or from light (phototrophs).

Table 1. Database of Biosignatures from Early to Mid-Archean Cherts.

Micro-environment	Sedimentary details	Location	Age (Ga)	Type of biomarker	Type of analysis	Interpretation	Sources
Lithosphere							
Pore water/mineral surfaces	Within shallow water sediments; N.B.: environment influenced by hydrothermal fluids	Kitty's Gap Chert, Pilbara	3.446	Coccoid colonies on volcanic detritus C isotopes −22 to −30 ‰	SEM Geochemistry Isotopes	Probably methanogens	Westall et al., 2006; Westall, 2003, 2004
	Carbonate-altered pillow lava rinds	Various localities Onverwacht Group Barberton	3.472–3.333	Corrosion features, carbon, C isotopes, +3.9 to −16.4‰	Thin section Isotopes	Biogenic corrosion	Furnes et al., 2004
	Chert/BIF	Marble Bar, Pilbara Towers (Dresser) Formation	3.46	Bacteriomorph filaments C/N ratios 2.4 to 19.4	SEM Geochemistry	Microbial activity	Orberger et al., 2006
Hydrothermal veins	Chert veins	North Pole, Pilbara	3.46	Bacteriomorph filaments and kerogen clots C isotopes, −36‰	Thin section Isotopes, Raman	Chemoautotrophic microorganisms	Ueno et al., 2004
	VMS deposits	North Pole, Pilbara	3.235	Bacteriomorph pyrite filaments	Thin section	Thermophilic, S-utilizing chemotrophs	Rasmussen, 2000
		Fig Tree Group Barberton	3.230–3.259	Organic carbon compounds C isotopes −25 to −30.2‰	Continuous scanning electron ionization mass spectroscopy Isotopes	Microbial	de Ronde and Ebbessen, 1996
Lithosphere/Hydrosphere Interface							
Deep sea (below wave base)	Bedded chert barite units; N.B.: also interpreted as shallow water sediments	Dresser (Towers) Formation North Pole, Pilbara	3.46	Bacteriomorph filaments C isotopes, −40 to −31‰	Thin section Isotopes	Organotrophic microorganisms	Ueno et al., 2001a, b; Nijman et al., 1998
	Carbonaceous cherts	Kromberg, Mendon Formations, Fig Tree and Moodies Groups, Barberton	3.416–3.2	Carbonaceous sediments Filamentous mat, C isotopes −40.8 to −9.5‰ S isotopes −4 to +1‰	Thin section Isotopes	Resedimented shallow water photosynthetic mat; SRBs, methanogens	Tice and Lowe, 2004; Kakegawa, 2001

	Rock type	Location	Age	Structure/Description	Methods	Interpretation	References
Shallow water (above wave base/photic zone)	Bedded carbonate/ cherts	North Pole, Pilbara	3.45–3.47	Stromatolite-like structures	Macroscopic, thin section Geochemistry	Microbial biofilms	Walter et al., 1980; Lowe, 1980; Walter, 1983; Nijman et al., 1998; Westall et al., 2002; Westall, 2003, 2004; van Kranendonk et al., 2003; Allwood et al., 2005a,b
	Bedded barite/chert	North Pole, Pilbara	3.45–3.47	S isotopes −16 to −2‰	Isotopes	SRBs	Shen et al., 2001
		North Pole, Pilbara	3.45–3.47	C isotopes −32.7‰ N isotopes −4.4‰ Kerogen ultrastructure	Isotopes HR-TEM	Biogenic kerogen	Skrzypczak et al., 2004
	Carbonaceous cherts	Fig Tree Group Barberton	3.300	Stromatolite-like structures	Macroscopic observation	Microbilite	Byerly et al., 1986
	Carbonaceous cherts	Variety of sites Onverwacht Group Barberton	3.472–3.333	Thin filaments <0.2 to 2.5 μm diam. C isotopes −38.5 to −16.7‰	Thin section Isotopes Geochemistry	Possibly microbial Microbial mats, photosynthetic	Walsh, 1992, 2004; Walsh and Lowe, 1985, 1999
		Josefsdal Chert Onverwacht Barberton	3.472–3.333	Filamentous microbial mat, filaments 0.3 μm diam and colony of rods/vibroids 1 μm diam, 2–3.8 μm length C isotopes −22.7‰	Thin section HR-SEM Geochemistry, Isotopes	Evaporitic, anaerobic photosynthetic microbial mat and SRB colony	Westall, 2003, 2004; Westall et al., 2001, 2004
		Kitty's Gap Chert Panorama Formation Pilbara	3.446	Biofilm consortium or filaments (0.25 μm) Coccoids (0.5 and 0.8 μm) and rods (1 μm) C isotopes −30 to −26‰	Thin section HR-SEM Geochemistry Isotopes	Anaerobic photosynthetic microbial biofilm	Westall, 2003, 2004; Westall et al., 2004, 2006
Open Ocean Planktonic	Metaturbidites?	Isua, SW Greenland	>3.7	C isotopes −18‰ (range −26 to −10‰)	Isotopes	Aerobic photosynthesis	Rosing, 1999; Rosing and Frei, 2004
		Isua, SW Greenland		N isotopes −3 to −1‰	Isotopes	No clear signal	van Zuilen et al., 2005
	Shales	Fig Tree Formation Barberton	3.3	C isotopes (range −38.5 to −16.7‰)	Thin section Isotopes Geochemistry	Photosynthetic microorganisms	Walsh and Lowe, 1999

HR-TEM, high-resolution transmission electron spectroscopy.

Likewise, inorganic or organic compounds can be used as the source of carbon. One of the oldest microbial groups, the lithoautotrophs, obtains its energy and carbon from inorganic sources, making use of favourable redox contrasts, i.e., a negative reduction potential (ΔG_r), to support metabolism [*Nisbet and Sleep*, 2001]. The Early-Mid Archean environment would have presented many such possibilities for redox reactions in association with rocks and debris of basic to ultrabasic volcanic composition (with some rhyolites and dacites) having chemically reactive surfaces, as well as elements, gases, sulphate and nitrate produced volcanically, and in association with metal sulphide deposits. The lithoautotrophic microorganisms that extract their energy and carbon from these inorganic sources provide the first step in the food chain. Their by-products and organic "remains", released after cell death, can then be used by other organotrophic microorganisms, the by-products of which, in turn, can be used by other microorganisms with yet another type of metabolism. The metabolic diversity exhibited by prokaryotes is so large that they have inhabited nearly every environment on Earth that possesses liquid water [*Madigan et al.*, 2000; *Newman and Banfield*, 2002]. Potential microbial habitats in the Early-Mid Archean period are summarized in Figure 1.

From a bacterial perspective, microenvironments can be broadly grouped as follows [*Nisbet and Sleep*, 2001; *Westall*, 2003; *Nisbet and Fowler*, 1999, 2004], recognising however, that even within these groups there will be small-scale nuances: (1) subsurface sediment pore space and fractures in sedimentary or volcanic rocks, including the surfaces of individual particles or rocks (volcanic or sedimentary) and vadose zones in terrestrial environments; (2) hydrothermal environments (in different depths of water, including subaerial systems); (3) sediment surfaces in deep (subwave-base), shallow, or littoral (and possibly even subaerial) environments.

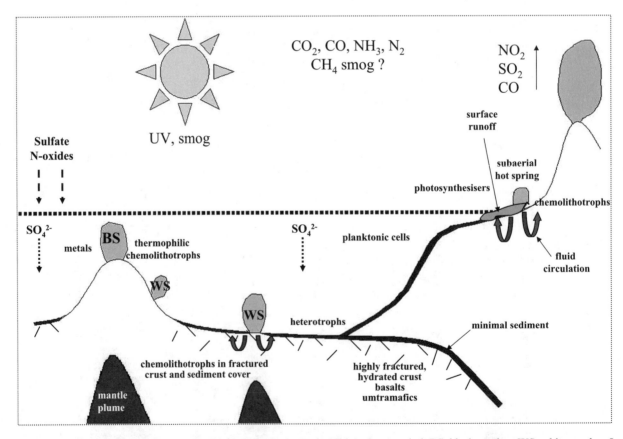

Figure 1. Sketch showing the potential microbial habitats in the Early-Mid Archean period. BS, black smoker; WS, white smoker. In the deep sea area, the potential habitats include the fractured, hydrated crust; the thin layer of sediments covering the crust; the black and white smokers; and the water column. In the shallow water regions around the emerged land masses (mainly volcano tops and emerged platforms), again sediment and fractured crust provide potential habitats, but the surfaces (subaqueous and subaerial sediment, lava, mineral) exposed to the sunlight represent habitats for potential photosynthesizers. Sketch inspired by *Nisbet and Sleep* [2001].

EARLY-MID ARCHEAN BIOSIGNATURES

In this review, we summarise the biosignature data related to Early-Mid Archean life in Table 1. A previous review [*Schopf and Walter*, 1983] evaluated the data published prior to 1983, hence we concentrate on later studies. The biosignatures are classified based on the microenvironment represented by the sediments in which they were found: (1) the lithosphere environment, including intra-sediment pore spaces and mineral surfaces, in/on the glassy rinds of pillow lavas, and within low-temperature hydrothermal veins; (2) interfacial environments between the lithosphere and the hydrosphere, e.g., on sediment surfaces in the deep sea environment (material transported from shallow water basins and the photic planktonic zone in the oceans), as well as in the photic zones of shallow water to littoral environments, and (3) the open ocean planktonic realm.

Lithosphere Environments

Intra-sediment pore space. Westall [2003, 2004] and *Westall et al.* [2004, 2005] investigated a shallow water tidal flat deposit from the 3.446 Ga Kitty's Gap Chert in the Pilbara. The surfaces of the volcanic shards and minerals in the pore space areas of these samples hosted several species of coccoid bacteria. These dividing cells, 0.5 μm and 0.8 μm in size, produced exopolymer-coated colonies that contained hundreds of individuals and were up to tens of micrometres in diameter (Figure 2). The microenvironment, i.e., their localisation on mineral substrates below the photic zone (i.e., subsurface) and their isotopic signature (δ^{13}C of −25.9‰ to −27.8‰) suggest that they represent probable lithotrophs that may be related to methanogens. The microbial activity of these microorganisms presumably contributed to the

alteration of the protoliths particles [*Westall et al.*, 2004, 2006; *Orberger et al.*, 2006].

Subsurface sediments/hydrothermal veins. Strong hydrothermal activity during the Early-Mid Archean is documented by silicified volcanic and sedimentary lithologies, chemical sediments, and contemporary and post-depositional chert veining [*Paris et al.*, 1985; *Duchač and Hanor*, 1987; *de Wit and Hart*, 1993; *de Vries*, 2004; *van Kranedonk*, 2004; *Orberger et al.*, 2005] as well as silicified microorganisms [*Walsh*, 1992, 2004; *Westall*, 1999, 2003, 2004, 2005; *Westall et al.*, 2001, 2002, 2004, 2005]. In the North Pole area of the Pilbara, for instance, the volcanic and sedimentary layers are cut by hydrothermal chert veins containing kerogenous clots and possible microfossil filaments, around tens of micrometres in size, that are characterised by average δ^{12}C ratios of about −36‰, suggesting fractionation by chemoautotrophic microorganisms [*Ueno et al.*, 2001a,b, 2004]. These researchers, however, based their interpretations of the microbial fossils on the criteria and descriptions of *Schopf* [1993] and suffer from the same criticisms. Although the kerogen may be biogenic in origin, the morphological structures probably represent redeposited amorphous carbon. Note, another group that investigated the chert veins in the North Pole area found that silicified volcanic shards, trapped in the vein matrix and representing entrained portions of country rock, had δ^{15}N ratios of +0.8‰ to −7.4‰ [*Pinti et al.*, 2001; *Pinti and Hashizume*, 2001]. These values were interpreted as an indication of the presence of chemoautotrophic microorganisms at the time of formation. Also in the Pilbara, hematitic vermicular and filamentous structures occur in a micro-BIF from the nearby 3.46 Ga Towers Formation, Marble Bar [*Orberger et al.*, 2005]. This deposit formed by the reaction of hydrothermal fluids with underlying

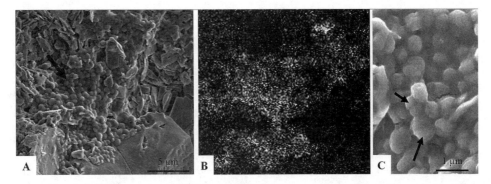

Figure 2. High-resolution SEM micrographs of a lightly HF-etched cut rock surface in 3.446 Ga sediments from the Kitty's Gap Chert, Pilbara [*Westall et al.*, 2006] showing a "monolayer" biofilm of probable coccoidal microfossils (sample coated with 4 nm Au/Pt). (A) General view of the biofilm; (B) EDX carbon map showing that the probable microfossils are carbonaceous; (C) detail showing that the colony consists of two different species, one with a median diameter of 0.5 μm and the other with a median diameter of 0.8 μm (arrows).

basalts. The microstructures and the C/N ratios of the carbonaceous component (2.4–19.4) are interpreted to indicate microbial activity.

Lithosphere/Hydrosphere Interfacial Environments

Deep sea (i.e., below wave-base) environment. Early-Mid Archean sediments deposited below wave-base have been described from only one locality in the Pilbara region: *Ueno et al.* [2001a,b] observed large carbonaceous structures (tens of micrometres in size, see Figure 3; the structures were similar to those in similarly aged 3.46 Ga cherts—although of hydrothermal origin—described by *Schopf* [1993]) in bedded cherts and barite sediments from the 3.49 Ga Dresser Formation in the North Pole Dome area. The carbon isotope ratios of these sediments ranged –40‰ to –31‰. The authors noted the influence of hydrothermal fluid activity in these sediments [*Ueno et al.*, 2004] and hypothesized that the carbonaceous structures and isotope signal were produced by organotrophic organisms. These very same sediments are interpreted by most other workers as strongly hydrothermally influenced, shallow water deposits [e.g., *Nijman et al.*, 1998]. The different depositional setting, however, has no bearing on the microbial interpretation. Although the actual carbonaceous structures described by *Ueno et al.* [2001a,b] probably, like those of *Schopf* [1993], represent redeposited amorphous organic carbon, the carbon could originally have been of biogenic origin [cf. *Pinti et al.*, 2001; *Orberger et al.*, 2005; *Westall*, 2003, 2004].

Mid-Archean rock formations from the Barberton greenstone belt comprise carbonaceous shales from the Kromberg Formation, Upper Mendon Formation, and Fig Tree and Moodies Groups. Isotope ratios of these shales range from –9.5‰ to –40.8‰ [*de Ronde and Ebbessen*, 1996; *Walsh and Lowe*, 1999; *Kakegawa*, 2001; *Tice and Lowe*, 2004], suggesting that the biological component was formed by (anoxygenic) photosynthetic microorganisms in a shallow water environment [*Tice and Lowe*, 2004]. The shallow origin of the carbonaceous component combined with the deep water sediment character may simply result from the transportation and sedimentation of pieces of microbial mat (Figure 4) from nearby shallow water zones. In another study of Fig Tree and Moodies Group shales, *Kakagawa* [2001] interpreted the C (–31‰ to –17‰) and S (–4‰ to +1‰) isotopes as indicating the presence of lithotrophic methanogens, and possibly heterotrophic sulphate-reducing bacteria.

Volcanogenic massive sulphide deposits (VMS) related to shallow (<100 m), black smoker, seafloor vents occur in both the Pilbara and Barberton terrains. Thin pyritized filaments (0.5–2 μm in diameter), embedded in hydrothermal chert (Figure 5) from a bituminous VMS from a 3.235 Ga deposit near the North Pole Dome were described by *Rasmussen*

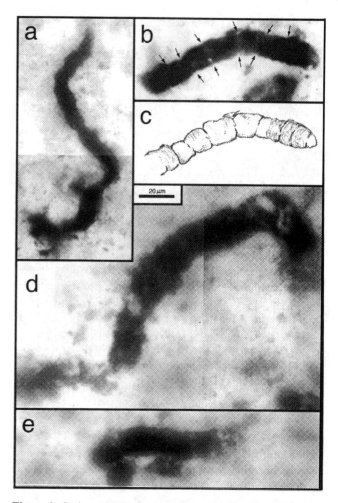

Figure 3. Carbonaceous, filamentous bacteriomorph structures in 3.46 Ga cherts from bedded chert/barite units in the North Pole Dome, Pilbara (light-microscope observations of thin sections, from *Ueno et al.* [2001a]). Although the structures are probably not the remains of microorganisms, the kerogen may have been originally derived from biogenic carbon.

[2000], who interpreted them as representing thermophilic, S-utilising chemotrophs. In the Barberton greenstone belt, a 3.230–3.259-Gyr-old VMS from the Fig Tree Group studied by *De Ronde and Ebbessen* [1996] contains organic carbon compounds up to 600 atomic mass units, as well as a $\delta^{13}C$ signature of –25‰ to –30.2‰ that they interpreted as being of microbial origin.

Another example of probable microbial protolith corrosion was observed in the glassy rinds of 3.480–3.322 Ga pillow lavas from the Barberton greenstone belt previously described by *Furnes et al.* [2004]. Although microfossils were not directly observed in the glassy pillow lava rinds, their presence was inferred from the occurrence of carbon in

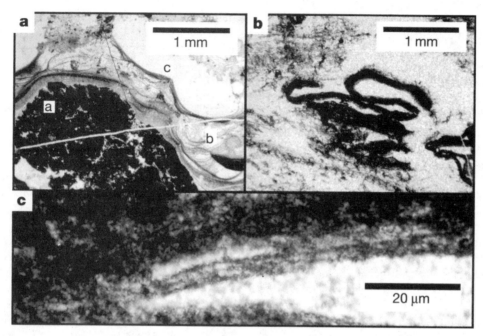

Figure 4. Black and white banded chert containing fragments of laminated carbonaceous material representing probable shallow-water microbial mats that were redeposited in deeper water (subwave base) (light-microscope observations of thin sections, from *Tice and Lowe* [2004]). The mats were probably produced by anoxygenic photosynthetic microorganisms. Carbonaceous laminae drape coarser particles (upper left), whereas probable mat fragments have been rolled up by the activity of currents (upper right). The lower micrograph shows possible microbial filaments within the mat fragments.

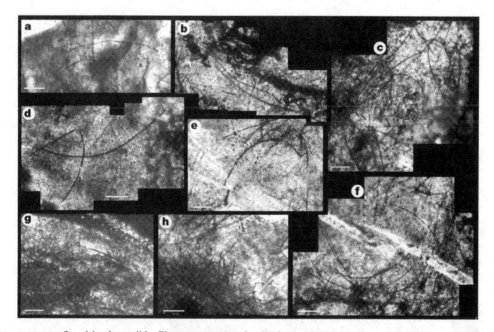

Figure 5. Photomontage of pyritized possible filamentous microfossils from a 3.230 Ga massive sulfide deposit, North Pole Dome, Pilbara (light-microscope observations of thin sections, from *Rasmussen* [2005]). The microfossils probably represent thermophilic chemolithotrophs that inhabited subseafloor hydrothermal environments.

the corrosion pits and the carbon isotope ratios ranging down to −16.4‰ of the carbonate-replaced, carbonaceous rinds.

Shallow water (i.e., within wave-base/photic) to littoral environments. Deposits from shallow water environments are far more common than deeper water settings in the Early-Mid Archean greenstone belts in both the Pilbara and Barberton regions. Macroscopic, domal to conical, stromatolite-like constructions have been described from the North Pole Dome area in the Pilbara and from the Barberton greenstone belt [*Walter et al.*, 1980; *Lowe*, 1980; *Walter*, 1983; *Byerly et al.*, 1986; *Nijman et al.*, 1998; *Westall et al.*, 2002; *Westall*, 2003, 2004; *Walsh*, 2004; *Allwood*, 2005a,b]. Those in the <3.446 Ga Strelley Pool Chert (North Pole Dome), for instance, appear to have formed on a carbonate platform, the oldest such platform identified to date [*Allwood et al.*, 2005a,b]. The similarity of such structures to those from the Late Archean and Proterozoic led most workers to hypothesise their construction from oxygenic photosynthetic bacteria, similar to those formed in modern environments [e.g., *Monty*, 1967; *Walter*, 1976]. On the other hand, *Lowe* [1994], *Buick and Dunlop* [1981], *Buick et al.* [1995], and *Grotzinger and Rothmann* [1996] noted that abiogenic mechanisms can produce similar features; *Lowe* [1994] concluded that the Early-Mid Archean stromatoloid structures were, in fact,

abiogenic structures. However, *Westall et al.* [2002], *van Kranendonk et al.* [2003], and *Allwood et al.*, [2005a,b] observed that, from the physical aspects of certain stromatolite structures in 3.45–3.47-Gyr-old cherts in the North Pole area of the Pilbara, they must have been constructed by sticky organic films. For example, the small stromatolite in Figure 6 consists of domed, contorted thin laminae, sandwiched between layers of hydrothermal silica that are not deformed. The laminae exhibit plastic deformation, which suggests that they were once formed of soft and pliable but robust layers of material [*Westall et al.*, 2002]. Their deformation may have been caused by the weight of the silica precipitated on top of the dome. Furthermore, the carbonate crusts of films from the Strelley Pool Chert (North Pole) have a geochemical signature identical to that of other Later Archean microbial carbonates [*van Kranendonk et al.*, 2003; *Allwood et al.*, 2005b].

As noted by *Krumbein* [1983], stromatolites are basically microbial mats that can be both planar as well as vertical in expression. More often than not, it is the associated polymer or EPS that is implicated in the fossilization of the mats [*Westall et al.*, 2000]. In this respect, well-developed biofilms and microbial mats can form relatively large features that may be visible in thin section (Figure 7A) [e.g., *Walsh*, 1992, 2004; *Walsh and Westall*, 2003; *Tice and Lowe*, 2004]. Optical microscope observations of microfossils in these deposits are,

Figure 6. Probable stromatolites from the Pilbara and Barberton [*Westall*, 2004]. (A,B) Section and plan view, respectively, of stromatolites in the <3.446 Ga Strelley Pool Chert, North Pole Dome. (C) Probable stromatolite with contorted, plastically deformed internal laminations from the 3.46 Ga Dresser Formation. (D) Low-amplitude stromatolitic domes from the 3.30 Ga Fig Tree Formation [cf. *Byerly et al.*, 1986].

however, rare: *Walsh and Lowe* [1985], *Walsh* [1992, 2004], *and Westall and Walsh* [2000] report long, thin (<0.2–2.5 μm diameter), filamentous microfossils (Figure 7B) from shallow water deposits of the Onverwacht Group in Barberton. These structures are clearly distinguishable from the much larger, very thick, amorphous, biogenic carbonaceous structures described by *Schopf* [1993] and *Ueno et al.* [2001a,b].

The most detailed investigations of planar or stratiform microbial mats in shallow water sediments made use of combined optical microscopy, high-resolution SEM, micrometre-scale geochemical analysis, and layer-scale (as opposed to bulk rock) carbon isotope investigations, to reveal the "construction" of microbial mats on sediment surfaces, including those in the littoral environment [*Westall*, 2003, 2004, 2005b; *Westall et al.*, 2004, 2005]. This kind of detail has demonstrated the interactions between microbial consortia within a biofilm, as well as their interactions with their immediate microenvironment. For example, *Westall, et al.* [2003, 2004] and *Westall* [2003, 2004] describe an Early-Mid Archean microbial mat consisting of 0.25-μm-thick filaments from the Josefsdal Chert in Barberton that formed under flowing water in a partly exposed, evaporitic littoral environment (Figure 8). This mat could have been formed by anoxygenic photosynthetic microorganisms ($\delta^{13}C$ of −22.7‰). It is closely associated with 1 μm wide and 2–3.8 μm colonies of long, rod- to vibroid-shaped bacteria (Figure 7). This is the earliest evidence for a subaerially exposed microbial mat. In a similar manner, *Westall et al.* [2005] documented a 3.446 Ga microbial mat consortium consisting of probable filamentous (0.25 μm diameter), cocci- (0.5 and 0.8 μm diameter), and rod-shaped bacteria (1 μm length) that formed in a tidal channel–mud flat environment in the Kitty's Gap

Chert from the Coppin Gap greenstone belt in the Pilbara. Given the microenvironment and the carbon isotope ratios ($\delta^{13}C$ of −26‰ to −30‰), anoxygenic photosynthetic microorganisms may have contributed to the formation of this mat as well.

The same bedded chert/barite units from the 3.449 Ga Dresser Formation of the North Pole Dome were analyzed by *Ueno et al.* [2001a,b,c], who proposed a deep water origin, and by *Nijman et al.* [1998] and *Shen et al.* [2001], who interpreted them as being shallow water deposits. *Shen et al.* [2001] obtained a $\delta^{34}S$ signature of −16‰ to −2‰ from these deposits that they interpreted as having been produced by sulfur-reducing bacteria, although the authors also note that other abiogenic processes can produce such a signature. Note that *Ueno et al.* [2001a,b,c] had hypothesized the presence of litho- or organotrophs based on the microstructures and the carbon isotopes of these same sediments. *Skrzypezak et al.* [2004] obtained a carbon isotope ratio of −32.7‰ and a nitrogen isotope ratio of −4.4‰ from a chert horizon in the lower part of the Dresser Formation (also called the Towers Formation). They made a high-resolution transmission electron microscopy study of the ultrastructure of the kerogen, from which they deduced that it was of biogenic origin.

The Open Ocean Planktonic Environment

The oldest deep water sediments are preserved as metaturbidites in the >3.7 Ga Isua Greenstone Belt [*Rosing*, 1999], although the interpretation that these rocks represent sediments is disputed (e.g., *Myers* [2001] regards them as mylonites). Carbon isotope ratios of −10‰ to −26‰ are interpreted as representing a biosignature from the sedimentation

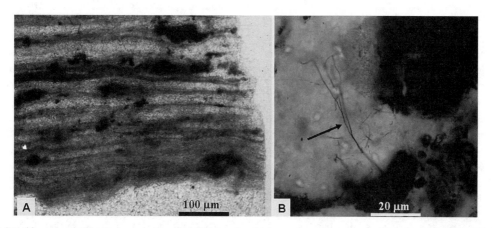

Figure 7. (A) Stratiform laminations representing a probable microbial mat in carbonaceous cherts in the 3.33–3.412 Ga Kromberg Formation as shown in Barberton light-microscope observations of thin sections [cf. *Walsh*, 1992, 2004; *Walsh and Lowe*, 1999]; (B) probable filamentous microfossil from the same sample (arrow) (light-microscope observations of thin sections, from *Westall* [2004]; cf. *Walsh* [1992]).

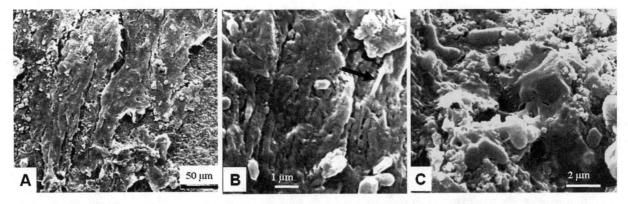

Figure 8. High-resolution SEM micrographs of a lightly HF-etched freshly broken rock surface (i.e., not an old fracture surface) coated with Au. Shown is a probable evaporitic, subaerial microbial mat from the 3.333–3.472 Ga Josefsdal Chert, Barberton [*Westall et al.*, 2001, 2004; *Westall*, 2003, 2004]. (A) Plan view of the desiccated, filamentous mat; (B) filaments (arrow) thickly coated with polymer, which may have been anoxygenic photosynthesizers; (C) rod- to vibroid-shaped microfossils probably associated with the filamentous mat.

of planktonic microorganisms, including oxygenic photosynthesizers [*Rosing*, 1999; *Rosing and Frei*, 2004]. *Van Zuilen et al.* [2003] verified that a part of the isotope signal with a value of −18‰ is original to the rock but the rest is due to contamination. However, the origin of this fractionation (biogenic or abiogenic) has yet to be verified. The values of the nitrogen isotopes contained in the metasedimentary graphite (−3‰ to −1‰) are likewise inconclusive with regards to biogenicity [*van Zuilen et al.*, 2005]. Based on the carbon isotope signature, *Walsh and Lowe* [1999] suggest that deep water deposits from Mid-Archean formations in the Barberton greenstone belt may contain the remains of open ocean planktonic photosynthetic microorganisms.

DISCUSSION

The Evidence for Early-Mid Archean Life

Whatever the limitations of the individual analyses presented above and in Table 1, there is an overwhelming consensus from the isotope, geochemical, mineralogical, and morphological data for the existence of life by at least ~3.5 Ga, preserved in the rocks of the Early-Mid Archean period in the Pilbara and in Barberton. The existence of life is documented by (1) its morphological remains including individual microfossils, their biofilms and microbial mats, and possibly by corrosion pits in rock surfaces, and (2) the carbonaceous and isotopic composition (including C, S, and N) of the rocks. Also, most of the data, when classified by microenvironment, are consistent with each other.

The high-resolution SEM observations of *Westall* [2003, 2004, 2005b] and *Westall et al.* [2001, 2005] demonstrate that the Early-Mid Archean microorganisms are individually

and collectively—i.e., microscopically, as in colonies, biofilms, and mats and macroscopically [*Byerly et al.*, 1986; *Walsh*, 1992, 2004; *Hofmann et al.*, 1999; *van Kranendonk*, 2003; *Furnes et al.*, 2004; *Allwood et al.*, 2005a,b]—identical to modern microorganisms. Despite the possibility of an abiogenic origin for the various (C, N, S) isotopic ratios measured, the values obtained are consistent with those produced by modern microorganisms. Moreover, when combined with the potential metabolic strategies available at a microenvironment scale (see below), the carbon isotope ratios, at least, concur with the metabolic possibilities. With respect to the S and N isotope ratios, the effects of metamorphism, albeit relatively low-grade, on those ratios has yet to be fully evaluated. To date, there is no known process that can produce the totality of structural and geochemical data, including possible concentrations of REEs and heavy metals, summarized in Table 1. We therefore conclude that the evidence for the existence of microorganisms, similar to modern prokaryotes, on the early Earth is overwhelming.

The carbonaceous detritus in sediments deposited in the deeper areas of basins consists of transported, shallow water, benthic photosynthetic microbial mats [*Walsh and Lowe*, 1999; *Tice and Lowe*, 2004], photosynthetic planktonic cells [*Walsh and Lowe*, 1999; *Rosing*, 1999; *Rosing and Frei*, 2004], as well as in situ colonies formed on the detritus [*Kakegawa*, 2001]. Torn fragments of microbial mats testify to transport from shallow waters [*Walsh and Lowe*, 1999; *Tice and Lowe*, 2004]; the clear identification of planktonic colonies, however, is very difficult because, in the modern environment, they are represented by flocs of "marine snow" that, once sedimented to the seafloor, provide nutrients for organisms living on and within the sediment [*Alldredge and Silver*, 1988; *Turley*, 2002]. Although *Kakegawa* [2001] based his interpretation of in situ microorganisms such as lithoautotrophs (e.g., methanogens) and heterotrophs (e.g., sulphide-reducing bacteria, or SRBs) on C and S isotope analyses, the observations of a direct relationship between microbial colonies and volcanic mineral surfaces [*Furnes et al.*, 2004; *Westall et al.*, 2004, 2005] indicates that his interpretation is plausible.

The Early-Mid Archean shallow water setting would offer a range of microenvironments that could support growth. The most obvious indications of microbial activity are the macroscopic stromatolites. Even though there is discussion regarding the biogenicity of some of these structures, e.g., the fine-scale internal laminations [*Walsh*, 2004], the presence of microfossils [*Westall*, unpub. data] and indications of sticky biofilms [*Westall et al.*, 2002; *van Kranendonk et al.*, 2003; *Allwood et al.*, 2005a,b] suggests that at least some of the stromatolites are probably of biogenic origin. Comparison of these early stromatolites with oxygenic photosynthetic (cyanobacterial) Late Archean and Proterozoic stromatolites prompted early investigators to implicate cyanobacteria in their formation [*Schopf and Packer*, 1987; *Schopf et al.*, 1993]. However, the growing evidence for environmental conditions with no or very little oxygen [*Rye et al.*, 1995; *van Kranendonk et al.*, 2003; *Orberger et al.*, 2005, *Westall et al.*, 2004, 2005] has prompted interpretations of anoxygenic photosynthesizers [*Kapler and Newman*, 2004] as the primary producers in these shallow water to littoral microbial mats [*Westall et al.*, 2001, 2004, 2005; *Westall*, 2004, 2005b; *Tice and Lowe*, 2004].

Metabolic Strategies

Given the microenvironments in which the fossil microorganisms occur and the type of sediments from which the relevant geochemical and isotopic data were obtained, it is evident that early life used a number of different metabolic strategies to obtain both energy and carbon from its immediate environment. The close association of microorganisms with rocky substrates, such as volcaniclastic sediments [*Westall*, 2003, 2004, 2005b; *Westall et al.*, 2005] or the vitreous rinds of pillow basalt lavas [*Furnes et al.*, 2004], suggests a chemolithotrophic metabolism. Ignoring the requirement to produce an entire cell from CO_2, this is the "simplest" metabolism in modern organisms and may have been the first metabolism used by early life [*Des Marais*, 2000]. Further refinements in interpretation can be made on the base of C and S isotope ratios, as well as microfossil morphology. Thus, the lithotrophs probably included methane-producing microorganisms (in the modern microbial world, these include methanogens belonging to the Archaea) and sulfate- (SO_4^{2-}) or sulfur- (S^0) reducing bacteria, which use organic compounds or H_2 as electron donors

and produce H_2S as a by-product. The occurrence of microbial mats in the photic zone [*Walsh*, 1992, 2004; *Tice and Lowe*, 2004; *Westall et al.*, 2001, 2004, 2005; *Westall*, 2003, 2004, 2005b] prompts the question as to whether their builders had actually developed photosynthesis. This advanced form of metabolism, in which energy for incorporating CO_2 into the cell is obtained from sunlight, allows a greater production of biomass [*Des Marais*, 2000]. Of the two types of photosynthesis, anoxygenic and oxygenic, the former is more simple than the latter, in which oxygen is produced as a by-product. Given the evidence for generally anoxic to very oxygen-poor conditions in that period, if the biofilms were produced by photosynthetic microorganisms, they must have used anoxygenic photosynthesis.

A number of biofilms possessing bacterial consortia have been identified in the Early-Mid Archean cherts [*Westall et al.*, 2001, 2004, 2005; *Westall*, 2003, 2004, 2005b]. One example from the Onverwacht Group in Barberton (3.5–3.3 Ga) [*Westall et al.*, 2001, 2004] consists of what are probably SRB directly overlying the surface of volcaniclastic sediments, which are, in turn, overlain by probable anoxygenic photosynthesizers. In such a consortium, the SRB would obtain their energy and carbon directly from the underlying rocky materials or from the photosynthesizers, in turn providing them with H_2S as a by-product of metabolism. The presence of evaporite minerals, such as gypsum, associated with the biofilm is evidence that a source of sulfate for the microbial metabolism was present, and pyrite crystals associated with the mat were probably formed as a result of the H_2S produced by the microbes.

Environmental Characteristics and Distribution of Early Life

Life in the Early-Mid Archean appears to have been widely distributed in deep water (subwave-base), shallow water, and even subaerial environments. Hydrothermal systems appear to have been abundant in all environments [e.g., *Paris et al.*, 1985; *Duchač and Hanor*, 1987; *de Wit and Hart*, 1992; *van Kranendonk et al.*, 2004; *Hofmann*, 2004]. The latter ranged from dilute flushing by hydrothermal fluids further away from the vents to more concentrated and probably hotter environments directly in the vicinity of the vents (e.g., the pyritized filaments described by *Rasmussen* [2000]).

In the absence of significant oxygen (and therefore ozone) in the Early-Mid Archean atmosphere, the observation of a microbial mat formed in a partially subaerial littoral environment [*Westall et al.*, 2005] provokes the question as to the resistance of the microorganisms to subaerial exposure and UV radiation. Microorganisms in modern environments with high insolation protect themselves by forming multilayered biofilms or mats in which the outer layers consist of dead cells that form a skin-like, protective layer [*Gorbushina*, 2003]. Also, rapid gene repair mechanisms are characteristic of radiation-resistant organisms, such as *Deinococcus radiodurans* [*Daly et al.*, 1994]. It is possible that the microbial communities in the Early-Mid Archean epoch survived by forming multilayered structures and developing gene repair mechanisms. It has been estimated that the amount of DNA-weighted UV radiation affecting early prokaryotes is on the order of about 54 W/m^2 [*Cockell*, 2001]. This is the same value for the surface of Mars today [*Cockell*, 2001]. Recent modeling, however, suggests that the high amount of water vapor in the atmospheres of the early terrestrial planets (Venus, Earth, and Mars) would have provided significant protection for microorganisms at the surface [*Lammer et al.*, 2005].

One of the important environmental characteristics of the Early-Mid Archean period is the prevalence of syndepositional and postdepositional hydrothermal activity. High heat flow from the mantle to the crust on the early Earth ensured thorough flushing and circulation of seawater through the crust [*Paris et al.*, 1985; *Duchač and Hanor*, 1987; *Hofmann*, 2004]. Evidence for this flushing comes from geochemical investigations of fluid inclusions [e.g., *de Ronde et al.*, 1994; *de Ronde and Ebbessen*, 1996] and geochemical tracing of element movement [*Orberger et al.*, 2005]. *Knauth and Lowe*, [2003] calculated that the Early-Mid Archean seawater was strongly influenced by hot, hydrothermal fluids and that the temperature of the global ocean was between 50 °C and 80 °C, and more probably in the upper reaches of this range. Estimations for hydrothermal fluid temperatures generally range from <150 °C to 220 °C [*de Ronde et al.*, 1994; *de Vries*, 2004], although *Brasier et al.* [2002] suggested >300 °C. However, most of the data suggest generally low-temperature hydrothermal exhalations <150 °C. The high global ocean temperatures and the pervasive nature of hydrothermal activity suggest that any life forms from this period must have been at least thermophilic (i.e., optimum growth temperature between 45 °C and 80 °C).

The Early-Mid Archean seawater was more enriched in minerals and had higher salinities compared to modern seawater [*Channer et al.*, 1997; *de Ronde et al.*, 1997], and a number of small, evaporitic deposits from both the Pilbara and Barberton have been described [*Lowe*, 1983; *Lowe and Worrell*, 1999; *de Vries*, 2004]. Although the importance of hydrothermal fluids on the formation of the brines and deposits is under discussion [*de Vries*, 2004], there is clear evidence of an evaporite suite deposited in and on a littoral microbial mat from the Barberton area [*Westall et al.*, 2005]. The evidently high salinities of the environment indicate that the microorganisms were at least halotolerant, if not halophilic (i.e., requiring NaCl for growth). The Early-Mid Archean environment is also supposed to have been slightly

acidic [*Grotzinger and Kasting*, 1994; *Orberger et al.*, 2005]. Under such conditions the microorganisms may have been acid-tolerant, but they would *not* have been comparable to contemporary acidophilic bacteria.

Microorganisms occurred in the pore spaces of subsurface sediments [*Westall et al.*, 2005], although probably not at depths as great as >900 m, as seen in modern deep sea sediments [*Parkes et al.*, 1994] because (1) such thicknesses of sediment have yet to be identified in Early-Mid Archean formations and (2) rapid silicification of the volcaniclastic sediments would have blocked the pore spaces. Carbon isotope analyses of the carbonaceous hydrothermal vein cherts may indicate the presence of deep-dwelling organisms (today going down to >3 km depth [*Moser et al.*, 2003]). *Pinti et al.* [2001] and *Orberger et al.* [2005] observed that the isotopic signature was preserved in chert shards within the vein-filling material, i.e., country rock, rather than in hydrothermal material. Although no microorganisms have yet been described from the vadose zone cracks that were abundant in some horizons of these formations [e.g., *de Vries*, 2004], there is evidence of potential endolithic (i.e., living in rocks) behavior of microorganisms corroding cavities in the surfaces of the pillow lavas [*Furnes et al.*, 2004].

Productivity and Biomass

Although life seems to have been widely distributed in the Early-Mid Archean epoch, it does not appear to have been volumetrically very important. The stromatolitic constructions, if they are at least partly of biogenic origin, are orders of magnitude smaller than those of the Late Archean/Proterozoic epochs. Thick accumulations of tabular microbial biofilms/mats are relatively rare [*Walsh*, 1999, 2004]. *Westall et al.* [2004, 2005] noted that, although common, the colonies of lithoautotrophic microorganisms in these sediments were very small and the biofilms relatively thin; as a consequence, total organic carbon contents are very low (~0.1%). Only the organic-rich shales deposited in deeper waters testify to a relatively abundant input of carbonaceous material from shallower water environments [*Walsh and Lowe*, 1999; *Tice and Lowe*, 2004] or from the planktonic photic zone [*Walsh and Lowe*, 1999; *Rosing*, 1999; *Rosing and Frei*, 2004; *Hofmann*, 2004]. The main limitation to productivity, and hence to biomass development, appears to be the fact that oxygenic photosynthesis and the corresponding aerobic heterotrophs had not yet developed [*Des Marais*, 2000].

Evolution

What is surprising about these remnants of the oldest life on Earth is (1) the modern-looking aspect of the individual microorganisms, their colonies, biofilms, and mats, (2) their diversity, and (3) their wide environmental distribution. This is not the record of primitive life forms; rather, it is the record of relatively highly evolved life forms. Granted that oxygenic photosynthesis had apparently not evolved by the Early-Mid Archean, the evidence for anoxygenic photosynthesis is somewhat circumstantial; however, the widespread development of microbial mats and stromatolites in environments favored by photosynthesizers (anoxygenic or oxygenic) today, the morphological similarity of the photic zone mats to those formed by modern photosynthetic microorganisms, and the carbon isotope signature, is strong support for such an interpretation.

The relative abundance, widespread distribution, and relatively evolved level of life at ~3.5 Ga implies that life must have been present much earlier, e.g., by 3.7–3.8 Ga, when the rocks of Isua and Akilia were formed. These rocks could potentially hold a record of this life, if the metamorphic overprinting can be disentangled from the potential biogenic signal. However, given the conclusions regarding the possibility of the evolution of anoxygenic photosynthesis (but not oxygenic photosynthesis) by 3.5 Ga, the claims of oxygenic photosynthesis by 3.7 Ga [*Schidlowski*, 1988, 2001; *Rosing*, 1999; *Rosing and Frei*, 2004] need to be reassessed.

The advanced level of evolution at 3.5 Ga has implications for the timing of the origin of life. Did life need to evolve before 4.0 Ga in order to reach the stage of anoxygenic photosynthesis by 3.5 Ga? If yes, this means that it survived throughout the period of heavy bombardment that lasted between 20 and 200 Myr and occurred between 4.0 and 3.85 Ga [*Maher and Stevenson*, 1988; *Sleep et al.*, 1989; *Nisbet and Sleep*, 2001; *Ryder*, 2002], despite the predictions of mass annihilation in planet-sterilising impacts [*Sleep et al.*, 1989; *Nisbet and Sleep*, 2001; *Ryder*, 2002].

Furthermore, the implication of the evidence for relatively evolved life at 3.5 Ga is that the oldest record of life on Earth is not the record of the progenote(s) or first life forms. That record has been lost on Earth. It may, however, exist elsewhere in the Solar System on a planet where plate tectonics has not destroyed the strata potentially holding the record of life's first steps. Mars is such a planet. The Southern Highlands of Mars comprise rocks of Noachian age (4.5 to ~3.5 Ga) that were formed when the environment on the planet was suitable for the appearance and early evolution of life.

CONCLUSIONS

Despite the controversies and false trails in the search for the earliest life on Earth, there exists a rich record in sediments from ~3.5 Ga from the Pilbara in Australia and Barberton in South Africa. Using the multitude of information available, which ranges from field to microscopic

observation and chemical and isotopic analyses, we have shown that life appears to have inhabited all the microenvironments available on the Early-Mid Archean Earth. These microenvironments include subsurface sediment pore waters, possibly low-temperature hydrothermal veins, mineral and rock surfaces, sediment surfaces, and possibly the planktonic realm. They occurred in the deep sea, in shallow water basins, and in littoral beach environments. By 3.5 Ga, it appears that life had already colonized exposed land surfaces, at least on the beach.

The Early-Mid Archean microorganisms were morphologically similar to modern organisms and behaved in the same way: building colonies, biofilms, and mats and interacting directly with their immediate substrate and with each other (in consortia). Their metabolic processes included chemolithotrophy, possibly methanogenesis, and possibly anoxygenic photosynthesis. Early life was diverse and included thermophilic, acid-tolerant, halo-tolerant to halophilic, and radiation-resistant species.

The diversity, relative level of evolution, and widespread distribution of life by 3.5 Ga implies that it must have evolved much earlier, possibly even before 4.0 Ga (in which case it was therefore not extinguished during the period of late heavy bombardment). However, no record of its appearance and early evolution remains on Earth; the Southern Highlands of Mars could potentially host this missing record.

Acknowledgments. We thank the reviewers for their useful comments.

REFERENCES

Alldredge, A.L., and M.W. Silver (1988), Characteristics, dynamics and significance of marine snow, *Prog. Oceanogr.*, 20, 41-82.

Allwood, A.C., M.R. Walter, M.J. van Kranendonk, and B.S. Kamber (2005a), 3.43 Ga Stromatolites, Rocky Shorelines and a Carbonate Platform: Strelley Pool Chert, Pilbara Craton, Western Australia, *Astrobiology*, 5, 184.

Allwood, A.C., M.R Walter, and M.J van Kranendonk (2005b), Stromatolite facies of the 3.43 Ga Strelley Pool Chert: Pilbara Craton, Western Australia, *Astrobiology*, 5, 184.

Arndt, N.T. (1994), Archean komatiites, in *Archean crustal evolution*, edited by K.C. Condie, pp. 11-44, Elsevier, Amsterdam.

Awramik, S.M., and J. Sprinkle (1999), Proterozoic stromatolites: the first marine evolutionary biota, *Historical Biology*, 13, 241-253.

Bazylinski, D.A. (1996), Controlled biomineralization of magnetic minerals by magnetotactic bacteria *Chem. Ecol.*, 132, 191-198.

Benning, L.G., V. Phoenix, N. Yee, M.G. Tobin, K.O. Konhauser, and B.W. Mountain (2002), Molecular characterization of cyanobacterial cells during silicification: A synchrotron-based infrared study, in *Proceedings of the Geochemistry of the Earth's Surface Conference, Honolulu, Hawai'i*, May 20-24, 2002, pp. 259-263.

Brantley, S.L., L. Lierman, M. Bau, and S. Wu (2001), Uptake of trace metals and rare elements from hornblende by a soil bacterium, *Geomicrobiol. J.*, 18, 37-61.

Brasier, M.D., O.R. Green, A.P. Jephcoat, A.K. Kleppe, M. van Kranendonk, J.F. Lindsay, A. Steele, and N. Grassineau (2002), Questioning the evidence for Earth's oldest fossils, *Nature*, 416, 76-81

Brocks J.J., G.A. Logan, R. Buick, and R.E. Summons (1999), Archean molecular fossils and the early rise of eukaryotes, *Science*, 285, 1033-1036.

Buick, R. (1990), Microfossil recognition in Archean rocks: an appraisal of spheroids and filaments from a 3500 m.y. old chert-barite unit at North Pole, Western Australia, *Palaios*, 5, 441-459.

Buick, R., D.I. Groves, J.S.R. Dunlop, and D.R. Lowe (1995), Abiological origin of described stromatolites older than 3.2 Ga; discussion and reply, *Geology*, 23, 191-192.

Burne, R.V., (1991/1992). Lilliput's Castles: stromatolites of Hamelin Pool, *Landscope*, 7, 34-41.

Byerly, G.R., M.M. Walsh, and D.L. Lowe (1986), Stromatolites from the 3300-3500 Myr Swaziland Supergroup, Barberton Mountain Land, South Africa, *Nature*, 319, 489-491.

Cady, S.L., and J.D. Farmer (1996), Fossilization processes in siliceous thermal springs: trends in preservation along the thermal gradient, in, *Evolution of Hydrothermal Ecosystems on Earth (and Mars)*, edited by G.R. Bock and J.A. Goodie, Ciba Symposium 202, pp 150-173, John Wiley, Chichester.

Cady, S.L., J.D. Farmer, J.P. Grotzinger, W.J. Schopf, and A. Steele (2003), Morphological biosignatures and the search for life on Mars, *Astrobiology*, 3, 351-368.

Channer, D.M.DeR., C.E.J. de Ronde, and E.T.C. Spooner (1997), The Cl⁻ Br I⁻ composition of ~3.23 Ga modified seawater: implications for the geological evolution of ocean halide chemistry, *Earth Planet. Sci. Lett.*, 150, 325-335.

Charaklis, W.G., and P.A. Wilderer (editors.) (1989), Structure and function of biofilms - Dahlem Workshop report, Wiley, Chichester.

Chekroun, K.B., C. Rodríguez-Navarro, M.T. González-Muñoz, J.M. Arias, G. Cultrone, and M. Rodríguez-Gallego (2004), Precipitation and Growth Morphology of Calcium Carbonate Induced by *Myxococcus Xanthus*: Implications for Recognition of Bacterial Carbonates, *J. Sed. Res.*, 74, 868-876.

Cockell, C.S. (2000), The ultraviolet history of the terrestrial planets – implications for biological evolution, *Planet. and Space Sci.*, 4, 203-214.

Daly, M.J., L. Ouyang, and K.W. Minton (1994), *In vitro* damage and *recA*-dependent repair of plasmid and chromosomal DNA in the radioresistant bacterium *Deinococcus radiodurans, J. Bacteriol.*, 176, 3508-3517.

Dauphas, N., M. van Zuilen, M. Wadwha, A.M. Davies, B. Marty, and P.E. Janney, (2004), Clues from Fe isotope variations of the origin of Early Archean BIFs from Greenland, *Science*, 306, 20077-2080.

de Ronde, C.E.J., M.T. de Wit, and E.T.C. Spooner (1994), Early Archean (>3.2 Ga) Fe-oxide–rich, hydrothermal discharge vents in the Barberton greenstone belt, South Africa, *Geol. Soc. Am. Bull.*, 196, 86-104.

de Ronde, C.E.J., D.M.DeR. Channer, K. Faure, C.J., Bray, and E.T.C. Spooner (1997), Fluid chemistry of Archean seafloor hydrothermal vents; implications for the composition of circa 3.2 Ga seawater, *Geochim. Cosmochim. Acta*, 61, 4025-4042.

de Ronde, C.E.J. and T.W. Ebbesen (1996), 3.2 billion years of organic compound formation near sea-floor hot springs, *Geology*, 24, 791-794.

de Ronde, C.E.J., M.J. de Wit, E.T.C. Spooner, D.M. deR. Channer, B.W. Christenson, C.J. Bray, K. Faure (2004), Ironstone pods in the Archean Barberton greenstone belt, South Africa: Earth's oldest seafloor hydrothermal vents reinterpreted as Quaternary subaerial springs: Comment and Reply. *Geology: Online forum*, e68-69.

Des Marais, D.J. (2000), When did photosynthesis emerge on Earth?, *Science*, 289, 1703-1705.

De Vries, S.T. (2004), Early Archean sedimentary basins: depositional environment and hydrothermal systems: Examples from the Barberton and Coppin Gap greenstone belts, PhD thesis, 159 pp, University of Utrecht, Geologica Ultraiectina, Utrecht, The Netherlands.

de Wit, M.J., and R.A. Hart (1993), Earth's earliest continental lithosphere, hydrothermal flux and crustal recycling, *Lithos*, 30, 309-336.

Duchač, K.C. and J.S. Hanor (1987), Origin and timing of the metasomatic silicification of an Early Archaean komatiite sequence, Barberton Mountain Land, South Africa, *Precambrian Res.*, 37, 125-146.

Fedo, C.M., and M.J. Whitehouse (2002), Metasomatic origin of quartz-pyroxene rock, Akilia, Greenland, and implications for Earth's earliest life, *Science*, 296, 1448-1452.

Ferris, F.G., T.J. Beveridge, and W.S. Fyfe (1986), Iron-silica crystallite nucleation by bacteria in a geothermal sediment, *Nature*, 320, 609-611.

Furnes, H., N.R. Banerjee, K. Muehlenbachs, H. *Staudigel, M. de Wit (2004),* Early Life Recorded in Archean Pillow Lavas, *Science,* 304, 578-581.

Garcia-Ruiz, J.M., S.T. Hyde, A.M. Carnerup, A.G. Christy, M.J. van Krankendonk, and N.J. Welham (2003), Self-assembled silica-carbonate structures and detection of ancient microfossils, *Science*, 302, 1194-1197

Gorbushina, A. (2003), Microcolonial fungi: Survival potential of terrestrial vegetative structures, *Astrobiol.*, 3, 543-554

Grotzinger, J.P., and J.F. Kasting (1993), New constraints on Precambrian ocean composition. *J. Geol.*, 101, 235-243.

Grotzinger, J. P., and D.H. Rothman, (1996), An abiotic model for stromatolite morphogenesis, *Nature*, 383, 423-25

Handley, K.M. (2004), *In situ* experiments on the growth and textural development of subaerial microstromatolites, Champagne Pool, Waiotapu, New Zealand, MSc thesis, 111 pp, University of Auckland, Auckland, New Zealand.

Hayes, J.M., I.R. Kaplan, and K.W. Wedeking (1983), Precambrian organic chemistry, preservation of the record, in *Earth's Earliest Biosphere*, edited by J.W. Schopf, pp. 93-134, Princeton Univ. Press, Princeton, New Jersey.

Heubeck, C., and D.R. Lowe (1999), Sedimentary petrography and provenance of the Archean Moodies Group, Barberton greenstone belt, in *Geologic evolution of the Barberton greenstone belt, South Africa*, edited by D.R. Lowe and G.R. Byerly, *Geol. Soc. Amer. SP*, 329, pp. 259-286, *Geol. Soc. Amer.*, Boulder, Colorado.

Hofmann, A. (2004), Sedimenatry vs hydrothermal processes for the origin of carbonaceous cherts in the Barberton greenstone belt, in *Field Forum on Processes on the Early Earth Abstract Volume, Kaapvaal Craton*, 4-9 July, 2004, compiled by Reimold, W.U. and Hofmann, A., pp. 35-37, Univ. Witwatersrand, Johannesburg, South Africa.

Hofmann, H.J., K. Grey, A.H. Hickman, and R.I. Thorpe (1999), Origin of 3.45 Ga coniform stromatolites in Warrawoona Group, Western Australia, *Geol. Soc. Am. Bull.*, 111, 1256-1262.

Holland, H.D. (1984), The chemical evolution of the atmosphere and oceans, Princeton Univ. Press, Princeton, New Jersey.

Holland, H.D. (1994), Early Proterozoic atmospheric change, in *Early Life on Earth*, edited by Bengston, S., *Nobel Symposium* No. 84., pp. 237-244, Columbia Univ.Press, New York.

Horneck, G., C. Mileikowsky, J. Melosh, J.W. Wilson, F.A. Cucinotta, and B. Gladman (2002), Viable transfer of microorganisms in the solar system and beyond, in *Astrobiology*, edited by G. Horneck and C. Baumstark-Kahn, pp. 57-78, Springer Verlag, Berlin, Germany.

Huber, H., M.J. Hohn, R. Rachel, T. Fuchs, V.C. Wimmer, and K.O. Stetter (2002), A new phylum of Archaea represented by a nano-sized hyperthermophilic symbiont, *Nature*, 417, 63-67.

Kakegawa, T. (2001), Isotopic signatures of early life in the Archean oceans: influence from submarine hydrothermal activities, in *Geochemistry and the Origin of Life*, edited by S. Nakashsima, S. Maruyama, A. Brack, and B.F. Windley, pp. 237-249., Universal Acad. Press, Tokyo, Japan.

Kappler, A., and D.K. Newman (2004), Formation of Fe(III)-minerals by Fe(II)-oxidizing photoautotrophic bacteria, *Geochim Cosmochim. Acta*, 68, 1217-1228.

Kashefi, K., and D.R. Lovley (2003), Extending the upper temperature limit for life, *Science*, 301, 934-936.

Kasting, J.F. (1993), Earth's early atmosphere. *Science*, 259, 920-926.

Knauth, L.P., and D.R. Lowe (2003), High Archean climatic temperature inferred from oxygen isotope geochemistry of cherts in the 3.5 Ga Swaziland Supergroup, South Africa. *Geol. Soc. Am. Bull.*, 115, 566-580.

Konhauser, K.O., T. Hamade, R.C. Morris, F.G. Ferris, G., Southam, R. Raiswell, and D. Canfield (2002), Could bacteria have formed the Precambrian banded iron formations?, *Geology*, 30, 1079-1082.

Konhauser, K.O., B. Jones, A.-L. Reysenbach, and R.W. Renaut (2003), Hot spring sinters: keys to understanding Earth's earliest life forms, *Can. J. Earth Sci.*, 40, 1713-1724.

Krumbein, W.E. (1983), Stromatolites - the challenge of a term in space and time. *Precambrian Res.*, 20, 493-531.

Krumbein, W.E., D.M. Paterson, and L.J. Stal, (eds.), 1994, Biostabilization of sediments: Oldenburg, Biblioteks und Informationssystem der Carl von Ossietzky Universität, 526 pp.

Kyte, F.T., A. Shukolyukov, G.W. Lugmaor, D.R. Lowe, G.R. Byerly (2003), Early Archean spherule beds: Chromium isotopes confirm origin through multiple impacts of projectiles of carbonaceous chondrite type, *Geology*, 31, 283-286.

Lammer, H., N. Yu., T. Kulikov, M. Penz, H.K. Leitner, N. Biernat, and V. Erkaev (2005), Stellar-planetary relations: Atmospheric stability as a prerequisite for planetary habitability, in: *A comparison of the dynamical evolution of planetary systems, Proceedings of the 6th Alexander von Humboldt colloquium*, edited by R. Dvorak and S. Ferraz-Mello, Kluwer, Dordrecht, The Netherlands, in press.

Lepland, A., G. Arrhenius, and D. Cornell (2002), Apatite in early Archean Isua supracrustal rocks, southern West Greenland: its origin, association with graphite and potential as a biomarker, *Precambrian Res.*, 118, 221-241.

Lepland, A., M. van Zuiland, G. Arrhenius, M.J. Whitehouse, and C.M. Fedo (2005), Questioning the evidence for earth's earliest life – Akilia revisited, *Geology*, 33, 77-79.

Lovelock, J.E. (1988). *The ages of Gaia*. W.W. Norton, New York.

Lowe, D.R. (1980), Stromatolites 3,400-Myr old from the Archean of Western Australia, *Nature*, 284, 441-443.

Lowe, D.R. (1983), Restricted shallow water sedimentation of zearly Archean stromatolitic and evaporitic strata of the Strelley Pool Chert, Pilbara Block, Western Australia, *Precambrian Res.*, 19, 239-283.

Lowe, D.R. (1994), Abiological origin of described stromatolites older than 3.2 Ga, *Geology*, 22, 287-390.

Lowe, D.R., and G.R. Byerly (1986a), Early Archean silicate spherules of probable impact origin, South Africa and Western Australia, *Geology*, 14, 83-86.

Lowe, D.R. and Byerly, G. R. 1986b. Archaean flow-top alteration zones formed initially in a low-temperature sulphate-rich environment, *Nature*, 324, 245-248

Lowe, D.R. and G.F. Worrell (1999), Sedimentology, mineralogy, and implications of silicified evaporates in the Kromberg Formation, Barberton Greenston Belt, South Africa, in *Geologic Evolution of the Barberton Greenstone Belt, South Africa. GSA SP.*, 329, edited by D.R. Lowe and G.R. Byerly, pp. 167-189, *Geol. Soc. Amer.*, Boulder, Colorado.

Lowe, D.R., G.R. Byerly, F.T. Kyte, A. Shukulyukov, F. Asaro, and A. Krull (2003), Spherule beds 3.47-3.24 billion years old in the Barberton Greenstone Belt, South Africa: a record of large meteorite impacts and their influence on early crustal and biological evolution, *Astrobiol.*, 3, 7-48.

Lyons, T.W., C.L. Zhang, and C.S. Romanek (2003), Introduction: Isotopic records of microbially mediated processes, *Chem. Geol.*, 195, 1-4.

Madigan, M.T., J.M. Martinko, and J. Parker (2000), Brock: Microbiology of Microorganisms, Prentice Hall Int., New Jersey.

Maher, K.A. and D.J. Stevenson (1988), Impact frustration of the origin of life, *Nature*, 331, 612-614.

McGregor, V.R. and B. Mason (1977), Petrogenesis and geochemistry of metabasaltic and metasedimentary enclaves in the Amîtsoq gneisses, West Greenland, *Amer. Mineralogist*, 62, 887-904.

Mojzsis, S.J., G. Arrhenius, K.D. McKeegan, T.M. Harrison, A.P. Nutman, and C.R.L. Friend (1996), Evidence for life on Earth before 3800 million years ago, *Nature*, 384, 55-59.

Monty, C.L.V. (1967), Distribution and structure of recent stromatolitic algal mats, eastern Andros island, Bahamas, *Soc. Géol. De Belgique Ann.*, 96, 585-624.

Moser, D.P., T.C. Onstott, J.K. Fredrickson, F.J. Brockman, D.L. Balkwill, G.R. Drake, S. Pfiffner, D.C. White, K. Takai, L.M. Pratt, J. Fong, B.S. Lollar, G. Slater, T.J. Phelps, N. Spoelstra, M. Deflaun, G. Southam, A.T. Welty, B.J. Baker, and J. Hoek (2003), Temporal shifts in the geochemistry and microbial community structure of an ultradeep mine borehole following isolation, *Geomicrobiol. J.*, 20, 517-548.

Myers, J.S. (2001), Protoliths of the 3.8-3.7 Ga Isua greenstone belt West Greenland, *Precambrian Res.*, 105, 129-141.

Newman, D.K., and J.F. Banfield (2002), Geomicrobiology: How molecular-scale interactions underpin biogeomolecular systems, *Science*, 296, 1071-1077.

Nijman, W., K.H. de Bruijne, and M. Valkering (1998), Growth fault control of Early Archaean cherts, barite mounds and chert-barite veins, North Pole Dome, Eastern Pilbara, Western Australia, *Precambrian Res.*, 88, 25-52.

Nisbet, E.G., and C.M.R. Fowler (1999), Archaean metabolic evolution of microbial mats, *Proc. R. Soc. Lond. B*, 266, 2375-2382.

Nisbet, E.G., and N.H. Sleep (2001), The habitat and nature of early life, *Nature*, 409, 1083-1091.

Nisbet, E.G., and C.M.R. Fowler (2004), The early history of life. In *Biogeochemistry*, edited by W.H. Schelsinger, pp 1-39, Treatise on Geochemistry, 8, Elsevier-Pergamon, Oxford, UK.

Orberger, B., V. Rouchon, F. Westall, S.T. de Vries, C. Wagner, and D.L. Pinti (2006), Protoliths and micro-environments of some Archean Cherts (Pilbara, Australia): Geological Society of America Bulletin, in Reimold, W.U. and Gibson, R., *Processes on the Early Earth, Geol. Soc. Amer. SP*, 405., in press.

Papineau, D., S.J. Mojzsis, J.A. Karhu, and B. Marty (2005), Nitrogen isotopic composition of ammoniated phyllosilicates: case studies from Precambrian metamorphosed sedimentary rocks, *Chem. Geol.*, 216, 37-58.

Paris, I., I.G., Stanistreet, and M.J. Hughes (1985), Cherts of the Barberton greenstone belt interpreted as products of submarine exhalative activity. *J. Geol.*, 93, 111-129.

Parkes, R.J., B.A. Cragg, S.J. Bale, J.M. Getliff, K. Goodman, P.A. Rochelle, J.C. Fry, A.J. Weightman, and S.M. Harvey (1994), Deep bacterial biosphere in Pacific Ocean sediments, *Nature*, 371, 410-411

Pavlov, A.A., J.F. Kasting, L.L. Brown, K.A. Rages, and R. Freedman (2001), Greenhouse warming by CH_4 in the atmosphere of early earth, *J. Geophys. Res.*, 105, 11981-11990.

Pflug, H.D. (1979), Archean fossil finds resembling yeasts, *Geol. Palaeontol.*, 13, 1-8.

Pflug, H.D. (2001), Earliest organic evolution, Essay to the memory of Bartholomew Nagy, *Precamb. Res.*, 106, 79-92.

Pflug, H.D., and H. Jaeschke-Boyer (1979), Combined structural and chemical analysis of 3,800-Myr-old microfossils, *Nature*, 280, 483-486.

Phoenix, V.R., D.G. Adams, and K.O. Konhauser (2000), Cell viability during hydrothermal biomineralisation, *Chem. Geol.*, 169, 329-338

Pinti, D.L., and K. Hashizume (2001), ^{15}N-depleted nitrogen in Early Archean kerogens: clues on ancient marine chemosynthetic-based ecosystems?, *Precambrian Res.*, 105, 85-88.

Pinti, D.L., K. Hashizume, and J. Matsuda (2001), Nitrogen and argon signatures in 3.8 to 2.8 Ga metasediments: Clues on the chemical state of the Archean ocean and the deep biosphere, *Geochim. Cosmochim. Acta*, 65, 2301-2315.

Rancourt, D.G., P.-J. Thibault, D. Mavrocordatos, and G. Lamarche (2005), Hydrous ferric oxide precipitation in the presence of non-metabolizing bacteria: Constraints on the mechanism of a biotic effect, *Geochim. Cosmochim. Acta*, 69, 553-577.

Rasmussen, B. (2000), Filamentous microfossils in a 3,235-million-year-old volcanogenic massive sulphide deposit, *Nature*, 405, 676-679.

Robbins, E.I. (1987), *Appelella ferrifera*, a possible new iron-coated microfossil in the Isua Iron-Formation, Southwestern Greenland, in *Precambrian Iron Formations*, edited by P.W.U. Appel and G.L. LaBerge, pp. 141-154, Theophrastes, Athens, Greece.

Robbins, E.I., G.L. LaBerge, and R.G. Schmidt (1987), A model for the biological precipitation of Precambrian Iron Formations –B: Morphological evidence and modern analogs, in *Precambrian Iron Formations*, edited by P.W.U. Appel and G.L. LaBerge, pp. 97-139, Theophrastes, Athens, Greece.

Robbins, E.I. and A.S. Iberall (1991), Mineral remains of early life on Earth? On Mars?, *Geomicrobiol. J.*, 9, 51-66.

Rosing M.T. (1999), ^{13}C depleted carbon microparticles in >3700 Ma seafloor sedimentary rocks from West Greenland, *Science*, 283, 674-676.

Rosing, M.T., and R. Frei (2004), U-rich Archaean sea-floor sediments from Greenland: indications of >3700 Ma oxygenic photosynthesis, *Earth and Planet. Sci. Lett.*, 217, 237-217244.

Rothschild, L.J., and R.L. Mancinelli (2001), Life in Extreme Environments. *Nature*, 409, 1092-1101

Ryder, G. (2002), Mass influx in the ancient Earth-Moon system and benign implications for the origin of life on Earth, *J. Geophys. Res.*, 107, 10.1029/2001JE001583.

Rye, R., P.H. Kuo, and H.D. Holland (1995), Atmospheric carbon dioxide concentrations before 2.2 billion years ago, *Nature*, 378, 603-605.

Schidlowski, M. (1988), A 3800 million-year isotopic record of life from carbon in sedimentary rocks, *Nature*, 333, 313-318.

Schidlowski, M. (2001), Carbon isotopes as biogeochemical recorders of life over 3.8 Ga of Earth history: Evolution of a concept, *Precambrian Res.*, 106, 117-134.

Schoenberg, R., B.S. Kamber, K.D. Collerson, and M. Moorbath (2002), Tungsten isotope evidence from ~3.8-Ga metamorphosed sediments for early meteorite bombardment of the Earth, *Nature*, 418, 403-405

Schopf, J.W. (1993), Microfossils of the Early Archean Apex Chert: New evidence of the antiquity of life, *Science*, 260, 640-646.

Schopf, J.W. (1998), Tracing the roots of the universal tree of life, in *The Molecular Origins of Life*, edited by A. Brack, pp. 336-362, Cambridge Univ. Press, Cambridge, UK.

Schopf, J.W., and M.R. Walter (1983), Archean microfossils: new evidence of ancient microbes, in *Earth's Earliest Biosphere*, edited by J.W Schopf, pp. 214-239, Princeton Univ. Press, Princeton, New Jersey.

Schopf, J.W., and B.M. Packer (1987), Early Archean (3.3 billion to 3.5 billion-year-old) microfossils from Warawoona Group, Australia, *Science*, 237, 70-73.

Schopf, J.W., A.B. Kudryavtsvev, D.G. Agresti, T.J. Wdowiak, and A.D. Czaja (2002), Laser-Raman imagery of Earth's earliest fossils, *Nature*, 416, 73-76.

Sephton, M.A., A.B. Verchovsky, P.A. Bland, I. Gilmour, M.M. Grady, and I. P. Wright (2003), Investigating the variations in carbon and nitrogen isotopes in carbonaceous chondrites, *Geochim. Cosmochim. Acta*, 67, 2093-2108.

Shen, Y., R. Buick, and D.E. Canfield (2001), Isotopic evidence for microbial sulphate reduction in the early Archaean era, *Nature*, 410, 77-81.

Skrzypezak, A., S. Derenne, F. Robert, L. Binet, D. Gourier, J.-N. Rouzaud, and C. Clinard (2004), Characterization of the organic matter in an Archean Chert (Warrawoona, Australia), 35th Lunar and Planetary Sci. Conference, Abstract # 1241.

Sleep, N.H., K.J. Zahnle, J.F. Kasting, and H.J. Morowitz (1989), Annihilation of ecosystems by large asteroid impacts on the early Earth, *Nature*, 342, 139-142.

Stahl, L.J. (1994), Microbial mats: Ecophysiological interactions related to biogenic sediment stabilization, in *Biostabilization of sediments*, edited by W.E. Krumbein, D.M. Patterson and L. Stahl, *et al.*, pp. 41-54, Biblioteks und Informationssystem der Carl von Ossietzky Universität, Oldenburgh, Germany.

Summons R.E., L.L. Jahnke, J.M. Hope, and J.H. Logan (1999), 2-Methylhopanoids as biomarkers for cyanobacterial oxygenic photosynthesis, *Nature,* 400, 554-557.

Thurston, P.C., and L.D. Ayres (2004), Archaean and Proteozoic Greenstone Belts: Setting and evolution, in *The Precambrian Earth: Tempos and Events*, P.G Eriksson*, et al.*, pp.311-334, Elsevier, Amsterdam, The Netherlands.

Tice, M., and D.R. Lowe (2004), Photosynthetic microbial mats in the 3,416-Myr-old ocean: *Nature*, 431, 549-552

Toporski, J.K.W., F. Westall, K.A. Thomas-Keprta, A. Steele, and D.S. Mckay (2002), The simulated silicification of bacteria – new clues to the modes and timing of bacterial preservation and implications for the search for extraterrestrial microfossils, *Astrobiol.*, 2, 1-26.

Turley, C. (2002), The importance of "marine snow", *Microbiol. Today*, 29, 177-179.

Tyler, S.A., and E.S. Barghoorn (1954), Occurrence of structurally preserved plants in pre-Cambrian rocks of the Canadian shield, *Science*, 119, 606-608.

Ueno, Y., S. Maruyama, Y. Isozaki, and H. Yurimoto (2001a), Early Archean (ca. 3.5 Ga) microfossils and ^{13}C-depleted carbonaceous matter in the North Pole area, Western Australia: Field occurrence and geochemistry, in *Geochemistry and the Origin of Life*, Eds. S. Nakashsima, S. Maruyama, A. Brack, and B.F. Windley, pp. 203-236, Universal Acad. Press, Tokyo, Japan.

Ueno, Y., Y. Isozaki, H. Yurimoto, and S. Maruyama (2001b), Carbon isotopic signatures of individual Archean microfossils (?) from Western Australia, *Int. Geol. Rev.*, 43, 196-212.

Ueno, Y., H. Yoshioka, S. Maruyama, and Y. Isozaki (2001c), Carbon and nitrogen isotope geochemistry of kerogen-rich ssilica dikes in the ca. 3.5 Ga North Pole area, Western Australia: sub-seafloor biosphere in the Archean, in *International Archean Symposium extended abstracts*, edited by K.F. Cassidy, J.N. Dunphy, and M.J. van Kranendonk, pp. 99-101, AGSO-Geoscience Australia 2001/37, Perth, Australia.

Ueno, Y., H. Yurimoto, H. Yoshioka, T. Komiya, and S. Maruyama (2002), Ion microprobe analysis of graphite from ca. 3.8 Ga metasediments, Isua supracrustal belt, West Greenland: Relationship between metamorphism and carbon isotopic composition, *Geochim. Cosmochim. Acta*, 66, 1257-1268

Ueno, Y., H. Yoshioka, S. Maruyama, and Y. Isozaki (2004), Carbon isotopes and petrography of kerogens in ca. 3.5 Ga hydrothermal silica dikes in the North Pole area, Western Australia, *Geochim. Cosmochim. Acta*, 68, 573-589

van Kranendonk, M.J., G.E. Webb, and B.S. Kamber (2003), Geological and trace element evidence for a marine sedimentary environment of deposition and biogenicity of 3.45 Ga carbonates in the Pilbara, and support for a reducing Archean ocean, *Geobiol.*, 1, 91-108.

van Kranendonk, M.J., and F. Pirajno (2004), Geochemistry of metabasalts and hydrothermal alteration zones associated with *c.* 3.45 Ga chert and barite deposits: implications for the geological setting of the Warrawoona Group, Pilbara Craton, Australia, *Geochem. Explor., Environ., Anal.*, 4, 253-278.

van Zuilen, M., A. Lepland, and G. Arrhenius (2002), Reassessing the evidence for the earliest traces of life, *Nature*, 418, 627-630.

van Zuilen, M., A. Lepland, J. Teranes, J. Finarelli, M. Wahlen, and G. Arrhenius (2003), Graphite and carbonates in the 3.8 Ga old Isua Supracrustal Belt, southern West Greenland, *Precambrian Res.*, 126, 331-348.

van Zuilen, M.A., K. Mathew, B. Wopenka, A. Lepland, K. Marti, and A. Arrhenius (2005), Nitrogen and argon isotopic signatures in graphite from the 3.8-Ga-old Isua Supracrustal Belt, Southern West Greenland, *Geochim. Cosmochim. Acta*, 69, 1241-1252.

Walker, J.C.G. (1977), *Evolution of the atmosphere*. Macmillan, New York.

Walsh, M.M. (1992). Microfossils and possible microfossils from the Early Archean Onverwacht Group, Barberton Mountain Land, South Africa, *Precambrian Res.*, 54, 271-293.

Walsh, M.M. (2004), Evaluation of early Archean volcanoclastic and volcanic flow rocks as possible sites for carbonaceous fossil microbes, *Astrobiol.*, 4, 429-437.

Walsh, M.M., and D.R. Lowe (1985), Filamentous microfossils from the 3500 Myr-old Onverwacht Group, Barberton Mountain Land, South Africa, *Nature*, 314, 530-532.

Walsh, M.M., and D.R. Lowe (1999), Modes of accumulation of carbonaceous matter in the early Archaean: A petrographic and geochemical study of carbonaceous cherts from the Swaziland Supergroup, in *Geologic evolution of the Barberton greenstone belt, South Africa*, eds. DR. Lowe, G.R. Byerly, Geol. Soc. Am Spec. Paper, 329, pp. 115-132, Geol. Soc. Amer., Boulder, Colorado.

Walsh, M.M., and F. Westall (2003), Archean biofilms preserved in the 3.2-3.6 Ga Swaziland Supergroup, South Africa, in *Fossil and Recent Biofilms,* edited by W.E. Krumbein, T. Dornieden, and M. Volkmann, pp. 307-316, Kluwer, Amsterdam, The Netherlands.

Walter, M.R. (1976), *Stromatolites*, Springer Verlag, Berlin.

Walter, M.R. (1983), Archean stromatolites: evidence of the Earth's earliest benthos, in *Earth's Earliest Biosphere,* edited by J.W. Schopf, pp. 187-213, Princeton University Press, Princeton, New Jersey.

Walter, M.R., R. Buick, and J.S.R. Dunlop (1980), Stromatolites 3,400-3,500 Myr old from the North Pole area, Western Australia, *Nature*, 284, 443-445.

Westall, F. (1999), The nature of fossil bacteria, *J. Geophys. Res., Planet.*, 104, 16,437-16,451.

Westall, F. (2003), Stephen Jay Gould, les procaryotes et leur évolution dans le contexte géologique. *Palevol*, 2, 485-501.

Westall, F. (2004), Early life on earth: The ancient fossil record, *in Astrobiology: Future Perspectives,* edited by P. Ehrenfreund, P., *et al.*, pp. 287-316, Kluwer, Dordrecht, The Netherlands.

Westall, F. (2005a), Life on the early Earth: A sedimentary view, *Science*, 434, 366-367.

Westall, F., (2005b), The geological context for the origin of life and the mineral signatures of fossil life, in *Lectures in Astrobiology*, Vol. 1, edited by M. Gargaud *et al.*, pp. 195-219, Speinger Verlag, Berlin, Germany.

Westall, F., L. Boni, and M.E. Guerzoni (1995), The experimental silicification of microbes, *Palaeontol.*, 38, 495-528.

Westall, F., and D. Gerneke (1998), Electron microscope methods in the search for the earliest life forms on Earth (in 3.5-3.3 Ga cherts from the Barberton greenstone belt, South Africa): applications for extraterrestrial studies, *SPIE: Instrum., Meth., Miss. Astrobiol.*, 3441, 158-169.

Westall, F., and M.M. Walsh (2000), The diversity of fossil microorganisms in Archaean-age rocks. In, *Journey to Diverse Microbial Worlds*, edited by J. Seckbach, pp. 15-27, Kluwer, Amsterdam, The Netherlands.

Westall, F., and R.L. Folk (2003), Exogenous carbonaceous microstructures in Early Archaean cherts and BIFs from the Isua greenstone belt: Implications for the search for life in ancient rocks, *Precambrian Res.*, 126, 313-330.

Westall, F., A. Steele, J. Toporski, M. Walsh, C. Allen, S. Guidry, E. Gibson, D. Mckay, and H. Chafetz (2000), Polymeric substances and biofilms as biomarkers in terrestrial materials: Implications for extraterrestrial samples, *J. Geophys. Res. Planets*, 105, 24,511-24,527.

Westall, F., M.J. De Wit, J. Dann, S. van Der Gaast., C. De Ronde., and D. Gerneke (2001a), Early Archaean fossil bacteria and biofilms in hydrothermally influenced, shallow water sediments, Barberton Greenstone Belt, South Africa, *Precambrian Res.*, 106, 94-112.

Westall, F., A. Brack, B. Barbier, M. Bertrand, and A. Chabin (2002), Early Earth and early life: an extreme environment and extremophiles - application to the search for life on Mars. *Proceedings of the Second European Workshop on Exo/Astrobiology Graz, Austria, 16-19 September 2002*, ESA SP-518, pp. 131-136, European Space Agency, Nordwijk, The Netherlands.

Westall, F., M.M. Walsh, J. Toporski, J. and A. Steele (2003), Fossil biofilms and the search for life on Mars, In, *Recent and Fossil Biofilms*, edited by W.E. Krumbein *et al.*, pp. 447-465, Kluwer, Dordrecht, The Netherlands.

Westall, F., O. Orberger, V. Rouchon, J.-N. Rouzaud, and I. Wright (2004), On the identification of Early Archaean microfossils in cherts from Barberton and the Pilbara, in *Field Forum on Processes on the Early Earth Abstract Volume, Kaapvaal Craton, 4-9 July, 2004*, compiled by Reimold, W.U. and Hofmann, A., pp. 94-97, Univ. Witwatersrand, Johannesburg, South Africa.

Westall, F., S.T. de Vries, W. Nijman, V. Rouchon, B. Orberger, V. Pearson, J. Watson, A. Verchovsky, I. Wright, J.-N. Rouzaud, D. Marchesini, and S. Anne (2006), The 3.466 Ga Kitty's Gap Chert, an Early Archaean microbial ecosystem, in Reimold, W.U. and Gibson, R., *Processes on the Early Earth, Geol. Soc. Amer.* SP., 405, in press.

Yushkin, N.P. (2000), Biomineral homologies, organismobiosis, and the problem of biomarkers, in Instruments, Methods, and Missions for Astrobiology III, Proceedings of SPIE, 4137, edited by R.B. Hoover, pp. 22-35, the International Society for Optical Engineering, Bellingham, Washington.

G. Southam, Department of Earth Sciences, University of Western Ontario, London, Ontario, Canada N6A 5B7. (gsoutham@uwo.ca)

F. Westall, Centre de Biophysique Moléculaire, CNRS, Rue Charles Sadron, 45071 Orléans cedex 02, France. (westall@cnrs-orleans.fr)

Biogeochemical Cycles of Sulfur and Nitrogen in the Archean Ocean and Atmosphere

Yanan Shen and Daniele L. Pinti

Département des Sciences de la Terre et de l'Atmosphère and GEOTOP-UQAM-McGill,
Université du Québec à Montréal, Montréal, Québec, Canada

Ko Hashizume

Department of Earth and Space Sciences, Osaka University, Osaka, Japan

Early geodynamic processes modeled the surface of the Earth, providing suitable environments and energy sources for the development of life. Studying the ancient biogeochemical cycles is thus a valuable way to understand the physico-chemical conditions of the early environments, such as the redox state of the primitive oceans and the atmosphere. To investigate early biogeochemistries, the measurement of stable isotopes of S, N, and C from ancient sedimentary rocks is crucial. Here we review the sulfur and nitrogen isotopic signatures measured in Archean rocks that suggest the antiquity of sulfate-reducing bacteria and chemolithotrophs living in a low-oxygen environment.

1. INTRODUCTION

The interest in studying ancient biogeochemistries is twofold. The first, the recognition and understanding of ancient primitive metabolic pathways, is imperative for the reconstruction of the tree of life. The second, the evolution of biogeochemical cycles, is directly related to the physico-chemical characteristics of the terrestrial reservoirs, such as the redox state of the atmosphere and the oceans. Understanding Archean metabolic pathways and biogeochemical cycles not only will yield coherent information about the state of the Archean Earth, but also will shed light on the geodynamic processes that modeled these ancient environments. In order to investigate early metabolisms, there are three possible pathways, including searching for microfossils, extracting molecular fossils, and measuring stable isotopes from ancient sedimentary rocks. Among them, the measurement of stable isotopes can be taken as a biomarker because biological isotopic fractionations of carbon, sulfur, or nitrogen usually have a signature distinct from that of non-biological processes [e.g., *Schidlowski et al.*, 1983; *Boyd*, 2001]. In this paper, we will briefly review the biochemical processes of isotopic fractionation during sulfate reduction as well as that of nitrogen in modern marine and hydrothermal settings. In combination with isotopic records from Archean sedimentary rocks, we will then discuss the antiquity of these metabolic processes and their implications for early evolution of the oceans and atmosphere.

2. SULFUR

2.1. Microbial Sulfate Reduction and Isotopic Fractionation

2.1.1. Biochemical processes. Microbial sulfate reduction is an energy-yielding metabolic process during which sulfate is reduced and sulfide is released, coupled with the oxidation of organic matter or hydrogen (H_2):

$$SO_4^{2-} + CH_2O \rightarrow H_2S + HCO_3^- \qquad (1)$$

$$SO_4^{2-} + H_2 \rightarrow H_2S + H_2O \qquad (2)$$

Archean Geodynamics and Environments
Geophysical Monograph Series 164
10.1029/164GM19

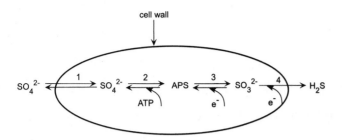

Figure 1. The biochemical pathway of dissimilatory sulfate reduction.

These energy-yielding reactions are carried out by several groups of prokaryotes that use them to obtain energy for growth and maintenance [e.g., *Widdle*, 1988]. The biochemical processes of dissimilatory sulfate reduction can be divided into two major steps (Figure 1), though the actual biochemical pathway may be more complicated and variable among microbial species [*Widdel and Hansen*, 1992]. The first step is endergonic reduction to sulfite catalyzed by soluble enzymes (steps 1–3) and this is reversible, which may allow kinetic isotopic effects to be expressed during sulfite reduction [*Rees*, 1973]. The second step involves exergonic sulfite reduction to sulfide catalyzed by membrane-bound enzymes (step 4) (Figure 1). It has been shown that the uptake of sulfate into the cell and then reduction by ATP sulfurylase is associated with only small fractionations [*Rees*, 1973]. However, reductive processes in the cytoplasm are accompanied by large fractionations, because APS reduction to sulfite and sulfite reduction to sulfide involve the breaking of S-O bonds. During these processes, the lighter sulfur isotope (^{32}S) tends to react faster than heavier sulfur (^{34}S) because the bonding energy of ^{32}S-O is less than for ^{34}S-O and thus breaking the ^{32}S-O bond is easier. Hence, the product H_2S is enriched in ^{32}S and depleted in ^{34}S.

2.1.2. Isotopic fractionation. Isotopic fractionation during sulfate reduction by pure cultures have shown large fractionations, ranging from 2‰ to 49‰ [e.g., Chambers and *Trudinger*, 1979; *Bollinger et al.*, 2001] (Figure 2). The considerable variations in isotopic fractionations may be controlled by many factors, such as sulfate concentration, electron donor, temperature, specific rate of sulfate reduction, and growth conditions [*Chambers and Trudinger*, 1979]. Importantly, *Detmers et al.* [2001] discovered that sulfate reducers that oxidized the carbon source completely to CO_2 show greater fractionation than those that release acetate as the final product of carbon oxidation during sulfate reduction. Moreover, *Detmers et al.* [2001] demonstrated that there is no relationship between phylogenetic positions, as measured on the 16S rRNA Tree of Life, and isotopic fractionation. Even the deepest branching shows fractionations similar to those of more recently evolved groups. Hence, it is likely that

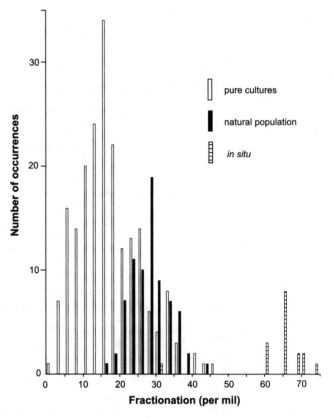

Figure 2. The histogram of sulfur isotope fractionation by pure cultures and natural populations of sulfate reducers (from *Shen and Buick* [2004]).

ancient sulfate-reducers fractionated isotopes in a similar way as their modern counterparts do.

Recent work has shown that natural populations of sulfate-reducers produce fractionations comparable to those of pure cultures, with an upper limit of 45‰ [*Habicht and Canfield*, 1997]. With non-limiting sulfate, minimum fractionations are rarely lower than 10‰ in sediments, unless excess organic substrate is added [*Böttcher et al.*, 1997]. So, it is likely that in natural environments with abundant supply of sulfate, the sulfides are isotopically fractionated by 10–40‰ relative to the sulfates.

However, the metabolic activities of sulfate-reducers in the deep biosphere might be different from those in near-surface environments. Coexisting dissolved sulfide and sulfate from hypersulfidic interstitial waters in deep subsurface ocean sediments show unusually large fractionations, up to 70‰ [*Wortmann et al.*, 2001], which is consistent with modeling result of pore-water sulfate in deep marine sediments [*Rudnicki et al.*, 2001] (Figure 2). Hence, we are left with the challenge to understand the mechanisms of isotopic fractionation during sulfate reduction in poorly known environments;

these may provide new insights into our interpretation of biogeochemical records of S isotopes over geological history.

2.2. Preservation of the Isotopic Record

Microbial sulfate reduction produces large amounts of free H_2S, which mostly reacts with heavy metal ions, mostly iron [*Berner*, 1984]:

$$Fe^{2+} + H_2S \rightarrow FeS + 2H^+ \qquad (3)$$

The metastable iron sulfides produced can react with intermediate sulfur species such as elemental sulfur or S_x^{2-} to form pyrite [e.g., *Wilkin and Barnes*, 1996]:

$$FeS + S^\circ \rightarrow FeS_2 \qquad (4)$$

$$FeS + S_x^{2-} \rightarrow FeS_2 + S_{(x-1)}^{2-} \qquad (5)$$

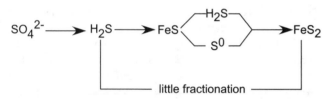

Figure 3. The schematic pathways of pyrite formation and associated isotopic fractionation.

Alternatively, pyrites can be formed through the oxidation of FeS, during which H_2S acts as the oxidizing agent [*Drobner et al.*, 1991; *Rickard*, 1997]:

$$FeS + H_2S \rightarrow FeS_2 + H_2 \qquad (6)$$

The isotopic fractionation associated with pyrite formation from dissolved sulfide at low temperatures is <1‰ [*Price and Shieh*, 1979]. Thus, primary isotopic signatures of sulfate reduction can be preserved in metal sulfides in sedimentary rocks (Figure 3). Therefore, the S-isotope signature of sedimentary sulfides can be a powerful tool for studying early life and environment.

2.3. Archean S-Isotopic Record

The oldest terrestrial S-isotopic records are from metamorphosed and deformed ferruginous rocks in the Isua Supracrustal Belt, Greenland (~3.8 Ga), which show a narrow range with an average of +0.5‰ ± 0.9‰ [*Monster et al.*, 1979; *Strauss*, 2003] (Figure 4). With few exceptions, sedimentary sulfides between 3.8 and 2.8 Ga are characterized by similarly narrow ranges, with isotopic compositions of −5‰ to +10‰ around an inferred d34S of coeval seawater sulfate of about +3–5‰ (Figure 4). In contrast, sedimentary sulfides in the ~2.7-Gyr-old Michipicoten and Woman River Iron Formations in Canada as well as in the Manjeri Formation in the Belingwe Belt in Zimbabwe are distinctly shifted to the negative, with $\delta^{34}S$ values as low as −19.9‰ and widely spread [*Goodwin et al.*, 1976; *Grassineau et al.*, 2001]

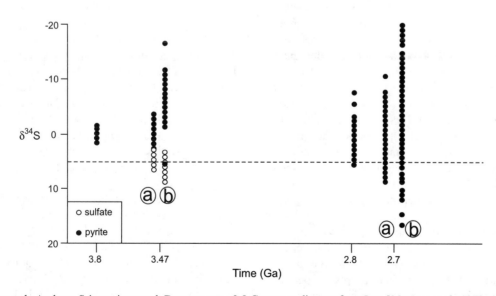

Figure 4. The early Archean S-isotopic record. Data source: ~3.8 Ga metasediments from Isua [*Monster et al.*, 1979; *Strauss*, 2003]; ~3.46 Ga North Pole barite deposite (a: *Lambert et al.* [1978]; b: *Shen et al.* [2001]); ~2.8 Ga sedimentary rocks [*Donnelly et al.*, 1977]; ~2.7 Ga sedimentary rocks (a: *Goodwin et al.* [1977]; b: *Grassineau et al.* [2000]). The dashed line represents a probable S-isotopic value of the early Archean seawater sulfate. Note that the geological time is not on scale.

(Figure 4). Thus, the $\delta^{34}S$ distributions of these sedimentary sulfides provide compelling evidence that microbial sulfate reduction had evolved by 2.7 Ga [*Goodwin et al.*, 1976; *Grassineau et al.*, 2001].

Though microbial sulfate reduction had evolved by 2.7 Ga, the interpretations of the small fractionations and narrow range of S-isotope values from older rocks remain controversial. The minimally fractionated early Archean sulfides have been attributed to either biological sulfate reduction at low sulfate concentrations, implying poorly oxygenated oceans, or a non-biological volcanogenic origin, implying sulfate reducers had not yet evolved [*Cameron*, 1982; *Walker and Brimblecombe*, 1985; *Canfield and Raiswell*, 1999]. By contrast, the competing model argued that the small fractionations of early Archean sulfides were formed by active biological sulfate reduction in warm and oxygenated Archean oceans [*Ohmoto and Felder*, 1987; *Ohmoto et al.*, 1993]. If the model of low sulfate concentration in the early Archean oceans is correct, some isolated sulfate-rich Archean basins must have produced large fractionations between sulfate and sulfides. Indeed, there are relics of some isolated sulfate-rich basins in the early Archean geological record [e.g., *Lambert et al.*, 1978; *Buick and Dunlop*, 1990]. Apparently, these isolated sulfate-rich sedimentary environments are excellent targets for investigating the metabolism of microbial sulfate reduction early in Earth's history.

2.4. A New Window on Antiquity of Microbial Sulfate Reduction

2.4.1. Geological setting. One of the best-known sites of sulfate-rich sedimentation in the early Archean is at an area in northwestern Australia named North Pole. The rocks there belong to the >3.46 Ga Warrawoona Group, a predominantly basaltic unit about 10 km thick [*Buick et al.*, 1995]. The rocks are remarkably well-preserved and have generally undergone only low-strain brittle deformation [*Dunlop and Buick*, 1981]. Pillow margins from drill-core contain prehnite and pumpellyite or epidote and actinolite as prograde mineral assemblages, indicating peak metamorphic grades of prehnite–pumpellyite to lowermost greenschist facies [*Dunlop and Buick*, 1981].

The best developed sulfate deposit is in the basal unit of the Dresser Formation, which is ~40 m thick. In this, and in all other sedimentary horizons within the Warrawoona Group at North Pole, original sedimentary minerals have been largely replaced by hydrothermal silica, now chert. However, primary textures and structures are in places perfectly preserved, defined by inclusions of original iron-titanium oxides, by metamorphic sericite and chlorite replacing original volcanic glass, or by microinclusions of dolomite after original carbonates [*Buick et al.*, 1981]. Large sulfate crystals

are now composed of barite ($BaSO_4$). Small sulfate crystals have been totally replaced by silica, now microquartz. Some of those intermediate in size have a barite core and a microquartz rind, and all of the large barite crystals have been silicified marginally to a depth of 1–2 mm [*Buick and Dunlop*, 1990].

Interfacial angle measurements on the diagenetic crystals show that they were initially composed of gypsum ($CaSO_4 \cdot 2H_2O$) that grew before lithification, as indicated by their cores of incorporated sediment [*Lambert et al.*, 1978; *Buick and Dunlop*, 1990]. Primary sulfate crystals precipitated in the back-barrier brine ponds, forming lenticular beds up to 0.5 km in lateral extent and up to 10 m thick. Individual layers of sulfate crystals (now dark-gray barite) are up to 15 cm thick, composed of bottom-nucleated sub-radiating fans of bladed hemipyramids [*Buick and Dunlop*, 1990]. That gypsum was indeed primary is confirmed by the presence on the draped interfaces of broken, eroded, and rounded crystals filling interstices between protruding crystal fans [*Buick and Dunlop*, 1990; *Shen and Buick*, 2004].

Barite replacement of gypsum occurred soon after diagenesis began and shortly after burial by the overlying basalts, as shown by the perfect preservation of crystal morphologies of the highly soluble phase gypsum. The most likely barium source was hydrothermal brines that percolated into the porous sediments from the surrounding hot basalt pile [*Lambert et al.*, 1978; *Buick and Dunlop*, 1990]. Several S-bearing species are associated with the North Pole barite deposit, including the bedded barite, the vein barite, and the pyrite laminae as well as the microscopic pyrites [*Buick and Dunlop*, 1990; *Shen et al.*, 2001; *Shen and Buick*, 2004].

2.4.2. S-isotopic results from North Pole barite deposits. The $\delta^{34}S$ data for sulfates are between +3.2‰ and +5‰ with an average of +4.3‰ [*Lambert et al.*, 1978; *Shen et al.*, 2001] (Figure 4). The two reduced phases of sulfur produce contrasting signatures. The macroscopic pyrite laminae show slightly ^{34}S-depleted values between −0.9‰ and −3.5‰ with an average of −2.4‰ [*Lambert et al.*, 1978; *Shen et al.*, 2001]. While these pyrites could represent unfractionated volcanogenic sulfur, they are consistently more ^{34}S-depleted than is typical of younger volcanogenic sulfide deposits. However, they are not sufficiently ^{34}S-depleted to be definitely biological, so their origin is still equivocal [*Buick and Dunlop*, 1990; *Shen and Buick*, 2004].

Most importantly, microscopic pyrites in the bedded barites are highly ^{34}S-depleted, with fractionations relative to coexisting sulfate ranging from 21.1‰ to 7.4‰, with an average of 11.6‰ [*Shen et al.*, 2001] (Figure 4). These microscopic pyrites in bedded barites at North Pole are intimately associated with organic substrate in the form of kerogen [*Buick and Dunlop*, 1990]. In addition, the $\delta^{34}S_{sulfide}$

values, from −1.3‰ to −16.8‰, show a clear negative trend relative to the original sulfate and a wide spread about the mean, typically as the result of biological sulfate reduction. Moreover, isotopic fractionations are within the range observed for modern sulfate-reducers metabolizing with >1 mM sulfate, for which typical fractionation is 10‰ to 26‰ [e.g., *Chambers and Trudinger*, 1979]. Therefore, the results from North Pole barite deposits demonstrate that microbial sulfate reduction had evolved by ~3.47 Ga [*Shen et al.*, 2001; *Shen and Buick*, 2004].

2.5. The Early Archean Atmosphere and Oceans

As the sulfur cycle is involved in atmospheric oxygen regulation, Precambrian sulfur isotopic records over geological time have been used to help understand the evolution of atmospheric oxygenation [e.g., *Canfield and Raiswell*, 1999; *Strauss*, 2004]. It is generally accepted that the burial of biogenic pyrites as well as organic matter in marine sediments and the weathering of pyrites on the continents may primarily control the balance of atmospheric oxygen in a geological time scale. The qualitative relationship between atmospheric oxygen and seawater sulfate is through the oxidative weathering of sulfide minerals on the continents and the subsequent delivery of dissolved sulfate to the oceans. Accordingly, less oxidative weathering of sulfides and thus low concentrations of seawater sulfate would be consistent with lower levels of atmospheric oxygen [*Hayes et al.*, 1992; *Canfield and Raiswell*, 1999; *Shen et al.*, 2003]. By contrast, a sulfate-rich Archean ocean would provide supportive evidence for high Archean atmospheric oxygen concentrations [*Ohmoto and Felder*, 1987; *Ohmoto et al.*, 1993].

As discussed above, most Archean sedimentary sulfides older than ~2.7 Ga are characterized by low fractionations (<10‰). By contrast, the isotopic data from North Pole barite deposits show a large spread of $\delta^{34}S$ values and large S-isotopic fractionations (Figure 4), and they provide strong evidence for that the metabolic process of biological sulfate reduction had evolved by 3.47 Ga. The differences between the North Pole deposit and other Archean sedimentary sulfides probably reflect environmental variations. The evaporite ponds at North Pole were localized oases of high sulfate concentrations maintained by evaporative concentration and hydrologic semi-isolation and containing organic electron donors derived from an indigenous stromatolitic microbiota [*Buick et al.*, 1981; *Buick and Dunlop*, 1990]. Thus, microbial sulfate reduction could operate under favorable conditions and consequently induce large isotopic fractionations. Accordingly, other Archean sedimentary sulfides characterized by small fractionations may have resulted from microbial sulfate reduction under low oceanic sulfate concentrations. The results from North Pole barite deposits

thus support models of low atmospheric oxygen during the early Archean.

Evidence for low atmospheric oxygen in the Archean also comes from multiple S-isotopic data. Recent isotopic measurements of the stable isotopes $\delta^{33}S$ and $\delta^{36}S$ from Precambrian sulfides and sulfates show anomalous mass-independent fractionation signature [*Farquhar et al.*, 2000; *Bekker et al.*, 2004; *Ono et al.*, 2003; *Mojzsis et al.*, 2003]. The photochemical experiments demonstrated that photochemical reactions of SO_2 and SO produce the S-isotopic signature similar to those observed from the Archean and early Proterozoic rocks [*Farquhar et al.*, 2001]. *Farquhar et al.* [2001] suggest that those photochemical reactions could operate significantly only when the atmospheric oxygen is low. Therefore, mass-independent S-isotopic records from the Archean and early Proterozoic rocks indicate that gas-phase atmospheric reactions may have played an important role in the Archean sulfur cycle and that the early Archean atmosphere was nearly free of oxygen [*Farquhar et al.*, 2000].

3. NITROGEN

3.1. Nitrogen in Silicate Rocks

Nitrogen occurs in silicate rocks mainly as ammonium (NH_4^+) in two forms: exchangeable ammonium, adsorbed on the charged surfaces of clay minerals and representing organic nitrogen that has been decomposed in sediments most recently; and fixed ammonium, which is strongly bound in the lattice structure of K-bearing minerals, where it can replace K^+ ions [*Honma and Itihara*, 1981]. It is generally assumed that the $^{15}N/^{14}N$ ratio (expressed as $\delta^{15}N$) of lattice-bound ammonium is mainly constrained by the N-isotopic composition of organic nitrogen compounds that have been decomposed and partially transformed into ammonium during early *diagenesis* in sediments. Thus the $\delta^{15}N$ of lattice-bound, or diagenetic, NH_4^+ has the potential to function as a paleoindicator [*Scholten*, 1991]. First studies on ammonium as a bioindicator in Precambrian rocks were carried out in the late '80s [*Schidlowski et al.*, 1983; *Zhang*, 1988; *Sano and Pillinger*, 1990] but not until recently did it become the focus of much attention [*Boyd and Philippot*, 1998; *Beaumont and Robert*, 1999; *Boyd*, 2001; *Pinti*, 2002; *Pinti and Hashizume*, 2001; *Pinti et al.*, 2001; *Jia and Kerrich*, 2004a; *Ueno et al.*, 2004; *Papineau et al.*, 2005; *Van Zuilen et al.*, 2005]. These studies added reliable data to the very poor data set of Precambrian N and initiated debates on the evolution of the nitrogen cycle on Earth [*Beaumont and Robert*, 1999; *Pinti and Hashizume*, 2001; *Marty and Dauphas*, 2003, 2004; *Jia and Kerrich*, 2004b].

3.2. The Nitrogen Biogeochemical Cycle in the Modern Ocean

For a better understanding of the N cycle in the past, one may need to look at the N cycling in modern marine environments (Figure 5). The processes of *ammonification* (Figure 5; "m"), *nitrification* ("n"), *denitrification* ("d"), and *assimilation* ("a") dominate the N cycle in the modern ocean. The kinetic fractionation of nitrogen resulting from this multi-step process produces an average $\delta^{15}N = 5$‰ for the dissolved organic nitrogen (DON; termed "Organic-N" in Figure 5) [*Sigman et al.*, 2000; *Sigman and Casciotti*, 2001].

Atmospheric N_2 enters the marine nitrogen cycle via the process of biological N_2 fixation ("f" in Figure 5). N_2-fixing organisms incorporate molecular nitrogen with little isotopic fractionation [$\varepsilon = -3$ to +1; *Fogel and Cifuentes*, 1993]. The energetically more favorable and more common form of N uptake is the consumption of fixed (or biologically available)

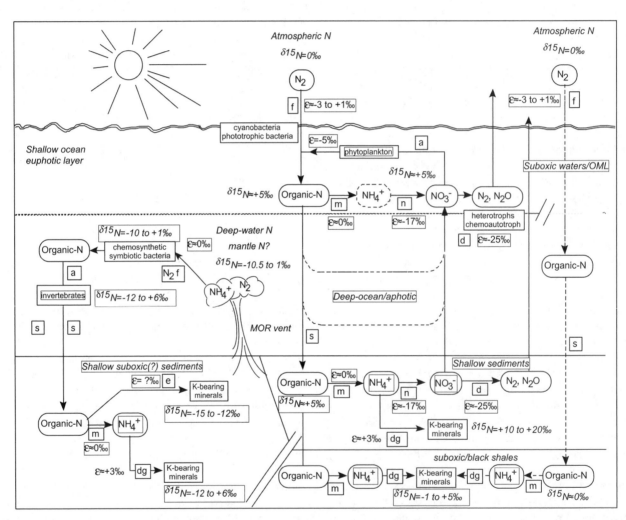

Figure 5. Schematic nitrogen cycling in marine and deep-sea hydrothermal environments. Isotopic shifts of nitrogen for the produced species are expressed by the symbol $\varepsilon = (k^{15}/k^{14} - 1)*1000$ and are from compilations of *Fogel and Cifuentes* [1993]; *Sigman and Casciotti* [2001]; and M. Lehmann (internal compilation of data). The measured $\delta^{15}N$ values (outlined in the figure) are from chemosynthetic prokaryotes [*Colaço et al.*, 2002; *Van Dover*, 2002; *Van Dover and Fry*, 1994; *Van Dover et al.*, 2001; *Kennicut and Burke*, 1995; *Conway et al.*, 1994]; invertebrates [*Burd et al.*, 2002; *Van Dover*, 2002; *Van Dover and Fry*, 1994; *Conway et al.*, 1994]; deep-ocean N_2 [*Kennicut and Burke*, 1995]; mantle nitrogen [*Marty et al.*, 1991; *Marty and Humbert*, 1997; *Marty and Zimmermann*, 1999]; and ammonium in marine sediments derived from biological fixation [*Rau et al.*, 1987] and from nitrification/denitrification processes [*Peters et al.*, 1978; *Sweeney et al.*, 1978; *Muzuka et al.*, 1991; *Holmes et al.*, 1996a, b; *De Lange*, 1998; *Cowie et al.*, 1998; *Milder et al.*, 1999; *Sadofsky and Bebout*, 2004]. Symbols: f, fixation; m, mineralization; n, nitrification; d, denitrification; a, assimilation (called also uptake); s, sedimentation; dg, diagenesis.

nitrogen. Primary producers utilize (assimilate) various forms of fixed N, but nitrate (NO_3^-) is by far the dominant species of fixed N in the modern ocean. The biological isotope fractionation associated with nitrate uptake (or nitrate assimilation; "a" in Figure 5) has been shown to be variable in laboratory experiments, but estimates for natural marine environments indicate a relatively constant N isotope effect of 4–6‰ [e.g., *Sigman and Casciotti*, 2001].

The transformation of organic-N back to nitrate is a multistep process. Organic matter degradation and mineralization (Figure 5; "m") refers to the enzymatic breakdown of organic matter into short-chain organic compounds (e.g., amino acids, proteins) and the subsequent release of ammonium (ammonification; "a"). Remineralization of organic N to ammonium is not associated with any significant nitrogen fractionation [$\varepsilon \leq -5$ to +0; *Sigman and Casciotti*, 2001]. Then, under aerobic conditions, NH_4^+ is rapidly oxidized to nitrite (NO_2^-) and subsequently to nitrate (NO_3^-) in a two-step process called nitrification (Figure 5; "n"). *Casciotti et al.* [2003] reported a significant isotopic fractionation for nitrification ($\varepsilon = -17‰$). Nitrate that has been produced during the remineralization of organic matter can be transported to surface waters, where it is available for re-uptake by phytoplankton.

In suboxic environments, nitrate is stepwise reduced to N_2O and/or N_2 by heterotrophic bacteria that can use NO_3^- as an alternative electron acceptor when O_2 is not readily available (denitrification; Figure 5, "d"). An important aspect of denitrification, and highly relevant for the marine N budget, is that it produces gaseous forms of N that escape to the atmosphere, making it the most important sink of fixed nitrogen in the ocean [e.g., *Codispoti et al.*, 2001; *Lehmann et al.*, 2004]. Similar to nitrate-assimilating algae, denitrifying bacteria preferentially use $^{14}NO_3^-$; the gaseous N_2 and N_2O products are thus depleted in ^{15}N, whereas the residual dissolved NO_3^- becomes enriched in ^{15}N. Yet, at least in the open marine environment, denitrification produces a much larger isotope effect ($\varepsilon = -25‰$) than nitrate assimilation [*Cline and Kaplan*, 1975; *Sigman et al.*, 2003; *Lehmann et al.*, 2004], which is responsible for the enrichment in ^{15}N of oceanic nitrate (relative to atmospheric nitrogen).

The mean $\delta^{15}N$ of nitrate in the ocean is +5‰. In most surface waters the nitrate supply from below is almost completely utilized by primary producers, with the consequence that the $\delta^{15}N$ of the sinking flux (and of sedimentary organic matter prior to diagenetic changes) will be close to 5‰. Indeed, marine sediments seem to have a relatively constant bulk $\delta^{15}N$ value of +5‰ to +7‰, similar to that of seawater nitrate [*Peters et al.*, 1978].

The fate of organic-N in sediments depends largely upon redox conditions. In aerobic surface sediments, NH_4^+ from organic matter degradation is readily converted to NO_3^-.

Deeper in the sediment column, where sediments are anaerobic (Figure 5), the ammonium is not rapidly oxidized by nitrifying bacteria (although anaerobic oxidation of ammonium is also possible). During *diagenesis* (Figure 5; "dg"), NH_4^+ is partially adsorbed to, or fixed in, clay minerals, such as illite or smectite [*Williams et al.*, 1995] or replaces K^+ in the lattice structure of biotite, muscovite, or K-feldspar [*Honma and Itihara*, 1981]. The isotopic ratio of fixed NH_4^+ in minerals depends on the balance between N fluxes involved in the sediments and at the sediment–water interface (nitrification, anaerobic ammonium oxidation, and the simple efflux of ammonium out of the sediments) but is most likely higher ($\delta^{15}N = +10‰$ to +20‰) than the original organic-N isotopic signature ($\delta^{15}N = +5‰$) because of the partial NH_4^+ oxidation within the sediment [*Lehmann et al.*, 2004]. It is generally assumed that diagenesis does not significantly change the N isotopic signature in sediments and that the $\delta^{15}N_{sediment}$ reflects the $\delta^{15}N$ of porewater NH_4^+ and, possibly, the $\delta^{15}N$ of sedimentary organic matter [e.g., *Boyd*, 2001]. Whether the isotopic composition of mineral-phase associated N also reflects primary isotope signals recorded during biosynthesis in the ocean water column remains unclear. Recent work by *Lehmann et al.* [2002] showed that a slight negative isotope shift of 3‰ occurs during microbial degradation of organic matter during suboxic early sedimentary diagenesis, but early diagenetic N isotopic shifts towards more positive values have also been reported [*Freudenthal et al.*, 2001].

The dominant source of energy driving biochemical cycles in the ocean is the sunlight. However in certain marine environments, such as deep-sea hydrothermal vents, microbial chemoautotrophic primary production can replace photosynthesis [*Conway et al.*, 1994 and references therein]. The term chemosynthesis, or more correctly chemoautolithotrophy, describes the synthesis of organic carbon compounds from CO_2 using energy and reducing power derived from the oxidation of inorganic compounds, such as H_2S, S_2O_3, CH_4, H_2 and NH_4^+, which support microbial chemosynthesis [*Jannasch and Mottl*, 1985]. In deep-sea hydrothermal vents at Mid-Ocean Ridges (MOR), reactions of seawater with crustal rocks at high temperatures produce the reduced chemical species used as the source of energy for reduction of CO_2 to organic carbon. Here, the predominant metabolic process could be N-fixation [Figure 5; *Kennicut and Burke*, 1995]. The N isotopic signature of the biomass produced by chemosynthetic processes is significantly depleted in ^{15}N. Typical $\delta^{15}N$ values of microbial mats at MOR range from −9.6‰ to +0.9‰ [Figure 6; *Van Dover and Fry*, 1994; *Noda et al.*, 2002; *Kennicut and Burke*, 1995; *Conway et al.*, 1994; and *Colaço et al.*, 2002]. Numerous invertebrate species that feed on vent-associated microorganisms display equally ^{15}N-depleted isotope signatures [$\delta^{15}N$ from −12‰ up to +6‰; *Van Dover and Fry*, 1994]. This organic matter may

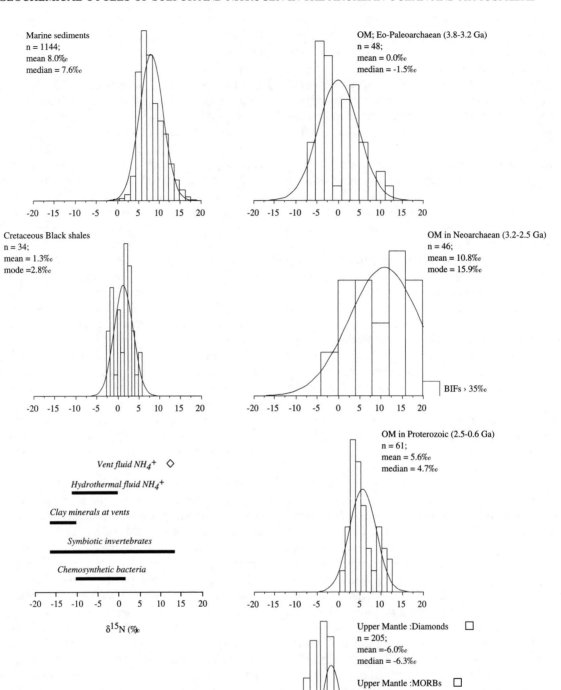

leave its light N isotopic signature in the NH_4^+ adsorbed to, or trapped in, vent-associated clay minerals [*Lilley et al.*, 1993] or by exchange with rocks during hydrothermal circulation [*Orberger et al.*, 2005a]. The NH_4^+ found in clay minerals of altered (i.e., ion exchange during hydrothermal water-rock interaction), volcanic rocks at the Suiyo Seamount, shows $\delta^{15}N$ values ranging from −14.7‰ to −10.5‰ [*Noda et al.*, 2002]. These values are in the range of those observed in chemosynthetic assemblages around hydrothermal vents, suggesting that primary organic-N isotopic signatures can be preserved even in heavily altered rocks. However, the strongly ^{15}N-depleted isotopic signature of chemosynthetic primary producers remains unexplained [*Kennicut and Burke*, 1995]. If metabolic processes were controlled by biological N_2-fixation, we would expect only small isotope N isotope fractionation (Figure 5). *Rau* [1981] suggested that the inorganic N source used by microbial producers in vent environments is mantle-N, which is known to be ^{15}N-depleted [−6‰ to −1.8‰; *Marty and Zimmermann*, 1999; *Sano et al.*, 1998; *Cartigny et al.*, 1997; 1998b; and Figure 6]. Given a negligible isotope effect for the fixation of N_2, a pure mantle N source can barely explain the observed negative $\delta^{15}N$ values of the chemosynthetic ecosystems [as low as −12‰; *Conway et al.*, 1994]. As a consequence, if the N_2 that is fluxing out of hydrothermal vents indeed reflects N from the upper mantle ($\delta^{15}N = -3 \pm 2$‰), either the isotope fractionation for N_2-fixation by chemolithotrophs is significantly higher than assumed previously, or a dissolved N pool other than N_2 (possibly NH_4^+) serves a primary N source for biosynthesis in vent systems. Few and contrasting isotopic analyses are available for N, N_2 or NH_4^+ at MOR vents. *Lilley et al.* [1993] reported a $\delta^{15}N_{(NH4+)}$ value of +12.4‰ for the Endeavor segment of the Juan de Fuca Ridge. Hydrothermal fluids from hot springs in Iceland show $\delta^{15}N$ values for nitrogen as low as −10.5‰ [*Marty et al.*, 1991]. Further studies of the biochemical cycles at MOR are needed to elucidate the

ultimate sources of N and the isotopic fractionation processes during chemosynthesis.

3.3. N Isotopic Fractionation in Archean Organic Matter and Ancient Metabolisms

Beaumont and Robert [1999] measured $\delta^{15}N$ of Precambrian kerogens from different localities and observed an isotopic shift from low $\delta^{15}N$ values down to −6‰ in the Archean to high $\delta^{15}N$ values up to +15‰ in the Proterozoic. The marked change of the N isotopic signature of organic matter between the Archean and the Proterozoic has been interpreted to be the result of major environmental changes. More precisely, the beginning of the Proterozoic corresponds roughly to the Great Oxidation Event (*Holland*, 2002; and references therein), where both the atmosphere and oceans became sufficiently oxygenated to allow for the biological production of NO_3^- and its use as a nitrogen source during biosynthesis. In the Archean, the fixation of atmospheric nitrogen (with little or no fractionation) may have dominated the biological cycle of N [*Beaumont and Robert*, 1999].

Since then, new nitrogen data have been obtained, complicating the scenario. Figure 6 shows the frequency distribution of the isotopic composition of organic-N extracted from kerogen and graphite preserved in cherts and banded iron formations (BIF) of Archean and Proterozoic ages and, for comparison, from organic matter in recent marine sediments and Phanerozoic marine black shales. We further compiled existing data on the N isotopic composition of organic matter (microbial mats, symbiontic invertebrates, zooplankton) from Karei, Indian Ridge; Lucky Strike, MAR; Gorda, Endeavour Site on the Juan de Fuca Ridge; Juan de Fuca flange; and Suiyo Seamount, Izu-Bonin; and in ammonium from various hydrothermal vent environments (Figure 6).

In this review, we treated Precambrian nitrogen isotopic data derived from "fossil" organic matter (kerogens) contained

Figure 6. Frequency distributions of the N isotopic composition of Precambrian organic matter (OM) from kerogens, compared to OM in terrestrial reservoirs of actual, subactual, or Phanerozoic age. Isotopic compositions of inorganic-N (metamorphic and magmatic) are also plotted. Gaussian curves superimposed on the frequency distribution represent the normal distribution centered on the mean value of each dataset (histograms and Gaussian curves calculated with StatView 5.0®). When compared to the observed distribution histograms, Gaussian curves are helpful in evidencing asymmetries or plurimodal distributions. Data sources: marine sediments [*Peters et al.*, 1978; *Sweeney et al.*, 1978; *Muzuka et al.*, 1991; *Holmes et al.*, 1996a, b; *De Lange*, 1998; *Cowie et al.*, 1998; *Milder et al.*, 1999; *Sadofsky and Bebout*, 2004]; Cretaceous marine black shales [*Rau et al.*, 1987]; Eoarchean and Paleoarchean OM [*Schidlowski et al.*, 1983; *Beaumont and Robert*, 1999; *Pinti et al.*, 2001; *Ueno et al.*, 2004; *Van Zuilen et al.*, 2005]; Neoarchean and Proterozoic OM [*Jia and Kerrich*, 2004a; *Beaumont and Robert*, 1999; *Schidlowski et al.*, 1983]; chemosynthetic bacteria [*Van Dover and Fry*, 1994; *Van Dover*, 2002; *Colaço et al.*, 2002]; chemosynthetic invertebrates [*Burd et al.*, 2002; *Van Dover*, 2002; *Van Dover and Fry*, 1994; *Conway et al.*, 1994]; ammonium from Iceland hot springs [*Marty et al.*, 1991]; and ammonium from MOR vents [*Lilley et al.*, 1993]. Data sources for upper mantle sampled by diamonds: *Cartigny et al.* [1997, 1998a], *Javoy et al.* [1986], *Boyd et al.* [1987], and *Boyd and Pillinger* [1994]; sources for sampling by MORBs: *Marty and Humbert* [1997], *Marty and Zimmermann* [1999], and *Sano et al.* [1998].

in rocks that have been slightly metamorphosed to prehnite–pumpellyite or lower greenschist facies. Nitrogen data derived from mica separates and/or from rocks that have been affected by higher degree of metamorphism (amphibolite facies) have been avoided because they are largely affected by "devolatilization", which produces a preferential release of ^{14}N from the host rock and a consequent increase of the $\delta^{15}N$ (see *Boyd and Philippot* [1998], *Pinti et al.* [2001], and *Dauphas and Marty* [2004] for a large treatment on the preservation of N in Precambrian rocks and minerals). In the case of Eoarchean rocks from the Isua Supracrustal Belt, West Greenland, a vigorous debate is underway on the biogenicity of the carbonaceous material preserved in these rocks [*Papineau et al.*, 2005; *Van Zuilen et al.*, 2002, 2005]. Here, we reported N data only from graphite preserved in a rock of Isua (sample AL43-3; *Van Zuilen et al.* [2005]), which seems to represent ancient biogenic material [*Rosing*, 1999].

Eo-paleoarchean (3.8–3.2 Ga) organic matter exhibits a bimodal distribution of its N isotopic composition, with $\delta^{15}N$ values being measured around −3.6‰ and +4.3‰, respectively (average $\delta^{15}N = 0.041‰$; Figure 6). Nitrogen from Neoarchean organic matter (3.2–2.5 Ga) is much more ^{15}N-enriched, with a $\delta^{15}N_{mean}$ value of +10.8‰. Some Neoarchean rocks, particularly BIFs, show an extreme enrichment in ^{15}N ($\delta^{15}N \geq +35‰$). The frequency distribution of organic matter for the Proterozoic shows again a decrease in the $\delta^{15}N$, with an average of +5.6‰, close to the value of nitrates in the modern ocean (Figure 5). Figure 7 shows the N isotopic composition measured in organic matter versus the age of deposition of the host rocks, from Eoarchean to the end of Proterozoic, and illustrates well the change from an Eo-paleoarchean dominated by ^{15}N-depleted organic matter, to largely positive $\delta^{15}N$ at the end of the Neoarchean and then again slowly reaching $\delta^{15}N$ values close to +5‰ at the end of Proterozoic.

The positive shift in the N isotopic composition of OM is reached before the Great Oxidation Event (Figure 7), and thus the direct relationship suggested by *Beaumont and Robert* [1999] between the oxygenation of the Earth and the isotopic change of N seems to be ephemeral. Yet, this shift is produced during the largest deposition of BIFs ever on Earth, between 2.7 and 2.6 Ga [*Isley and Abbott*, 1999; Figure 7]. The deposition of BIFs requires the presence of oxygen in the ocean, although in very limited amounts, likely 0.1–1% of the PAL [*Isley*, 1995]. If these conditions were sufficient to have nitrates in the ocean, also in limited amounts, we could speculate that these elevated values were obtained by local metabolic pathways, such as chemosynthesis in microbial mats, where N, as nitrate, was incorporated and used as an electron source. Another hypothesis is that, as observed sometimes throughout the recent Earth history, enhanced

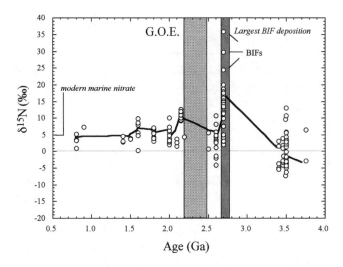

Figure 7. The N ($\delta^{15}N$) isotopic variations during Precambrian time plotted for organic matter. The curve represents the average secular variation of organic-N obtained with a smoothing method (computed with StatView 5.0®). The light gray area labeled G.O.E. indicates the time period covering the Great Oxidation Event, as suggested by *Holland* [2002] and *Bekker et al.* [2004]. The dark gray area indicates the time period when the largest ever deposition of Banded Iron Formations occurred on the Earth, following time series curves calculated by *Isley and Abbott* [1995].

denitrification may have shifted the $\delta^{15}N$ of oceanic nitrate towards elevated values, which are then recorded by organic matter in marine sediments [*Altabet et al.*, 2002; *Altabet and Francois*, 1994]. The degree of kinetic isotopic fractionation of N depends also on the capacity of supply nitrates as a reactant. In the first case (steady-state model) the isotopic shift of the N product will be much lower than in the case of a closed system, for which the isotopic shift can be described in terms of Rayleigh fractionation kinetics [*Sigman and Casciotti*, 2002]. In an Archean ocean containing limited amounts of nitrates, the *nitrification–denitrification–assimilation* process could have easily produced a higher isotopic shift than that observed in the modern ocean, because of the earlier scarceness of nitrates as a reactant (closed system's model). With the formation of a quantitative global nitrate pool in response to the oxygenation of the ocean at the beginning of the Proterozoic, the emergence of microbial processes that led to the enrichment in $\delta^{15}N$ in the nitrate pool and to a continuous supply of nitrates, the isotopic shift during *nitrification–denitrification–assimilation* would have decrease to the modern average value of +5‰ (Figure 7).

Jia and Kerrich [2004a, b] challenged recently the hypothesis of *Beaumont and Robert* [1999], suggesting that the high Archean $\delta^{15}N$ values reflect the presence of a ^{15}N-enriched residual atmosphere of chondritic composition

($\delta^{15}N = 30$–42‰), used by microorganisms in their metabolic synthesis. It is difficult to assume that a residual atmosphere of chondritic composition lasted half of the Earth's history without being recycled and its isotopic composition modified [*Dauphas and Marty*, 2004]. Major volatiles have been added to the Earth in the first 100 Ma of its history [e.g., *Morbidelli et al.*, 2000; *Dauphas and Marty*, 2002; *Pinti*, 2005]. The Hadean was a tectonically very active era, with the entire atmosphere–ocean system possibly recycled through the mantle in less than 150 Ma [*Pinti*, 2005]. Yet, the presence of a residual, slightly ^{15}N-enriched (+3 to +5‰) very early atmosphere is an attractive hypothesis that could explain, through the processes of biological fixation of atmospheric nitrogen, the high Eo-Paleoarchean $\delta^{15}N$ values centered on +4.3‰ (Figure 6), but certainly not the low values down to –7.4‰ (Figure 6; and *Pinti et al.* [2001]). The redox state of the depositional environment affects also the isotopic composition of sedimentary N. Organic matter preserved in an anoxic or sub-oxic marine environment, such as the Archean ocean, contains large amounts of N supplied by microbial fixation rather than derived from assimilation of nitrates and always exhibits low $\delta^{15}N$ [e.g., *Cowie et al.*, 1998]. A Phanerozoic analogue of anoxic Archean oceanic sediments is likely the Cretaceous black shales of *Rau et al.* [1987]. Yet, this organic matter exhibits $\delta^{15}N$ values that are, if at all, only slightly negative (Figure 6). Alternatively, the ^{15}N-depleted values measured in Eo-Paleoarchean organic matter may be explained by the significant contribution of biomass from chemosynthetic microbial activity, using light ammonium and/or mantle N as their nitrogen source [*Pinti and Hashizume*, 2001]. A dominant mantle N source in the ocean cannot be ruled out at that time. There are indeed numerous geochemical evidences that the chemistry of the Archean ocean was buffered by the mantle and sea-floor hydrothermal water–rock interactions [e.g., *Pinti*, 2005; and references therein].

The origin of the Archean rocks from which the organic matter was extracted (mainly cherts) suggests a genetic link with the mantle. Archean cherts are sedimentary rocks chiefly composed of micro- and cryptocrystalline quartz, often deposited in association with submarine exhalative hydrothermal activity [*de Wit et al.*, 1982; *Paris et al.*, 1985; *Sugitani*, 1992; *Kato and Nakamura*, 2003; *Orberger et al.*, 2005b]. In Figure 8, we plotted the argon and nitrogen isotopic composition of MORBs (which contain N and Ar extracted from the upper mantle) and those measured in the kerogeneous cherts—samples PANO C-85 and PANO D-136-00 (see *Pinti et al.* [2001] and *Orberger et al.* [2005] for details on these samples)—from dykes of North Pole representing hydrothermal veins at MOR [*Van Kranendonk et al.*, 2001; *Nijman et al.*, 1998]. Variations in the N and Ar isotopic signatures of MORBs result from the mixing of three

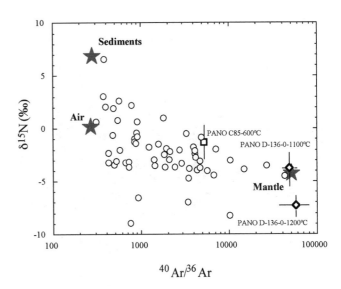

Figure 8. The nitrogen ($\delta^{15}N$) and argon ($^{40}Ar/^{36}Ar$) isotopic composition of upper mantle, as sampled by MORBs [white dots: *Marty and Dauphas*, 2003] plotted against the N and Ar composition of Paleoarchean kerogeneous black cherts PANO C-85 and PANO D-136-0, North Pole Dome, Pilbara, Western Australia (data from *Pinti et al.* [2001]; see *Orberger et al.*[2005b] for details on these cherts). The atmosphere, upper mantle, and recycled sediments end-members (gray stars) are plotted following values proposed by *Sano et al.* [1998].

components: recycled sedimentary N at subduction zones ($\delta^{15}N = +7$‰; $^{40}Ar/^{36}Ar = 295.5$), air contamination ($\delta^{15}N = 0$‰; $^{40}Ar/^{36}Ar = 295.5$), and an upper mantle component ($\delta^{15}N = -3 \pm 2$‰; $^{40}Ar/^{36}Ar$ ratio $\geq 40,000$) [*Sano et al.*, 1998]. The N and Ar isotopic composition of the cherts from North Pole is very close to that of pure mantle component (Figure 8). Recently, we performed additional N and Ar measurements on cherts within pillow basalts at the roots of the hydrothermal dykes of North Pole Dome, confirming the presence of this component with a $\delta^{15}N$ of -6.4‰ ± 1.7‰ and a $^{40}Ar/^{36}Ar$ ratio up to 30,000 (sample Pi-47-00; *Pinti et al.*, 2003, and in prep.). The isotopic similarity of N to mantle signatures is in agreement with the hypothesis of chemosynthetic bacteria using mantle N at hydrothermal vents, as primary contributors to the organic matter pool in the Eo-Paleoarchean.

The most convincing evidence of a dominant role of chemosynthesis as metabolic pathway in the Eo-Paleoarchean is when the N and C isotopic compositions of organic mater at that time are plotted together (Figure 9). Their isotopic composition is distinct from those measured in Phanerozoic black shales or in organic matter from recent marine sediments, but it is in the range of those measured in chemosynthetic communities. The highly ^{13}C-depleted signature of Archean organic matter may be explained by

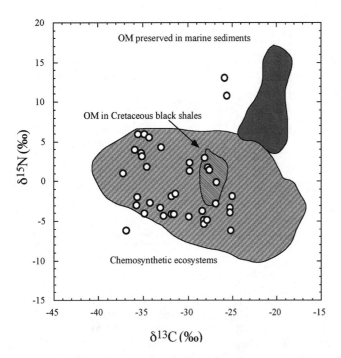

Figure 9. The nitrogen and carbon isotopic composition of Paleoarchean organic matter [data from *Beaumont and Robert*, 1999; *Ueno et al.*, 2004; *Pinti et al.*, 2001] against the isotopic composition of organic matter preserved in modern marine and deep-sea ecosystems. Data sources are: marine sediments [*Peters et al.*, 1978; *Muzuka et al.*, 1991; *Holmes et al.*, 1996a, b; *De Lange*, 1998; *Sadofsky and Bebout*, 2004]; Cretaceous marine sediments (black shales) [*Rau et al.*, 1987]; and chemosynthetic bacteria [*Van Dover and Fry*, 1994; *Colaço et al.*, 2002].

autothrophic fixation of seawater-dissolved inorganic carbon [*Conway et al.*, 1994] or, in some particular environment, from microbial methanotrophy [*Ueno et al.*, 2004].

The similarity between the Archean and the present-day N and C isotopic composition of hydrothermal vent organisms (Figure 9) supports the general idea that life at the hydrothermal vents represents the best modern natural analogue for life on the early Earth [*Pinti and Hashizume*, 2001]. On the other hand, we cannot exclude other forms of microbial life at that time. How Archean ecosystems functioned is still a matter of debate. *Nisbet and Fowler* [2004] suggested a diversification of the ecosystems as we observe it on today's Earth with oxygenic photosynthetic mats, stromatolites, and cyanobacterial plankton at the surface of the ocean, and anoxygenic photosynthesizers, methanogens at depth, and hyperthermophiles (cf. and likely chemosynthesizers) at MOR. The bimodal distribution of the N isotopic composition during Eo-Paleoarchean seems to support such a diversification of ecosystems; with anoxygenic photosynthesizers producing ^{15}N-enriched organic matter, via microbial fixation, and chemolithoautotrophs producing ^{15}N-depleted biomass.

There are also alternative pathways to denitrification, such as Anaerobic AMMonium Oxidation the so-called "anammox", that could have been predominant in earlier times [*Nisbet and Fowler*, 2004] and that could explain some of the observed values measured in the Archean kerogens.

We would like to spend few words on the high isotopic shifts observed in nitrogen from BIFs (Figure 7). *Beaumont and Robert* [1999] measured δ^{15}N values from +24.4 to +35.8‰ in the 2.69 Ga Michipicoten Iron Formation, Ontario, Canada, while *Pinti et al.* [2001] measured δ^{15}N values up to 21.4‰ in 2.49 Ga Dales Gorge member BIFs, Pilbara craton, Western Australia. These high N isotopic ratios are hard to explain by biological cycling of a residual ^{15}N-rich atmosphere or enhanced denitrification during first oxygenation of the ocean, at the end of the Archean. A possible mechanism to explain the contemporary deposition of iron oxides/oxy-hydroxides and an isotopic fractionation of residual dissolved nitrogen towards higher δ^{15}N values (which, through microbial uptake, would then leave its isotopic signature in the preserved biomass) could be the nitrate-dependent microbial oxidation of ferrous (II) to ferric (III) iron [*Straub et al.*, 1996]. This can take place in anaerobic conditions by light-independent, chemotrophic microbial activity [*Straub et al.*, 1996], by reactions such as:

$$10FeCO_3 + 2NO_3^- + 24H_2O \rightarrow$$
$$10Fe(OH)_3 + N_2 + 10HCO_3^- + 8H^+ \quad (7)$$

$$15FeO + 2HNO_3^- \rightarrow 15Fe_3O_4 + N_2 + H_2O \quad (8)$$

This process probably has a similar effect on the isotopic composition of residual nitrate as that of denitrification processes, but unfortunately there is not yet experimental knowledge on the isotopic fractionation during iron oxidation processes, to confirm this hypothesis.

4. CONCLUSIONS

In this review, we show how isotopic records of sulfur and nitrogen help us understand ancient metabolic processes as well as early ocean and atmospheric chemistry. Sulfur isotopic data from North Pole barites indicate that sulfate-reducing metabolism had already evolved by 3.5 Ga. The differences in the δ^{34}S between North Pole deposits and other Archean sedimentary sulfides probably reflect environmental variations. The North Pole evaporite ponds were localized oases of high sulfate concentrations. Thus, biological sulfate reduction could operate under favorable conditions, producing larger isotopic fractionations. Other early Archean sedimentary sulfides may have resulted from biological sulfate reduction under low ocean sulfate concentrations, implying a low oxygen environment.

Nitrogen isotopic composition of Paleoarchean kerogens from Pilbara and Barberton show a bi-modal distribution centered around the $\delta^{15}N$ values of $-3.6‰$ and $+4.3‰$, respectively. The occurrence of a global Archean anoxic ocean buffered by the mantle [*Pinti, 2005*] could explain these values: The oceanic N pool could have been dominated by ammonium derived from biological fixation of atmospheric N_2, operated by anoxygenic photosynthesizers and mantle N, possibly in the form of ammonium, metabolized by chemolithotroph organisms living at proximity of hydrothermal vents.

Acknowledgments. We are indebted to Moritz Lehmann for valuable discussions on the nitrogen cycle in the modern ocean and for improving the text. Harald Strauss and Jesper Nielsen are thanked for their constructive reviews. Canada Research Chairs Program and NSERC Discovery Grant funded YS research. French CNRS-PNP and GDR-Exobiology programs, and Japanese JSPS and Mitsubishi Foundation, funded DLP and KH research, respectively.

REFERENCES

Altabet, M.A., and R. Francois, Sedimentary nitrogen isotopic ratio as a recorder for surface ocean nitrate utilization, *Glob. Biogeochem. Cycle*, 8, 103-116, 1994.

Altabet, M.A., Higginson, M.J., and D.W. Murray, The effect of millennial-scale changes in Arabian Sea denitrification on atmospheric CO_2, *Nature*, 415, 159-162, 2002.

Beaumont, V., and F. Robert, Nitrogen isotope ratios of kerogens in Precambrian cherts: a record of the evolution of atmosphere chemistry?, *Precambrian Res.*, 96, 63-82, 1999.

Bekker, A., H.D. Holland, P.L. Wang, D. Rumble III, H.J. Stein, J.L. Hannah, L.L. Coetzee, and N.J. Beukes, Dating the rise of atmospheric oxygen, *Nature*, 427, 117-120, 2004.

Berner, R.A., 1984. Sedimentary pyrite formation: an update, *Geochim. Cosmochim. Acta*, 48, 605-615, 1984.

Bolliger, C., M.H. Schroth, S.M. Bernasconi, J. Kleikemper, and J. Zeyer, Sulfur isotope fractionation during microbial sulfate reduction by toluene-degrading bacteria, *Geochim. Cosmochim. Acta*, 65, 3289-3298, 2001.

Böttcher, M.E., A. Rusch, T. Hopner, and H.J. Brumsack, Stable sulfur isotope effects related to local intense sulfate reduction in a tidal sandflat (southern North Sea): Results from loading experiments, *Isot. Environ. Health Stud.*, 33, 109-129, 1997.

Boyd, S.R., Ammonium as a biomarker in Precambrian metasediments, *Precambrian Res.*, 108, 159-173, 2001.

Boyd, S.R., D.P. Mattey, C.T. Pillinger, H.J. Milledge, M. Mendelssohn, and M. Seal, Multiple growth events during diamond genesis - an integrated study of carbon and nitrogen isotopes and nitrogen aggregation state in coated stones, *Earth Planet. Sci. Lett.*, 86, 341-353, 1987.

Boyd, S.R., and P. Philippot, Precambrian ammonium biogeochemistry: a study of the Moine metasediments, Scotland, *Chem. Geol.*, 144, 257-268, 1998.

Boyd, S.R., and C.T. Pillinger, A preliminary study of N-15 N-14 in octahedral growth form diamonds, *Chem. Geol.*, 116, 43-59, 1994.

Buick, R., and J.S.R. Dunlop, Evaporite sediments of early Archaean age from the Warrawoona Group, North Pole, Western Australia, *Sedimentology*, 37, 247-277, 1990.

Buick, R., J.S.R. Dunlop, and D.I. Groves, Stromatolite recognition in ancient rocks; an appraisal of irregularly laminated structures in an early Archean chert-barite unit from North Pole, Western Australia, *Alcheringa*, 5, 161-179, 1981.

Buick, R., J.R. Thornett, N.J. McNaughton, J.B. Smith, M.E. Barley, and M. Savage, Record of emergent continental crust ~3.5 billion years ago in the Pilbara craton of Australia, *Nature*, 375, 574-577, 1995.

Burd, B.J., R.E. Thomson, and S.E. Calvert, Isotopic composition of hydrothermal epiplume zooplankton: evidence of enhanced carbon recycling in the water column, *Deep-Sea Res. Part I-Oceanogr. Res. Pap.*, 49, 1877-1900, 2002.

Cameron, E.M., Sulphate and sulphate reduction in early Precambrian oceans, *Nature*, 296, 145-148, 1982.

Canfield, D.E., and R. Raiswell, The evolution of the sulfur cycle, *Am. J. Sci.*, 299, 697-723, 1999.

Cartigny, P., S.R. Boyd, J.W. Harris, and M. Javoy, Nitrogen isotopes in peridotitic diamonds from Fuxian, China: the mantle signature, *Terra Nova*, 9, 175-179, 1997.

Cartigny, P., J.W. Harris, and M. Javoy, Eclogitic diamond formation at Jwaneng: No room for a recycled component, *Science*, 280, 1421-1424, 1998a.

Cartigny, P., J.W. Harris, D. Phillips, M. Girard, and M. Javoy, Subduction-related diamonds? The evidence for a mantle-derived origin from coupled $\delta^{13}C$-$\delta^{15}N$ determinations, *Chem. Geol.*, 147, 147-159, 1998b.

Casciotti, K.L., D.M. Sigman, and B.B. Ward, Linking diversity and stable isotope fractionation in ammonia-oxidizing bacteria, *Geomicrobiol. J.*, 20, 335-353, 2003.

Chambers, L.A., and P.A. Trudinger, Microbiological fractionation of stable sulfur isotopes: A review and critique, *Geomicrobiol. J.*, 1, 249-293, 1979.

Cline, J.D., and I.R. Kaplan, Isotopic fractionation of dissolved nitrate during denitrification in the eastern tropical North Pacific Ocean, *Mar. Chem.*, 3, 271-299, 1975.

Codispoti, L.A., J.A. Brandes, J.P. Christensen, A.H. Devol, S.W.A. Naqvi, H.W. Paerl, and T. Yoshinari, The oceanic fixed nitrogen and nitrous oxide budgets: Moving targets as we enter the anthropocene?, *Sci. Mar.*, 65, 85-105, 2001.

Colaço, A., F. Dehairs, and D. Desbruyères, Nutritional relations of deep-sea hydrothermal fields at the Mid-Atlantic Ridge: a stable isotope approach, *Deep-Sea Res. Part I-Oceanogr. Res. Pap.*, 49, 395-412, 2002.

Conway, N.M., M.C. Kennicutt, and C.L. Van Dover, Stable isotopes in the study of marine chemosynthetic-based ecosystems, in *Stable isotopes in ecology and environmental science, Methods in Ecology*, edited by K. Lajtha and R.H. Michener, pp. 158-186, Blackwell, Oxford, 1994.

Cowie, G., S. Calvert, G. De Lange, R. Keil, and J. Hedges, Extents and implications of organic matter alteration at oxidation fronts in turbidites from the Madeira abyssal plain, in *Proceedings of the Ocean Drilling Program, Scientific Results*, edited by

P.P.E. Weaver, H.-U. Schmincke and J.V. Firth, pp. 581-589, College Station, TX, 1998.

Dauphas, N., and B. Marty, Heavy nitrogen in carbonatites of the Kola Peninsula: A possible signature of the deep mantle, *Science*, 286, 2488-2490, 1999.

Dauphas, N., and B. Marty, Inference on the nature and the mass of Earth's late veneer from noble metals and gases, *J. Geophys. Res.*, 107, 5129-5136, 2002.

Dauphas, N., and B. Marty, "A large secular variation in the nitrogen isotopic composition of the atmosphere since the Archaean?": response to a comment on "The nitrogen record of crust-mantle interaction and mantle convection from Archaean to Present" by R. Kerrich and Y. Jia, *Earth Plan. Sci. Lett.*, 225, 441-450, 2004.

De Lange, G.J., Oxic vs. anoxic diagenetic alteration of turbiditic sediments in the Madeira abyssal plain, eastern North Atlantic, in *Proceedings of the Ocean Drilling Program, Scientific Results*, edited by P.P.E. Weaver, H.-U. Schmincke and J.V. Firth, pp. 573-580, College Station, TX, 1998.

Detmers, J., V. Brüchert, K.S. Habicht, and J. Kuever, Diversity of sulfur isotope fractionation by sulfate-reducing prokaryotes, *Appl. Envir. Microbiol.*, 67, 888-894, 2001.

de Wit, M., Hart, R., Martin, A., and P. Abbott, Archean abiogenic and probable biogenic structures associated with mineralized hydrothermal vent systems and regional metasomatism, with implications for greenstone belt studies, *Econ. Geol.*, 77, 1783-1802, 1982.

Donnelly, T.H., I.B. Lambert, D.Z. Oehler, J.A. Hallberg, D.R. Hudson, J.W. Smith, O.A. Bavinton, and L. Golding, A reconnaissance study of stable isotope ratios in Archaean rocks from the Yilgarn Block, western Australia. *J. Geol. Soc. Australia*, 24, 409-420, 1977.

Drobner, E., H. Huber, G. Wachtershauser, D. Rose, and K.O. Stetter, Pyrite formation linked with hydrogen evolution under anaerobic conditions, *Nature*, 346, 742-744, 1990.

Dunlop, J.S.R., and R. Buick, Archaean epiclastic sediments derived from mafic volcanics, North Pole, Pilbara Block, Western Australia, *Spec. Publ. Geol. Soc. Australia*, 7, 225-233, 1981.

Farquhar, J., H. Bao, and M.H. Thiemens, Atmospheric influence of Earth's earliest sulfur cycle, *Science*, 289, 756-758, 2000.

Farquhar, J., S. Airieau, and M.H. Thiemens, Observation of wavelength-sensitive mass-independent sulfur isotope effects during SO2 photolysis: Implications for the early atmosphere. *J. Geophy. Res. – Planets*, 106, 32829-32839, 2001.

Fogel, M.L., and L.A. Cifuentes, Isotope fractionation during primary production, in *Organic Geochemistry*, edited by E.M.H. and S.A. Macko, pp. 73-98, Plenum Press, New York, NY, 1993.

Freudenthal, T., T. Wagner, F. Wenzhofer, M. Zabel, and G. Wefer, Early diagenesis of organic matter from sediments of the eastern subtropical Atlantic: evidence from stable nitrogen and carbon isotopes, *Geochim. Cosmochim. Acta*, 65, 1795-1808, 2001.

Goodwin, A.M., J. Monster, and H.G. Thode, Carbon and sulfur isotope abundance in Archean Iron-Formations and early Precambrian life, *Econ. Geol.*, 71, 870-891, 1976.

Grassineau, N.V., E.G. Nisbet, M.J. Bickle, C.M.R. Fowler, D. Lowry, D.P. Mattey, P. Abell, and A. Martin, Antiquity of the biological sulphur cycle: evidence from sulphur and carbon isotopes in 2700 million-year-old rocks of the Belingwe Belt, Zimbabwe, *Proc. Royal Soc. London B*, 268, 113-119, 2000.

Habicht, K., and D.E. Canfield, Sulfur isotope fractionation during bacterial sulfate reduction in organic-rich sediments, *Geochim. Cosmochim. Acta*, 61, 5351-5361, 1997.

Hayes, J.M., Lambert, I.B., and H. Strauss, The sulfur-isotopic record, in *The Proterozoic Biosphere*, edited by J.W. Schopf, and C. Klein, pp. 129-132, Cambridge University Press, Cambridge, 1992.

Holland, H.D., Volcanic gases, black smokers, and the great oxidation event, *Geochim. Cosmochim. Acta*, 66, 3811-3826, 2002.

Holmes, M.E., P.J. Müller, R.R. Schneider, M. Segl, J. Pätzold, and G. Wefer, Stable nitrogen isotopes in Angola Basin surface sediments, *Mar. Geol.*, 134, 1-12, 1996a.

Holmes, M.E., P.J. Muller, R.R. Schneider, M. Segl, and G. Wefer, Spatial variations in euphotic zone nitrate utilization based on $\delta^{15}N$ in surface sediments, *Geo-Mar. Lett.*, 18, 58-65, 1996b.

Honma, H., and Y. Itihara, Distribution of ammonium in minerals of metamorphic and granitic rocks, *Geochim. Cosmochim. Acta*, 45, 983-988, 1981.

Isley, A.E., Hydrothermal plumes and the delivery of iron to Banded Iron Formation, *J. Geol.*, 103, 169-185, 1995.

Isley, A.E., and D.H. Abbott, Plume-related mafic volcanism and the deposition of banded iron formation, *J. Geophys. Res.*, 104, 15461-15477, 1999.

Jannasch, H.W., and M.J. Mottl, Geomicrobiology of deep-sea hydrothermal vents, *Science*, 229, 717-723, 1985.

Javoy, M., F. Pineau, and H. Delorme, Carbon and nitrogen isotopes in the mantle, *Chem. Geol.*, 57, 41-62, 1986.

Jia, Y., and R. Kerrich, Nitrogen 15–enriched Precambrian kerogen and hydrothermal systems, *Geochem. Geophys. Geosyst.*, 5, Q07005, doi:10.1029/2004GC000716, 2004a.

Jia, Y., and R. Kerrich, A Reinterpretation of the crustal N-isotope record: evidence for a ^{15}N-enriched Archean atmosphere?, *Terra Nova*, 16, 102-108, 2004b.

Kato, Y., and K. Nakamura, Origin and global tectonic significance of Early Archean cherts from the Marble Bar greenstone belt, Pilbara Craton, Western Australia, *Precambrian Res.*, 125, 191-243, 2003.

Kennicut, M.C.I., and R.A.J. Burke, Stable isotopes: clues to biological cycling of elements at hydrothermal vents, in *Deep-sea hydrothermal vents*, edited by D. M. Karl, pp. 275-287, CRC Press, Boca Raton, TX, 1995.

Lambert, I.B., T.H. Donnelly, J.S.R. Dunlop, and D.I. Groves, Stable isotopic compositions of early Archaean sulphate deposits of probable evaporitic and volcanogenic origins, *Nature*, 276, 808-811.

Lehmann, M.F., S.M. Bernasconi, A. Barbieri, and J.A. McKenzie, Preservation of organic matter and alteration of its carbon and nitrogen isotope composition during simulated and in situ early sedimentary diagenesis, *Geochim. Cosmochim. Acta*, 66, 3573-3584, 2002.

Lehmann, M.F., D.M. Sigman, and W.M. Berelson, Coupling the $^{15}N/^{14}N$ and $^{18}O/^{16}O$ of nitrate as a constraint on benthic nitrogen cycling, *Mar. Chem.*, 88, 1-20, 2004.

Lilley, M.D., D.A. Butterfield, E.J. Olson, J.E. Lupton, S.A. Macko, and R.E. McDuff, Anomalous CH_4 and NH_4^+ concentrations at an unsedimented mid-ocean-ridge hydrothermal system, *Nature*, 364, 45-47, 1993.

Marty, B., and N. Dauphas, The nitrogen record of crust-mantle interaction and mantle convection from Archean to present, *Earth Planet. Sci. Lett.*, 206, 397-410, 2003.

Marty, B., E. Gunnlaugsson, A. Jambon, N. Oskarsson, M. Ozima, F. Pineau, and P. Torssander, Gas Geochemistry of Geothermal Fluids, the Hengill Area, Southwest Rift-Zone of Iceland, *Chem. Geol.*, 91, 207-225, 1991.

Marty, B., and F. Humbert, Nitrogen and argon isotopes in oceanic basalts, *Earth Planet. Sci. Lett.*, 152, 101-112, 1997.

Marty, B., and L. Zimmermann, Volatiles (He, C, N, Ar) in mid-ocean ridge basalts: Assessment of shallow-level fractionation and characterization of source composition, *Geochim. Cosmochim. Acta*, 63, 3619-3633, 1999.

Mojzsis, S.J., C.D. Coath, J.P. Greenwood, K.D. McKeegan, and T.M. Harrison, Mass-independent isotope effects in Archaean (2.5 to 3.8 Ga) sedimentary sulfides by ion microprobe multicollection, *Geochim. Cosmochim. Acta*, 67, 1635-1658, 2003.

Monster, J., P.W.U. Appel, H.G. Thode, M. Schidlowski, C.M. Carmichael, and D. Bridgwater, Sulfur isotope studies in Early Archaean sediments from Isua, West Greenland: Implications for the antiquity of bacterial sulfate reduction, *Geochim. Cosmochim. Acta*, 43, 405-413, 1979.

Morbidelli, A., J. Chambers, J.I. Lunine, J.M. Petit, and F. Robert, Source regions and timescales for the delivery of water to the Earth, *Meteor. Planet. Sci.*, 35, 1309-1320, 2000.

Muzuka, A.N.N., S.A. Macko, and T.F. Pedersen, Stable carbon and nitrogen isotope compositions of organic matter from Site 724 and Site 725, Oman margin, in *Proceedings of the Ocean Drilling Program, Scientific Results*, edited by W.L. Prell and N. Niitsuma, pp. 571-586, College Station, TX, 1991.

Nisbet, E.G., and C.M.R. Fowler, The early history of life, in *Biogeochemistry*, edited by W.H. Schlesinger, Vol. 8 *Treatise of Geochemistry*, pp. 1-39, Elsevier-Pergamon, Oxford, 2001.

Noda, M., T. Kakegawa, H. Naraoka, K. Marumo, and T. Urabe, Geochemistry of phosphorous and nitrogen in volcanic rocks altered by submarine hydrothermal activities at the Suiyo Seamount, Japan, *Eos Trans. AGU*, 83, Fall Meet. Suppl., Abstract V72A-1291, 2002.

Ohmoto, H., and R.P. Felder, Bacterial activity in the warmer, sulphate-bearing, Archaean oceans, *Nature*, 328, 244-246, 1987.

Ohmoto, H., T. Kakegawa, and D.R. Lowe, 3.4-billion-year-old biogenic pyrites from Barberton, South Africa: sulfur isotope evidence, *Science*, 262, 555-557, 1993.

Ono, S., J.L. Eigenbrode, A.A. Pavlov, P. Kharecha, D. Rumble, J.F. Kasting, and K.H. Freeman, New insights into Archean cycle from mass-independent sulfur isotope records from the Hamersley basin, Australia, *Earth and Planet. Sci. Letts.*, 213, 15-30, 2003.

Orberger, B., J.-P. Gallien, D.L. Pinti, M. Fialin, L. Daudin, D.R. Grocke, and J. Pasava, Nitrogen and carbon partitioning in diagenetic and hydrothermal minerals from Paleozoic Black Shales, (Selwyn Basin, Yukon Territories, Canada), *Chem. Geol.*, 218, 249-264, 2005a.

Orberger, B., V. Rouchon, F. Westall, S.T. de Vries, D.L. Pinti, C. Wagner, R. Wirth, and K. Hashizume, Micro-facies and origin of some Archaean cherts (Pilbara, Australia), in, *Processes on the Early Earth*, edited by U. Reimold and R.L. Gibson, in press, Geol. Soc. Am. Spec. Pap., Boulder, CO, 2005b.

Papineau, D., S.J. Mojzsis, J.A. Karhu, and B. Marty, Nitrogen isotopic composition of ammoniated phyllosilicates: case studies from Precambrian metamorphosed sedimentary rocks, *Chem. Geol.*, 216, 37-58, 2005.

Paris, I., I.G. Stranistreet, and M.J. Huges, Cherts of the Barberton Greenstone Belt interpreted as products of submarine exhalative activity, *J. Geol.*, 93, 111-130, 1985.

Peters, K.E., R.E. Sweeney, and I.R. Kaplan, Correlation of carbon and nitrogen stable isotope ratios in sedimentary organic matter, *Limnol. Oceanogr.*, 23, 598-604, 1978.

Pinti, D.L., The isotopic record of Archean nitrogen and the evolution of the early earth, *Trends in Geochemistry*, 2, 1-17, 2002.

Pinti, D.L., The formation and evolution of the oceans, in *Lectures in Astrobiology*, vol. 1, edited by M. Gargaud, B. Barbier, H. Martin and J. Reisse, pp. 83-107, Springer-Verlag, Berlin, 2005.

Pinti, D.L., and K. Hashizume, ^{15}N-depleted nitrogen in Early Archean kerogens: clues on ancient marine chemosynthetic-based ecosystems? A comment to Beaumont, V., Robert, F., 1999. Precambrian Res. 96, 62-82, *Precambrian Res.*, 105, 85-88, 2001.

Pinti, D.L., K. Hashizume, and J. Matsuda, Nitrogen and argon signatures in 3.8 to 2.8 Ga metasediments: Clues on the chemical state of the Archean ocean and the deep biosphere, *Geochim. Cosmochim. Acta*, 65, 2301-2315, 2001.

Pinti, D.L., K. Hashizume, P. Philippot, J. Foriel, and P. Rey, Nitrogen quest in Archean metasediments of Pilbara, Australia, *Geochim. Cosmochim. Acta*, A287, 2003.

Price, F.T., and Y.N. Shieh, Fractionations of sulfur isotopes during laboratory synthesis of pyrite at low temperatures, *Chem. Geol.*, 27, 245-253, 1979.

Rau, G.H., Low ^{15}N/^{14}N in hydrothermal vent animals: ecological implications, *Nature*, 289, 484-485, 1981.

Rau, G.H., M.A. Arthur, and W.E. Dean, ^{15}N/^{14}N variations in Cretaceous Atlantic sedimentary sequences: implication for past changes in marine nitrogen biogeochemistry, *Earth Planet. Sci. Lett.*, 82, 269-279, 1987.

Rees, C.E., A steady-state model for sulphur isotopes fractionation in bacterial reduction process, *Geochim. Cosmochim. Acta*, 37, 1141-1162, 1973.

Rickard, D., Kinetics of pyrite formation by the H_2S oxidation of Fe(II) monosulfide in aqueous solutions between 25°C and 125°C: the rate equation, *Geochim. Cosmochim. Acta*, 61, 115-134, 1997.

Rosing M.T., ^{13}C-Depleted carbon microparticles in 3700-Ma sea-floor sedimentary rocks from West Greenland, *Science*, 283, 674-676, 1999.

Rudnicki, M.D., H. Elderfield, and B. Spiro, Fractionation of sulfur isotopes during bacterial sulfate reduction in deep ocean sediments at elevated temperatures, *Geochim. Cosmochim. Acta*, 65, 777-789, 2001.

Sadofsky, S.J., and G.E. Bebout, Nitrogen geochemistry of subducting sediments: New results from the Izu-Bonin-Mariana margin and insights regarding global nitrogen subduction, *Geochem. Geophys. Geosyst.*, 5, doi:10.1029/2003GC000543, 2004.

Sano, Y., and C.T. Pillinger, Nitrogen isotopes and N_2/Ar ratios in cherts: An attempt to measure time evolution of atmospheric δ^{15}N value, *Geochem. J.*, 24, 315-325, 1990.

Sano, Y., Takahata, N., Nishio, Y., and B. Marty, Nitrogen recycling in subduction zones, *Geophys. Res. Lett.*, 25, 2289-2292, 1998.

Schidlowski, M., Hayes, J.M., and I.R. Kaplan, Isotope inferences of ancient biochemistries: Carbon, sulfur, hydrogen, and nitrogen, in, *Earth's Earliest Biosphere: Its Origin and Evolution*, edited by J.W. Schopf, Princeton University Press, Princeton, NJ, pp. 149-186, 1983.

Scholten, S.O., The distribution of nitrogen isotopes in sediments, *Geologica Ultraiectina*, 81, 101, 1991.

Shen, Y., R. Buick, and D.E. Canfield, Isotopic evidence for microbial sulphate reduction in the early Archaean era, *Nature*, 410, 77-81, 2001.

Shen, Y., A.H. Knoll, and M.R. Walter, Evidence for low sulphate and anoxia in a mid-Proterozoic marine basin, *Nature*, 423, 632-635, 2003.

Shen, Y., and R. Buick, The antiquity of microbial sulfate reduction, *Earth-Sci. Rev.*, 64, 243-272, 2004.

Sigman, D.M., and K.L. Casciotti, Nitroge319n isotopes in the ocean, in *Encyclopedia of Ocean Sciences*, edited by J.H. Steele, K.K. Turekian and S.A. Thorpe, pp. 1884-1894, Academic Press, London, 2001.

Sigman, D.M., R. Robinson, A.N. Knapp, A. van Geen, D.C. McCorkle, J.A. Brandes, and R.C. Thunell, Distinguishing between water column and sedimentary denitrification in the Santa Barbara Basin using the stable isotopes of nitrate, *Geochem. Geophys. Geosyst.*, 4, doi: 10.1029/2002GC000384, 2003.

Straub, K.L., M. Benz, B. Schink, and F. Widdel, Anaerobic, nitrate-dependent microbial oxidation of ferrous iron, *Appl. Envir. Microbiol.*, 62, 1458-1460, 1996.

Strauss, H., Sulphur isotopes and the early Archaean sulphur cycle, *Precambrian Res.*, 126, 349-361, 2003.

Strauss, H., 4 Ga of seawater evolution: Evidence from the sulfur isotopic composition of sulfate. *Geol. Soc. Amer. Bull. Spec. Papers*, 379, 195-205, 2004.

Sugitani, K., Geochemical characteristics of Archean cherts and other sedimentary rocks in the Pilbara Block, Western Australia: evidence for Archean seawater enriched in hydrothermally-derived iron and silica, *Precambrian Res.*, 57, 21-47, 1992.

Sweeney, R.E., K.K. Liu, and I.R. Kaplan, Oceanic nitrogen isotopes and their uses in determining the source of sedimentary nitrogen, *Dept. Sci. Ind. Res. New Zealand Bull.*, 220, 9-26, 1978.

Ueno, Y., H. Yoshioka, S. Maruyama, and Y. Isozaki, Carbon isotopes and petrography of kerogens in similar to 3.5-Ga hydrothermal silica dikes in the North Pole area, Western Australia, *Geochim. Cosmochim. Acta*, 68, 573-589, 2004.

Van Dover, C.L., Trophic relationship among invertebrates at the Karei hydrothermal vent field (Central Indian Ridge), *Mar. Biol.*, 141, 761-772, 2002.

Van Dover, C.L., and B. Fry, Microorganisms as food resources at deep-sea hydrothermal vents, *Limnol. Oceanogr.*, 39, 51-57, 1994.

Van Dover, C.L., T.L. Harmer, Z.P. McKiness, and C. Meredith, Biogeography and ecological setting of Indian Ocean hydro-thermal vents, *Science*, 294, 818-823, 2001.

Van Kranendonk, M.J., A.H. Hickman, I.R. Williams, and W. Nijman, *Archaean geology of the East Pilbara granite-greesntone terrane, Western Australia- A field guide*, Geol. Survey W. Australia, Perth, 2001.

Van Zuilen, M.A., A. Lepland, and G. Arrhenius, Reassessing the evidence for the earliest traces of life, *Nature*, 418, 627-630, 2002.

Van Zuilen, M.A., K. Mathew, B. Wopenka, A. Lepland, K. Marti, and G. Arrhenius, Nitrogen and argon isotopic signatures in graphite from the 3.8-Ga-old Isua Supracrustal Belt, Southern West Greenland, *Geochim. Cosmochim. Acta*, 69, 1241-1252, 2005.

Wada, E., T. Kadonaga, and S. Matsuo, ^{15}N abundance in nitrogen in naturally occurring substances and global assessment of denitrification from isotopic viewpoint, *Geochem. J.*, 9, 139-148, 1975.

Walker, J.C.G., and P. Brimblecombe, Iron and sulfur in the Prebiologic ocean, *Precambrian Res.*, 28, 205-222, 1985.

Widdel, F., Microbiology and ecology of sulfate- and sulfur-reducing bacteria, in *Biology of Anaerobic Organisms*, edited by A.J.B. Zehnder, pp. 469-585, John Wiley & Sons, New York, NY, 1988.

Widdel, F., and T.A. Hansen, The dissimilatory sulfate- and sulfur-reducing bacteria, in *The Prokaryotes*, edited by A. Balows, H.G. Trüper, M. Dworkin, W. Harder, K.H. Schleifer, vol. 1, pp. 583-624, Springer-Verlag, New York, NY, 1992.

Wilkin, R.T., and H.L. Barnes, Pyrite formation by reactions of iron monosulfides with dissolved inorganic and organic sulfur species, *Geochim. Cosmochim. Acta,* 60, 4167-4179, 1996.

Williams, L.B., R.E. Ferrell Jr., I. Hutcheon, A.J. Bakel, M.M. Walsh, and H.R. Krouse, Nitrogen isotope geochemistry of organic matter and mineral during diagenesis and hydrocarbon migration, *Geochim. Cosmochim. Acta*, 59, 765-779, 1995.

Wortmann, U.G., S.M. Bernasconi, and M. Böttcher, Hypersulfidic deep biosphere indicates extreme sulfur isotope fractionation during single-step microbial sulfate reduction, *Geology*, 29, 647-650, 2001.

Zhang, D., *Nitrogen Concentrations and Isotopic Compositions of Some Terrestrial Rock,* Ph.D. Thesis, pp. 157, The University of Chicago, Chicago, IL, 1988.

K. Hashizume, Department of Earth and Space Sciences, Faculty of Science, Osaka University, 1-1 Machikaneyama, Toyonaka, 560-0043 Osaka, Japan.

D.L. Pinti and Y. Shen, Département des Sciences de la Terre et de l'Atmosphère and GEOTOP-UQAM-McGill, Université du Québec à Montréal, C.P. 8888, Succ. Centre-Ville, Montréal, Québec, H3C 3P8 Canada. (shen.yanan@uqam.ca)